Geology and the Environment in Western Europe

A coordinated statement by
The Western European Geological Surveys

Edited by

G. INNES LUMSDEN

and

THE EDITORIAL BOARD OF THE DIRECTORS OF THE WESTERN EUROPEAN GEOLOGICAL SURVEYS

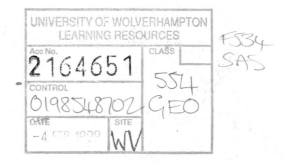

CLARENDON PRESS · OXFORD
1994

Oxford University Press, Walton Street, Oxford OX2 6DP

Oxford New York Toronto
Delhi Bombay Calcutta Madras Karachi
Kuala Lumpur Singapore Hong Kong Tokyo
Nairobi Dar es Salaam Cape Town
Melbourne Auckland Madrid
and associated companies in
Berlin Ibadan

Oxford is a trade mark of Oxford University Press

Published in the United States
by Oxford University Press Inc., New York

A catalogue record for this book is available from the British Library

Library of Congress Cataloging in Publication Data

Geology and the environment in Western Europe: a coordinated
statement by the Western European geological surveys: edited by
G. Innes Lumsden and the editorial board of the directors of the
Western European geological surveys.
Includes bibliographical references and index.
1. Environmental geology—Europe. 2. Geology—Europe. 3. Natural
resources—Europe. I. Lumsden, G. Innes (George Innes)
QE38.G46   1991   554—dc20   91-28274
ISBN 0-19-854870-2 (Pbk)

Printed in Hong Kong

The frontispiece is a motif from 'The Verdal Slide in 1893' painted by Herman W. Anker. This quick-clay slide is one of the biggest natural catastrophes to have occurred in Norway. Five and a half million cubic metres spilt out and converted 11 000 km^2 into a quagmire in half an hour. In all 112 people were killed of the 250 who lived in the area. Many dramatic eye-witness accounts of the slide and the rescue work are collected in a short book entitled *Description of the catastrophe in Verdal on Thursday night 19 May 1893* edited by Olai Hartmann, and published by Stenkjers Dagblad.

The painting belongs to Trondhjems Kunstforening.

Foreword

Gro Harlem Brundtland, Prime Minister of Norway

The World Commission on Environment and Development called for urgent action to safeguard the security, well-being, and very survival of the planet. A change to sustainable development is absolutely necessary to ensure that the needs of the present generation are met without compromising the ability of future generations to meet their needs. All sectors in our societies must participate in this effort if we are to save our common future.

Scientists have an important role to play in these efforts. Scientific research is continually increasing the stock of cumulative knowledge about the global environment. Our science and technology give us the potential to look deeper into and better understand natural systems.

Scientists have a responsibility to convey this knowledge both to the public and to the political community. This volume is one example that scientists are now living up to this responsibility. It presents a broad picture of international cooperation involving Geological Surveys from most parts of Western Europe. It provides an interesting account of the geological framework of western and southern Europe. In so doing, it reminds us that an understanding of basic science is absolutely necessary in determining what we are doing to our environment—and how we can minimize or even reverse the effect of ecological deterioration.

Geologists have for long been in the forefront of the discovery and exploitation of non-renewable resources. These resources underpin not only our economies, but modern civilization itself. But this exploitation has also led to many of the present environmental problems, such as the increased greenhouse effect from the burning of fossil fuels, or the toxic contamination of some areas as a result of the mining of heavy metals.

This volume makes use of a series of case histories—almost all involving the work of the Surveys—to illustrate the way in which man's activities can have profoundly adverse effects on the natural geological environment. It also demonstrates the natural hazards which can impede man's developments or which can be a danger to life.

Its pages contain salutary lessons for Europe in their documentation of the impact of mining, the pollution of vital groundwater resources, land stability and erosion, the contamination of soils and waterways, and the relatively unconstrained use of non-renewable resources. Once again, we are reminded that environmental degradation does not respect national boundaries, and that any attempt to ameliorate these effects can only be carried out on an international basis.

Geosciences can be the key to overcoming or avoiding many of the problems which face us in dealing with the need to build the economy whilst safeguarding the environment. The discovery of more natural gas will increase the supply of cleaner fuel; geothermal power is even more environmentally friendly and, in some areas at least, may hold the key to more acceptable sources of energy.

The study of hydrogeology, fundamental to the development of geothermal energy, is also important in dealing with the increasing concerns about global climatic change, and in particular, the impact of these changes on groundwater supplies. Geology can also provide a better guide to global change by giving us a better understanding of the way such changes have occurred in the past. Here too, this volume is likely to be of wide interest in its documentation of environmental change since the climax of the last ice age some 18 000 years ago.

I am confident that this book will be of great interest not only to geologists, but to the wider community concerned with environment and development.

Preface

For some years now, the Directors of the national Geological Surveys of the countries constituting Western Europe have recognized that the comprehensive accumulation of data and expertise in their organizations is not being applied to the best advantage in tackling the environmental problems facing mankind. Much has now been done internally and between the organizations to improve communication, to minimize problems of language and varying practice, and to organize the data so that the scale of the available database can be recognized, the variations and shortcomings from country to country can be identified, and the access to and retrieval from that international database can be optimized. It is now felt that all the relevant data and expertise available in the Geological Surveys could be readily brought to bear on any national or international environmental study or problem in Western Europe and adjacent and associated territories.

By far the greater problem is ensuring that the non-geologists, the politicians, the decision makers, the industrialists, the planners, and the laymen recognize the scale and relevance of the data and expertise which exist, and that they understand when and how to achieve their most effective application to everyday problems in the environment, whatever their size. All too often geology is not applied when it should be, and problems become disasters or they are overcome or ameliorated at unnecessary cost. This book is an attempt to improve the communication between the geologist and the non-geologist, to demonstrate by example the types of problem and situation to which geological input is relevant, and to impress the non-geologist with the scale of national and international input which can be sought from the Geological Surveys. It has been pitched deliberately at a level to be understood by the intelligent layman, and yet the geological content will be important for students of environmental geology and for earth scientists of other disciplines. It emphasizes the function of a national Geological Survey as it operates in the area between government, industry, and academia, providing independent and impartial advice without prejudice. It is expected that the book will be used by most readers for reference rather than read in its entirety. For this reason there is a deliberate overlap and even repetition in places.

This is the first time that the 21 Geological Surveys of Western Europe have attempted to cooperate in producing a major publication. The Directors wisely appointed a small Editorial Board and then agreed the framework of chapters proposed and the selection of chapter coordinators reflecting the strengths of their parent Surveys. More than 230 contributors from Geological Surveys and associated organizations then offered text and illustrations which could have filled several volumes. All the contributions were of great value, but some have had to be reduced to little more than a mention in order to achieve the desired length and balance of the publication. Every contributor is acknowledged by number throughout the text, and readers wishing to expand on any topic are advised to refer directly to the contributor concerned for further information. The chapter and section coordinators, each an expert in the area covered, have assembled the contributions in both text and illustration form, and edited these to formulate each chapter and section for further consolidation and integration by the Editorial Board. As a result of much effort and excellent collaboration the final text was completed on schedule. In writing the first edition in English every effort has been made to preserve the meaning and style of each contributor. Further editions in other languages are planned.

The book demonstrates the ability of the large number of earth scientists located in the Geological Surveys of Western Europe to work together in maximizing the value, interpretation, and application of the comprehensive European geological database for the benefit of all Europeans. It is now for the decision makers in the community to ensure that that ability is recognized and used with maximum effect.

Oxford
April 1991

G. I. L.
Editor-in-Chief

Contents

The Editorial Board

of the
Directors of the Western European Geological Surveys

Chapter Coordinators

The Western European Geological Surveys: Abbreviations

Austria	Geologische Bundesanstalt	GBA
Belgium	Service Géologique de Belgique	SGB
Cyprus	Geological Survey Department	GSD
Denmark	Danmarks Geologiske Undersøgelse	DGU
Finland	Geologian Tutkimuskeskus	GSF
France	Bureau de Recherches Géologiques et Minières	BRGM
Germany	Bundesanstalt für Geowissenschaften und Rohstoffe	BGR
	German States represented by Geologisches Landesamt Nordrhein–Westfalen	GLA/NW
Greece	Institute of Geology and Mineral Exploration	IGME
Greenland	Grønlands Geologiske Undersøgelse	GGU
Iceland	Orkustofnun (National Energy Authority)	NEA
Ireland	Geological Survey of Ireland	GSI
Italy	Servicio Geologico d'Italia	SGI
Luxembourg	Service Géologique du Luxembourg	SGL
The Netherlands	Rijks Geologische Dienst	RGD
Norway	Norges Geologiske Undersøkelse	NGU
Portugal	Serviços Geológicos de Portugal	SGP
Spain	Instituto Tecnológico Geominero de España	ITGE
Sweden	Sveriges Geologiska Undersökning	SGU
Switzerland	Service Hydrologique et Géologique National	SHGN
Turkey	Maden Tetkik ve Arama Genel Müdürlüğü	MTA
United Kingdom	British Geological Survey	BGS

1

Objectives of the book

The Editorial Board

1.1. INTRODUCTION [107]*

Future environmental changes may very well have unforeseen impact on human welfare and may represent the most challenging problems ever faced by mankind. Whether these changes are natural processes that are part of our evolving globe or environmental changes induced by man, they require our direct attention in order to ensure the survival of the human race.

The environment has become one of the major pre-occupations of governments. The evidence is to be seen in the growing emphasis on recognizing and understanding global changes, and particularly those influencing our climate. Examples that spring to mind are the greenhouse effect, the destruction of the ozone layer, potential sea-level variations, and also the whole range of problems related to industrial activities and to the adverse effect on our cultural heritage. The problem is that many solid-earth processes, whether natural or anthropogenic in origin, usually proceed at a pace that is almost imperceptible to human observation. However, such processes are also at the root of most natural disasters which, though usually restricted in their impact and damage, tell us of the fragile nature of our environment. These effects are the subject of research and monitoring on a global scale, but the underlying causes have hitherto received very little attention from policy-makers worldwide.

Another insidious problem is the accumulation of polluting material above and below the Earth's surface, which can have catastrophic impacts on the environment and thus on mankind. The solid Earth, as a source of renewable and non-renewable resources, must receive more attention by scientists and policy-makers alike on global, regional, and local scales, in order to ensure the economic health of all countries. Population growth and the desire for higher living standards lead to increased industrialization. This, in turn, requires economic and policy decisions that take into consideration the essential relationship between scientific research and informed decision-making.

The Western European Geological Surveys (WEGS) with their geoscientific databases can supply the necessary scientific input to prudent policy-making. The Surveys, many founded a long time ago, have an official responsibility for the collection of geoscientific data, and for making these and their interpretation readily available to the public in the form of reports, maps, and databases. Control, management, and availability of these databases is of prime importance for the future, and the role of geology in risk-prediction is clearly underlined.

As instruments for the study and control of the physical geological environment, the national Geological Surveys are, as a group, at the centre of all actions that concern that environment. In Western Europe, 21 Surveys have been grouped informally since 1971 in the WEGS which serves as a meeting and discussion forum. With the gradual increase in European cooperation, the WEGS is naturally beginning to play an essential role in the field of the European environment and the control of the related risks.

The idea of producing a synthesis volume concerning these matters was born out of the experience acquired by a wide range of Survey staff in varied European settings. One of the clear conclusions of such experience is, for instance, that most of the lessons learned in one country can be applied to almost any other country. This book is aimed at the widest possible readership; at the layman as well as the specialist. Everyone should be able to find food for thought in its content, in the fact that future patterns of environmental work/studies will have to be multidisciplinary, and that priorities will have to be clearly defined.

1.2. A GENERAL OBJECTIVE: THE ENVIRONMENT

1.2.1. Definition

According to a commonly used definition, the environment is the combination of natural (physical, physiological, and biological) and cultural/sociological conditions that may act on living organisms and human activities. Geology belongs to the environment. The geological environment refers more specifically to the Earth and water systems, and it may be affected by the two major risk categories: those linked to natural hazards and those arising out of development activities.

The Geological Surveys, together with other institutions, are deeply involved in these topics. They relate to these hazards through their responsibilities for geological infrastructures and various types of applied studies. Geological mapping, large-scale geological syntheses, and the management of surface and sub-surface data banks are an integral part of the geological infrastructure. Applied studies fall into two main categories: the exploration, development, and protection of natural resources, and all work related to the development and management of the surface and sub-surface.

1.2.2. A growing implication

All countries today are both economically and socially involved in problems related to the environment, whether in fighting against natural or man-induced catastrophies, or in controlling development, where production is the first objective. Even though some governments with a heightened 'ecological awareness', such as those of The Netherlands, Switzerland, and West Germany, have been leading in this

---

* Numbers in square brackets throughout the text refer to the contributors as listed in Appendix 2.

Fig. 1.1   The sea on a calm day, but the clouds are gathering. (Photo: Geological Survey of Denmark)

domain, the alarm is usually raised by international organizations, which then attempt to marshal combined efforts on trans-national problems.

The problem is focused in the 'Brundtland Report' to the World Commission on Environment and Development with the title *Our common future*. This report clearly outlines the dangers for the human race if due measures are not taken to redress man's destructive impact on the environment in which future generations will live.

Certain of the more important traditional and recent national and international actions are worth mentioning and are indicative of the total effort undertaken.

At the national level most countries are currently involved in environmental management and protection, with varying degrees of involvement or efficiency.

In Germany, for instance, monitoring activities (water, air, soil) are mainly in the hands of the water and environment boards of the Länder, whereas national support for broader research and inventories is coordinated by the corresponding federal office.

In The Netherlands, a specific research institute, the National Institute of Public Health and Environment Hygiene (RIVM) has been set up to deal with all aspects of environment studies and protection.

In France, the water agencies (Agences de l'eau) were set up in the 1960s and have a key role in the monitoring of surface water quality and for financing depollution investments. Their areas of competence do not correspond to administrative boundaries, but rather to physical limits, and include one or several water-basins.

The list of national actions could be extended much further.

At the international level, many actions are coordinated by the European Community, and the creation in 1990 of the European Agency for the Environment will play a major role by strongly emphasizing the will of the EEC to improve data collection and environment management in the European Community. On the other hand, there is a traditional involvement of the EEC in support of inventories and research projects in the field of environment. There is every reason to believe that this involvement will increase, and the general move towards greater concern with the human environment will be of increasing importance to all European (EEC and non-EEC) governmental authorities.

Some programmes concern definite areas more specifically. This is the case with the Mediterranean Strategy and Action Plan (MEDSAP), created in late 1988 by the EEC (DG XI), which is aiming at the protection of the Mediterranean Sea and its catchment basins.

At the world level, many other programmes could be mentioned, all supported by the United Nations or the World Bank. The United Nations Organization monitors a specific programme on the environment (UNEP) and sponsors the International Decade for Natural Disaster Reduction (IDNDR) covering the period 1990–2000.

The World Bank is promoting action plans for the environment at a national scale in several countries, mostly in Africa, and also has a project for cleaning Asian capitals. The need for adequate data led the World Bank to set up an environmental information system (ENVIS) which started at the end of 1987.

Also at the global scale, the International Geosphere–Biosphere Programme (IGBP) Global Change will incorporate facts and data of all geoscientific areas in order to estimate the influence of natural processes together with anthropogenic factors on the future development of our climate.

The involvement of the Geological Surveys in such programmes is a matter of tradition, and has little bearing on the ministry or government department under which they function. Still, it is quite obvious that future developments will demand an even larger degree of cooperation between organizations and states. It will be necessary to arrive at a commonly accepted basis for the scientific research of problems related to the environment, leading to the creation of suitable systems for the prevention and remedy of environmental risks.

1.3. BACKGROUND: THE WESTERN EUROPEAN GEOLOGICAL SURVEYS (WEGS)

1.3.1. The foundations

The WEGS association was created in 1971 as an informal discussion group without legal responsibility. The idea was to constitute a forum for discussion of problems of common interest, to benefit from mutual experiences, and to facilitate the exchange of information and the taking of common actions.

The WEGS functions by means of an annual meeting of the Directors of the various Surveys, and through the actions of Working Groups they have set up on subjects of common interest. During the 1980s the following topics were studied by Working Groups:

(1) applied mineralogy;

(2) the market for geological maps;

(3) collaboration with developing countries;

(4) energy;

(5) remote sensing

(6) applied geochemistry and mapping;

(7) applications of computers;

(8) the environment;

(9) particular problems of the marine environment.

Today, only groups (5) to (9) are still active. The Working Group on the environment plays a key role as it is concerned

with standards in environmental matters seen from the viewpoint of the work of Geological Surveys. It also has been given a coordinating role for the synthesis of 'geology and the environment', the subject of this work.

In certain cases, the groups also liaise with other, existing, international groups. For example, the Chairman of the group working on the applications of computers participates in the meetings of the International Consortium of Geological Surveys for Earth Computing Sciences (ICGSECS).

After having been in existence for almost two decades, the inherently flexible WEGS network has proved its efficacy for creating a basis for relations and contacts, and for generating programmes with several partners.

1.3.2. The multidisciplinary approach

The diversity of the Working Groups underlines the multidisciplinary nature of the work related to the geological environment undertaken by the Geological Surveys. Physics, chemistry, mathematics, and geology are combined in the study of the complex phenomena which constitute the subject area of the earth sciences. Increasingly, disciplines that previously had little relevance to the work of a Geological Survey, such as biology, ecology, hydrology, and climatology, now play a significant role. Topography is commonly a component of the studies, as it frequently reflects aspects of the phenomena under review.

The competence of the Geological Surveys is thus manifold, covering the surface, the sub-surface and the associated fluids. The common roots lie in several fields of activity, but, as emphasized above, the Surveys also have strongly varied fields of interest.

1.3.3. Diversity of the assignments

The various Geological Surveys have a great diversity and range of assignments, some of which are very specific:

- The Swiss National Hydrological and Geological Survey is unique in having the official task of operating the national monitoring network for surface water with respect to both quantity and quality.
- The services of Germany, both the BGR (p. xii) and those of the Länder, are the only ones having the responsibility for soil mapping and the major associated scientific programmes.
- The regional survey of the Piemonte (Italy) is alone in having a mandate in meteorology.

It should furthermore be noted that certain Surveys are responsible to the ministry or department of the environment.

The activities of geological surveying and the construction of surface and sub-surface data banks are common to all Surveys. Most of them deal with exploration for minerals, whereas only one (France) is concerned with research in mineral beneficiation. Only four Surveys (UK, The Netherlands, Germany, and Denmark) have an official role in surveying, data bank management, and administration control in support of exploration of hydrocarbons.

As indicated above, the German Surveys (BGR and Länder) are the only ones which deal with soil mapping and soil research, and in Switzerland the monitoring of surface water is in the hands of the National Survey, which carries out both geological and hydrological surveys. In contrast, most surveys deal with groundwater at the national, regional, and local level, and the impact on it of such as waste disposal. A much smaller number (including Germany, UK, and France)

Fig. 1.2 Precipitation and wave-action erode the sea-cliff, and threaten coastal inhabitants and activities. (Photo: Geological Survey of Denmark)

deal with radioactive waste disposal alongside the legal administrative bodies concerned.

Whilst the maintenance of the basic geoscientific infrastructure is common to all Surveys, in the area of applied studies an enormous range of subjects has been developed in various combinations and to varying degrees in each Survey.

This diversity in activities has led to the recognition that the Western European Geological Surveys could combine their efforts on geological syntheses of matters of fundamental importance to mankind. The pace of development and changes in the environment, however, have sharply focused the need for unified action in an area which is now clearly identified as a priority.

In writing the present synthesis, those Surveys that have responsibilities in the specific fields of the environment were made priority contributors to the relevant chapters. In wider terms, it appears that no Geological Survey can act alone to evaluate the problems related to the environment. In fact, one of the major difficulties is the coordination of the various

inputs from all relevant organizations. Suggestions have been made nationally and internationally for the solution of this problem, but so far they are most certainly insufficient. One of the aims of this work is to show to what extent the Geological Surveys can be the central coordinators for this necessary synchronization of effort.

Such an objective demands a strongly inter-ministerial approach, which is already available to some extent in the relationships between Geological Surveys, as is shown by the diversity of their respective administrative supervision (Table 1.1).

1.3.4. Organization

The scale at which the problems are tackled is also an important factor. In this context, the relevance of the national coordinating role of a Geological Survey must be stressed. This is emphasized by its strong regional presence and its direct contacts with the appropriate regional authorities.

Table 1.1 Ministries controlling Geological Surveys[1]

Ministry or Department	Country
Science and Research	Austria, United Kingdom[2]
Economic Affairs	Belgium, Germany (BGR and five Länder Surveys), The Netherlands
Industry and/or Energy	Finland, France, Greece, Iceland, Ireland, Norway, Portugal, Spain, Sweden, Turkey
Environment	Denmark, Germany (four Länder Surveys), Italy, Switzerland (federal survey)
Agriculture and Natural Resources	Cyprus
Public Works	Luxembourg
Regional Government	Switzerland (cantonal surveys), Italy (regional surveys)

[1] As at the 1 September 1990
[2] Strictly speaking under the Natural Environment Research Council rather than the Department of the Environment, so the UK could also be placed under the heading 'Environment'

In this respect also, the situation differs from one country to another. It ranges from the centralized situation in metropolitan France, with 22 regional offices relating to a central Geological Survey, to the strong regional structure in Germany, with nine autonomous Länder Surveys independent of the federal Survey. In all cases, the national/regional relationships are strong, whether in terms of specific regional problems, in the choice of geological infrastructure, or in the selection and equipping of pilot sites.

1.4. PROBLEMS DEALT WITH IN THE SYNTHESIS

The study of environmental problems is the general framework in which the present synthesis is placed. But, more precisely, it is the idea of 'risk', such as vulnerability or exhaustion of a resource, or the degradation of man's environment, that lies at the core of the debate. The problems posed by 'risk' must be answered by specialists, preferably by means of an integrated multidisciplinary approach.

1.4.1. Hazard and risk [152]

The term 'hazard' refers to natural phenomena (seismic hazard, landslide). The notion of 'risk' covers more specifically an economic appraisal of potential damage to man or his economic activities. In fact, it covers two quite distinct aspects:

(1) the probability that a phenomenon will occur;

(2) the consequences when such a phenomenon does occur.

The first aspect corresponds to a physical reality, such as the development of pollution or a landslide. The second aspect has a socio-economic dimension embracing the impact on the physical environment.

The aim of these comments is to highlight the complexity of the notion of 'risk', both in its physical and in its socio-economic dimensions. To these first-order factors must be added two secondary elements. These are the factors of space (spatial extent of the phenomenon) and time (duration of the impact). The synthesis aims at a better understanding of this complex system of relationships. However, these subdivisions should not mask the fact that such effects are commonly cumulative, in particular for the hazards induced by man, and

that in many cases all gradations between short- and long-term can exist.

1.4.2. Expert evaluation

The study of hazards, whether natural or induced by man, requires specialized evaluation. The study of the physical geological environment, under its apparent unity, covers an enormous diversity of problems, which can be grouped into three major units:

(1) the solid environment, which is specific because of the type of rocks involved, their (varied) ages and their properties;

(2) the fluid (gases and liquids) environment, the importance of which is fundamental, either as a pervasive element of the solid environment, or as a mode of transfer, particularly of pollution;

(3) the interaction between the solid and the fluid environments.

A knowledge of the solid/liquid interactions at all scales is a fundamental prerequisite for understanding the phenomena that affect the environment, and the Geological Surveys have the specialized expertise available in all fields concerned.

Although the input from each specialized area is essential, the role of the overall coordinator in combining these expert fields is fundamental to the understanding of any geological problem.

1.4.3. Integrated approach to the environment

The effort in assembling all knowledge on a subject is not limited to the study of a given problem alone; rather it has to be part of the integrated approach to environmental problems that bears on the physical environment as a whole. This integration also plays a role in the various phases of a risk study, such as diagnosis, prevention, monitoring, and remedy after the problem has begun to develop.

The use of computers has opened up a vast opportunity for the multidisciplinary approach to problems, both in storing data, and in their processing and use for specific aims. This powerful tool has made it possible to have an integrated approach, even though the environmental problems themselves and the work on them may be widespread.

Fig. 1.3 The flooding of a coastal lowland shows what could happen as a result of a rise in sea-level related to the 'greenhouse effect'. (Photo: Geological Survey of Denmark)

Climatology has adopted this approach long since, and recently with very powerful computing tools. It is now of vital importance that the work in support of controlling and developing the geological environment to the maximum benefit of mankind follows this example, which implies quite significant investments—of the appropriate order of magnitude as required by the goal—and the necessity to work in the long term.

1.5. TARGETED READERS AND THE ORGANIZATION OF THIS BOOK

This work addresses itself in the first place to the non-specialist and, in a general sense, to all people having to deal with environmental matters. Thus, as citizens, we are all concerned with this book.

Each type of hazard is described in detail and illustrated by carefully selected case histories. The latter are appropriate for each country, whether or not the cases themselves are geographically located in the country involved.

Finally, proposals for further action and research are presented, showing the dedication of the Geological Surveys to contribute to the further mastery of environmental problems.

1.5.1. Informing the decision makers

The wish to address the non-specialist arises out of the need to inform decision makers. This requires a clear explanation of the functions and fields of competence of the Geological Surveys in matters concerning the environment.

This aspect is clearly shown by Chapter 2, the only part of the book which approaches fundamental geology. A practical presentation of geology is aimed for here, putting the accent on the causes of natural catastrophies and on the vulnerability of natural resources in a general sense.

The specialist can, through this same text, become aware of unfamiliar hazards, or find approaches in the collaborative

Fig. 1.4 Basalt cliffs under attack from the North Atlantic. (Photo: Geological Survey of Denmark)

areas between hazard types and methods of approach. On the other hand, the layman who has a direct relationship with the environment should find food for thought, concerning both his individual behaviour and that of his fellow citizens as a whole.

1.5.2. Specialized chapters

In order to reach all the different sectors of the public domain, it was decided to present each type of hazard individually in Chapters 4 and 5. Chapter 6 is then devoted to outlining the more global aspects of the role of the geosciences in planning and development work.

Chapter 3 is concerned with the various natural resources, from the viewpoint of their geological relationships, and from that of their exploitation and management. Chapter 4 gives information about the various potential hazards from natural phenomena and Chapter 5 describes many hazards arising out of development on and in the geological environment.

Each sub-chapter is organized around a presentation of significant case histories which should serve to illustrate the diagnosis of the main problems that are posed for each hazard type, as well as the points of similarity displayed by them. In each case, the socio-economic aspects are also discussed.

1.5.3. Case histories

The case histories have been selected so as to provide the best illustration of the specific types of problems involved. With respect to the necessary adjustment to local conditions (e.g. geology, climate), the reader can easily make the connection between the detailed study of a case history in another country with similar situations in his own surroundings.

The case histories originally submitted have, in most cases, been shortened considerably or presented only in abstracts. Readers who are interested in obtaining further information regarding individual case histories are advised to contact the authors as identified in Appendix 2.

1.5.4. Perspectives and lines of action

Proposals for future action form the logical end-point of this work. After diagnosing the problems and risks, it is important to propose common actions which could go some way towards coping better with the environment in the future. Without anticipating the conclusions of Chapter 7, a few remarks are apposite here.

It is evident that many of the identified problems are common to the experience of all Surveys. This leads to two areas of common interest:

(1) the search for uniformity in collecting, presenting and accessing the data;

(2) the use of common methodologies in approaching problems, covering diagnosis, prevention, monitoring, and reduction.

The interfaces between the various components of the physical environment also form a unifying factor, which militates in favour of standardized procedures in methodology, in the modelling of air/soil/water interactions at surface and underground, and in the experiments carried out on pilot sites, to name but a few subjects.

In encouraging the uniformity of approach, attention should be drawn to complementary areas of applied science, to the many different disciplines involved, and to the varying degrees of awareness of the problems amongst decision makers and others, upon whose actions the quality of life depends.

In a general sense, this synthesis should facilitate the understanding of the action necessary in the face of the ever-increasing impact of problems in the environment.

2

Introduction to the geology of western and southern Europe

A. Autran

2.1. PREAMBLE: SOME BASIC PRINCIPLES IN EARTH SCIENCES

2.1.1. Geology [65]

Geology is most commonly defined as the science that aims to describe and understand the genesis of the materials that make up the Earth and to study past and present transformations. Etymologically, geology denotes the science of the Earth, although the term 'earth sciences' is more generally used today because the range of scientific disciplines has developed considerably over the past century and now covers most of the sciences, including the humanities. By applying and adapting the laws of physics and chemistry to terrestrial phenomena, two major branches of earth sciences, geophysics and geochemistry, have been developed in order to quantify and qualify the processes that control the 'life' of the Earth.

All the terms in use cover key concepts that comprise the basis of modern geology: the rocks themselves (and hence the component materials, mineral resources, and water resources), the transformations they have undergone (and hence the dynamics of their evolution), and the fossils they contain (and hence the study of earlier forms of life).

Let us try, however, to look at some other areas, more specifically with a view to the environment, the subject with which we are concerned here. The study of air and atmosphere is not strictly a part of geology, although climatic conditions do govern the rhythm of erosion: substances transported by the wind, for example, are re-deposited on the ground and in water and form part of the geological cycle.

Biology and the humanities are clearly distinct from the earth sciences, and yet they are the basis for all studies of traces of past living forms. Similarly, ecology covers a complex system of interactions between the living environment, fluids (air and water), the mineral components physically and chemically present, and the topography, itself conditioned by the nature of the rocks, the type of deformation they have undergone, and the cause of erosion.

Thus the earth sciences are not isolated in an ivory tower but are an integral part of the whole range of scientific disciplines that involve the very core of our existence. The geologist is concerned with the evolution of terrestrial phenomena over time periods that may be very long, but also with specific episodes of crisis; his scale of observation ranges from a crystal, to a continent, to the Earth as a whole, and he must therefore manage to synthesize complex and multiform data and make increasing use of modern computer technologies.

2.1.2. Rocks

Geologists normally distinguish three main types of rock, each of which provides information on the various physical and chemical processes that created them: sedimentary rocks, igneous rocks, and metamorphic rocks. The latter are metamorphosed as the result of the transformation of sedimentary or igneous rocks by complete recrystallization and deformation due to their slow transport to great depths within the Earth's crust, where heat, increasing pressure, and migration of fluids (e.g. water, CO_2) produce changes in the initial mineral components but generally preserve the chemical composition. Igneous rocks result from fusion into a silicate melt (=magma) of the rocks that form the mantle and external crust of the Earth. Where the temperatures and pressures are high enough, the metamorphic rocks themselves can undergo partial fusion.

These rock liquids (magmas) easily migrate upwards due to their low viscosity and density. They can remain in the upper crust, forming intrusive or plutonic rocks that crystallize into coarse-grained aggregates as they slowly cool (e.g. granites), or they can erupt from volcanoes at the surfaces in the form of volcanic rocks that everywhere crystallize into fine-grained aggregates or glass (e.g. basalt). Magmatic rocks therefore form the essential agent that transports chemical components from the Earth's interior toward the surface, and it is these rocks that also formed the Earth's early crust. The Moon provides a remarkable frozen image of this process, having been cooled for the past three thousand million years and having at no time been affected by water or wind at the surface. Magmatic rocks are generally classified according to the amount of silica they contain, rocks with a low silica content being referred to as basic or mafic (e.g. peridotites, basalts) and rocks with a high silica content being referred to as acidic (e.g. granites).

Sedimentary rocks appeared on Earth after the other two types, as a result of surface dismantling of the different rock types by physical and chemical disaggregation at the contact with the atmosphere, followed by transport either by water in rivers or by the wind, to sites of deposition, very commonly in the sea but also in intracontinental fresh-water lakes, sabkhahs, and deltas. Deposition is governed by the laws of physical hydrodynamics and by chemical or biochemical processes, the latter being responsible for most calcareous rocks, for example.

2.1.3. Time [44]

The most important factor for the geologist is time. It should be borne in mind that the creation of the Earth dates back some 4.5 thousand million years. The famous 'Big Bang' by which the universe is considered to have been created is some three times older. An understanding of time implies two very important geological concepts: first the ability to measure it, and second the ability to follow its evolution in order to comprehend the dynamics of the phenomena involved.

Two main tools are used for this purpose, based on the facts that some elements are radiogenic and that the species that make up the living world have evolved considerably with time and continue to do so. Some elements contain natural radioactive isotopes that disintegrate in the course of time to a chain of other elements. This property of natural radioactive transmutation is the basis for the measurement of geological time, or *geochronology*. The method involves measuring the present composition of a rock for a given radioactive element and its degenerative components, and comparing this composition with the calculated initial composition at the moment when the elements in question became incorporated in the rock and began to disintegrate within it. Such is the case with uranium, which is gradually transformed to lead. When used on certain selected minerals, the uranium–lead method of age determination will give the age of rocks some thousands of millions of years old.

At the other end of the time scale, radiogenic carbon (carbon 14) is used with reference to non-radiogenic carbon in order to date recent geological formations over periods ranging from several thousand to some tens of thousands of years. This method can be used on fossilized vegetal remains or on the carbon gas dissolved in underground aquifers in order to establish the time of origin. For the same very recent past, some other methods often give very precise ages useful for deciphering environment change during the last few thousands of years. Some of these are also isotopic, using measurements of short-lived nuclides from the natural decay series of uranium and thorium. They show for example the short time (100–10 000 years) required for the extraction of basalt from the mantle and its eruption at the surface, or the age of Quaternary coastal reefs. Others are directly related to biological processes such as dendrochronology (the study of growth rings in trees) or the amino acid method.

Primitive life forms were present on Earth some 3.6 thousand million years ago. Fossils of living species are commonly used to date geological formations dating back some 500 to 550 Ma*. Over this long period of evolution of living organisms, man made his appearance only some 2.5 Ma ago.

Because modern techniques of absolute age determination have evolved only over the past few decades, the specific time scale and succession of geological events which formed the existing world are, as used by geoscientists today, based to a major extent on knowledge acquired over almost 200 years of study of the living environment in the form of fossil remains (Fig. 2.1).

Apart from their usefulness in dating sedimentary rock strata (enabling the division of geological history into *stratigraphic eras* and *stages*: Fig. 2.1), associations of fossil species give clear data to reconstruct the past environment in which such animals and plants lived: for example, deep open sea or coastal area, continental lacustrine, and warm or cold sea water. Enormous progress has been made since the late eighteenth century when the principles of stratigraphical palaeontology were established. By about 1850, all the major groups of fossils had already been discovered. From 1945, the requirements of oil exploration resulted in the very rapid development of micropalaeontology: so it is possible to exploit data obtained from a multitude of very small organisms. Recent biometrical techniques in micropalaeontology allow very precise dating in suitable sediments, better than that achieved by isotopic geochronology.

* The symbol Ma is used to denote million of years.

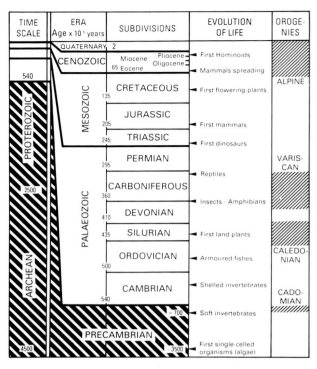

Fig. 2.1 Time-scale and main stratigraphical divisions (isotopic ages from IUGS (Commission of Stratigraphy) and G. S. Odin (1990) *Geochronique 35*).

2.1.4. Dynamics—principles of plate tectonics

About 2 000 Ma years ago the Earth began to function as a physical and chemical system comparable to that observed today. Since then, mutiple modifications of the form of continental masses and oceans have taken place at the Earth's surface. Work by geologists over the past 80 years has made it possible to reconstruct these successive past geographies with a fair degree of accuracy, at least as for the last 300 Ma or so.

The interior of the Earth, at depth, can now be described with increasing accuracy thanks to recent work by geophysicists, using such data as the Earth's magnetic field and the propagation of shock waves from earthquakes. In recent years, such work has been increasingly based on measurements made from satellites (including measurements of the gravity field, of the precise shape of the earth, and of the ocean floor topography) or from oceanographic ships (essentially concerned with exploration of the ocean floor).

Intensive exploration of ocean areas over the past 30 years has completely changed our view of the Earth. It shows that the ocean floors, representing about 65 per cent of the surface area of our planet, have been renewed in less than 200 Ma as a result of gigantic recycling of matter between the ocean floor and the deep layers of the Earth. This mechanism is now becoming better understood from measurements and observations by geochemists, who analyse the composition of the various types of rock that rise to the surface out of volcanoes, and who can reproduce in the laboratory the very high pressures and temperatures that prevail in the first few hundred kilometres below the Earth's surface.

Such data indicate that the Earth's interior comprises three concentric spheres of material that differ considerably in composition, the density increasing substantially with depth (Fig. 2.2).

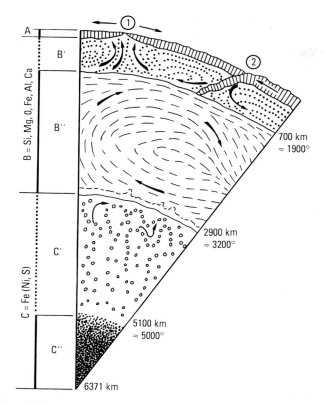

Fig. 2.2 Section through the Earth showing its chemical and mechanical zonation (temperature in °C).

A: Rigid *lithosphere* comprising oceanic or continental crust upon the uppermost rigid part of the mantle (see Fig. 2.4).

B: *Mantle* (silicates and oxides); B′: upper mantle, partly molten and ductile (asthenophere), with large convection cells; B″: crystalline lower mantle (rigid).

C: *Core* (mainly iron); C′: external liquid core, with fluid currents producing the Earth's magnetic field; C″: internal solid core.

1. Spreading ridge
2. Subduction zone

The *crust* forms the external envelope of the Earth. It comprises the oceanic crust, 4–6 km thick and of relatively uniform basaltic composition, and the continental crust, of very varied composition (igneous, metamorphic, and sedimentary rocks), very commonly between 30 and 40 km thick, though up to 70 km thick beneath major mountain belts and thinner below the great sedimentary basins.

The *mantle* extends from the base of the crust to a depth of 2900 km. It is much more homogeneous in composition, composed essentially of magnesium silicates and lesser amounts of iron, calcium, and aluminium. This is a solid and crystalline zone, in which very slow movement of matter occurs. As a function of temperature, which rises with depth, the upper mantle, crystalline and rigid in its upper part, undergoes partial fusion from a depth of 10–300 km below the base of the crust. Thus, a thick zone appears in the mantle within which a silicate liquid (basalt) is dispersed in small proportions. The presence of this liquid means that the mantle loses its rigidity, resulting in a zone of easy deformation upon which the crust and outer mantle form a kind of hard shell around the Earth. This shell is known as the *lithosphere* (or sphere of stone) and it can slide, moved even by very weak forces, over the slightly molten mantle referred to as the *asthenosphere* (or zone of no resistance). In this way, three zones of differing rigidity are superimposed on zones of

differing chemical and mineralogical composition (the oceanic crust, the continental crust, and the mantle); the depth of the limits between these zones (the external lithosphere, the middle asthenosphere, and the rigid deep mantle) is related to the distribution of temperature and pressure in the mantle.

The *core* in the centre, at a depth of 2900–6400 km, is assumed to consist predominantly of iron, nickel, and sulphur.

Dissipation of the Earth's heat, from the internal core to the external rigid lithosphere, generates within the mantle an extensive and slow stirring of matter, in the form of convection currents. Geophysicists and seismologists have in recent years learned to observe and describe the position of the hot rising zones and cold descending zones in these huge convection cells within the mantle. These mantle dynamics are responsible for complex movements in the continental and oceanic lithospheres that can be observed and measured today, but the presence of which in earlier periods of the Earth's history can also be reconstructed by geological analysis.

Some major principles of the earth sciences are the basis of the currently accepted *theory of plate tectonics or global geodynamics* of the planet. The theory was developed around 1970, and explains the various aspects of the geological history of the Earth's surface, including:

- formation of oceanic and continental crust;
- distribution of oceans and continents;
- present-day velocity of horizontal plate movement;
- formation of mountain ranges and sedimentary basins;
- formation of volcanic rocks and granites;
- distribution of temperatures at depth;
- transformation of rocks by deformation and metamorphism;
- successive geographies encountered in the course of geological time;
- climatic variations.

The basic principles of plate tectonics are as follows:

1. The lithosphere is composed of a small number of rigid plates (currently twelve in number) bounded by borders where most of the deformation occurs. A plate can be formed either from continental crust or oceanic crust or, more generally, from both at once. The Pacific plate, for example, is now composed only of oceanic lithosphere, whereas the African plate comprises the continent of Africa, the eastern half of the central and southern Atlantic Ocean, and the western half of the Indian Ocean (Fig. 2.3).

2. Three types of plate boundary are distinguished:

(a) Boundaries at which oceanic crust is created by the rise to the surface of huge amounts of basalt formed within the mantle below the lithosphere (Fig. 2.4). This creation of crust increases the width of the area occupied by ocean floors by some 1–20 cm per year. Plates diverging from these *spreading ridges* move apart from one another at equal velocities.

(b) Boundaries at which the plates move closer to one another. It is at these boundaries that material disappears and balances the new material of the oceanic crust created at spreading ridges, the surface area and volume of the Earth being fixed, at least at the time-scale of this process. This disappearance of oceanic surface results from the oceanic lithosphere descending into the mantle by the process of *subduction*.

(c) Boundaries at which two adjacent plates slide along a major vertical fault, such as the famous San Andreas Fault in California and the Anatolian Fault between the Arabian and Eurasian plates in Turkey.

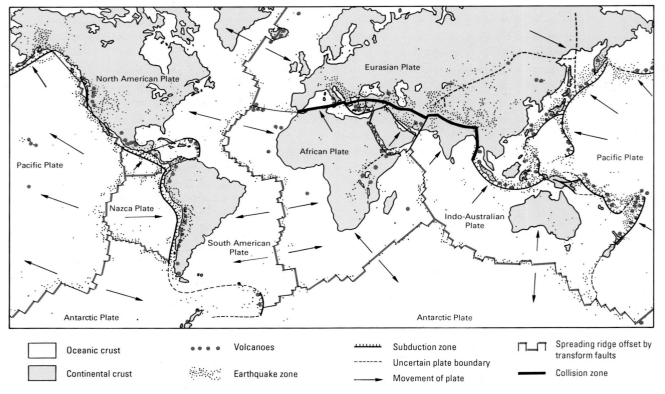

Fig. 2.3 Boundaries of the present 12 major plates on Earth.

3. The plates are in constant movement dependent on the velocity of movement apart and subduction into the mantle of their oceanic portions. The continents borne by the plates are in this way constantly displaced in relation to one another: at some point in time they were joined in a single mass (such as Pangaea, 300 Ma ago (Fig. 2.10)); at others they were dispersed and separated by several oceans, as is the case today. This whole process is known as *continental drift*.

The cyclical surface dynamics of our planet (each cycle lasting about 200 Ma) comprise several major stages and some critical surface sites:

1. *Oceanic rifting*: above rising convection currents, the continental crust tends to become thinner, crack, and open up. This rift allows the surface eruption of basaltic volcanism of deep origin, or oceanic volcanism. This type of oceanic rift, the first stage of creation of a new ocean, is well illustrated by the present-day Red Sea and the Afar region.

2. *Oceanic expansion*: basaltic effusion develops in the axis of the initial rift forming a spreading-ridge where oceanic crust is continuously created. This new crust is like a rigid floor sliding over the underlying viscous mantle (the asthenosphere), and gradually separates the two continental masses. This process is known as oceanic expansion, and can be observed today in the Atlantic Ocean and on land in Iceland. The margins of the American, African, and European continents, that are moving increasingly apart in this way, are referred to as passive continental margins.

3. *Oceanic subduction*: the oceanic crust formed over the spreading-ridge moves progressively away from the 'hot zone' of the ridge where the basalt pours out and cools. The thickness and density of the oceanic lithosphere thus increases with age and distance from the ridge by cooling. This mechanism causes deepening of the ocean floor, and creates abyssal plains 5000 m beneath the sea surface, whereas the submarine

mountain ranges formed by the hot oceanic spreading ridges are covered by less than 2500 m of water.

The dense, cooled, oceanic lithosphere finally plunges back into the viscous mantle (or asthenosphere), where it is subducted at a velocity comparable to that at which the oceanic crust over the ridge of the same ocean is being simultaneously created. This phenomenon is known as oceanic subduction, and is marked by the creation of deep trenches in the oceanic floor (such as the Mariana trench, the Japan trench, or the Peru trench). In Europe only one small subduction area exists, beneath southern Greece and Sicily (Fig. 2.11, p. 19).

Major thermal disequilibrium accompanies this subduction at depth. Progressive reheating of the subducted oceanic crust results in metamorphic alteration and partial fusion of that crust at the same time as partial fusion of the mantle. The magmas thus created rise to the surface and progressively build new continental crust over an area about 100 km wide above the subduction zone.

Where subduction occurs below a piece of oceanic plate it forms *island arcs*, strings of volcanic islands bounding subductive oceanic trenches, such as the Antilles arc and the many island arcs in the northern and western Pacific.

Where the oceanic plate plunges below a continental part of the adjacent plate, the magmas rise through the continental crust, heat it, and form a range of mountains known as the *active margin*. This type of process can be observed in the Andes, from Ecuador to Chile.

4. *Oceanic closure*: after a stage of opening up, of variable duration, all oceanic plates begin the inevitable process of disappearance by subduction. In this way, the lithosphere of the Atlantic Ocean, the creation of which continues over the vast mid-Atlantic ridge, begins to disappear into the zone of subduction in the Antilles, at which point it is at its oldest (180 Ma). The lithosphere of the Indian Ocean also disappears

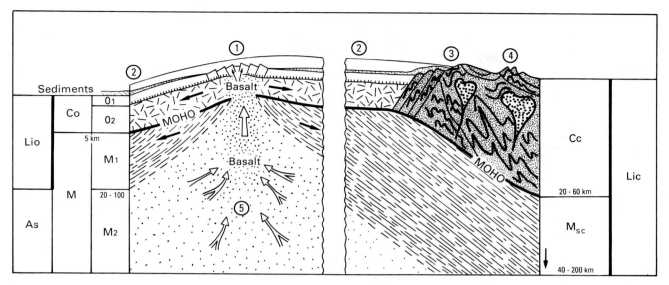

Fig. 2.4 The two types of lithosphere. To the right, continental lithosphere (Lic); to the left, creation of new oceanic lithosphere (Lio) at an oceanic spreading ridge (1).
1. Axial spreading ridge with its rift valley in which new oceanic crust (Co) is created by basalt influx from zone 5 beneath, where it collects from partial melting of mantle M2 in rising asthenosphere (As).
01. Upper oceanic crust (basalt).
02. Lower oceanic crust (mafic and ultramafic rocks). Co crust has a constant thickness of 5 km.
2. Abyssal plains (~5000 m) with thin sedimentary cover.
3. Continental rise or passive margin of the continental crust (Cc); the thickness of the rigid lithosphere upper mantle (M1 or Msc) increases away from the spreading ridge (1) due to its progressive decrease in temperature.

along the island arcs of Malaysia and Indonesia. On either side of these subduction zones, there is necessarily a point where two continental portions of converging plates begin to move closer together; this movement continues until the plunging oceanic portion disappears completely and the ocean closes.

5. *Continental collision*: once oceanic closure is complete, the continental portion of the plate, from which the ocean has disappeared, reaches the subduction zone. The very light continental crust cannot be absorbed into the mantle and subduction ceases. However, convergence of the two plates does not stop immediately, because it is maintained by more general dynamics related to movement of the other plates.

The two continental lithospheres juxtaposed along the missing ocean (= *suture zone*) are in this way pushed towards one another by their respective plates and collide. During this collision (the duration of which can be some tens of millions of years) fundamental tectonic processes result in the formation of an *orogenic belt* and in the progressive building of great mountain ranges. In this way, the Alps and the Himalayas were created, both the result of disappearance of the Tethys Sea and of collision between the continental parts of two plates: the Eurasian and African plates for the Alps, and the Eurasian and Indian plates for the Himalayas (Fig. 2.3). In the course of continental collisions, the two blocks of old continental crust undergo major changes such as folding, metamorphic recrystallization, superimposition of thick sheets of continental crust (overthrusts), and genesis of granite. The resulting considerable thickening of the continental crust (from 35 to 60 or 70 km) causes disequilibrium between the crust and the dense mantle on which it floats, and it is essentially this disequilibrium that provokes the rise of mountainous relief features.

Finally, the speed at which the two plates converge decreases, the system becomes blocked, and new plate boundaries appear elsewhere.

This brief description of the theory of plate tectonics illustrates the four major stages of the Earth's geodynamic cycle:

(1) opening up of an ocean and transfer of material (basalt) from the mantle towards the oceanic crust;

(2) oceanic expansion;

(3) closure of the ocean in the area of the subduction zone, at which point further transfer of material occurs from the oceanic crust and mantle towards the surface, creating a new continental crust;

(4) collision of blocks of continental crust, and building of mountain ranges.

General and repeated processes therefore work together to build the Earth's surface. Geologists and geophysicists have now deciphered some of them, and can identify their trace in the deformed sedimentary, magmatic, and metamorphic rocks that make up the continent of Europe.

Systematic exploration of the ocean floor by the International Ocean Drilling Program has now validated this theory and has revealed a major factor in the understanding of the Earth's evolution: *the ocean floor, representing two-thirds of the Earth's surface, is everywhere composed of young surfaces, the oldest dating back 200 Ma*, created by magmatic flow along ocean ridge zones and disappearing continuously in subduction zones or destructive margins, within which the material is re-incorporated into the deep mantle of the Earth. It is this continuous recycling that, by magmatic processes, progressively transfers matters from the mantle to the surface, resulting in the formation of continental masses. These rocks are lighter than the oceanic crust, float on the mantle, and cannot be re-incorporated into it. *The continents are in this way the only record of the Earth's history.*

2.1.5. The Earth today [65]

What today's geology shows us is not only a synthesis of thousands of millions of years of the Earth's history, both at the surface and below, but also the very recent effect over several million years of very varied climatic influence which resulted in the relief forms, geography, shorelines, and river systems that can be observed today.

Two hundred years' observation of the various types of rock that make up the continents, and 60 years of careful and increasingly accurate analysis of the causes and processes of their association, have convinced geologists that these geological events have been consistently repeated for at least the past thousand million years. Detailed analysis of each type of phenomenon in action at the present time, forms a fundamental branch of geology/geophysics as practised by every geologist. Wherever he may be he can observe these processes at the scale of the Earth as a whole or on a very local scale (such as an active volcano or the side of a valley). It is this analysis that makes it possible to comprehend the causes of processes that are either currently active or date from the recent past. Geologists use these 'contemporary causes' to determine the history of past events by identifying significant traces of such features and processes preserved in fossil form in the rocks.

All outcropping rocks and all rocks intersected by boreholes, trenches, and tunnels have been analysed in this way for several generations, resulting in progressive precision and validity of observation. This vast stock of data is now handled by the Geological Surveys of individual countries in the form of factual and bibliographic geological data banks and geological maps. These maps provide a synthetic picture of the history of geological processes in a given region and show the surface distribution and geometrical organization of the various rock types. Examined together with the topographic data, this understanding of superimposed or adjacent masses makes it possible to extrapolate the structures of buried portions at depth. Such maps are constantly used at the planning stage, for all types of surface and subsurface development and in exploration for minerals, water, and energy resources. In many cases they are difficult for non-geologists to read and comprehend, essentially because of the great amount of information they contain. For this reason, the national Geological Surveys and certain specialist consultancy bureaux undertake to 'translate' geological maps and data for given types of application (such as groundwater management, mineral exploration, or urban development) by making a thematic examination of one aspect of the data, selecting those factors and implications that are most pertinent for the type of application in question.

2.2. AN OVERVIEW OF THE GEOLOGICAL BUILDING OF EUROPE

The continent of Europe is part of the record of the Earth's continental crust-building process over the past 3000 Ma. Many areas of ocean have been created and destroyed by the successive collisions of older continental margins. From the oldest continental crust (in Finland) to the youngest ocean crust in existence (still being created today in the Tyrrhenian Sea or the Aegean), geologists have deciphered about nine orogenic periods which have given rise to high mountain ranges and have changed the successive geographies of continents and seas.

The present geography of Europe is very recent in geological time. The northern Atlantic Ocean only began to open in the Cretaceous, 100 Ma ago, and the separation of Europe and North America dates back only some 50 Ma. The Alpine orogeny was responsible for building a mountain range less than 60 Ma ago. The last million years represent a period of great human importance: human societies first appeared in the Palaeolithic, during which four great ice ages substantially changed the geography of Europe. The mountain ranges in existence during these ice ages were approximately the same as today, but the sea level underwent great variations of up to 150 m, continuously modifying the coastlines, particularly in flat-lying areas. The climate and vegetation underwent four corresponding changes, with contrasts as great in each place as that visible today between northern Scandinavia and the Mediterranean or northern Africa.

The interest in describing the geological evolution of Europe is far from purely academic. This evolution was responsible for the very diverse geology encountered throughout the countries and to a great extent for the relief at the surface, a factor of obvious importance for natural borders and for the distribution of lines of communication. It is also responsible for the vulnerability of certain regions, for the specific location of natural mineral resources, and for the natural hazards (seismic or volcanic activity, earth movements) that penalize some regions.

Two other basic principles of geological analysis, the concepts of 'cover' and 'basement', are essential for the understanding of the geology of Europe: younger sedimentary material forms a 'cover' over older and deformed rocks that have been recrystallized by metamorphism and injected with igneous rocks, forming a 'basement' below the more recent cover.

Basements are created in the course of each orogeny by an erosional and peneplanation phase that immediately succeeds the thickening of the continental crust during the 'collision stage'. The products of the dismantling of emergent land masses, and in particular of mountainous zones progressively eroded and levelled, are transported by rivers towards seas and oceans where they fill large sedimentary basins. Some such basins form directly over the mountain range that preceded them, as is the case of the basins less than 10 Ma old formed over the Alpine range in the Po–Adriatic region of Italy or the Aegean Sea in Greece. These are the first intra-range cover rocks, commonly deformed in the course of compression related to the end of the orogeny (Fig. 2.11, p. 18).

True post-orogenic *cover rocks* appear during the phase of extension at continental scale that everywhere succeeds the phase of continental collision and mountain-building. In very large sedimentary basins such as that of northern Europe (the North Sea and Germany), the fill process can last for 500 Ma, in this case successively forming the cover of the Precambrian in Scandinavia to the east, the Caledonian range in the United Kingdom and Norway to the west, and the Hercynian range in the United Kingdom, France, Belgium, and Germany to the south (Fig. 2.12, p. 18). This type of deposition over a fairly stable continental platform or over a continental shelf bounding the oceans, generally comprises large amounts of limestone and occurs in shallow water, as can be seen today in the North Sea and the Adriatic.

A second type of thick sedimentary deposit is trapped at the foot of the continental rise at the junction with the oceanic crust and the passive continental margin, at a depth of 5000 m. At the centre of the oceans themselves, far from continental detrital influx, very little sedimentary material is deposited and it is of biological origin (limestone or silica) (Fig. 2.4).

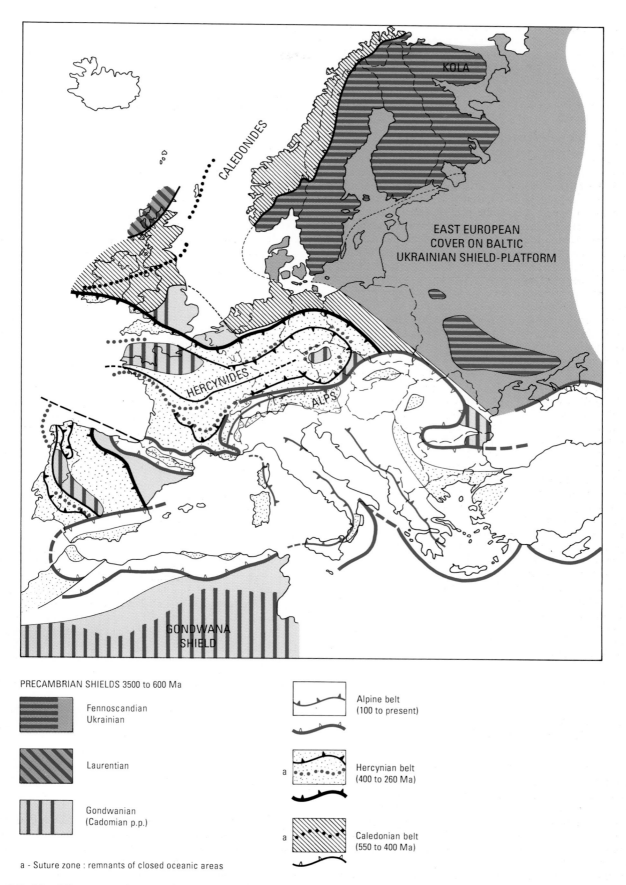

Fig. 2.5 The different ages of basement in Europe: Precambrian (3500–600 Ma); Caledonian belt (550–400 Ma); Hercynian–Variscan belt (400–260 Ma).

Using all the data accumulated over many years and applying the extensive and sophisticated techniques and expertise now available, geologists are in a position to understand the orogenies of the past and to reconstruct the process of the formation of sedimentary basins with considerable accuracy and reliability.

2.2.1. The basement of Europe: its long and complex evolution

As shown in Fig. 2.5, the basement is visible in outcrop in many European countries, from the extreme north of Scandinavia to southern Spain. It is also present in hidden form throughout Europe, buried below more recent rocks.

The European basement is commonly subdivided into eight major cycles grouped as two entities: the Precambrian and the Palaeozoic. The Primary Era, also referred to as Palaeozoic, began about 570 Ma ago. All Earth history prior to that is referred to as Precambrian by reference to the Cambrian, the first stage of the Palaeozoic. (The term Cambrian comes from Cambria, the Latin word for Wales.) It was during the Cambrian (or slightly before) that the first faunas appeared which can be reliably used to date geological time. For the Precambrian, only radiometric age determination can be used.

The Precambrian [15] [170] [148]

Precambrian basement is found to the north and south of the great North Sea–North Germany–Russia sedimentary basin (Figs 2.5, 2.11, pp. 15 and 20). In the north, a succession of orogenic cycles between 3500 and 900 Ma ago constructed a resistant shield that has remained a single block from the Atlantic to the Urals. This is the *Fennoscandian Shield*, which continues beneath the younger sedimentary cover of the Russian Platform to reappear in the Ukraine. It is formed by four major orogenic belts, becoming progressively younger westwards, which, between 2900 and 900 Ma, enveloped nuclei of Archaean relics 3500–3100 Ma old in Finland and the Kola Peninsula in northern Russia (Fig. 2.6).

The last major event to affect this ancient shield, was the Grenville–Sveconorwegian orogeny at about 1000–900 Ma. At that time, Fennoscandia collided with the Laurentian Shield (Fig. 2.7) (named after the St Lawrence River), that occupies 80 per cent of Canada, all of Greenland and parts of the northern USA, forming a single immense continental plate. At the end of the Precambrian, Fennoscandia was separated from the Laurentian Shield by the formation of an ocean occupying a similar position to the present-day Atlantic and known as Iapetus, which closed again 400 Ma ago to give rise to the Caledonian orogeny. The only relics of the Laurentian Shield in Europe are the Lewisian rocks of northwest Scotland, which were separated from the rest of the shield at the opening of the modern Atlantic Ocean 50 Ma ago.

The Precambrian to the south of the North European sedimentary basin, is very different. It belongs to another great ancient shield, that of Africa and South America, known as Gondwana, but is found only as fragments, incorporated into

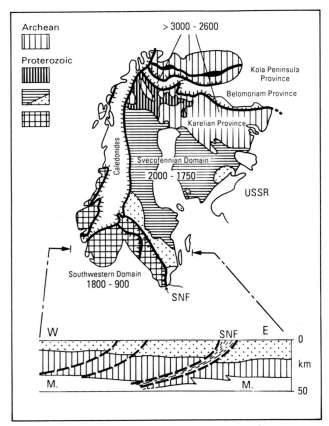

Fig. 2.6 The polyorogenic organization of the Fennoscandian Shield with its Archaean northern part and the surrounding Early and Middle Proterozoic orogenies. The seismic section of the crust across the southern domain shows the crustal strips thrust eastwards during the last Sveconorwegian orogeny. M: Mantle; vertical lines: lower dense crust; dots: upper crust; SNF: Sveconorwegian front (from EUGENO-S project: *Tectonophysics* 1989 vol. 162, n°1).

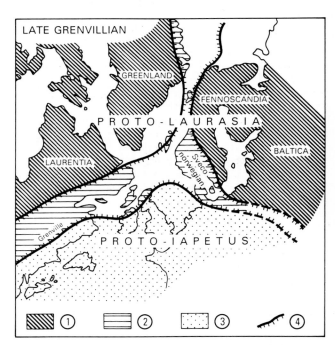

Fig. 2.7 Upper Proterozoic (about 900 Ma) collage of Laurentian and Fennoscandian shields during the Grenville–Sveconorwegian orogeny, resulting in a large continental plate—Protolaurasia (after Ziegler 1978).
1. Continental cratons > 1700 Ma.
2. Active fold belt 1200–900 Ma.
3. Oceanic area < 1000 Ma.
4. Deformation fronts of the active fold belt.

the Variscan–Hercynian and Alpine orogenic belts in Spain, Brittany, southern England, and Bohemia (Fig. 2.5).

Only two orogenic cycles have been identified in these shield areas, one at about 2000 Ma and one at the end of the Precambrian, between about 700 and 550 Ma ago. This latter is the Pan-African orogeny (known in Europe as the Cadomian), so called because, at about 600 Ma, this episode was responsible for welding together the various older components into a single unit: the Gondwana supercontinent.

A change in the dynamics of Earth's evolution that took place at the begining of the Proterozoic (around 2200 Ma) is well illustrated in the Fennoscandian Shield, and also the progressive formation of continental crust by the extraction of magma from the mantle to form igneous rocks, as in Archaean time, and the recycling of older continental crust by deformation, metamorphism, and fusion to form new magmas during succeeding orogenic cycles.

Between 3500 and 2000 Ma the very high temperature basic magmatism deriving directly from the mantle was still very much predominant, giving the 'greenstone belts'. The first cover sediments appeared after the *Lopian orogeny* (2900–2600 Ma). Mineral resources typical of the oldest Precambrian (Ni, Cr, Pt, Cu) are being actively explored in Finland.

In the large area (almost 2000 km across), affected by the *Svecofennian orogenic cycle* (2000–1750 Ma), the identification of relics of oceanic crust and island arcs, first created during the Early Proterozoic, and sedimentary basins has made it possible to reconstruct a geodynamic setting involving the various stages of the plate tectonic regime, with the opening of oceanic basins, followed by their closure and the collision and overriding of their bordering continental margins. Mineral resources typical of recent oceanic and island arc settings are well represented in Finland and Sweden (the Cu–Zn–Co deposits of Outokumpu, Skellefte, Orijärvi, and Bergslagen districts).

Lastly, still farther west, in southern Sweden and Norway, the *South-west orogenic cycle*, lasting 900 Ma, consisted of a succession of varied magmatic episodes, and the formation and filling of sedimentary basins that became welded to the western edge of the Svecofennian shield in broad north–south strips, particularly during the final collision (the Sveconorwegian orogeny at about 1000–900 Ma).

Geophysical exploration during the last ten years has shown the existence of thick (40–50 km) continental crust throughout the shield and generally a very low heat flow. Crustal thickening by overthrusting during collision has also been 'imaged' in the Svecofennian and Sveconorwegian cycles (Fig. 2.6). Metamorphic and intrusive igneous rocks now predominate at the surface as a result of erosion of the entire superstructure of these old mountain belts. Successive palaeolatitudinal positions of this ancient shield show that continental drift was a continuous process 3000 Ma ago. This Precambrian shield, in common with all continental parts of the Earth suffered great geographical changes of position (Fig. 2.8).

The Palaeozoic

At the end of the Proterozoic, a broad ocean, Iapetus, separated the three Precambrian shields, Laurentia, Baltica, and Gondwana (Fig. 2.9). The complete closure of the oceanic areas (from 700–400 Ma) separating these ancient continental masses led to the Caledonian and then the Variscan (or Hercynian) orogenies at respectively 500–400 Ma and 400–280 Ma, forming two separate belts in Europe, but superimposed one upon the other in the Appalachians of western North America. There was no Atlantic Ocean at that time! (Fig. 2.10).

The older, *Caledonian orogenic belt*, between Laurentia and Baltica, is well represented in Ireland, the UK, Norway, Sweden, and Svalbard. This orogeny was responsible during the Devonian (about 400 Ma ago) for the re-establishment of an immense continent extending from Gulf of Mexico to the Urals, known now as Laurussia. The younger *Hercynian belt*, between Gondwana and Laurussia, now forms the basement throughout southern and central Europe, in the Iberian peninsula, France, southern England, Germany as far as the Bohemian massif, and almost most of the basement now incorporated in the Alpine chain (Fig. 2.5).

These two Palaeozoic orogenies, in contrast to those of the Precambrian, were not responsible for the formation of much new continental crust. Through deformation and metamorphism, they essentially reworked the pre-existing Precambrian crustal material involved in collision.

Throughout the length of its axial zone, the *Caledonian orogenic belt* includes the mafic and peridotite rock assemblages characteristic of oceanic lithosphere (ophiolites). These are relics of the Iapetus ocean floor that separated North America–Greenland from Scandinavia at the end of the Precambrian and disappeared at about 420 Ma during the Caledonian collision. This was an event of exceptional intensity, the material constituting the belt being transported bodily eastward as nappes overriding the Fennoscandian basement in Norway and Sweden for hundreds of kilometres. On the opposite western edge of the belt, westward moving nappes overrode the Greenland basement, as can be seen in Scotland and eastern Greenland. The present Atlantic Ocean opened 320 to 350 Ma later more or less on the site of Iapetus, starting at the beginning of the Tertiary. It can thus be seen that the suture between Laurentia and Baltica had never healed perfectly, but rather was the site of three successive oceanic orogenic zones, each suffering a cycle of opening, closure, and collision.

The *Variscan, or Hercynian belt* has had a particularly complex history. Relics of oceanic crust are rare, occurring only in Galicia (north-western Spain), at Lizard Point in Cornwall (south-western England), in the Belledonne massif in the French Alps, and in Bohemia–Poland. Granitic intrusions, on the other hand, are abundant and varied. The piled-up nappe structure of the belt has been revealed only recently by careful investigation of the timing and conditions of metamorphism and its evolution through time. Seismic reflection profiles (Fig. 2.13, p. 23) have confirmed the huge size of these overthrust slices of crust, formed during collision between 390 and 300 Ma. Thus, contrary to the traditional opinions held only twenty years ago, when it was contrasted with them, the Variscan chain, with translation in opposite directions, northward onto Laurussia and southward onto Gondwana, evolved in a very similar way to the Caledonian and Alpine chains.

At the end of the Variscan cycle, the two great Precambrian shields of Gondwana and Laurussia were welded together, and to the east, the third of the great Asiatic shields, that of Siberia and China, was also welded to Europe across the Urals, the Tien Shan, and the Quin Lin belts, making up the continuation of the Variscan belt into Asia. At the end of the Palaeozoic there thus existed only a single continental mass on the Earth's surface—*Pangaea*—surrounded by a single immense ocean—Tethys (Fig. 2.10).

This configuration was not to remain stable, and from the Permian onwards, more particularly during the Triassic and Jurassic (245–180 Ma) a new zone of oceanic lithosphere

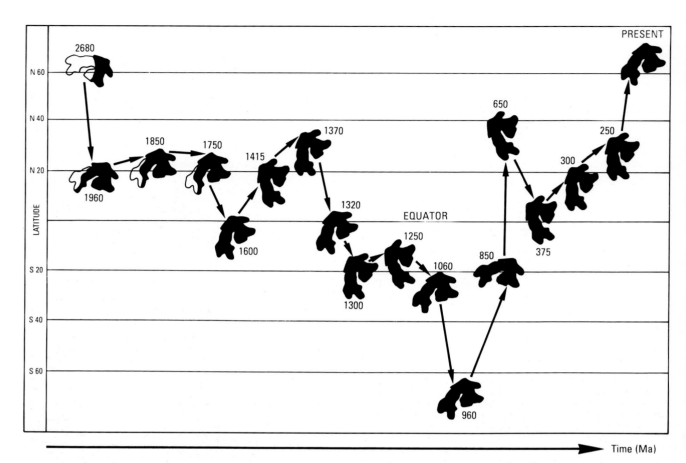

Fig. 2.8 Drift history of the Fennoscandian Shield, from the Archaean to the present. Fennoscandia is plotted at the correct palaeolatitudes and in its correct orientation (with respect to present orientation). To show Fennoscandia at its successive positions through time it has been shifted arbitrarily to the right (longitudes are not shown as they cannot be determined by palaeomagnetic methods). The variation in size of Fennoscandia is due to its growth by lateral accretion during the Svecofennian and Sveconorwegian orogenies. Black areas indicate the extent of the Shield at the period given in Ma. Palaeolatitudes are identified by measuring the magnetic field of the Earth frozen in rocks which appeared during the successive stages of growth of the Precambrian continental crust. (Modified by Trond Torsvik and Mary von Knorring from Pesonen, L. J., Tovsvik, T. H., Elming, S. Å., and Bylund, G. (1989). Crustal evolution of Fennoscandia—palaeomagnetic constraints. *Tectonophysics*. **162**, 27–49.)

formed more or less on the site of the collison zone (suture) of the Hercynian belt. The associated intracontinental extension gave rise to the sedimentary basins in which the Mesozoic and Cenozoic cover rocks of Europe were deposited.

2.2.2. The younger sedimentary basins

The cover rocks of Europe occur in several great sedimentary basins covering rocks from the older mountain chains and abutting against Hercynian rocks that were emerging from the sea only in part at that time (Fig. 2.11). Some of these basins were subsequently deformed and incorporated into the Alpine range from 100 Ma onward. Later, during the Tertiary, smaller basins were created over the Alpine range and at its margins.

Following the Hercynian orogeny, a phase of general dismantling resulted in the formation of small localized basins on Pangaea, filled with sediments during the Permian (280–245 Ma), in the final stage of the Palaeozoic. They are followed by Mesozoic (245–65 Ma) and Cenozoic (65–0 Ma) platform basins and by the development of new passive margins on the two sides of the E–W trending new Tethys Ocean, and subsequently of the N–S Atlantic Ocean. This

mid-European basin resting between the three Precambrian shields has the more complex history and structure (Fig. 2.11 and 2.12), and traps major energy resources (see Chapter 3).

Large sedimentary basins were filled by a variety of sediments (clay, limestone, and sandstone) that either pass from one to another in a complex system of lateral and vertical facies variations, or show breaks in sedimentation of variable amplitude. These sedimentary sequences are generally only slightly altered and are therefore well suited for studying depositional mechanisms and subsequent processes that affected them, such as diagenesis and compaction. It is also in these sequences that study of the past living-environment revealed by macro-fauna and flora, and micro-fauna and flora, can best be made.

It would be a mistake to conclude, however, that these sedimentary sequences have remained unaffected by geodynamical processes. Apart from the sedimentary dynamics already mentioned, and partly conditioned by the re-activation of structures acquired in the Hercynian, they are affected by Alpine movements over 500 km from the mountain range. In this way, a considerable number of faults have brought geological strata of differing ages into contact. Subsidence troughs have formed in which particular types of sedimentation occur, but they show very attenuated fold structures.

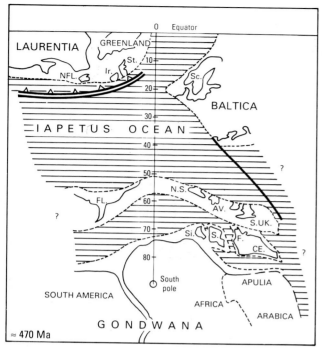

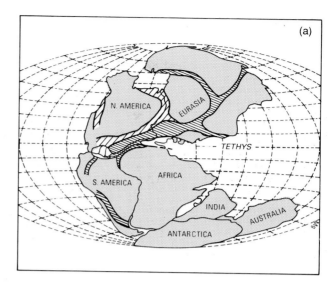

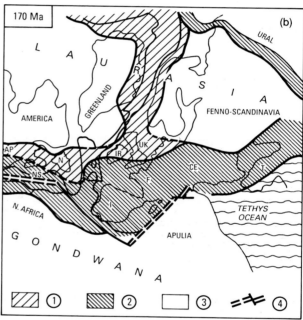

Fig. 2.9 Probable organization of major continental plates and oceanic areas at the beginning of the Caledonian–Hercynian orogenic period (c. 470 Ma). This sketch has been constructed using palaeomagnetic data for this period which give the latitudinal positions of the continental components (data from Peroud and Van der Voo (1984). *Geology*, **84**, 579). All the continental components of the future Europe were at this time in the southern hemisphere, located around the three major Precambrian shields separated by the Iapetus ocean; these were Scotland and N. Ireland (ST and IR) along the Laurentian Shield, and Scandinavia (SC), Denmark and Russia on the Baltic Shield. On the border of the Gondwana Shield the situation was similar to the present complex geography in south-east Asia, with several oceanic areas separating fragments (small plates) of the Gondwana border: one major fragment comprising southern England (S.UK) and Ireland, the Avalon Peninsula (AV) in Newfoundland (NFL), Nova Scotia (NS), and to the west the present eastern border of the USA (Florida Fl). Smaller plates, nearer to Africa, include South Iberia (Si), Spain and southern France (S), northern France (F), and Central Europe (CE). The final welding together of all these fragments by displacements and collisions between 460 Ma and 270 Ma is depicted on Fig. 2.10 (oceanic areas at that time are horizontally ruled).

Processes of re-activation and readjustment continue today and their study is referred to under the term neotectonics.

2.2.3. The Alpine mountain belt

Figure 2.11 shows the distribution of distinctive zones of the Alpine orogenic belt that mark the Mediterranean and its periphery. The Alpine arc crosses eastwards through the Balkans, Turkey, Iran, Afghanistan, and into the Himalayan range. Much better preserved than the Hercynian, the Alpine orogeny, with some regional variants, extends around the Earth.

The Alps are essentially the result of collision between the European and the African basement. Superposition of the borders of these two great blocks gave rise to two major mountain ranges. The northern or Alpine branch was translated to the north, whereas the southern or Dinaride–Magrebide branch was translated to the south over the old African basement. Each of the branches comprises an

Fig. 2.10 Pangaea: collage of all the continents at the end of the Hercynian orogeny and final organization of the Caledonian-Hercynian belt: (a) general reconstruction of Pangaea showing the location of the Palaeozoic orogenic belts; (b) enlarged sketch of the European part of Pangaea showing the Caledonian (1) and Hercynian (2) belts between Precambrian shields (3) and the position of the new spreading ridges (4) that initiated the beginning of the Alpine cycle.

Ap: Appalachians; N-S: Nova Scotia, N: Newfoundland; M: Morocco; I: Iberia, F: France; IR: Ireland, UK: United Kingdom, CE: Central Europe; T: Turkey (from Olivet *et al.* (In press). *Synthèse géologique des Pyrénées*. BRGM-ITGE).

external part (in the direction of translation), which is less altered and which forms a transition with stable peripheral regions, and an internal part which has been subjected to the most intense alteration (in the form of folding and metamorphism). It is this internal part which contains the blocks of basement: Hercynian, Cadomian, or even older.

The recent seismic soundings across the chain in the Pyrenees and the French/Swiss/Italian part of the Alps, give very spectacular images for this collision zone at the scale of

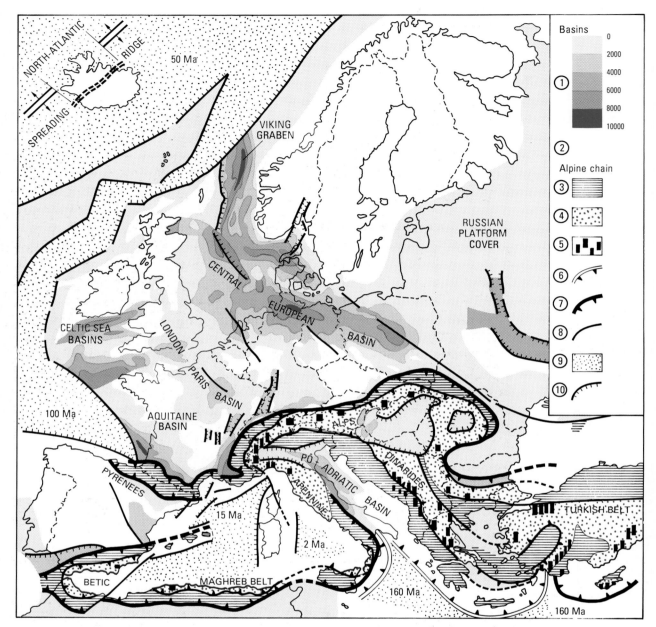

Fig. 2.11 Alpine chain and sedimentary basins in Europe.

Sedimentary basins. 1. These are indicated by the thickness of the post-Permian deposits that filled them. The major central European basin is shown, with its extensions on the Atlantic passive continental margin, in the North Sea basin and in the Celtic Sea. 2. Basement to the sedimentary cover, outcropping around the basins.

Alpine chain. 3. External zone. 4. Internal zone. 5. Remnants of oceanic crust (ophiolites). 6. Subduction zone. 7. Opposed thrust directions in the Alpine margins. 8. Faults. 9. Areas of oceanic crust with their age of formation. 10. Continental rise (see Fig. 2.4), a zone of transition from continental to oceanic crust.

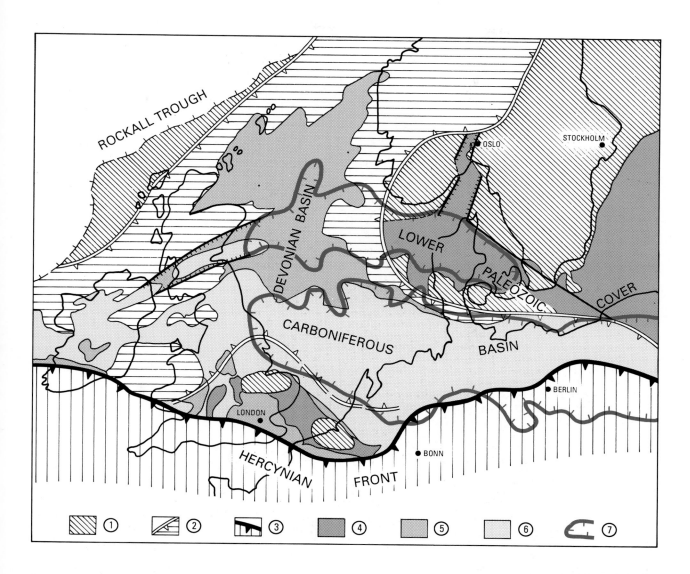

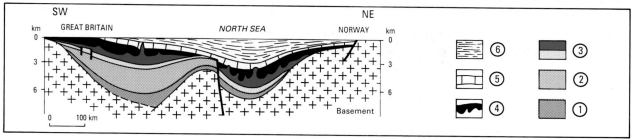

Fig. 2.12 The early, deep North Sea and Central European Palaeozoic basins (after Ziegler 1978). The geological reconstruction of these completely concealed deep basins is based upon seismic and drilling data from oil exploration carried out during the last 20 years. The map shows the three successive basement-forming orogenic belts and their cover-rocks:

(a) the Precambrian of the Baltic Shield and the London–Brabant high (1) overlain by Cambrian to Silurian cover (4);

(b) the Caledonides (2) and the external edge of the Hercynian belt (3) overlain by Devonian (5) to Carboniferous (6) cover;

(c) all three types of basement overlain by the thick Permian evaporitic (Zechstein) succession (7), at the base of the Mesozoic–Cenozoic basin-fill (the latter is shown on Fig. 2.11).

The section shows the lateral displacement of the maximum thickness of basin-fill deposits during the successive sedimentary episodes.

1. Devonian; 2. Carboniferous with coal basins; 3. Permian with evaporites giving diapiric structures; 4. Triassic, Liassic, Jurassic, and Lower Cretaceous; 5. Chalk of Upper Cretaceous; 6. Cenozoic.

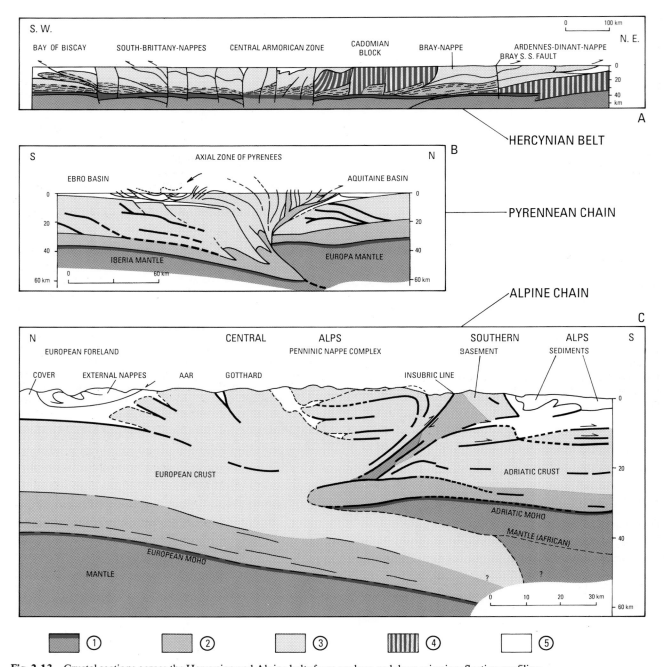

Fig. 2.13 Crustal sections across the Hercynian and Alpine belts from geology and deep seismic reflection profiling:

(a) Section across the *Hercynian belt* in France, from the Pyrenees in the south to northern France (Ecors programme 1984–88). This section, after Matte and Hirn (1988). *Tectonics* **7**, No. 2, shows the opposed vergence of the nappes and the central older block of the Cadomian orogeny (630–550 Ma, North Brittany block), transported northward by the Bray nappe.

(b) This section of the *Pyrenean belt*, obtained through a Franco-Spanish cooperative programme (after Choukroune and Ecors team (1989); *Tectonics* **8.1**, 23) also shows the divergent outward movement of nappes and the resulting 'flower structure' and thickening of the crust.

(c) Deep structure of the *Alps* in a N–S section across Switzerland (Swiss Programme PNR 20 after Frei, W., Heitzmann, P., Lehner, P., Muller, S., Olivier, R., Pfiffner, A., Steck, A., and Valasek, P. (1989) *Nature* **340**, 544–7). This section shows the regular subduction of the European crust beneath the Alps. This crust is indented and scraped by the southern Adriatic (or African) crust with a wedge of mantle at its base. This late configuration postdates earlier polyphase Alpine deformation which began in the Cretaceous (~ 100 Ma).

1. Mantle with Moho zone at its top; 2. Lower reflective crust; 3. Crustal basement; 4. Block of Cadomian crust (profile A); 5. Cover formations (foreland).

the upper lithosphere as a whole. They show the scraping, splitting, thrusting, and superposition of huge nappes of the continental crust or of its cover, expelled in the two opposite directions (Fig. 2.13). Thus older proposals and careful reconstruction of the 'anatomy' of the chain by geologists, based only on observations of surface outcrops, are now validated to a depth of 30–50 km or more.

This schematic scenario shows the great geological diversity of Europe. There is no doubt that future scientific developments will make it possible to refine the models proposed today and even to modify or change them in some cases. The next step is to assess the impact that this geological evolution has had on the formation and location of mineral resources in Europe.

2.3. GEOLOGY AND MINERAL RESOURCES [44] [65]

2.3.1. Types of mineral resources and their modes of formation

The present-day geology of Europe is the result of more than three thousand million years of complex evolution. The rocks now visible at the surface bear the mark of successive events which affected them in the past. Some of these events reflect the direct origin, at the time of their formation, of the mineral substances in such rocks; others caused the trapping or later concentration of mineral substances.

It must be borne in mind that the way in which mineral deposits are formed is, like the geological formations that surround them, subject in the course of time to processes of deposition, alteration, and dissolution by mechanisms that are repeated during geological time. Hence a deposit emplaced at a given time and during a geological cycle can be modified or even disappear in the course of later cycles. Schematically, four main genetic mechanisms for mineral deposits can be distinguished:

1. Some mineral deposits are directly associated with the sedimentation of diverse varieties of rock deposited in lakes, seas, and lagoons and with the later evolution of such sediments during compaction and induration, in all cases associated with displacement of interstitial water. These are known as *exogenic* deposits.

2. Other deposits are created during crystallization of the magmas at the origin of some igneous rocks. These are referred to as *endogenic* deposits.

3. The greatest number and the greatest variety of mineral deposits result from the alteration of pre-existing rocks or are deposited directly in the cracks or veins in such rocks. The mineral substances are precipitated from aqueous fluids charged with dissolved chemical components that circulate in rocks after their consolidation under the effect of pressure and heat. They do not necessarily bear any genetic relationship to the host rock. These are known as *hydrothermal* deposits. Present-day hydrothermal upwellings (thermal springs) represent the outlets of deep hydraulic circulations.

4. *Superficial alteration* of rocks under aggressive climatic conditions, such as those encountered in the intertropical zone, leads to the elimination by dissolution of many chemical elements in the rocks, and in some cases a superficial residue forms, different from the initial rock,

and commonly very rich in minerals of economic value, such as kaolin, aluminium bauxite, and nickeliferous minerals. The deposits belonging to this group, formed at the surface of the continents, are very sensitive to erosion and are therefore generally of recent age.

Exogenic deposits are related to the depositional dynamics of sediments and associated fluids. A sedimentary rock (such as sand, limestone, sandstone, salt or potash, coal) can represent a mineral deposit in itself if it forms a resource of economic value. Its exploitation can be of immediate or potential interest. In other cases, the deposit is contained in a sedimentary rock, either as an integral part of it (high-quality material within generally low-quality material), as is the case of clay deposits or deposits of white limestone (for use in paints), or as a concentration within it, for example oolitic iron deposits composed of small, rounded granules of iron oxide in a sandy or clayey setting (the word oolite indicates eggs of stone), or phosphate deposits.

A particular category of sedimentary deposits is represented by certain fluids that form a mineral deposit in themselves. Such deposits are either trapped in the sediments (such as oil, natural gas, and some types of groundwater occuring in what are known as captive aquifers) or are renewed at a variable rate, as is the case with most groundwater resources.

Endogenic deposits are more specifically related to igneous rocks, whether these are emplaced at the surface (volcanic rocks), at modest depth, or at greater depth (plutonic rocks, named after Pluto, the Greek god of the infernal regions). In the last case, the deposits generally show crystallization into coarser minerals. For the mineral deposits, as for the host rocks, the temperature, pressure, and composition of the fluids dissolved in the magmas or heated by them in the surroundings, are key factors governing genesis. For this reason, there is a very wide range of metalliferous deposits, each representing a different temperature of emplacement and type of magmatic rock. As in the case of sedimentary rocks, the host rocks (granite, basalt, or porphyry) can themselves be economic targets in places as ornamental stone or raw materials for industry (rockwool, cement, etc).

The boundaries between these two main categories, and also hydrothermal deposits, are not always clear, and intermediate stages are encountered. This is particularly related to fluid movements prevailing in the crust (hydrothermalism): there are many intermediate stages between surface fluids (rainfall and run-off that infiltrates at depth) and the waters that either rise from great depth and accompany the emplacement of igneous rocks, or are caught up in convection currents and rise after being heated and charged with dissolved metals at great depth. Thus we describe these kinds of hydrothermal deposit with igneous or sedimentary rocks and related dynamics, as shown schematically on Fig. 2.14.

These three main categories of deposit were formed during most of the geological periods defined previously. Thus a description of mineral deposits based on genesis would seem appropriate. The location of the main depositional zones cited is shown in Fig. 2.15.

2.3.2. Mineral deposits formed by endogenic and high-temperature hydrothermal processes

The formations identified in the European basement (of Precambrian and Palaeozoic age) comprise a vast suite of volcanic and plutonic rocks with which mineral deposits are commonly associated. However, mineral deposits of this type

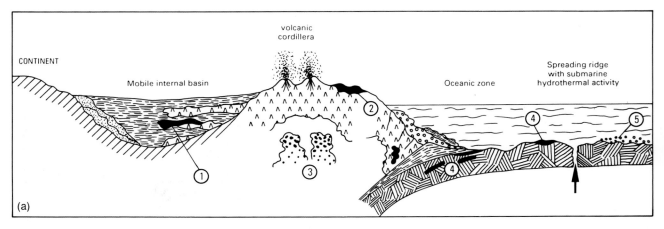

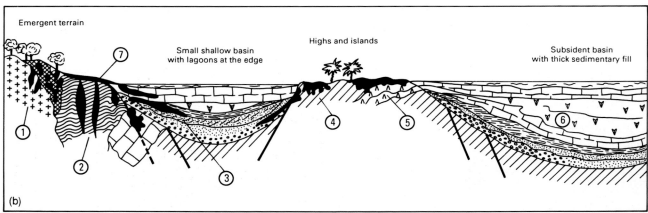

Fig. 2.14 Schematic location of the main types of mineral deposit in widespread geological settings (from *La terre notre planète* (ed. Soc. Géol. de France, 1980).

(a) Main types of deposit associated with sea-floor spreading and subduction zones: 1. Submarine mineralization associated with volcanic activity (Cu–Pb–Zn massive sulphide bodies), (see also setting 4); 2. Near-surface hydrothermal activity in volcanic arcs, depositing Au, Ag, Pb, Zn; 3. Disseminated Cu–Mo–Au porphyry mineralization; 4. Cu–Cr mineralization associated with the formation of oceanic crust and hydrothermal activity; 5. Ni–Mn–Cu nodules on abyssal ocean floors.

(b) Main types of deposit of continents and sedimentary basins: 1. Mineralization associated with granite (Wo, Au, Sn, U) or mafic rocks (Cr, Pt, Ni, Cu); 2. Mineralized hydrothermal veins (Pb, Zn, Cu, Ag, Sb, F, Ba); 3. Sedimentary mineralization (Pb, Ag, Cu, Fe, Mn, magnesite); 4. Basin-margin sedimentary mineralization (F, Ba, Pb, Cu); 5. Deposits formed by concentration during weathering (bauxite, Ni-laterite); 6. Evaporites (gypsum, potash, halite, Li, B); 7. River or beach placers (Au, Ti, Zr).

are not exclusive to these rocks and are also encountered in mountain ranges of the Alpine cycle. The nature of these mineral deposits varies depending on the chemical composition (acid or basic) of the rocks in which they are generated and the geological setting of their emplacement. Three main types are distinguished:

(1) sulphide bodies containing the main non-ferrous metals (lead, zinc, copper) and related principally to acidic volcanic rocks;

(2) mineralization (uranium, tin, tungsten, gold, antimony, arsenic, mercury) related to the emplacement of granites and to associated faulting;

(3) mineralization (chromite, platinoids, cobalt, nickel, titanium) associated with basic rocks.

Mineral deposits belonging to the first category form bodies some thousands to some tens of million m^3 in volume often with massive sulphides. They have formed in diverse geological settings, though everywhere in relation to submarine hydrothermal activity. They show many analogies with

presently forming metallic deposits, discovered recently on the ocean floor, along the spreading ridges where the oceanic crust is created and suboceanic hydrothermalism is very active.

In all these deposits, non-ferrous metals are associated with iron sulphide (pyrite), which is generally predominant. Sulphur is therefore the major element present: although in many cases it can be of economic interest (for example, in the manufacture of sulphuric acid), it is also frequently a penalizing factor presenting an environmental hazard.

This type of deposit is encountered in the old Precambrian rocks of Finland and Sweden, the Cadomian of France, the Caledonian range of Scandinavia, in the Hercynian of Germany and France, and particularly in the great Huelva–Alentejo belt in the southern Iberian peninsula. Some large iron oxide deposits, such as Kiruna in Sweden, also belong to this type.

Deposits associated with granite occur either in disseminated form within the granite (tin, tungsten, uranium), in masses near the contact between the granite and the country rock (iron, tungsten), or more commonly in long, narrow vein

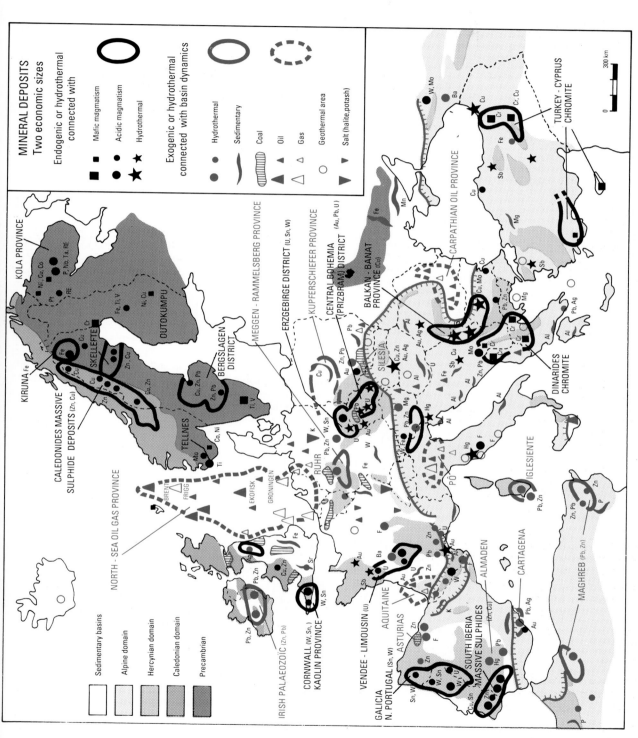

Fig. 2.15 The main mineral deposits of Europe. Some major metallogenic provinces or districts are indicated. Alpine domain is bounded by the thick thrust-line (in colour).

structures. These latter are deposits of uranium, tin, tungsten, zinc, lead, gold, baryte, and fluorite, covering a fairly wide range of depositional temperatures, and emplaced in relation to the various episodes of granite genesis and more particularly to the final episodes of an orogenic cycle.

The basic rocks with which the mineral deposits of Europe are associated correspond principally to fragments of oceanic lithosphere (ophiolites) that have become included in the continental crust due to the action of geological dynamics (plate tectonics). In some cases, deposits are associated with other kinds of basic igneous rocks, for example some of the nickel sulphide, chromite, and iron deposits in the Precambrian of Finland. The ophiolites principally contain copper sulphide deposits accompanying iron pyrite (as in Cyprus, Norway, and Turkey), chromite deposits (as in Cyprus, Turkey, Greece, and Albania) and, although to a lesser extent, platinoids, and deposits of asbestos and talc. The ophiolites of Europe are principally related to the Alpine cycle though they are also known from the Hercynian and Caledonian (e.g. chromite and platinum deposits of Shetland).

2.3.3. Mineral deposits generated during formation of sedimentary basins

To begin with it is useful to distinguish between solids and fluids. Solids are closely associated from the genetic viewpoint with the rocks that envelop them. In the case of fluids, it is the reservoir storage factor that predominates. Like endogenic rocks, rocks formed by exogenic processes can represent mineral deposits in themselves. This is the case of sand and gravel for concrete, clays for the tile, brick, and ceramics industries, limestone for lime and cement, lime sulphates such as gypsum and anhydrite used as raw materials in the manufacture of plaster, silica sands used in the glass industry, and various types of stone, including marble, used in the construction industry.

In the category of metallic mineral deposits, several major types are represented in Europe. Many oolitic iron deposits occur in the Palaeozoic (such as the deposits dating from 500 and 440 Ma in Brittany and Portugal) and the Mesozoic (such as the deposits dating from about 180 Ma in Lorraine, Luxembourg, and Germany). As regards non-ferrous metals, copper abounds in the clays (shales) of the Permian (280–245 Ma) in central Europe (Germany and Poland), and these occur in the famous facies known as the 'Kupferschiefer'.

The abundance of lead and zinc deposits makes Europe one of the most important provinces in the world for these metals. Such deposits are encountered in the Triassic (245–205 Ma) of Italy, Austria, Poland, and France and in the Jurassic (205–135 Ma) of France, around massifs belonging to the Hercynian cycle or in the Alpine Chain. Deposits of similar type are encountered in the Cambrian (570–500 Ma) of Sweden, Sardinia, and Spain, in the Carboniferous (345–280 Ma) of Ireland, and in the Cretaceous (130–65 Ma) of Spain.

Mediterranean Europe contains many aluminium (bauxite) deposits that abound in the Palaeozoic of Turkey and particularly in the Cretaceous of the entire north coast of the Mediterranean, in Spain, southern France, Italy, Hungary, Yugoslavia, Greece, and Turkey, where the re-sedimentation of primary residual alumina deposits has protected them from erosion.

The first European industrial revolution was associated with development of the iron and steel industry, based on the iron ore resources already mentioned above and on the solid

fuels (coal and lignite) with which Europe is abundantly provided.

Coal deposits are of continental sedimentary origin, and are derived from thick accumulations of vegetal fragments. In Europe they appeared during the late Carboniferous and early Permian (330–280 Ma), when Europe was in equatorial latitudes, under conditions that enabled the accumulation of thick masses of vegetal material. This accumulation took place in two types of site: in lakes, and in coastal swamps. The former were the sites of numerous small deposits upon the eroded Hercynian belt in Portugal, Spain, France, Germany, Austria, and Bohemia, and also in extensive lacustrine areas such as the Lorraine–Saar–Saale basin and the Briançonnais–Valais basin, now incorporated in the Alpine chain. The coastal swamps were located along the borders of the Hercynian mountain belt near marginal seas. These deposits are partly marine and the coal within them is generally more abundant and more continuous; in the north they extend from Ireland and the UK through northern France, Belgium, and the Ruhr to Silesia in Poland, and in the south they are present in Asturia, in northern Spain. The deepest of these, in the North German–North Sea basin, are the parent rocks of many of the oil and natural gas deposits of the region.

Brown coal or *lignite* is of more recent origin, derived from vegetal material laid down in the Cretaceous and Tertiary. The deposits are less deeply buried and have been formed essentially by compaction with little heating due to burial. Because of the ease of extraction by opencast mining, the large lignite deposits of north Germany and Austria, France, Spain, and Czechoslovakia are an important source of fuel for power stations and of raw material for the chemical industry.

This review of 'solid' deposits of exogenic origin cannot be complete without mention of *salt* and *potash*, the first of major importance for the glass and chloride industry and for human consumption, the second being one of the key components in the fertilizer industry. Waste from these industries in many cases has a marked negative impact on the environment.

The importance of *fluid deposits* requires little elaboration. They are part of our everyday existence, whether in the form of the petrol we put in our cars or of the water we drink.

The *hydrocarbons* (oil and natural gas) are found exclusively in sedimentary basins, and are present in variable quantities in most of the European sedimentary basins. Their existence depends upon several factors:

(1) the presence of source rock, i.e., rocks that were deposited in a non-oxidizing environment and contained large quantities of organic material;

(2) burial sufficiently deep for the rocks to be heated to the point where the solid organic material is transformed into oil and/or gas;

(3) the deposition of reservoir rock above the source rock; these are permeable rocks of high porosity, such as limestone and sandstone, into which the hydrocarbons created in the underlying or laterally adjacent source rock can migrate and be stored;

(4) trapping by an impermeable cap rock (clay, natural salt) overlying the reservoir rock and preventing the escape of the hydrocarbons towards the surface.

Detailed knowledge of the history of the filling of and deformation of the sedimentary basins is essential in exploration for hydrocarbons. The large deep basins of the North Germany–North Sea area contain source rocks, reservoir rocks and cap rocks at several different levels, so that a number of deposits are present, in places overlying one another. The extent of the basin beneath the North Sea and on the continental shelf off the British Isles and Norway has, during the last twenty years, led to a considerable development of exploration and to the exploitation of off-shore deposits beneath as much as 200 m of water (see section 3.3.1).

Water is also a mineral resource. It can be trapped in sedimentary aquifers with no outlet, referred to as captive aquifers, and generally flows more or less slowly within the host rock. The rate of flow can vary from a few millimetres per day to several metres and even several tens or hundreds of metres per day. All types of geological environment (see section 3.2), whether basement or cover, can harbour aquifers created by the combined action of rock porosity (the interstices between the component grains) and the porosity caused by fractures in the rock. Water is the raw material for thermomineral waters and for geothermal energy.

This brief presentation of the diversity of mineral resources shows that mineral deposits are specifically related to particular genetic processes and geological settings. Similar specific relationships are encountered in recent geological evolution and the environmental consequences, a theme that will be examined below.

2.4. RECENT GEOLOGICAL EVOLUTION AND THE ENVIRONMENT

The complex geological history described above does not mean that the present situation is either fixed or definitive. The history of the Earth shows a continuous evolving process, but we have great difficulty in appraising or envisaging changes in the future dynamics at the scale of human life, so ridiculously short when compared to the scale of geological time. Furthermore, the normal pace of this evolution is such a continuous and slow-moving process, currently involving variations of centimetric order per century, that only recent improvements in metrology can improve our overall assessments. The average rate of evolution is interspersed with paroxysmal events that cause abrupt changes and in many cases cause massive loss of life. Such events are due either to internal phenomena such as volcanic eruptions and earthquakes, or to external phenomena such as climate. The location of the former and the impact of the latter are closely related to the geological setting, which directly governs the dynamics of volcanic eruptions or earthquakes and indirectly induces the zoning of climatic effects. In the case of volcanic eruptions and earthquakes, continuing geological dynamics are the cause. In the case of climatic phenomena, the cause is to be found in the balanced exchanges between the oceanic and continental areas of the planet, dependent on the slow modification of geography and more locally on the variety of rocks, the geological setting, and the type of relief that results.

2.4.1. Active and stable zones

In terms of the geological time-scale, there is no permanence of stable zones in given geographical locations, although there is a specific distribution within each segment of geological time. All of what is now referred to as the European basement consists of zones that were highly active at a given earlier moment in geological history.

The main zone of instability today is to be found in the area of the Alpine chain, the evolution of which is still continuing. This is shown on Fig. 2.16 by the distribution of volcanic and seismic activity typically concentrated in the Alpine domain.

Alpine movements are part of the general closure between the two great plates of Eurasia and Africa, the end result of which is that the Mediterranean will gradually disappear over the course of some tens of million years. Today only a small part of the former Tethys Ocean persists in the eastern Mediterranean. A subduction zone, the only one in Europe, beneath Sicily–Greece–southern Turkey, is responsible for some of the seismicity and present-day volcanism in this area. But in other parts of the Alpine chain, the collision stage is largely past and the present boundaries of the Eurasian, African, and Arabian plates are long continental faults along which horizontal movement takes place. The zones of most intense seismicity, most hazardous for human activity, are associated with these faults, such as in Turkey (the Anatolian and the Dead Sea–Lebanon Faults), in Italy, Yugoslavia, Greece, southern Spain, and Portugal. The map in Fig. 2.16(b) clearly shows this relation in the linear disposition of earthquake epicentres. A third factor in instability and seismic activity is directly related to the areas of high relief and their continuing uplift, which generate extensional movements in the upper crust.

Phenomena related to such instability are observed not only in the Alpine orogenic zones themselves, and in the diversity of their present-day geographical distribution, but also in the counter-effects of Alpine movements which continue to affect neighbouring geological units. This applies principally to Hercynian Europe and to peripheral sedimentary basins.

The other cause of instability, much more modest, is related to the dynamics of the mid-Atlantic ridge, where ocean crust is currently being formed and which is affected by transverse fractures known as transform faults. From this ridge, the Atlantic Ocean is widening at a rate of several centimetres a year and the associated dynamics are generating volcanic activity particularly well expressed in Iceland and in the Azores.

Although no zone can be regarded as completely stable, it can be seen clearly on Fig. 2.16(a) that northern Europe is a relatively rigid structure that has barely been affected either by the counter-effects of Alpine folding or by processes originating from the mid-Atlantic ridge. In contrast, the impact of external dynamics is far from negligible, which brings us naturally to the subject of climatic influence. The few intense earthquakes known in this area (Kattegat, Bothnian Gulf) are probably more related to renewed movement of major faults on the shield during the very rapid uplift (~800 m in 6000 years) that succeeded the unloading caused by the melting of the last ice cap of the Würm age.

2.4.2. Quaternary climatic evolution and its consequences: the ice ages

The geological history of the world, and of Europe in particular, shows to what extent the climate in any given region has changed in the course of geological time as a result of the combined action of plate movement and variations in terrestrial insolation.

Mention was made above of a belt of Cretaceous aluminium (bauxite) deposits on the northern periphery of the Mediterranean. Genesis of this type of deposit is related to a humid

(a)

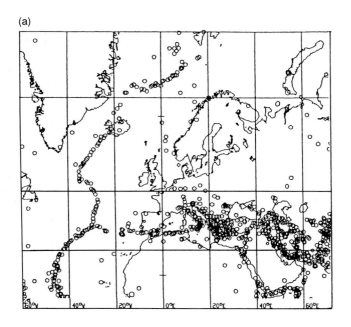

(b)

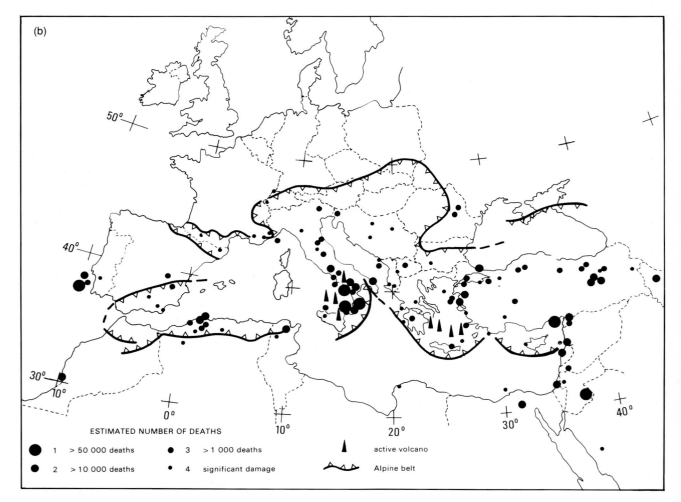

Fig. 2.16 Seismicity and active volcanism in Europe.
(a) Map showing instrumentally determined locations of epicentres of earthquakes of magnitude > 4.6 between 1974 and 1984 (source: BGR, Hannover, Germany). The Mid-Atlantic Ridge and the Alps, highly active zones, show up very clearly;
(b) main historical earthquakes on the Alpine zone in the last 1000 years (taken from *Géochronique* 1982). The active volcanoes in Italy and Greece can also be identified.

tropical climate, and it is now necessary to move more than 5000 km southward to find this type of climatic setting. In this case, it is the northward drift of the plate over 100 Ma which has transported the area of the present Europe from tropical to temperate climates. This does not imply a major global change in the latitudinal climatic zones. Even more indicative is the Sahara, which was an ocean of greenery 6000 years ago and is a desert today. Study of the geological period known as the Quaternary, the most recent in relation to the present day (that is to say the last two million years), makes it possible to assess the variability of climate on a very short time-scale, without significant modification of the latitudinal position of the studied area.

The ice ages [42]

From the beginning of Earth history seven ice ages have been detected. In each case a large continental area was situated at one of the poles, favouring the storage of huge quantities of water as ice and associated cooling of the oceans. We have seen that the dynamics of plates govern the variations in climate on the Earth, by the changes in geographical position and dimensions of continental and oceanic surfaces changing their positions with respect to the equator and the poles. These positions control the thermal regulation of the atmosphere through its coupling with oceanic currents, themselves modified by the shapes of the oceanic areas. The beginning of the present cool period 3 Ma ago, could result from the stable position of the Antarctic continent over the south pole from 30 Ma ago and the N–S arrangement of continental masses which precludes thermal exchange between the three main oceans except in southern latitudes.

Over the past 2.4 Ma, northern and central Europe has gone through six major glacial periods, during which the ice caps advanced some thousands of kilometres. The ice of the Riss ice age (dating back 350 000 to 120 000 years) covered an area extending from Scandinavia to The Netherlands (Fig. 2.17). The last ice age (the Würm, dating from 70 000 to 10 000 years) affected northern Europe and two-thirds of the United Kingdom.

Each of the great ice ages lasted for some tens of thousand years, and was separated from the following ice age by a warm period, referred to as an interglacial. Furthermore, each ice age was subdivided into ten shorter cycles, each lasting for about 10 000 years, and representing advance and retreat of the ice cap. Still further subdivisions can be introduced into these shorter cycles, and the last cold spell, for example, dates from the fifteenth to the nineteenth century, the so-called 'Little Ice Age'. In the present geographical situation, which has been stable for 3 Ma, A. L. Berger has been able to show that the alternating glacial and interglacial periods of the last million years have been controlled by variations of insolation resulting from slight variations in the astronomical orientation of the Earth's axis of rotation and orbital position around the Sun (Milankovitch cycles). Thus assuming there is no anthropic modification of climate, it is possible to predict the climatic changes for the next 100 000 years, with very probably a new ice age in about 60 000 years time (Fig. 2.17).

Movements of the ice cap are directly related to climatic evolution. Today we are in an interglacial period, but during each glacial period the movement of ice caps have many direct and indirect effects. So-called isostatic mechanisms cause depression of the basement under the effect of the weight of ice. Since the disappearance of the last ice cap 10 000 years ago, Scandinavia has risen in altitude by some 700 m, and the trend continues.

At the periphery of ice caps, there is a vast area of permafrost (permanent frozen underground) extending to a distance of 1000 km from the face of the cap. During the Riss ice age (350 000–120 000 years ago), the permafrost reached the northern coast of the Mediterranean. The thickness of this permafrost can vary considerably, but today it is 450 m thick at Svalbard and almost 1.5 km thick in Siberia. Considerable erosion is caused by the action of meltwater and ice cap movement, and sub-glacial canyons can be up to several hundred metres deep.

Sea level also varies in relation to the advance of the ice, which traps a considerable portion of the water of the oceans in solid form. Since the end of the last ice age (the Würm), the sea level has risen by about 120 m. Figure 2.17 shows the resulting coastline variations in western Europe.

At the European scale, the ice ages were critical to the formation of the present-day physiography of central and northern Europe, now characterized by a temperate climate. The effect of climate on geological deposits is marked, and this can be interpreted with increasing precision with decreasing geological age. Quaternary deposits can be studied and interpreted in great detail, specifically via the study of fossil pollen which makes it possible to reconstruct the vegetation of each given region and therefore the successive local climates. In the recent post-glacial period, it has been possible to establish subdivisions of some 1000 years. Each different climate has its specific characteristics in relation both to insolation and to the nature and cadence of precipitation, with resulting specific aspects (such as snowfall, flooding, or desertification) on a given region, although many effects are induced by interaction between climatic factors and the geological setting.

The short time-scale climatic changes [35]

For our concern with human activities on the Earth, we need to obtain precise information on the rhythms and amplitudes of climatic variation at the scale of human activities, i.e. 10 to 100 years. Thus a fine record of climatic change on a long time-scale (>100 000 years) would provide the backdrop against which human intervention in the global climatic environment must be assessed. The temporal resolution required for the proper study of the recent past, the present and the immediate future, however, demands a record of climatic changes over intermediate and short time-scales (10–1000 years) that is capable of resolving events in terms of years, decades, or at the most centuries. Such high-resolution records are rare, but they do exist; the principal types in sedimentary rocks are the annual deposits of calcium carbonate in corals and stromatolites,* and the finely laminated (varved) sediments of some lakes, marginal basins and deep fjords. Ice cores recovering the whole thickness of polar ice caps also offer continuous recording up to 150 000 years BP. By the same method, high-altitude glaciers have given a detailed record of temperature, precipitation, and aerosol trapping for the past 1000 years, showing the clear marks of the Mediaeval Climatic Optimum (between 1000 and 1500), the Little Ice Age (between 1500 and 1900) and the gradual warming that has characterized the globe since the mid-nineteenth century (Fig. 2.18(a)).

The increasing modern deficit of water on land has stimulated studies concerning changes in the water circulation in the oceans and atmosphere, and water resources of continents on the global scale during various time intervals.

---

* Reef-like rock generated by rhythmic microbial growth.

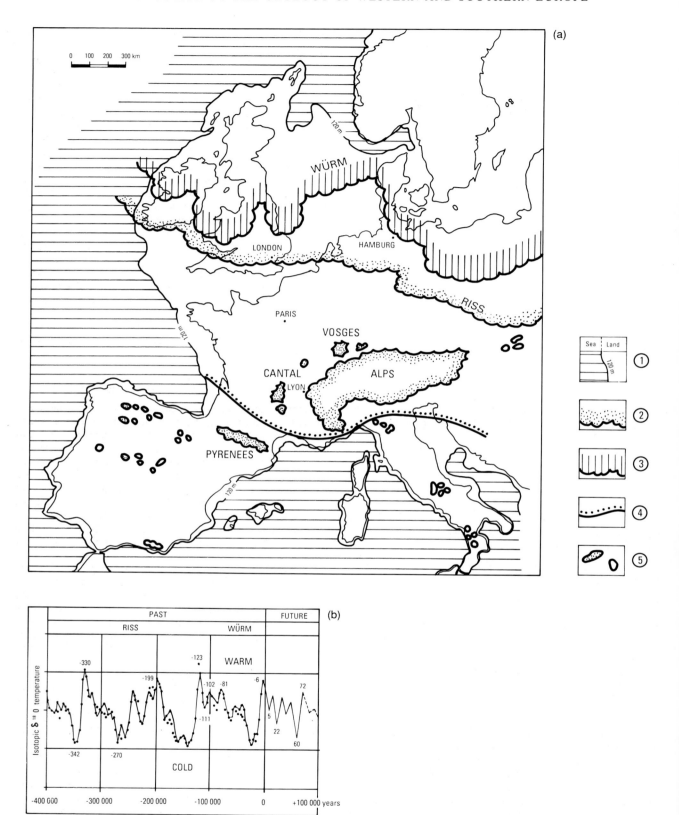

Fig. 2.17 (a) Europe during the last two glaciations, the Riss and the Würm.

1. Coastline during period of low sea-level caused by glaciation; 2. Maximum extension of Riss ice cap (200 000 BP); 3. Maximum extension of ice cap during Würm (18 000 BP); 4. Southern boundary of permafrost (after Maarleveld, J. C. (1976). *Biuletyn Peryglacjalny*. **26**, 57–78); 5. Mountainous areas with permanent ice.

(b) Past and future climatic variations = Milankovitch astronomical cycles (insolation) calibrated by temperatures recorded in recent marine sediments (after Berger, A. (1980) *Vistas in astronomy*. **24**, 103). Ages of maxima and minima are in thousands of years; temperature calculated from isotopic composition of oxygen in microfossil marine tests.

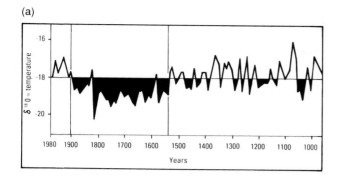

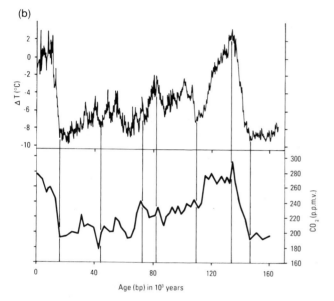

Fig. 2.18 Two kinds of continuous records of climatic change parameters in ice cores. (a) through 1000 years in ice core of Quelcaya Cap, Peru, at 5670 m altitude: variation of the concentration of $\Delta^{18}O$ isotope, directly correlated with mean temperature at surface. Three major periods are well illustrated: the present warm period (from 1900), the cold period between 1900 and 1550 (Little Ice Age) and the warmer period during the Middle Ages (from Thompson, L. G. and Mosley-Thomson, E. in Berger, A. and Labeyrie, D. (1987). *Abrupt climatic changes*. Reidel, Dordrecht)); (b) through 160 000 years in ice core at Vostock station (Antarctica). Here the two curves, with a sampling step of approximately 5000 years, show very good correlation between temperature variations (ΔT°C: isotopic temperature by reference to mean present surface temperature—55.5 °C) and in lower curve, CO_2 content (in parts per million of volume) of the air bubbles trapped in ice (from Barnola, J. M., Raynaud, D., Korotkevich, Y. S., and Lorius, C. (1987). *Nature* **329**, 408–14).

The tools to investigate the past climatic changes in detail are available nowadays. But the reasons for these changes are not yet well understood. The objective of the major international scientific programme PIGB is to make an interactive budget of the thermal, gas, water, and biological consumption and production in atmosphere, oceans, and continents, in order to model climatic changes and to forecast climate evolution in the near future. In this challenge, geologists and geochemists working together with biologists and physicists will contribute significantly by reconstructing past changes.

Understanding the evolution of precipitation in the short-term may contribute to the verification of palaeoenvironmental data used in global numerical models. For instance, significant increases in mid-latitude precipitation and concur-

rent decreases in low-latitude precipitation have occurred during the last 30–40 years.

The second major feature characterizing the modern climate is the global heating of 0.5 °C observed since the last century and its possible relationship with the enormous input of carbon dioxide into the atmosphere from the burning of fossil fuels (anthropic greenhouse effect). The CO_2 content of the atmosphere has increased from a pre-industrial value of ca. 280 ppmv (part per million in volume) in 1800 to 350 ppmv today. It has been demonstrated from the measurements of CO_2 in the bubbles of air trapped in ice cores, that the total CO_2 in the atmosphere has also fluctuated on glacial timescales. The changes in the atmospheric CO_2 content during the last 150 000 years show a direct relationship to the temperature changes of the atmosphere (Fig. 2.18(b)). Simulation models indicate that a doubling of the CO_2 content (ca 560 ppm) would involve a 1.5 to 4.5 °C mean temperature increase before the mid-twenty-first century. An increase of the CO_2-induced warming of the climate would have catastrophic social and economic impact, especially for agriculture and forests, but also for flat coastal areas which could be submerged by a sea level rise.

Human activity may already have influenced climate through forest burning and other forms of deforestation or desertification.

2.4.3. Geology and present-day geography [65] [42]

Whatever the climate, its ultimate effect is the generalized peneplanation of topography. In general, the rate at which mountains are destroyed by erosion or basins are filled with sediments is a few decimetres every 1000 years. In detail, the situation differs between young mountain ranges such as the Alps and old ranges such as the Massif Central of France, and rates of sedimentation vary considerably from one basin to another. Over a 1000 year period, therefore, there is a global range of several centimetres to several metres in the case of erosion and sedimentation, and great variability in time within a given zone, related to successive climatic conditions, for example.

This trend towards the levelling of the topographic features occurs in very diversified forms depending on the geological setting, the nature of the rocks (the component minerals and the degree of cohesion), the dynamics which affected them in the past (fracturing and folding), and current neotectonics. In similar geological environments, erosion and alteration of the rocks can evolve very differently, depending on the characteristics of climate.

The topographic forms visible today, the distribution of the drainage system, and the zoning of vegetation, are thus the result of interaction between geological and climatic factors. This observation is not purely academic, but very logically leads to the analysis of problems of vulnerability associated with natural factors. Detailed analysis of relief forms, geology, and climatic factors is essential for the diagnosis of problems and the design of remedial measures. Analysis of relief forms and climate are the basis for the monitoring of flooding, and, together with the geology, for the study of basin banks.

Any study of natural hazards undertaken with a view to risk or crisis management must incorporate a combined assessment of the geomorphology, geology, and climate. Apart from the physical reality of the phenomena involved, they are of particular importance for decisions related to the distribution of inhabited areas and areas scheduled for development, such distribution being largely conditioned by physical factors such as relief, climate, and location of natural resources.

2.5. GEOLOGY: A KEY TO UNDERSTANDING LAND-USE [65]

Land-use is the term used to denote the distribution of inhabited areas and centres of economic activity. This discipline also governs the selection of zones which should be preserved because of their natural beauty, vulnerability, or scientific interest. At all scales, the siting of inhabited areas must take account of the geology. At the scale of an individual dwelling or building, the selection and calculation of foundations, conditions of drainage, and structural support are essentially based on the nature and characteristics of the underlying rocks. At the scale of a zone scheduled for development geological factors of greater or lesser magnitude must be taken into account, including the presence of permeable or impermeable zones, large fractures, zones subject to the risk of settling, natural or artificial cavities, and the risk for contamination of groundwater resources.

All these characteristics are related to the present-day configuration of the rocks in the study area and to phenomena associated with the internal and external dynamics to which they have been subjected. Their understanding is therefore of prime importance. Obviously, the greater the extent of a development or the degree of safety imposed, the more important it will be to take account of geological factors. The Seveso EEC directive on listed establishments has resulted in a new range of geological control studies. Improved use of geological information in zones scheduled for development would make it possible to avoid unforeseen problems in many cases and thus reduce site development costs.

The ground on which we walk is the zone of interaction between geology and climate. Under the effect of climate, rocks are altered in ways dependent on their nature and on the degree of aggressiveness of the climate in question. The result is a fringe of alteration, of which the topsoil is only the superficial part. The thickness of this fringe is very variable, ranging from some tens of centimetres to several metres in temperate climates and to several tens of metres in tropical climates. The preservation of this fringe and of the topsoil at the surface, is highly dependent on the climate and on the rate of erosion. In scarp zones, the rock is commonly bare and there is little or no protection from a layer of soil. The preservation of soil itself is related directly to the vegetation cover above the zone of alteration.

As a result of development works and agricultural activity, man plays a very important role in the equilibrium of the Earth's surface layer. Without suitable management of agricultural land and forestry, this equilibrium becomes delicate and even damaged, as is seen in the badlands in various parts of the world. The concept of short-term yield for a given economic activity must therefore be integrated within a long-term view designed to preserve the equilibrium, the natural capital given to us, or even to improve it.

The superficial layer of the Earth must not be looked at in isolation. By definition, it represents the interface between the atmosphere and the geological environment: this layer is therefore the recipient of precipitation and of the matter borne by precipitation, and is also the site of transfer and exchange with the geological substrate. All pollution derived from human activity, whether domestic or industrial, ends up in the soil and in the hydrological cycle. The fight against pollution thus depends essentially on controlling the actual source of pollution, but it is also limited by our ability to control its spread and to mop it up where necessary. Whether in the case of isolated pollution (e.g. from a domestic urban waste tip or an industrial waste dump) or of more generalized pollution (e.g. by nitrates or pesticides), it is the hydrological component of the geological environment that is the recipient.

The understanding of this environment and of the interaction between 'solid' rock and the fluids within it is essential, therefore, if we are to establish the appropriate management structures to warn of, monitor, control, and correct pollution. The more vulnerable the area (e.g. the existence of rocks particularly permeable to fluids, densely inhabited zones, or an aquifer important for human consumption), the stricter must be the control and the more effective the management.

It is essential therefore, to take account of the vulnerability of an area when planning the management of land-use, whatever the scale, from the smallest community up to the largest development zone. The most difficult part is probably in relation to the general equilibrium, for which it is necessary to decide between space which should be preserved and space planned for development, involving a balance between development priorities and the necessity of safeguarding our biological environment, the Earth. No action in this context can be neutral, the effects are both cumulative and generalized.

The general principles and comments made in this chapter are an introduction to the themes developed in much more detail in Chapters 3-6, the location and role of natural resources, the importance of natural hazards and their management, the impact of development and of human activity, and the necessity for integrated planning at various scales. This chapter attempts to show how geology is important in this context, both as regards its dynamics and as regards the environment on which the effects of climate and of human activity are imposed.

RECOMMENDED FURTHER READING

Ager, D. V. (1980). *The geology of Europe.* (McGraw-Hill, New York).

Auboin, J. (1980). Geology of Europe: a synthesis. *Episodes.* **80**, 3–8.

Berger, A. (1980). The Milankovitch astronomical theory of palaeo-climates: a modern review. *Vistas in Astronomy.* **24**, 103–22.

Broeker, W. and Denton, G. (1990). Les cycles glaciaires. *Pour la Science.* No. 149, pp. 62–70. What drives glacial cycles. *Scientific American.* No. 1, pp. 43–50.

Caron, J. M., Gauthier, A., Schaaf, A., Ulysses, J., and Wozniak, J. (1989). *Comprendre et enseigner la planète Terre.* (Ophrys, Gap).

Collective Authors (1978–89). *Mineral deposits of Europe.* (Institution of Mining and Metallurgy and Mineralogical Society, London) (for detailed references see section 3.5).

Collective Authors (1980). *Geology of the European countries.* Published in cooperation with the Comité National Français de Géologie (C.N.F.G.) on the occasion of the 26th International Geological Congress. Dunod, Paris. 4 vols.

Denton, G. H. and Hughes, T. J. (eds) (1981). *The last great ice sheets.* (Wiley, New York).

Dunning, F. W., Adams, P. J., Thackray, J. C., Van Rose, S., Mercier, I. F., and Roberts, R. H. (1981). *The story of the Earth.* (Geological Museum, HMSO, London).

Embleton, C. (ed.) (1984). *Geomorphology of Europe.* (Weinheim, Deerfield Beach Florida).

Gee, G. D. G. and Sturt, B. A. (eds) (1986). *The Caledonide orogen-Scandinavia and related areas.* (Wiley, Chichester).

Goguel, J. (1980). *Géologie de l'environnement.* (Masson, Paris).

Haenel, R. and Staroste, E. (1988). *Atlas of geothermal resources in the European Community, Austria and Switzerland.* (Schäfer, Hannover).

Illing, L. V. and Hobson, G. D. (1981). *Proceedings of the second conference on Petroleum geology of the continental shelf of North-West Europe.* (Heyden, London, on behalf of the Institute of Petroleum, London).

Jones, P. D. (1990). Le climat des mille dernières années. *La recherche.* **21**, 304–12.

Levin, H. L. (1985). *Essentials of Earth science*. (Saunders, Philadelphia).

Perrodon, A. (1983). *Dynamics of oil and gas accumulations*. Mémoire Centres de Recherches Exploration Production Elf-Aquitaine No. 5.

Ziegler, P. A. (1978). North-western Europe: tectonics and basin development. *Geologie en Mijnbouw*. **57**, 589–626.

Ziegler, P. A. (1982). *Geological atlas of western and central Europe*. (Elsevier, Amsterdam).

3

Natural resources in the geological environment

B. Kelk

Resources and reserves

The dictionary definition of *resource* as 'something in reserve or ready if needed' has been extended for mineral resources to comprise all materials surmised to exist, having present or future values. In geological terms, a resource is a concentration of naturally occurring solid, liquid, or gaseous materials in or on the Earth's crust in such a form that economic extraction of a commodity is currently or potentially feasible.

Material classified as a *reserve* is that portion of an identified resource producible at a profit at the time of classification. *Total resources* are materials that have present or future value and comprise identified or known materials plus those not yet identified, but which on the basis of geological evidence are presumed to exist.

Public attention is focused usually on current economic availability of mineral materials (reserves). Long-term public and commercial planning, however, must be based on the probability of geological identification of resources in as yet undiscovered deposits, and of the technological development of economic extraction processes for presently uneconomic deposits. Thus, all the components of *total resources* must be continuously reassessed in the light of new geological knowledge, of progress in science and technology, and of shifts in economic and political conditions.

Whilst it has been common to think in terms of resources in the national context, there has been an increasing shift to international trade in very many of these commodities. This trade, well-established in hydrocarbons and metalliferous minerals, is now developing even in the common industrial minerals such as aggregates (see section 3.4). The economics of transportation are therefore playing an increasing role in the determination of the 'value', or economic feasibility, of any resource.

Another requirement of long-term planning is the weighting of total or multi-commodity resource availability against a particular need. To achieve this the general classification system must be uniformly applicable to all commodities so that data for alternate or substitute commodities can be compared.

To serve these planning purposes *total resources* are classified both in terms of economic feasibility and of the degree of geological assurance. The factors involved are incorporated in Fig. 3.1 to provide a graphic classification of total resources.

Thankfully, after decades or centuries of neglect, it is now accepted that the environment has also to be taken into account in both the short- and long-term planning. Whilst not affecting the *total resource* figures, environmental considerations can clearly diminish the calculation of economically workable *reserves*.

3.1. OVERVIEW [209]

The innumerable, and diverse, practical applications of geology based on the Earth's resources of water, rocks, minerals, fossil fuels, and space affect each and every member of the population. Many of these applications are so much a part of everyday living that, except for the most obvious, like the energy minerals, they are generally taken for granted and little heed is taken of the impact of their use on the longer-term reserves or on the environment.

The more economically advanced the country or region is, the more widespread and pervasive are the by-products of its geological resources. Thus, whilst most people will recognize the need for constructional and industrial minerals (including sand, gravel, clay, limestone, and gypsum) in making the fabric of a house (for bricks, tiles, concrete, plastic, etc.), few will recognize that such commodities are widely used in virtually every room. For example, calcium carbonate minerals (from limestone, chalk, and dolomite) are to be found in pharmaceutical products (toothpaste and floor cleaners), in catering (such as in some brands of white bread and in dog biscuits) and as a white pigment, filler, or extender in paints, plastics, and paper.

In the well-developed regions of the world the daily *domestic* consumption of water is in the range of 150 to 500 litres per person, whilst, in the UK for example, the annual usage of sand and gravel is equivalent to 50 to 60 barrowloads for every man, woman, and child. Whilst recycling of many of the resources, such as iron and silica sand, is practised increasingly, the demand on these finite resources is maintained.

Most of the geological resources, such as coal and hydrocarbons, are in limited supply, whereas water—perhaps the most important natural resource of all—is both limited in quantity and increasingly threatened in quality. The exploitation of the natural resources, so crucial to the everyday life and economic well-being of Western Europe, becomes increasingly difficult—as the population increases, as urban development spreads, and as the major resources dwindle and new, often smaller ones are developed. Increasingly, care has to be taken that the exploitation of new resources does not sterilize another—or in the case of water, contaminate or pollute it. The sterilization of a resource is often brought about by temporally varying economic pressures, as the explcitation of, for example, an underground coal mine is terminated by the lower cost of imported coal. Unfortunately, once an under-

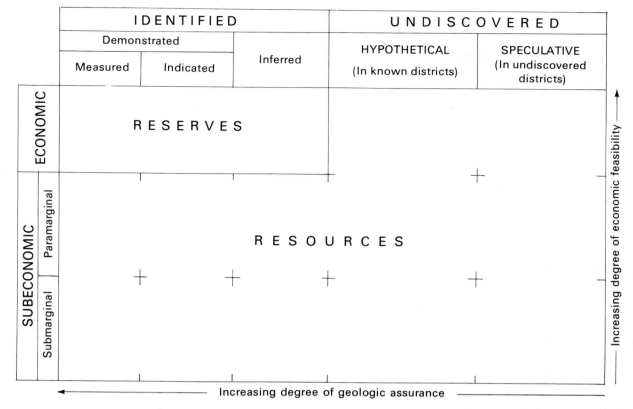

Fig. 3.1 Classification of resources as published in *US Geological Survey Bulletin* 1450-A.

ground mine has been closed, the extremely high cost of a potential later re-development virtually prohibits that possibility.

As all who live in or visit the major cities of Western Europe know, space is a commodity in short supply. The space for car parking is often severely limited whilst hectares are dedicated to the storage of bulk materials like hydrocarbons. Increasingly, therefore, attention is turning to the use of underground space both to alleviate the problems at the surface and to make use of the very stable temperature and humidity regimes of the subsurface natural and man-made cavities.

Fortunately, the resource scope of Western Europe has been notably enhanced in the last few decades by the search for, and exploitation of, offshore minerals, particularly oil and gas. This development will continue no doubt, and be supplemented by the exploitation of other mineral resources from the sea-bed and underlying rocks. However, exploitation offshore carries many risks both financially and environmentally; these are referred to elsewhere in this book.

The relatively densely populated region of Western Europe has been at the forefront of development since well before the Industrial Revolution and has, therefore, a long history of exploitation of its indigenous resources. That long history of exploitation brings with it an equally long history of consequential degradation to our environment; for example, by the undermining of buildings and services, and uncontrolled tipping of many types of toxic and undesirable wastes, and by the pollution of soil, aquifers, and atmosphere. The disruption, damage, and cost to living resources, the ecology and the human population are all too obvious. The demand for restoration and maintenance of environmental quality now presents the resource industry and society generally with a challenge to manage successfully the exploitation and use of resources in a manner which avoids undesirable and harmful consequences to the environment.

3.1.1. Economics of resource exploitation

In minerals we have a series of distinct commodities, some hard to find and of great financial and political importance like oil, others occurring almost ubiquitously so that they are produced in great volume yet seldom moved far. At the other end of the spectrum are the high-value precious metals and gems, which can be moved around the world with little increase in price. Between these two extremes is the whole range of metals and more specialized industrial minerals, each with its own market and commodity economies that govern its price and distribution.

Except where there are strong political considerations, a geological resource is exploited only when it can provide an adequate return on the investment. The most important factors in this context are: the extent to which the product can be sold (revenue), the cost of extracting or otherwise working the resource (including the quality and extent of the resource), the cost of transport to the market, the environmental constraints on its exploitation, the proximity to necessary infrastructure, and the political considerations of taxation and planning.

Some commodities, such as oil and metals, have prices set by central trading bodies or markets which thereby create a general equivalence of prices. In the absence of a 'London Metal Exchange' or similar open arrangements for the sale of other commodities such as industrial minerals, the same mineral or product can sell at very different prices from one customer to another depending on individual contracts. However, commodities traded through international exchanges are much more subject to the attention of financial speculators and hence to very notable fluctuations in prices over time.

There are also significant differences between different resources in the speed with which a discovery can be brought on-stream and, indeed, in the costs of having to stop and restart. For example, 10 years elapse, on average, from the discovery of a metal ore deposit to the commissioning of the mine and plant, whilst an oil well can be producing in a matter of months. Similarly, an oil well can be shut down and restarted at little cost or damage, whereas a deep mine, once closed, may be extremely expensive to re-open.

The economics of exploitation are discussed in more detail in Section 3.4. Whilst the specific concern there is industrial minerals, the principles apply to all the resources, though some tend to be more subject to political influences than others.

3.1.2. The wise use of resources

From the preceding paragraphs it can be seen that the whole topic of the exploitation of our natural resources is one of balance and risk. It is a balance of economic exploitation on the one hand against the longer-term availability of reserves on the other; of one resource against another; and of the maintenance and growth of our economies against the risk to the environment caused by resource exploitation. These problems of balance and the mitigation of environmental risk all demand good management, and certainly much better management than has been in the past. That management itself must be based on a full knowledge of all the factors involved.

Concern about a country's natural resources has been the main reason for the establishment of many national and state Geological Surveys. Initially, the concern may have been metalliferous minerals or coal, but later, industrial minerals, water, and environmental impact reached the awareness of politicians and the programmes of the Surveys.

In dealing with resources and, indeed, with the environmental consequences of their exploitation, the Geological Surveys undertake four key roles.

(1) the provision of information and advice to government, industry, and the public, about the national geology and its contained resources;

(2) the provision of resource-specific services to government, often on a contracted basis. These may include the collation and maintenance of databases and inventories of a country's resources and related environmental aspects, as well as the execution of prospecting and reconnaissance surveys as a stimulus to the resource industry;

(3) research and development, such as resource genesis, the methodology of assessment and exploration, techniques of beneficiation, and aspects of environmental impact;

(4) the collection and collation of national and international statistics of the production, export and import of major resource commodities. This may be associated with the representation to government of related resource-industry interests.

The following sections describe much of the first three roles but as the fourth is a more general 'overview' service, it is appropriate to comment further on it here.

In some countries, such as the USA, a specific Bureau of Mines exists to carry out much of this role. In Western Europe, whilst the activity may be more formalized in one country than another, many Geological Surveys provide at least some element of this service. The general public evidence of the role is given by publications such as *World Mineral Statistics* by the British Geological Survey. However, the supporting databases, and the expertise and advice of the Geological Survey geoscientists, are commonly used to enable government departments and ministers to establish policies for resource management and, where appropriate, support for industry.

This chapter is structured to provide a framework for discussion of each of the main geologically-based resources. Though not every Western European Geological Survey has the responsibility for providing advice on all such resources (for example, only in Germany is the Soil Survey function combined with a Geological Survey), it is appropriate to address them all here. It is not possible to treat all the resources in exactly the same manner. Nevertheless, the following sections generally address the distribution of the resources, their applications, the means by which exploration and evaluation are carried out, the methods of exploitation and beneficiation, and the related problems of environmental impact. Variations in national policies relevant to mineral resources are to be expected, but attention is drawn to the ever-increasing international trade in commodities, and the concomitant need for the greater coordination, at least in Western Europe, of information about them and the underlying geology, and the legislation controlling their exploitation. Only by that degree of coordination can our natural resources be used wisely, to strengthen our economies and enhance our way of life, and at the same time conserve the most precious resource of all—the environment.

3.2. GROUNDWATER [39]

3.2.1. Introduction

Water is one of the very rare minerals which are mobile and which exist in the atmosphere, on the surface, and in the subsurface. These peculiarities, along with its well-known physical, chemical, and biological properties, account for the interest in this vital substance, but are also, unfortunately, responsible for the difficult nature of hydrogeology and the strict requirements of water regulations.

Groundwater, the only natural resource which is renewable at a human scale, is the subject of the hydrologist's studies. It is but one stage of the great hydrological cycle and it cannot be well understood without a proper scientific approach to other components of the cycle i.e. climatic conditions, infiltration, and run-off. Furthermore, adequate groundwater management should not be isolated altogether from surface water appraisal from the scientific, technical, and economic point of view. This close association between hydrology (at the surface) and hydrogeology (underground) is not recognized administratively in most countries, and usually the national Geological Surveys are concerned with the latter only. The one exception is Switzerland, where the National Hydrological and Geological Survey has responsibility for both aspects.

Water is of fundamental importance in Western Europe with its dense population and high level industrialization. Resource aspects only are considered in this section, but the important influences of pollution are discussed extensively in Chapter 5.

Europe's average rainfall is 734 mm per year, from which evapotranspiration is 415 mm and run-off 319 mm. This means an average of 0.3 m^3 of water per square metre of land,

Table 3.1 Present and future water resources in Western Europe, according to J. Margat (Défense et illustration des eaux souterraines en Europe. *Hydrogéologie*, No. 2, 1989, BRGM, Orléans)

Country	(1) Theoretical total water resource (renewable) billion m³/year	(2) Population 1985 (millions) (WB atlas 1987)	(3) Per capita resource ratio 1985 (1)/(2) m³/head/year	(4) Pop. 2020 (average assessment millions)	(5) Per capita resource ratio 2020 (1)/(4) m³/head/year
Germany (West)	161.00	61.06	2 637	54.66	2 945
Austria	90.30	7.55	11 960	7.36	12 269
Belgium	12.50	9.85	1 269	10.05	1 244
Denmark	13.00	5.15	2 524	4.85	2 680
Spain	111.30	38.73	2 874	45.38	2 453
Finland	113.00	4.92	22 967	5.04	22 421
France	185.00	55.13	3 356	58.35	3 171
Greece	58.65	9.94	5 900	10.74	5 461
Ireland	50.00	3.56	14 045	5.16	9 690
Iceland	170.00	0.24	708 333	0.30	566 667
Italy	187.00	56.95	3 284	57.50	3 252
Luxembourg	5.00	0.37	13 514	0.34	14 706
Norway	413.00	4.14	99 758	4.25	97 176
Netherlands	90.00	14.49	6 211	14.86	6 057
Portugal	65.60	10.20	6 431	12.17	5 390
UK	120.00	56.73	2 115	56.32	2 131
Sweden	180.00	8.33	21 609	7.82	23 018
Switzerland	50.00	6.42	7 788	5.92	8 446

which places Europe in the middle ranks of world countries as far as renewable water resources are concerned.

This generalization hides the considerable disparity in resources amongst the countries as demonstrated by Table 3.1 which shows, for example, that a citizen of Iceland has available 558 times as much water as a Belgian. Such variation leads to significant consequences for the economy of each country and for the well-being of its inhabitants.

3.2.2. Resource assessment

Two elements govern the availability of groundwater: the presence of mountains and the permeability of the surface deposits; national, and even regional variations relate directly to their effects. Thus, in France, an average of 54 per cent (a European record) of all running water is groundwater at some stage, but infiltration to the subsurface varies from 1 to 2 per cent to more than 90 per cent according to the scale of relief and the nature of the geological strata present.

Areas such as Scotland, Scandinavia, and north-west Iberia, where high precipitation (over 1500 mm per year) coincides with mountains composed of granitic and metamorphic rocks and cold or temperate climate, have strong surface run-off. In contrast, the chalk plains and plateaux of central Europe and the karst regions of the Mediterranean countries are areas of predominant underground flow. The water policy and the involvement of Surveys dealing with hydrogeology should relate directly to the proportion of groundwater in the total supply. In reality, the situation is much more complex and varies greatly from one country to the other.

European groundwater resources are now well known, thanks to the work of the national Surveys (specialist inventories, reviews, balances, models, and mapping) and other associated organizations. Two synoptic presentations are available: the UNESCO 1:1 500 000 international hydrogeological map (not yet complete for the most southerly zone) and the EEC 1:500 000 water resources map (restricted to the countries which were members in 1982).

The geological characteristics of the formations give them specific technical behaviours and lead to different resource management methods (see Fig. 3.2). Case histories are given below to demonstrate the situation with respect to most examples of water-bearing formations.

Alluvium

Alluvial deposits, commonly restricted to narrow belts along watercourses, but locally widespread (the Po, Rhine, Rhône delta, etc.) are favoured for the tapping of large yields under easy conditions. They are also zones of intense conflict with other uses, particularly the impact of aggregate extraction which is widespread throughout Europe. Alluvium also provides a means of ready exchange between surface and groundwater which, along with the filter effects of the banks, are particularly important features in global water management. The natural filtering properties of these deposits are also used to achieve adequate qualities of groundwater from artificial recharge devices.

Valley alluvium in the Var Plain, Alpes Maritimes, France
[38] [39] [45]

Although small, only 24 km long by 1 km wide, the alluvial plain of the lower Var valley (Fig. 3.3), close to Nice and the other tourist centres of the Côte d'Azur, constitutes a remarkable teaching model. It has been possible to demonstrate the notion of an 'aquifer system' there, through thirty years of monitoring and investigations of the

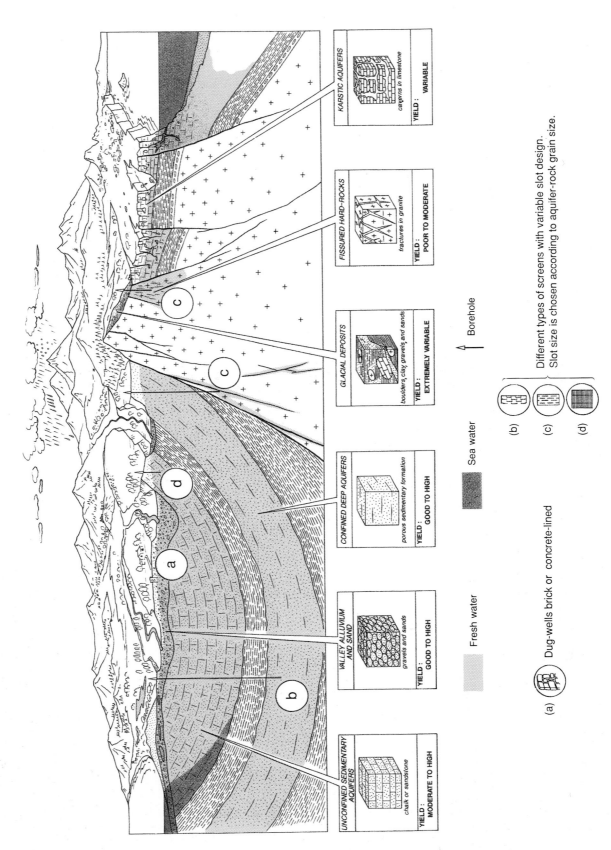

Fig. 3.2 Main types of aquifer (not to scale).

UNCONFINED SEDIMENTARY AQUIFERS
chalk or sandstone
YIELD : MODERATE TO HIGH

VALLEY ALLUVIUM AND SAND
gravels and sands
YIELD : GOOD TO HIGH

CONFINED DEEP AQUIFERS
porous sedimentary formation
YIELD : GOOD TO HIGH

GLACIAL DEPOSITS
boulders, clay gravels and sands
YIELD : EXTREMELY VARIABLE

FISSURED HARD-ROCKS
fractures in granite
YIELD : POOR TO MODERATE

KARSTIC AQUIFERS
caverns in limestone
YIELD : VARIABLE

Fresh water

Sea water

Borehole

(a) Dug-wells brick or concrete-lined

(b)
(c)
(d)
Different types of screens with variable slot design.
Slot size is chosen according to aquifer-rock grain size.

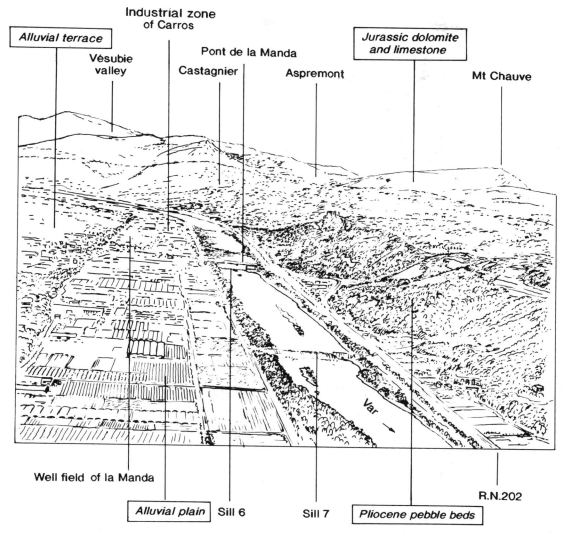

Fig. 3.3 The Var plain, aerial view.

relations between the river and the alluvial aquifer, and between the alluvial aquifer and its enclosing alluvial deposits. The considerable economic interests invested in this small area have generated problems over the years. These are now partly resolved at least, thanks to progressively accumulated hydrogeological knowledge. The long-term piezometric monitoring, effected by BRGM's Provence–Alpes–Côte d'Azur Agency, was particularly useful in aiding the departmental administration in its decision-making.

The Var, which formerly defined the boundary between France and the County of Nice, is like its Ligurian neighbours in being both an Alpine and a Mediterranean river. This duality endows it with a very variable régime, with a low rate of flow at its mouth of 15 m³s⁻¹ and a five-year flood of 1300m³s⁻¹.

The high altitude of its drainage basin (the Mercantour mountains reach 3000 m) makes it an area of high rainfall and niveous régime, often augmented by Mediterranean spring rainfall, which, added to the meltwater, can create sudden floods. Autumn rainfall gives rise to a second flood-prone period. This régime, combined with the variety of rock types that underlie the basin, gives rise to rapid erosion, limited now by protective measures in the upper reaches, and high turbidity. From the air, approaching Nice airport, which is built on the delta of the Var, a plume of muddy water covering several square kilometres can be seen extending into the Mediterranean.

During the last several million years, and 'recently' during the ice ages, when sea level was 120 m lower than at present, the Var built up a considerable alluvial and deltaic system. The material torn from the crystalline rocks of the Mercantour, and from the Alps, forms an assortment of stony facies much sought after by users of aggregate. It is this that has led to most of the problems that will be mentioned later.

The tectonic trench, with transverse fractures, down which the Var flows is composed of limestone and dolomite, and suffered karstification before the deposition of alluvium. The first deposits were 200 m of Pliocene pebble-beds, which were followed by varied deposits of Pleistocene to Recent alluvium. Upstream, the plain consists almost entirely of coarse material, but clayey intervals near the coast separate a lower, confined aquifer from the unconfined shallow aquifer. It is from the latter that most of the water is taken.

In their natural state, the high-water and low-water channels were essentially the same, as the low-level water followed the channels eroded by the preceding high-level flow. Dyking, begun in 1844, followed by the creation of so-called 'sealing compartments' or silt traps, were responsible for the present shape of the channel, and for the existence of a usable surface covering about 30 km². The proximity of Nice, the tourist centres of the coast, and the residential area of the hinterland, expanding rapidly during the 1960s, together led to a series of user conflicts, remarkable for their concentration and for the size of the investments involved in so small an area.

In the area of silty alluvium, the plain first became the market garden of Nice. The aquifer, tapped by seven well-fields, contributed progressively to the supply of water for almost half a million people, while the coarse alluvium in the river bed, and therefore the public domain, was regarded as a virtually free and inexhaustible reserve of concrete aggregate and roadway ballast. The flat ground, so rare in the area, also attracted industrial activities. The stage was thus set for the intense tensions to arise that led to user conflicts.

Dredging the active bed for gravel rapidly lowered the stream water surface, which, while tending to re-establish itself naturally between the pits, was always at a lower level because the Var and its tributaries no longer transported the coarse material necessary to fill the hollows. The alluvial aquifer, in direct connection with the stream water surface, was thus lowered by as much as 9 m over twenty years, which meant that the pumps on several hundred market gardens could no longer extract water from the boreholes. In order to remedy this, the administration first financed the deepening of the wells, a perfectly adequate measure since the aquifer is several tens of metres thick. However, those utilizers of the aquifer who extracted water for public use as drinking water became worried that the resource was seriously threatened.

BRGM's Provence–Alpes–Côte d'Azur Agency (Regional Geological Survey as it was known then), installed a system of observation wells in order to quantify the drawdown and to provide information, in the form of piezometric maps and variation diagrams, for the arbitration of disputes.

The administration then decided to raise the stream water surface artificially by installing sills. Electrical simulations (mathematical models were still very little used) showed that sills a few metres high, at 1 km intervals, would enable restoration of the piezometric level. The users, who had formed a user group, were required to pay the cost of constructing the sills, through a tax levied per tonne of gravel extracted. In return, exploitation of the gravel pits was permitted in the troughs between each sill.

When the first sills were completed, water-level gauges showed a marked, rapid rise in the water table, which in the middle of the reaches between the sills, rose to the level of the Var. However, satisfaction was not to last long: only a year later, owing to drainage of the aquifer from below the sills, the water table fell back to the level of the next downstream stretch of water, where it had been previously. Fine sediment brought down by floods had been trapped in the troughs between the sills, effectively sealing the bed of the controlled river. The limitation of exchanges between the river and the aquifer to times of flood only, with lateral flooding, caused not only a lowering of the groundwater level (which, in view of the thickness of the aquifer, would probably have been no more than a nuisance) but could also have limited recharge in the long term, so threatening the supply of potable water to the public. Was the next thing going to be a sealed-off Var, flowing on a bed of clay and completely isolated from the aquifer?

It was possible to introduce such a sealed zone into a mathematical model developed by BRGM. This made it possible to produce a representative image of the alluvial groundwater, from the point where the river emerges from the mountains to its mouth on the coast. The only boundary conditions that were imposed were those of rate of exchange with the aquifer, except downstream, where the potential was fixed at sea level. After calibration this model enabled a faithful representation of the piezometric surface (errors of less than 1 m over a total difference of elevation of 120 m) to be made; it was thus possible to show indirectly that there was a lateral input of some hundreds of litres per second from parts of the karstified Upper Jurassic, contributing significantly to the recharge of the aquifer.

This tool was used several times to study various problems of control and management of the aquifer. It was thus possible, taking into account water abstraction which increased from 30 million to 50 million m³ yr⁻¹, to show that it was essential to preserve the recharge zones in the upstream part of the aquifer, and to create no

further sealed troughs. Consequently, the extraction of gravel from the bed of the Var has been totally prohibited since 1983.

It was possible by simulation to forecast the impact of the sills and of various hypothetical possibilities of cleaning the river bed between the sills. Simulation of a number of well-fields and undeveloped areas, in some instances on precise scales, using in such cases more accurate multi-stage models, enabled estimation of the potential for increasing abstraction consistent with respecting the requirements for maintaining water quality.

It was thus shown that the effective management of an alluvial system requires that the parts played by the watercourse, its banks (a very particular interface because of the effects of sealing) and the adjacent water-bearing deposits (which in places extend the limits of the system considerable distances, thus increasing the potential resources) all be taken into account.

Glacial deposits

In addition to the moraines of the last stages of glacial retreat, which have an important role in the regulation of springs in the hydrogeology of the Alpine countries, the glacial deposits of the plains are of particular interest in northern Europe. Because of the intrinsic heterogeneity of the medium, careful exploration is necessary to detect the 'washed' material which will constitute the favoured zones for tapping the water.

Uppsala aquifer—Sweden [177] [178] [171]

The city of Uppsala and its suburbs is today one of the areas in Sweden experiencing rapid growth. The city centre is situated 65 km NNW of Stockholm (Fig. 3.4). The number of inhabitants is 160 000. Uppsala is known as a university and service city. Industry is diversified and, with a few exceptions, rather small.

The municipal water supply is based on groundwater, both natural and, lately, artificial. Uppsala, like many of the old Swedish towns in the southern and eastern parts of the country, was founded on an esker that offered good foundation facilities as well as strategic advantages. In addition, the eskers are known to be very good aquifers generally. The Uppsala esker is one of the largest in Sweden.

The part of the esker that is important for the public water supply is to a large extent situated on the western slope of a fault system in a deep basin in the Archaean bedrock. As a result of ice movements during the last ice age, till was deposited on the bedrock. At the end of this period, a great river under the ice transported loose material to the river mouth at the ice edge and, as the ice melted, deposited the coarse fractions there as a huge tail. Silt and clay settled further out in a diffuse plume. This all happened under water. Due to seasonal climatic variations, a varved glacial clay filled the distant parts of the basin, above the till. Slowly, the land started to rise from the sea, and fine material was washed out from till islands and the esker, forming a post-glacial clay in the still water-covered parts of what was now an archipelago. The result, seen today, is the Uppsala clay plain with low hills of bare rock or till and with the esker on its western border. In the fault area, the esker material is locally at least 150 m deep. Along the esker, clay depths of more than 100 m have been found. The River Fyris that now flows through the city, is a small remnant of the once mighty ice river. A sketch map of the geology and the water supply schemes is shown in Fig. 3.4.

Hydrogeologically, the esker is of the drainage type, in that it transports water not only from the esker itself but also from the surroundings. During some periods of the year, the river drains the esker, resulting in the occurrence of springs in the esker sides. On the other hand, induced infiltration occurs in parts of the esker during periods of low groundwater levels in the aquifer. Compared to similar but larger aquifers in central Europe, the turnover time of water is short. This is an important factor in maintaining the supply and quality of the groundwater. The threats to the groundwater are many: great parts of the city and main roads are located on top of the

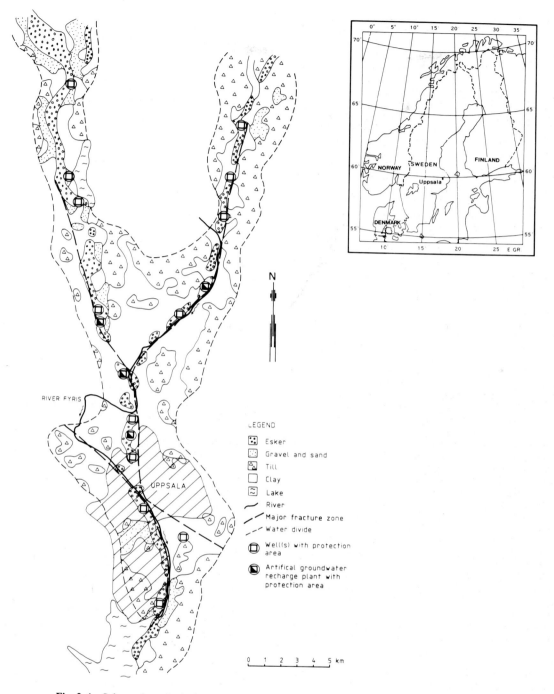

Fig. 3.4 Schematic geological map and water supply schemes of the Uppsala area of Sweden.

unprotected esker and heavy farming is common on the plain, where the clay provides good agricultural conditions.

The qualities of the Uppsala esker as a water reservoir of major interest were discovered in the Middle Ages, when the town inhabitants used groundwater emerging from springs in the esker flanks. The first pipeline system was built in the 1640s to supply the needs of the royal castle located on top of the esker. Construction of the modern supply system started in 1872. Water was initially pumped from the major springs, and the demand was only a few hundred m³ per day, as compared to an average today of 60 000 m³. Water quality has always been excellent.

As the city continued to grow, wells drilled in the glaciofluvial material were connected to the supply system, using geological and hydrogeological information from the Geological Survey. In the 1950s, calculations showed that the amount of available groundwater in the esker was less than the demand. Lowering of the groundwater levels in Uppsala caused settling problems in the central part of the city. Investigations showed that the hydraulic capacity of the esker was sufficient for increased pumping, but not the available groundwater. The idea of supplementing the amount of groundwater with artificial recharge using water from the River Fyris then arose. A field test was carried out, using an old gravel pit in the esker, combined with tracer tests in order to check not only the direction of the flow but also the velocity of the recharged water. Although the picture was distorted, the tests gave positive results, and basins for artificial recharge were constructed. The scheme, including the drilling of new production wells, was completed in the mid-1960s. Due to the continuing growth of Uppsala, another, similar scheme was put into operation in the mid-1970s. Again, the geological maps and other information from the Geological Survey were used with great success. Recent water production and artificial recharge rates in Uppsala can be seen in Fig. 3.5.

The time lapse for the artificially recharged groundwater to reach the production wells is 6–8 months. The groundwater produced does not need any kind of purification or other treatment since the quality of the distributed water is still as good as it was before the artificial recharge started. The material of the esker acts like a surface waterworks, and without the use of any chemicals it purifies the infiltrated water. In order to prevent bacteria from growing in the reticulation system, a minor amount of chlorine is added to the distributed water. The combined use of natural and artificial groundwater thus safeguards the modern public water supply.

In the late 1980s, the question of the hydraulic qualities of the esker arose again. Uppsala needs still more water to cope with the predicted demands, and it is vital to know the limits of hydraulic capacity and possible artificial recharge in order to avoid a potential reduction in quality. A mathematical model is used in order to answer the following questions:

(1) is there any foreseeable impact on water quality?

(2) is it possible to maintain the groundwater levels within the set limits, especially in the central parts of the city, in order to avoid settling and pollution problems?

It is clear that the accuracy of any such model, and the potential for avoiding drastic changes in groundwater levels and associated hazards, will relate directly to the detail of knowledge available about the geological strata concerned.

In addition to these investigations and calculations, the setting up of protection areas is important for a good public water supply. Such areas, consisting of boreholes and inner and outer protection zones, have been established since 1 January 1990. Restrictions within each area vary with the activities involved (for such as transport, handling of chemicals and sewage, farming, and energy wells) and the protection area is ratified by the County Government Board. A good knowledge of the geology and of the velocity of the groundwater in the aquifer is essential if the setting of the boundaries of areas and zones is to be successful.

As a whole, the relevant municipal authorities have a firm grip of the Uppsala main aquifer and its immediate surroundings, but supplementary knowledge of the total recharge area is needed. In order to study the groundwater drained by the esker from the distant parts of this area, the Department of Water and Waste Water, the Department of Environmental Health in Uppsala, and the Geological Survey of Sweden have put an investigation network into operation. Data from the National Groundwater Network, operated by the Survey in the long-term, are used. A similar, joint network, to study the groundwater regime in the recharge area, is under discussion.

The Geological Survey was relocated from Stockholm to Uppsala in 1978, and it is only in recent years that the interchange between the municipality and the authority has started to flourish. Those involved are convinced that this is of great joint benefit. The city now has geoscientific expertise and information readily available, and the Survey is learning a lot about the demands for its competence and services, and how to collaborate with municipal departments. The 'Uppsala project' serves as a fine model for future Survey enterprise in being involved in similar areas of activity.

Fissured hard rock

The development in the 1960s and 1970s of economically viable and very quick methods of drilling (i.e. hammer drilling using compressed air and tungsten carbide bits) in igneous and metamorphic rocks has led to the discovery and exploitation of modest but widely distributed resources held in fractures in these ancient formations. By contrast, a distinctive hydrogeological domain, as in the Canary Islands, constituting

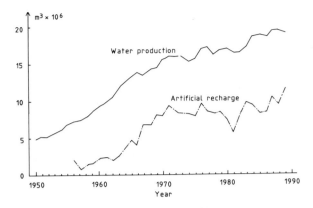

Fig. 3.5 Water production and artificial recharge in Uppsala, Sweden, 1950–89.

younger volcanic rocks, highly fractured, in places sealed by weathering processes, and with scoriaceous layers, is commonly exploited by the construction of drainage tunnels.

Fractured media in the Armorican Massif, France [39]

The drought which affected the north-west of France and the south of Great Britain in 1976 provided an opportunity to demonstrate the water-bearing properties of aquifers in hard (igneous and metamorphic), tectonically fractured rocks.

Until then the water requirements of the region had been met by two types of surface resources. Before the advent of a public water supply system, wells, drains, and galleries had been the main sources of drinking water, tapping the resources of grits and surficial weathered rocks. The increased demands of post-war expansion, however, meant that the yield was unable to meet the needs of communities, so government departments, assisted by their respective research departments, developed a policy of building dams, sometimes transporting water over great distances from the hilly hinterland to the more highly populated coastal regions. This gave rise to the image of Brittany as France's 'slate roof', good only for receiving rainwater to be stored either in domestic rainwater tanks or in reservoirs erected by water authorities.

Following the 1976 drought this image of Brittany was called into question by the Geological Survey. The drought affected the surface mainly as the groundwater had been well recharged the year before. The principal problems faced by this agricultural and stock-raising region were the lack of moisture in the soil and the low incidence of surface storage for animals. Thousands of boreholes were sunk by a spontaneous generation of amateur drillers. At the same time, numerous publicly owned boreholes were drilled, under the supervision of professional hydrogeologists, to provide additional supplies for various communities.

By law in France, the task of collecting geological sections and technical data from boreholes falls to the national Geological Survey. The scientific analysis of these data provides a solid foundation for the study of basement hydrogeology. In Brittany alone, nearly 16 000 boreholes were sunk, 8000 of which were recorded in the subsurface database.

The structural data from the boreholes are unbiased because the boreholes were sited by the drillers according to the clients' requirements, rather than by geologists. When analysed statistically these data reveal 'provinces' within which a certain amount of homogeneity exists between boreholes, apparently related more to structure than to the composition of the rocks. At about the same time hydrogeology was becoming more practically orientated, enabling yields to be improved on the local scale through the more careful siting of boreholes and better supervision of the work.

General studies based on large amounts of borehole data and individual methodological or technical studies were combined as

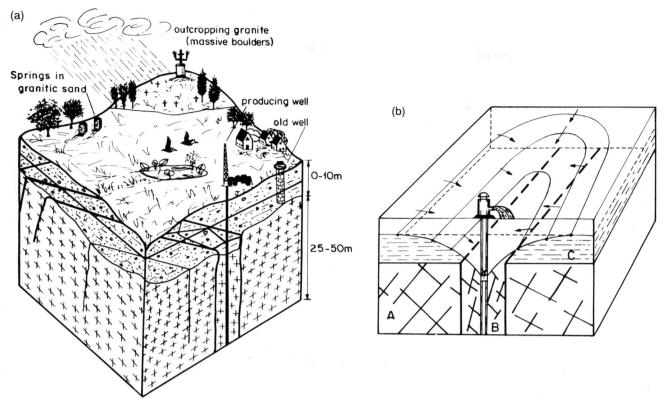

Fig. 3.6   (a) Geological model. Two-layer system: fractured rocks and weathered rocks. (b) Hydrodynamic model. The fracture zone B drains the poorly-fractured rocks A and the weathered zone C, thus contributing to the withdrawal-recharge equilibrium. Note the elliptical shape of the depression cone.

part of the Regional Geological Surveys' programmes for Brittany, Loire valley, and Basse-Normandie. During the 1980s a research programme, 'Hydrogeology of fractured media', was set up, giving support to the work carried out by the public services and the activities of the hydrogeologists in the regional survey teams.

The resource was globally evaluated by numerous studies based on an assessment of the reduction of discharge from the drainage basins, and revealed that granites have a greater storage capacity than schists. Because of the oceanic climate, aquifer recharge can be relied upon in nine out of ten years, as winter rainfall is the main source supplying the aquifers through surficial deposits. The mean effective rainfall is of the order of 500 mm in the Finistère Department, and 200 mm farther inland in the Ile-et-Vilaine Department. As capacity of the fractured reservoir is not very large, these amounts of water recharge it very quickly. The real storage capacity, however, lies in the weathered rocks. The exceptional lack of effective rainfall during the 1988–9 winter deprived these reservoirs of their recharge, yet only two boreholes dried up.

In total, the capacity of the water-bearing body is large enough to guarantee water supply on an annual basis. It is nevertheless true that from over 90 per cent of the boreholes with potential average yields of several m^3 h^{-1}, an average of only 15 m^3 d^{-1} is pumped. A recent evaluation—although prior to the 1989 drought during which irrigation by pumping was resorted to—indicates that the crop-growers and stock-rearers withdraw an estimated 30 million m^3 y^{-1}, for a total renewable resource (effective rainfall) of close on 90 million m^3 y^{-1}. The groundwater resource is still underexploited: the limiting factor is probably more in the nature of the aquifer (local availability only) than in the general water balance.

From research projects and statistical studies it has been established that the main aquifers are fracture zones, sometimes several kilometres long and in places partly sealed with weathering products, in others enlarged by chemical solution. They possess both storage and production capacities (see Fig. 3.6(a) and (b)).

Moreover, the surficial weathered zones and the various sands and colluvial deposits provide storage and delayed recharge, but could not produce the required yield alone. The appropriate management of the two elements together, as a true two-layer system, provides substantial resources. Hydrodynamically, the fracture zones are continuous aquifers with two watertight boundaries. The water in these zones can thus flow in two directions (see Fig. 3.6(b)).

Exploration methods have made appreciable progress thanks to various multi-technique programmes on pilot sites. Reconnaissance by satellite images followed by detailed photography remains the basic everyday tool for determining the structural framework. Measuring the radon gas emanations reaching the surface also enables faults to be located and assessments to be made of the degree of fracturing and, thus, the potential yield. In certain rock types radon levels may be enhanced when the degree of fracturing is high. The resistivity of rocks often relates to the amount of clay materials connected with alteration along fractures. Such alteration zones are not very permeable and often give low measurements for radon.

Private initiatives at the level of farm requirements are now giving way to professional hydrogeological studies directed towards identifying supplies for small towns, factories and agricultural co-operatives.

With a pH often less than 6 and a high iron content, the chemistry of water has to be monitored, but pollution is the main limiting factor with respect to quality. Excessive grazing and over-application of nitrogenous fertilizers have resulted in many of these aquifers having nitrate contents that render them useless for potable water. Nevertheless, two reasons for hope lie in the characteristics of the aquifer:

1. Its low handling capacity and rapid rate of recharge suggest that relatively rapid recovery of the zones reserved for potable water supply is possible.

2. Natural denitrification by bacteria deriving their energy from pyrite, which is abundant beneath the oxidized weathered zone, has been proved, so that water of adequate quality may be available from 'polluted' areas.

Confined groundwater

Management of confined groundwater reserves will have to be quantitative, with protective steps against over exploitation and allocation of the resources to priority uses. As the status of a confined aquifer is given by stratigraphic conditions and not by the reservoir properites, every kind of aquifer, if overlain by impervious layers, is potentially confined or 'artesian'. Because of poor recharge at their boundaries or possible leakage through overlying strata of low permeability, confined aquifers have to be used with caution. Water is produced by pressure-release rather than by abstraction. The time factor is the essential consideration of managers dealing with the decline of pressure, as an unsteady state governs the hydraulics of unconfined aquifers.

Nevertheless, confined aquifers can be exploited intensively during short periods as well as used at a lower level over a long period. Given these choices, mathematical models have to be used to identify pumping strategies to optimize the location of wells, abstraction rates, and the greatest benefit to society.

The hydrogeological relations of the emergency water supply of the Grand Duchy of Luxembourg [138] [139]

Since 1965 about 70 per cent of the water consumption of the Grand Duchy of Luxembourg has been derived from underground sources and about 30 per cent from surface waters. The surface waters come exclusively from the barrage of the 'Haute-Sûre' in Esch/Sûre. The total water consumption of the whole country is about 120 000 m^3 d^{-1}. The surface waters are treated and distributed to the main reservoirs of the syndicates of local communities or directly to local communities by the Syndicat des Eaux du barrage d'Esch/Sûre (SEBES). The water treatment plant produces about 35 000 m^3 d^{-1}, with a maximum of 70 000 m^3 d^{-1}.

The lac d'Esch/Sûre is to be emptied in 1991 for repair of the dam wall. The repairs will last for about seven months; five weeks are needed to empty the dam and a further two months to refill it. The water supply will be out of service for about nine months as the treatment plant works effectively with a water level somewhat below maximum. The government has decided to exploit the existing groundwater reserves to ensure the continuous provision of drinking water. This choice has been made after ruling out in 1985 eight other solutions based on surface water use (dams existing or to be built).

A plan to develop groundwater resources from several sites was then adopted. The aquifers involved are the Luxembourg Sandstone of Lower Liassic age and the Triassic Border Facies of Bunter, Muschelkalk and Keuper age. Three abstraction sites will be adjacent to the distribution pipes of the SEBES so that this existing network can be used to transport the groundwater. A fourth site will be close to a pumping station of the Syndicat des Eaux du Sud, a main client of the SEBES (Fig. 3.7).

Test wells were drilled at Everlingen Schaedhaff, Trois-Ponts (Hagen), and Cloche d'Or. Cores were taken of the aquifers, pumping tests were carried out, and the boreholes were used to examine geological and hydrogeological relationships.

As long ago as 1965, the Geological Survey proposed the drilling of a number of wells to study the hydrogeology of the formations overlying the impervious basement. The objective was to study the variation of lithology and permeability of the various aquifers, and of the quality of the groundwater in various areas of the country,

and to identify aquifers which could be exploited in case of emergency.

The first exploratory well was drilled in 1968 at Mersch. It started in the Keuper, passed through the entire Triassic sequence and reached the Devonian basement at −313 metres. Several potential aquifers were located in beds of sandstone and conglomerate.

A second well was drilled in 1972 close to the reservoir at Rehberg. It started in the Domerian, and reached the Rhetian at 343.50 metres and the Devonian basement at 694 metres. A pumping test produced 150 m^3 hr^{-1} with a drawdown of 174 metres from the Luxembourg Sandstone between 219 m and 310 m. The quality of the water was good in contrast to that from the lower Triassic aquifer which was unacceptably hard as a result of the presence of gypsum.

In 1979 a third well farther north near Bettborn started in the Muschelkalk and reached the Devonian basement at a depth of 100 metres. It produced an artesian flow of about 18 m^3 hr^{-1} of water of good quality.

The results of the exploratory project confirmed that in certain areas large quantities of groundwater of acceptable quality are available for exploitation. The plan to use these while the dam at Esch/Sûre was being repaired was adopted in1986 with an estimated cost of 700 000 000 Lfr (about 16 700 000 ECU). During 1987–8, 19 production wells were drilled at a final diameter of 600 mm. At the same time 18 observation wells were completed to monitor groundwater level variations during the successive pumping tests and later during the exploitation. Geophysical logging was used throughout to monitor the geological and hydrogeological conditions as each hole was drilled.

Eventually it was decided because of cost to limit production (and consumption) to 50 000 m^3 d^{-1} and the site at Cloche d'Or was abandoned. Mathematical models of the groundwater were used to plan the management of each site, to develop a rotational pumping scheme for all the wells at each site, and to allow continuous monitoring of the water levels. The new data collected during the short period of intensive exploitation has allowed the recalibration of the models which can now be used to predict long-term production rates, to monitor and evaluate the quality of the water, and to manage the effective use of the resources in the future.

Karstic aquifers

Karstic aquifers are found in limestone terrains where fissures develop and grow naturally by chemical solution. They are confined to countries which have carbonate platforms of coastal, often reef, origin. These range from the Palaeozoic (Carboniferous Limestone of Ireland) to the Mesozoic (Jurassic and Lower Cretaceous rocks deposited in the former Tethys ocean). Karst reservoirs are generally low in capacity but, locally, high yields of several m^3 s^{-1} are available (the principal supplies for Rome and Vienna, for example). The karst regime resembles that of surface water and is particularly vulnerable to pollution. Large research and development schemes have been carried out in places, aimed at regulating the yield, as at the Lez Spring at Montpellier in France, or at avoiding saline intrusion in the coastal zone as in Greece and at Port-Miou, France.

In the first case, the aim of this work was to tap water at a deeper level than that of the springs, in order to enhance the yield in periods of low water; this practice requires favourable structural conditions, with a good reservoir development below the level of the springs. In the second case the intrusion of sea-water is controlled by favourable structural conditions, especially karstic channels. In both cases, a very accurate knowledge of the subterranean channels is required: frogmen are often the colleagues of hydrogeologists! Another solution is to prospect for water as far as possible in the hinterland, before any saline contamination has taken place. The following case history demonstrates the potential.

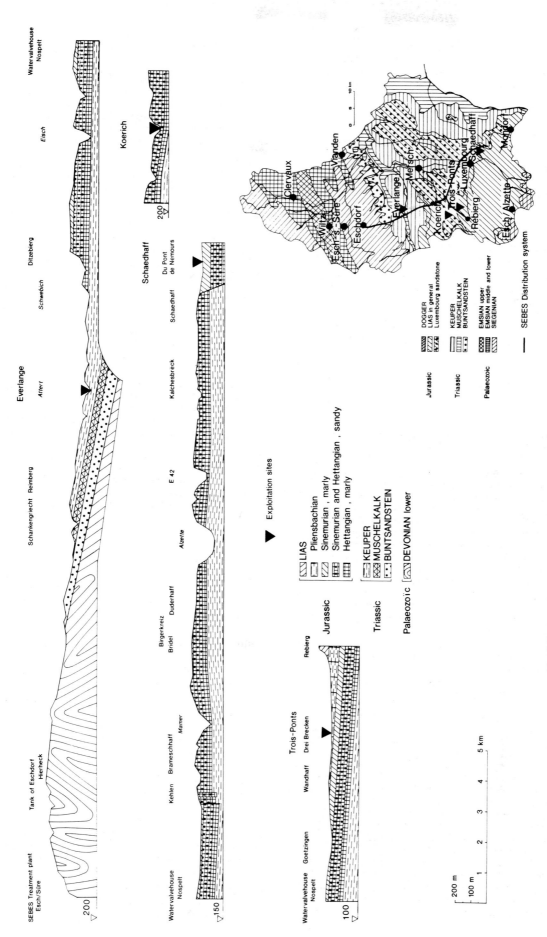

Fig. 3.7 Location of Luxembourg water supply network and alternative supply solution.

Hydrogeological investigation of the Agios Nikolaos area of Crete Island [117] [118]

Introduction

Brackish water bubbles up from springs along about 300 m of Cretaceous limestones at Almyros, about 1 km south of Agios Nikolaos in eastern Crete. Despite this indication of potential, the aquifers of the area remained undeveloped and the need for water supply was partially met by the transportation of fresh water by vehicles from other areas of the island. Irrigation remained at a primitive stage. To develop the promising tourist and agricultural potential necessitated the identification and exploitation of all freshwater resources in the area.

Initially, drilling programmes were carried out by private firms, but these met with little success mainly because they failed to overcome problems of water quality. Eventually, in 1976, the Institute of Geology and Mineral Exploration planned and executed a detailed hydrogeological investigation of the area, including the brackish springs of Almyros.

The case

The geology of the area is predominantly folded massive limestone and dolomite faulted against and overthrust onto schists and phyllites.

The following stratigraphical units are present in the catchment area of the Almyros springs (Fig. 3.8):

(1) aluvial fill of valleys, mainly gravel, sand and silt;

(2) Pliocene to Pleistocene marls and silts;

(3) Kritza flysch;

(4) late Tertiary limestone breccias, cropping out mainly in the hilly coastal area west of Agios Nikolaos. Their northern limit is located at the Almyros springs. The whole unit is overthrust onto the Tripolitza allochthonous series;

(5) the Tripolitza allochthonous series of dolomite and limestone of Jurassic to Cretaceous age. They crop out as klippen above a phyllitic series containing quartzites. They have been metamorphosed, forming a tectonic window. This unit has been overthrust onto the Plattenkalk series;

(6) the Plattenkalk series which appears to be autochthonous. It consists of cherty, well bedded to platy limestone.

The main tectonic feature of the area is a system of north-easterly faults and folds, culminating in the large anticline of Lassithi plain. Another major tectonic structure is the eastward-dipping Drassi syncline. The Lassithi anticline consists of Plattenkalk sediments and the Drassi syncline of Tripolitza sediments. Between these two units there is an area of heavily faulted phyllite quartzite, which curves from Kritza to Kakonia and seems to be the boundary of sea water intrusion.

In recognition of the complicated geological structure of this karst region, it was decided to carry out an integrated hydrogeological investigation before the implementation of a drilling programme for exploitation purposes.

The work started with detailed geological mapping of the area. Interpretations of satellite images and air photographs with respect to karst hydrology were carried out. Water level recorders were established in the Almyros springs. The flow of the springs was measured on a weekly basis. Sampling and chemical analyses were also done on a monthly basis. Exploratory wells were constructed. The fluctuation of the karstic groundwater level was recorded in a number of exploratory wells automatically, or manually on a weekly basis.

Electrical and electromagnetic measurements were carried out in the area to detect the main fault through which groundwater flows to the Amyros springs as well as the stratigraphy and structures which seem to control the karst water system. Temperature logging and other geophysical measurements were made in the exploratory wells. As far as the temperature logging is concerned, the distribution of temperature was used to point out the zones of high permeability of the karstic formations and the recharge zones of the system.

Deuterium and oxygen-18 data were used to identify the origin of recharge of the Almyros springs. Tritium data were used to estimate the age of the groundwater.

As a result of all the hydrogeological research it was established that:

1. The total flow rate of the Almyros springs fluctuates between 1.5 and 3.5 $m^3 s^{-1}$ as a consequence both of the changes of hydraulic pressure of the karst system and of the tidal variations in sea level. It is noticed that the increase of the yield of the springs happens slowly, even after a heavy precipitation. This means that the karst system is very extensive or that at least a part of it (Lakonia area) behaves like an elastic system exerting pressure on the interface between fresh and sea water.

2. The part of the karst system below sea level has a high capacity for water storage and this allows the deep intrusion of sea water inland. As a result freshwater is floating on salt water upstream from the springs.

3. The chloride content of the water of the springs ranges between 2700 and 3000 ppm, which indicates that the mean contribution of sea water to the springs is about 13.5 per cent.

4. The annual volume of flow of freshwater from the main point of discharge is estimated to be 77.3×10^6 m^3 and the total discharge from all the springs is estimated to be about 100×10^6 m^3 per year. This corresponds to a total recharge area ranging between 130 and 160 km^2.

5. The main source of recharge of the springs is the Tripolitza limestones forming the mountains of Selena and of Tsivi. Only a few small springs occur on the above mountains. The surface run-off is very low. So, most of the precipitation, which varies with elevation from about 400 mm yr^{-1} near the sea to about 1100 mm yr^{-1} on the mountains higher than 1000 m, infiltrates and penetrates to the deep aquifer which feeds the Almyros springs.

The flow-paths of the groundwater towards the springs, and the boundary of sea water intrusion were determined.

Using all the information and data acquired during the hydrogeological research project, a drilling programme to exploit the karst groundwater resources of Almyros was planned and implemented.

The first boreholes were located at low elevations upstream from the springs. They yielded a good quantity of water with 600 ppm chloride content. These boreholes are used today for irrigation purposes. The next boreholes were located at higher elevations, but always at sites where karst groundwater could be reached at reasonable depths. These have been successful in providing Agios Nikolaos with plenty of water of good quality. Finally, boreholes sunk in the Drassi area yield large quantities of water with 30 ppm chloride content.

Conclusions

This hydrogeological investigation is considered to be successful, not only because it provided the necessary data to support a production drilling programme, but also because it answered various scientific and theoretical questions about the hydraulics of the Almyros springs, the exploration of karst regions, and the methods and techniques to be applied in similar cases.

If the Institute of Geology and Mineral Exploration had taken action at an earlier stage, the previous drilling programmes would not have been implemented. Not only would considerable unnecessary costs have been avoided, but the large socio-economic benefits would also have accrued from the earlier development of the area.

Today the semi-arid environment of Agios Nikolaos has given way to ever-increasing green belts, the local economy is accelerating and a significant tourist and agricultural development is obvious in the area already.

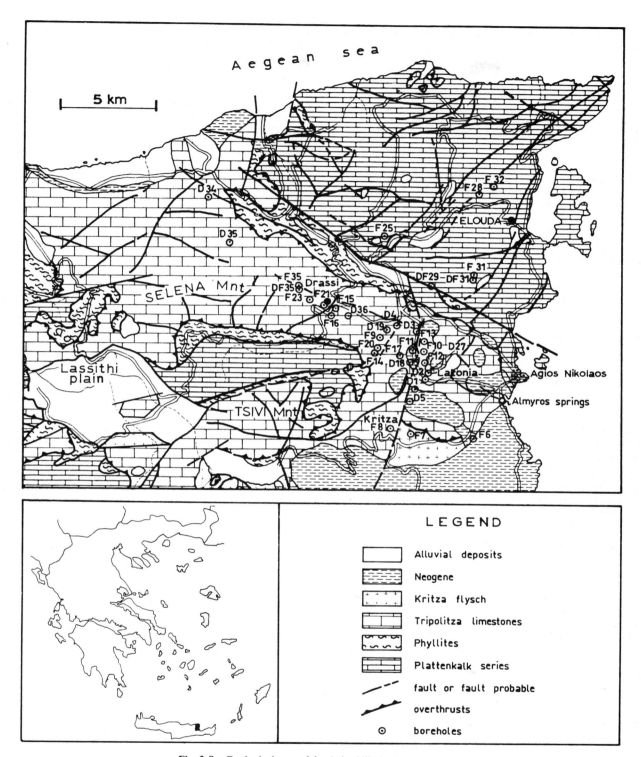

Fig. 3.8 Geological map of the Agios Nikolaos area (Crete).

This case is a fine example of cooperation between two Geological Surveys, IGME-Athens and BGR-Hannover. A collaboration project on Groundwater Technology was established between IGME and BGR in 1981. Its objective was to improve the utilization of various methods and techniques in hydrogeological research in karst regions. Some of them were applied in the Agios Nikolaos area and the results were included in the final evaluation.

Sedimentary basins

Aquifers in sedimentary basin are the most extensive of all and have been subjected to much exploration, which has resulted in the assembly of large data banks of information acquired from many boreholes.

The Triassic and Lower Liassic sandstones and the Chalk of central Europe, and the Tertiary sands and molasses in southern Europe especially, provide the largest aquifers, many of which extend across national boundaries. Some of these aquifers are confined by being overlain by impermeable beds. The contained groundwater is under pressure, difficult to recharge, but relatively protected from pollution. Unconfined aquifers on the other hand are readily recharged, but are highly vulnerable to localized and widespread pollution. The following case history is concerned with the Chalk, the greatest European unconfined aquifer of all.

The hydrogeology of the Chalk in the United Kingdom: the evolution of our understanding [199] [231]

Introduction

The Chalk is an important aquifer in the UK, France, Belgium, and Denmark (Fig. 3.9). In the UK it provides about 55 per cent of the groundwater used for water supply. There are two components to the porosity and permeability of this soft limestone. The first is the matrix, or intergranular component, and the second is the fissure network. Most of the fissures are micro-fissures less than 1 mm wide that impart a permeability of 10^{-7} to 10^{-5} ms^{-1}. This primary fissure component is developed by weathering and solution processes, creating a network of major, or macro-fissures. The porosity of the macro-fissure component is only of the order of 10^{-4} and most of the specific yield of the Chalk of 10^{-2} is due to the porosity of the micro-fissures. In the saturated zone, solution processes enhance transmissivities up to 2.5×10^{-2} m^2 s^{-1}. Generally, most of the water flowing in this zone is in the upper 60–100 m. The macro-fissures act as a water distribution system, gaining water from the matrix which is the storage reservoir.

By 1945 groundwater development in the English Chalk had led to localized falling water levels, reduced stream flows, and, in places, saline intrusion. Legislation was necessary to control abstraction and over the years the development of the aquifer has been related to increasing legal controls: the Water Act of 1945 promoted an era of resource assessment; the Water Resources Act of 1963 led to an era of groundwater management, and the Control of Pollution Act in 1974 heralded the current emphasis on risk contamination and groundwater quality.

Pre-1945

Collection of data on the Chalk aquifer by the Geological Survey of Great Britain began to increase in the mid-1930s as a direct consequence of a major drought in 1933. During the Second World War, development of the aquifer continued and it was realized that greater control of abstraction would be required in order to manage the aquifer properly.

Ground subsidence in London and its relationship to the uncontrolled decline in artesian pressure* in the Chalk aquifer beneath London was recognized before the war. In the early 1940s attempts were made to estimate the rate of infiltration into the Chalk.

Era of water resource assessment

The Water Act of 1945 introduced the first legal control of groundwater abstraction. In order to implement the policy it was necessary to establish a reliable database and identify well sites, and to collect information including well construction details, rest and pumping water levels, yields, and groundwater quality data. This task was entrusted to the Geological Survey. An assessment of the quantities of water taken from the Chalk consequently became available for the first time in 1948.

The next stage in understanding the groundwater resources was an attempt to assess the rate of infiltration using meteorological data to calculate aquifer recharge. This work has continued and has been supplemented with base flow studies.

The development of techniques for analysing pumping tests enabled the recognition of the relationship between yield–drawdown curves in Chalk wells and variations in transmissivity. This also confirmed that higher transmissivities (2 to 4×10^{-2} m^2 s^{-1}) occur in valleys whereas lower values (1 to 2×10^{-4} m^2 s^{-1}) are common elsewhere.

Development in downhole wireline geophysical logging demonstrated the role of secondary permeability and fissure flow within the Chalk. Logging also gave the means of recognizing diagnostic lithological features.

Era of groundwater management

This era began with the Water Resources Act 1963 and the inception of the catchment-based river authorities which were responsible for pollution prevention, fisheries, land drainage, and water-resources.

Regional groundwater schemes were introduced to provide water supplies but also to supplement and protect low summer river flows at the expense of high winter run-off. Emphasis was based on regional groundwater management in conjunction with surface water resources using river basins as the basic unit. These programmes enabled river flow to be augmented by groundwater and groundwater storage to be supplemented by artificial recharge. In addition, saline intrusion was controlled by seasonal pumping from boreholes at different distances from the sea, thereby making the maximum use of groundwater storage. For example, in the South Downs of Sussex coastal boreholes are used when groundwater levels are high in winter while production boreholes further inland are used during the summer months to minimize saline intrusion. This procedure enables 94 per cent of the infiltration in a drought year to be licensed for abstraction.

Digital and analogue computer models have also been used widely to understand the Chalk aquifer. Despite its fissured nature, numerical models have been successfully constructed for regional studies which match observed conditions of groundwater drawdown and stream depletion.

Era of groundwater quality

The Control of Pollution Act 1974 promoted a greater interest in pollution and aquifer protection than before. It also became apparent that different flow paths available within the Chalk aquifer offered different transport mechanisms to groundwater and potential pollutants. The role of the unsaturated zone came under close scrutiny. Initial analyses of thermonuclear tritium in rivers derived from Chalk springs showed surprisingly low concentrations. Dry coring of the Chalk revealed that most of the tritium was stored in the matrix in the upper part of the profile and that infiltration was moving through the unsaturated zone at about 1 m yr^{-1} by a process of piston displacement. About 15 per cent of the total appeared to move more rapidly through the unsaturated zone directly to the water table via the fissure system.

Research showed that the median pore throat size of chalk averages about 0.5 microns. Only about 1 per cent of the bulk volume can thus drain by gravity and this is primarily from the micro-fissure network. It therefore appears that recharge of the unsaturated zone is not by rapid transit of water in macro-fissures, but by piston displacement in micro-fissures and the larger pores of the matrix. The solutes in the micro-fissures diffuse into the matrix and move down by continuous exchange between the matrix and the micro-fissures. Once infiltration exceeds a threshold value (about 4 mm d^{-1}) flow also occurs in the macro-fissures.

---

* The Chalk aquifer is mostly unconfined, but it is confined in the centres of the London and the Paris basins beneath impervious layers of Tertiary strata.

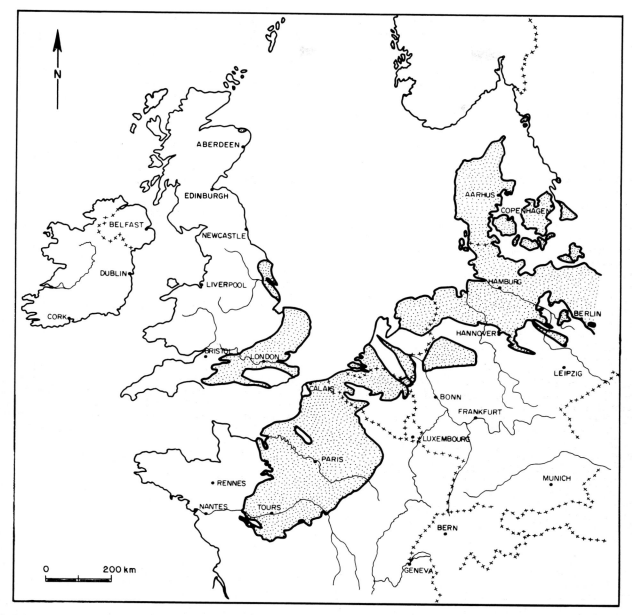

Fig. 3.9 Distribution of Chalk aquifers in Western Europe, (after Ziegler, P. A. (1982)). *Geological atlas of Western and Central Europe.* (Shell Inter. Pet. Maat.)

Investigation into rising nitrate levels drew further attention to the apparent anomaly between the traditional view of rapid transit of water infiltrating through the Chalk's unsaturated zone to the water table (about 1 m d^{-1}) and the environmental tritium studies which showed very slow rates of movement (about 1 m yr^{-1}). Extensive field studies involving dry coring of the Chalk and analysis of pore waters confirmed that the movement of nitrate, as with other solutes, is retarded because of diffusion into the matrix of the Chalk. Much of this work has been carried out by the Geological Survey.

In the saturated zone, in addition to the physical displacement of water downgradient in the matrix, there is a continuous exchange of water between the matrix and the fissures. Although the bulk of the water in the matrix is immobile, it is involved in the overall water transport since it is moving by continuous exchange of molecules by means of diffusion, and the total water storage is steadily displaced down the hydraulic gradient.

Because of the importance of the Chalk as a source of water supply, increasing consideration is being given to aquifer protection policies. These specify areas around Chalk sources with radii of between 1.75 and 2.5 km for rates of pumping of less than 6 × 10^{-2} m^3 s^{-1} to in excess of 0.18 m^3 s^{-1} respectively. The size of the protection area is designed to safeguard against microbial contamination.

Conclusions

Although our knowledge of the Chalk has increased greatly in recent years, topics that will need to be addressed in the future include:

(1) the nature and composition of the matrix;

(2) the influence on aquifer properties of diagenetic and structural change;

(3) the nature and distribution of clay minerals and organic compounds within the Chalk;

(4) water–rock reactions and their bearing on the chemical composition of the groundwater;

(5) the careful design of aquifer tests to interpret the nature of flow in a dual porosity aquifer.

Just as much of the research on the Chalk in the past has been led by the Geological Survey, it is anticipated that the Geological Survey will retain a major role in the future.

Thermal and mineral water

Thermal and mineral waters, which are of great historical importance and economic value, are not associated with specific rock types but are mainly related to structural conditions and the style of evolution of mountain ranges. The main task of the thermal hydrogeologist is to acquire detailed understanding of the structures involved, whether they be Hercynian with essentially sodic–carbonate water, or Alpine with chloride–sulphate water and sulphurous emanations. The main tools involved in the successful increased exploitation of the resources to produce high yields by deep tapping, and in improving the protection against surface pollution, are developed by Geological Surveys—basic geoscientific expertise, systematic geological mapping, and data acquisition.

Protection of mineral and medicinal springs in Stuttgart-Bad Cannstatt [81] [82] [83]

Introduction

Stuttgart, capital of the State of Baden-Württemberg, is the centre of an important area of economic agglomeration and has one of the richest networks of mineral springs in Europe. From a total flow of mineral water of $300 l s^{-1}$, $230 l s^{-1}$ are kerbed and $160 l s^{-1}$ are officially certified as being medicinal. The water is used by 1.6 million visitors every year for healing and relaxation. Adequate water protection and availability of water are accepted goals of the city's planning policy within the overall development of an industrial conurbation.

Geological conditions (Fig. 3.10)

Stuttgart is situated on a sequence of Mesozoic hard rocks comprising the Lower or Black Jurassic (Hettangian, Lower Simenurian), the Upper, Middle, and Lower Keuper, and the Upper Muschelkalk. A borehole for thermal waters found Lower Muschelkalk, Buntsandstein, and, at a depth of 450 m, basement rocks. The solution of gypsum and halite in the strata has led to subsidence and the formation of cavities which are commonly filled by valley alluvium. In places, the process continues and new cavities are formed from time to time.

For several hundred thousand years, carbonated mineral and thermal waters have reached the surface along tectonic faults and zones which have been opened and enlarged by leaching phenomena. Very thick sinter, known as freestone (Stuttgarter Travertin or Sauerwasserkalk) has been precipitated from these waters at times. With increased deepening of the Stuttgart sedimentary basin, areas of upwards water flow and sinter zones repeatedly changed. The mineral waters originate mainly from the Mesozoic Muschelkalk strata, in which they were mineralized by the solution of halite, gypsum, limestone, and dolomite. The provenance of the carbonic acid is believed to be a deep source of volcanic exhalations.

The network of springs

Mineral waters of varied composition and originating from Muschechalk strata, reach the surface over an area of about 3 km².

From the west, mineral water flows down dip from an open area of karstic outcrop with very little cover. With 1.1 to 1.3 g l⁻¹ of dissolved solid matter and 14–16 °C of temperature, it can be classified as a weakly warmed, poorly carbonated mineral water of low concentration. In contrast, highly concentrated mineral water of up to 6 g l⁻¹ of dissolved matter enters an area flowing up dip from the south where there is thick cover. It is highly carbonated (up to 2 g l⁻¹ CO_2) and subthermal with a temperature of 17.5–20.5 °C. The regeneration of this extraordinarily productive water current ($150 l s^{-1}$) has not yet been explained. Between the two main streams, mixing produces a flow of intermediate concentration.

The total weight of dissolved halite, gypsum, dolomite, and limestone produced within the network is 78 t d⁻¹, and the amount of free carbonic acid, partly degassed by pressure decrease as it reaches the surface, is 23 t d⁻¹.

The original seepages have been replaced by wells 20–140 m deep from which the flow of mineral water is legally controlled.

Some of the mineral water flowing from the west passes under the inner city and industrialized areas of Stuttgart, where the groundwater regime (mainly the low-concentrate mineral waters) has been contaminated to some extent by highly volatile chlorine hydrocarbon compounds. Strict controls have been introduced and the level of contamination has been reduced to less than 5 µg l⁻¹.

According to their content of radioactive isotopes, waters coming from the west are young while waters coming from the south are very old. Contents of up to 0.95 Bq l⁻¹ of tritium (^3H) prove a mixing process of younger and highly concentrated older water. Consequently, the danger of pollutants infiltrating the medicinal springs cannot be precluded.

Plans have been made for the entire area of inflow of mineral waters around Stuttgart-Bad Cannstatt to be made a protected area of more than 10 000 km², divided into qualitative and quantitative protection zones.

Protection of mineral water at building sites

All building projects within the urban district are already being examined. The risks they present are analysed and both exploration and construction phases are being controlled. The following conditions have been set for all sites:

1. Flat-lying and pile foundations must be located above the strata bearing the mineral waters, and they should not have any water-retaining installations.

2. Connections between different groundwater levels, allowing access of pollutants to greater depths, and connections which permanently drain mineral water and cause pressure decrease or degassing of carbon dioxide are not allowed.

3. Water-soluble components of building materials must not get into groundwater.

4. Intensive site exploration must be carried out to preclude construction-induced risks to mineral water. This will include hydrochemical tests, and tests for carbon dioxide and man-made pollutants. During deep construction work, contingency plans must be available for immediate sealing of the subsoil if necessary. On individual complicated sites proof and hydraulic modelling will have to demonstrate that potential problems can be controlled.

For individual buildings, problems related to variations in the subsoil can be overcome by carrying out appropriate site exploration. This is not the case for linear transport developments. The construction of these under the city of Stuttgart has proved to be difficult and expensive because of the varied subsurface conditions. The construction of several municipal tunnels as well as of the high-speed line for the German Federal Railways (Deutsche Bundesbahn) under the city has proved to be geotechnically difficult. Because of high water pressure in the Gipskeuper at the valley margins, the lines have to be accommodated in several small-diameter single-track tunnels and within the urban area four parallel tunnels have to be driven through mostly water-saturated valley deposits and collapsed rock-masses of the Middle Keuper. Fault

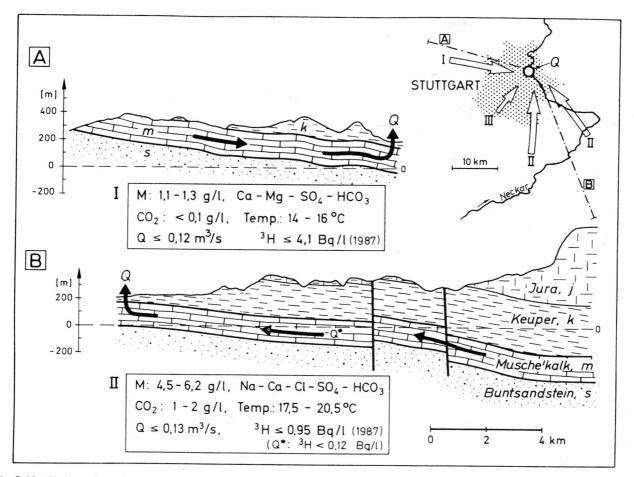

Fig. 3.10 Hydrogeology of the network of springs (Q) around Stuttgart-Bad Cannstatt (Germany) containing significant data on the main inflows from west (I) and south (II). A secondary flow (III), a mixture of (I) and (II), comes from the south-west. M = Mineralization: concentration of dissolved matter and characteristic components; ³H = Tritium in Becquerel per litre.

zones, vulnerable with regards to mineral water protection, have to be crossed. To avoid the existing infrastructure of the built-up area, the tunnels will have to be at greater depth thus jeopardizing the natural protection of the mineral waters provided by overlying strata. Such construction work requires the use of freezing and compressed-air techniques in order to ensure the protection of the mineral waters.

3.2.3. Exploitation methods

The methods used for exploiting groundwater are usually directly linked with the type of underground reservoir involved much more than with deliberate technical choices. Nevertheless, the economics of a water supply project also dictates the type of installation. For example, in an alluvial plain aquifer, an individual well yielding one litre per second may coexist with collector wells supplying 200 l s⁻¹ or more for a town or a large factory. Types of development over the last 30 years can be classified as follows:

1. Alluvium: large-diameter wells, bucket-sunk wells and, locally, collector wells (constructed according to various European patents) are often grouped in sets, along river banks, where they operate as natural filtering units for the surface water.

2. Sedimentary rocks in artesian basins such as that of Aquitaine: standard boreholes, one to several hundred metres

deep, sunk by rotary drills fitted with efficient anti-sand screens and filter gravel, or other drilling methods such as reverse circulation for the large diameter boreholes.

3. Karst: for large communities, such as the city of Rome, very high yields of up to 1 m³ s⁻¹ are got from natural springs, supplemented in some places with galleries; shafts and galleries are constructed to tap water below the natural spring level, as in Montpellier (France). Tapping of smaller but none the less important yields is done by borehole, especially in the palaeokarsts of the plains (Germany, Belgium, Ireland) and in partly karstic formations. Karstic channels and the porosity of fractures and joints improve the yields from the latter. Modern techniques for drilling standard boreholes (replacement of percussion by down-hole hammer drilling) has enabled small-scale economic exploitation of these aquifers and the development of irrigation works, as in Greece, Spain, and France.

4. Fractured media: shallow boreholes, on average about 50 m deep, drilled by down-hole hammer are ideal for domestic or farming use in areas which are poorly supplied by the public networks, as in Scandinavia and the Armorican Massif of Brittany (France). When yields are sufficient, the water from such boreholes is incorporated into the distribution supply, as is the case for the water of the green rocks of Troodos, in Cyprus.

5. Glacial deposits: the heterogeneity of these deposits ranging from large boulders to fine-grained poorly consolidated material, has long posed drilling problems. These

difficulties are now overcome by the association of the down-hole hammer and driven pipes: the hammer breaks the boulders and the casing holds the poorly consolidated material.

6. Thermal and mineral water: due to the pollution of spring water at the surface, water tapping at depth is now widely used. The requirement to maintain the quality of deep water and to protect it from microbial pollutions, has led to the use of advanced drilling techniques (continuous disinfection while drilling, the use of special materials, and so on), which are successful only when planned and supervised by professional hydrogeologists.

3.2.4. Management policies

Geological Surveys are not responsible for the management of water resources, but they make two important contributions:

(1) collecting, processing and holding for consultation the data required for management;

(2) bringing technical and scientific assistance to the management, in particular on resource and storage protection.

Research and development in these areas should influence the attitudes of decision makers and politicians as to the importance of groundwater and will contribute to the development of appropriate legislation.

Data management

Most European countries now have data banks related to subsurface information, and in particular to groundwater for which detailed geological knowledge is of fundamental importance particularly with respect to reservoirs. The most up-to-date hydrogeological studies, including essential physical and chemical elements, could not be successful without reference to the detailed geological infrastructure provided by the data banks of many national Surveys.

Supervision and monitoring

The supervision of water resources is in the hands of many varied organizations throughout Western Europe. In many cases, water is in the hands of *ad hoc* organizations which are concerned with a general approach to groundwater protection more than with the origin of the water. Quantitative monitoring of the resource of piezometric networks is entrusted either to specialized organizations, as in Germany and The Netherlands, or shared between miscellaneous departments and region or basin administrators. On that score, the French Geological Survey benefits from government support by a *lettre de mission* (detached service warrant) entrusting it to collect and organize groundwater information and make it available to the public. In the UK all boreholes for water over 15 m in depth, together with the results of any tests, have to be notified to the British Geological Survey and entered into the publicly accessible national data bank.

The density of stations in any network for observation or control is related to the style and scope of investigation rather than institutional choice. In The Netherlands for example, the high density of water level control points (more than 15 000) is justified to control the delicate freshwater/salt water balance (unconfined aquifer) over practically the whole country. In contrast, the information required for managing the confined aquifer of the 'lower sands' of Aquitaine (France) is supplied by only 27 points. It should also be added that, in

this case, the rationalized reduction of the number of monitoring points is the result of 30 years of experience. The logged data, particularly from the observation wells, enables a piezometric map of the aquifer to be drawn up to provide a basic reference for modelling. When a suitably representative model has been developed, points showing variations accurately are readily identified and pollution trends can be monitored accurately. Nevertheless, high-density observation networks may be a tradition in national or regional monitoring and provide easy access to detailed piezometric data.

Technical and scientific assistance to management

Considering the great variety of tasks, ranging from basic hydrogeological mapping to the production of models for groundwater management, the potential contribution available from the data banks and expertise of national Geological Surveys is obvious. Bearing in mind the close relationship between surface and groundwater, and the influences affecting quantity and quality, the value of this contribution can be maximized only if it is coordinated with the input of many non-geological bodies.

Quantitative problems, such as the over-pumping of aquifers, the allocation of resources to various sectors of the community, and the rising levels of groundwater in some urban environments, are monitored and studied in most countries. The principal contribution of the Geological Survey is mapping and data acquisition and ideally precedes the work of water engineers to whom basic information is supplied. In addition, the detailed mapping of aquifers and the presentation of data through Geographical Information Systems, are available for multidisciplinary analysis and land-using planning.

Modelling

Modelling, carried out by the various techniques developed in many public hydrogeological surveys, is the most effective tool in hydrogeological studies. Examples are models for the management of groundwater, especially that which is over-exploited or threatened—the marine intrusion in The Netherlands, the rising water tables in Great Britain, and pollution control in Alsace (France). Research on the development of these tools is nevertheless often carried out in universities or even in specialized centres.

Research and methodology

Research in the groundwater field is very diversified and carried out in many technical institutes and laboratories. The contribution made by the Geological Surveys varies greatly from country to country. In Germany, it is partly decentralized in various organizations of the Länder. In contrast, it is concentrated more in the national Geological Surveys in France, Denmark, and Spain. In Great Britain, many organizations contribute to the hydrogeological research alongside the BGS, and in The Netherlands, the research is commonly carried out elsewhere than in the Geological Survey.

Global resource management

The widespread availability of groundwater offered by nature, and the existence of highly sophisticated tools for exploration and development, should bring about a high rate of exploitation. Unfortunately, this does not always seem to be so. In Fig. 3.11, comparison is made between two ratios: abstraction

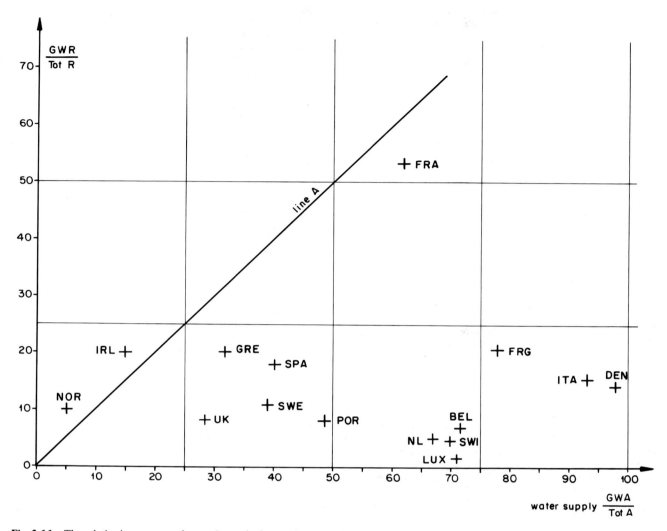

Fig. 3.11 The relative importance of groundwater in the supply, versus the importance of groundwater as a natural resource: line A, with slope 1/1, shows the balance.
GWR: Groundwater resource = relative importance of groundwater in the resource.
Tot R: Total resource
GWA: Groundwater abstraction = relative importance of groundwater in the supply.
Tot A: Total abstraction

of groundwater versus satisfaction of total water demand, and the relative share of groundwater in the total resource. Comparisons between the importance attached to groundwater can be made for each country, regardless of size.

It is not the countries with an abundance of groundwater that use it the most, but rather the contrary. The Benelux countries, Italy, and Denmark demonstrate this, especially for the public supply. On the other hand, France, which has abundant resources of groundwater, uses it to a very modest degree, as do the few countries lying above line A (slope is 1:1) in the diagram. The advantages of groundwater, its constant availability and quality, are thus little appreciated still in economic and political spheres. Although out of sight and out of mind, groundwater has frequently provided the vital reserve source during periods of drought. Despite this and the knowledge of the fundamental interactions and exchanges between surface and groundwater, the case for groundwater and its value has yet to be made. In this respect the task of Geological Surveys and associated bodies has hardly begun.

On the other hand, groundwater resources are not inexhaustible, even in the case of temperate climate where they are considered to be renewable. Although less than 20 per cent of the total resource is used, local over-pumping is common. Excessive lowering of the water table—100 m and more in Andalusia, Spain for example—may cause intrusions of saline water in coastal areas (see Section 5.5).

An increase in the demand for water comes, above all, from areas of agricultural irrigation in southern Europe. Despite the alarmist forecasts of the 1960s and 1970s, urban consumption is practically stable (better management, reduction in water losses), and industrial consumption has been sharply reduced as a result of modernization. In the flat regions, which are difficult to irrigate by dams and canals, the farmers take water directly, through boreholes, from the major sedimentary and alluvial aquifers. In western France, for example, this type of consumption has grown tenfold in regions of maize and sunflower cultivation. This causes extreme conflicts in the development of water supplies. It is a cause for reflection that, through irrigation for the production of cereals and animal feed, 2000–3000 litres of water are required to produce one kilogram of pork!

The quantities of groundwater taken for farming are so great that, locally, surface streams may stop flowing. This does not surprise the hydrogeologist and provides him with the best evidence to demonstrate the connection between surface and groundwater, which is so important in reverse. The effective management of the resource at an appropriate scale will be possible only when legislation brings the control of the use of groundwater into the public domain alongside surface water bringing forward the concept, therefore, that all water is from a common source.

Aquifers can be considered to be underground regulating reservoirs storing rainwater accumulated during years of high precipitation, for use to supplement surface supplies during years of low precipitation. The capacity and control of such reservoirs is far greater than can be achieved by the harnessing of river supplies by damming.

In addition to the advances of scientific knowledge and the development of sophisticated techniques for the analysis and forecasting of groundwater flow, the concept of the common source for all water is by far the most important advance in recent decades. Hydrogeologists must now convince other professions concerned with water supply that this concept is essential to securing adequate supplies of water for man in the long term. Unfortunately, the vast global water cycle is increasingly susceptible to pollution, a topic dealt with in more depth in Section 5.5.

RECOMMENDED FURTHER READING

Bodelle, J. and Margat, J. (1980). *L'eau souterraine en France* (Masson, Paris).

Castany, G. (1982). *Principes et méthodes de l'hydrogéologie.* (Dunod, Paris).

Collective Authors (1988). Karst hydrogeology and karst environment protection. *Proceedings of the 21st IAH Congress Guilin, China, 1988.* (Geological Publishing House, Beijing).

Collective Authors (1989). *Hydrogeologie.* No. 2, 1989, Europe—BRGM (special issue about groundwater in Europe).

Collective Authors (1990). Chalk. *Proceedings of the International Chalk Symposium, Brighton 1989.* (Thomas Telford, London).

Custodio, E. and Llamas, M. R. (1983). *Hidrologia subterranea.* 2nd edn. (Omega, Barcelona).

Custodio, E., Gurgui, A., and Ferreira Lobo, J. P. (eds). (1988). *Groundwater flow and quality modelling.* (Reidel, Dordrecht).

Drever, J. L. (1988). *The geochemistry of natural waters.* 2nd edn. (Prentice Hall, Englewood Cliffs, NJ).

Driscoll, F. G. (1987). *Groundwater and wells,* 2nd edn. (Johnson Division, St. Paul, Minnesota).

Erhard-Cassegrain, A., and Margat, J. (1982). *Introduction à l'économie générale de l'eau.* (Masson, Paris).

Prohic, E., Zoetl, J., Tanner, M. J., and Roche, B. N. (eds). (1989). *Hydrogeology of limestone terranes.* (Heise, Hannover).

Robins, N. S. (1990). *Hydrogeology of Scotland.* (HMSO, London).

3.3. ENERGY RESOURCES

Whilst water, air, and food are essential to man's survival, nothing has stimulated his advance more than the availability and application of an ever-growing range of energy resources. In his earliest days, the available energy was transmitted through himself or his harnessed animals, but the energy available from burning wood soon enabled him to keep warm, to ward off predators, and eventually to make better use of other natural resources.

Successively, as man has discovered new energy resources, and how best to use them, so civilization has taken further steps forward. The ability to produce coke and apply its great heat was a major factor in the Industrial Revolution. Likewise, the enormous advances in transport achieved this century have relied almost totally on the realization of the potential of hydrocarbons and their wide exploitation.

In the last few thousand years, as man has discovered new forms of energy, the demand for it has always risen to meet the increasing population and the additional applications. The more traditional energy resources, therefore, have seldom been abandoned.

Overlap in the applicability and use of all the energy resources means that there has been, and still can be, significant change in fuel usage from time to time as the economics of the exploitation of the various resources change. This overlap has allowed very different regional, or national, energy-use patterns to develop, even within Europe, for a variety of historical, geographical, geological, meteorological, and social, as well as economic, reasons. Therefore, whilst the changes in the prices of the different fuels drive changes in those patterns they do so in different ways in the various scenarios. Increasingly, the environmental concerns are important factors in the continuous development of energy use patterns.

Whilst the energy resources are by no means all geologically based, the major forms now in great demand are contained in the Earth's crust, whether arising from the past decay of organic matter (as coal and oil), or as minerals (such as uranium). Furthermore, of the 'newer' or 'alternative' energies, heat generated by geological processes (geothermal energy) is already contributing to the modern aims of minimizing environmental distress often associated with resource exploitation. In the following pages these geologically-based energy resources are described and the environmental implications of their exploitation are discussed.

3.3.1. Hydrocarbons [207] [209]

Introduction

Commercial accumulations of oil and natural gas are distributed throughout western Europe (Fig. 3.12). They occur in sedimentary strata ranging from Cambrian to Cenozoic and in various geological settings.

Europe has a long history of the use of petroleum; natural petroleum was used for lighting and lubrication as early as 1546 in Lower Saxony. By the middle of the eighteenth century oil sands were mined for oil in the Pechelbronn area of the Rhine Graben, whilst exploration of underground oil by drilling began in 1958 at Wietze, north of Hannover, where oil was encountered in economic quantities. Sixteen years later regular production started at this site. In Britain, James 'Paraffin' Young initially exploited seepage oil from a Derbyshire coalmine before opening a distillation plant for cannel coal and later oil shales at Bathgate, Scotland, in 1851.

The bulk of western Europe's current oil production and most of its remaining reserves are located offshore, with most exploration activity and hydrocarbons production concentrated in the North Sea. Offshore exploration in the North Sea began in the early 1960s, following the discovery in 1959 of the giant onshore Groningen gas field in The Netherlands, and success came in 1965 with the discovery of the West Sole gas

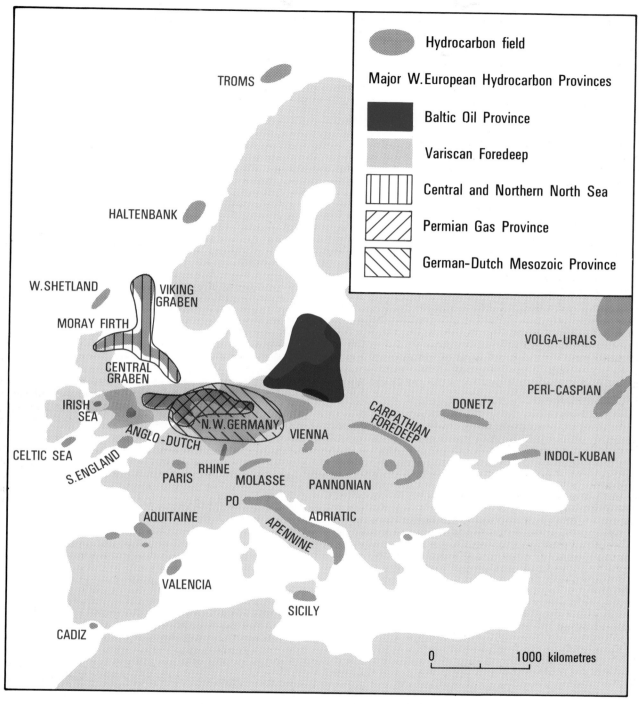

Fig. 3.12 Location of major hydrocarbon areas in Europe.

field in the UK sector. The first commercial oil discovery in the North Sea was made in 1969, with the discovery of the Ekofisk field, in the Norwegian sector. However, oil had already been encountered in 1966, by the A-1 well drilled in the Danish sector. At the time this discovery was considered to be non-commercial, but about 25 years later there are now plans to exploit this as the Kraka Field.

In the last decade the hydrocarbon output of Western Europe (Table 3.2) has risen to account for over 6 per cent of the world's production of crude petroleum, and just over 9 per cent of the world's natural gas production (1989 figures). Output from offshore fields now dominates such statistics.

Whilst exploration for hydrocarbons tends to stimulate local reaction, the environmental effects are largely transitory except for those associated with refineries and storage depots. Conversely, the widespread benefits and secondary products from these natural resources are legion.

Resource evaluation

Accumulations of hydrocarbons

Oil and gas are formed by the natural modification of materials produced by living organisms. Where sufficient organic material is deposited in a non-oxidizing environment

Table 3.2 Western European oil and gas production 1987–89. Figures from *World Mineral Statistics 1985–89* (British Geological Survey 1990)

County	1987		1988		1989	
	Crude petroleum (Mt)	Natural gas (Mm³)	Crude petroleum (Mt)	Natural gas (Mm³)	Crude petroleum (Mt)	Natural gas (Mm³)
United Kingdom	117.66	47.6	109.46	15.76	87.40	44.76
Norway	49.52	29.87	56.18	29.75	74.89	31.96
The Netherlands	4.29	74.25	3.91	65.61	3.81	71.57
Denmark	4.55	—	4.70	—	5.42	—
Italy	3.91	13.32	4.81	16.63	4.58	16.98
Germany (West)	3.80	15.87	3.94	14.78	3.77	14.65
France	3.23	3.88	3.39	4.64	3.24	4.41
Turkey	2.63	—	2.56	—	2.84	—
Spain	1.64	0.75	1.48	0.92	1.04	1.55
Austria	1.06	1.17	1.18	1.27	1.16	1.28
Greece	1.21	—	1.11	—	0.91	—
Ireland	—	1.67	—	2.02	—	2.30
Belgium	—	0.04	—	0.03	—	0.02

a source rock is created. Source rock can develop in both marine and non-marine environments but the chances of survival of any resulting hydrocarbons are greater in the marine environment. Once a source rock is deposited, sufficient heat is necessary (mostly arising from depth of burial and geothermal gradient) to transform the solid organic matter to oil and gas. The fluid hydrocarbons are displaced from source rocks by changes in the confining pressures which result from deep burial or folding, and migration through the adjacent strata ensues.

Subsequently, for hydrocarbon accumulations to occur there must be both reservoir and cap rocks. The shape of the cap must prevent lateral and upward escape of the oil or gas; such features (see Figs. 3.13 and 3.14) are known as traps. The reservoir rock's essential characteristic is an adequate porosity to hold the fluid hydrocarbons and, ideally, a good permeability to allow those hydrocarbons to flow into production wells when they are drilled. Much of the porosity, and hence storage capacity, is provided by the intergranular spaces within the rocks, but joints and cracks can greatly increase the total permeability.

Sedimentary rocks provide the most common reservoirs, particularly the coarser grained types of sandstone, limestone, and dolomite. Much less commonly, shales or even igneous and metamorphic rocks may be sufficiently cleaved or fractured to provide adequate openings for oil or gas storage. As the hydrocarbons are extracted from a reservoir (with a consequent lessening of reservoir pressure) there is evidence that the porosity and permeability decrease in response to the greater effect of the confining pressures on the slightly compressible rocks. Reservoir rocks range in age from Precambrian to Plio-Pleistocene but in Western Europe the most economic are of Permian to Tertiary age.

The cap rock is the barrier to the upward escape of oil and gas and is commonly a shale, salt, gypsum, or anhydrite. Essentially it has low permeability, and ideally it is flexible enough to have withstood tectonic movements without fracturing.

Whilst traps may vary widely in detail, there are a few common factors involved; for example, a suitable relationship between reservoir rock and fracture-free vertical or lateral seal(s). For effective trapping one or more of the seals must provide, singly or in combination, an upward convex barrier. There are two main types of traps (Fig. 3.13): structural and stratigraphic, though these may also occur in combination. Figure 3.14 depicts a number of possible stratigraphic traps in sandstones.

Accumulations of oil and gas are by no means all of commercial interest. Even when adequately sealed, subsequent tectonic movements and erosional processes can destroy some of the traps. In many oil provinces a whole stack of source rocks, reservoirs, and traps, occur together with frequent structural breaks and unconformities.

Hydrocarbon provinces in Western Europe

The main hydrocarbon deposits of the world are all associated with major sedimentary basins. Within Western Europe, the intracratonic Northwest European Basin—which extends over an area of 1 500 000 km²—contains a number of significant hydrocarbon provinces (see Fig. 3.12): (1) the Baltic oil province, (2) Varsican foredeep, (3) Permian gas province, (4) Dutch–German Mesozoic oil province, and (5) central and

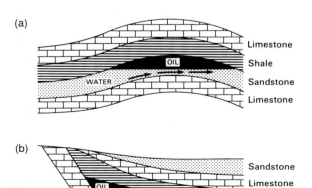

Fig. 3.13 (a) structural and (b) stratigraphic traps.

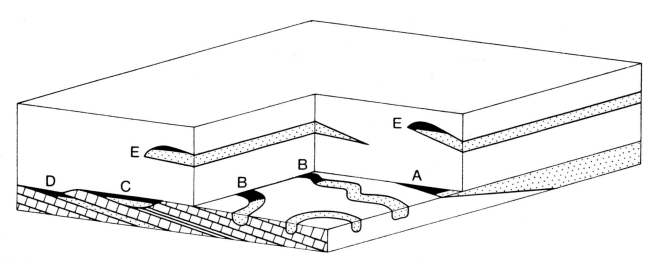

Fig. 3.14 Stratigraphic traps in sandstones. A: supra-unconformity pinchout; B: supra-unconformity channels; C: sub-unconformity strike-valley sand trap; D: sub-unconformity truncation trap; E: shoestring marine sand bars.

northern North Sea. Some of these are stacked on top of one another.

The oil and gas fields of Austria, Germany, France, and Spain have been consistent small producers. The increase in oil prices led to renewed exploration of these well known areas and to the discovery of new hydrocarbon fields. Further south, the Mediterranean region consists of several deep basins and worthwhile hydrocarbon resources exist in a number of them. The continental shelves of the Gulf of Valencia, the Italian Adriatic, the Pelagian Sea, part of the Aegean Sea, and the Nile delta have been notable in this respect.

The petroliferous sedimentary basins of Europe differ widely in their geological character, in the age of their sedimentary fill, its composition and thickness, and in the degree and style of their deformation. The hydrocarbon potential, which is related to these factors, also varies from basin to basin and province to province. On a global scale, one of the most productive areas for hydrocarbons is the North Sea (Fig. 3.15), where the UK, Norway, The Netherlands, and Denmark have benefited from the discovery of oil and/or natural gas offshore. The geological setting of the North Sea and its hydrocarbon accumulations is the subject of this section which will also highlight some of the published contributions on the geology made by the Geological Surveys of the adjacent countries.

The North Sea

The North Sea spans two distinct hydrocarbon provinces, namely the gas province of the Southern Permian Basin and the oil and gas provinces of the central and northern North Sea. The principal characteristics of the geology of these provinces and the habitat of their hydrocarbon accumulations are as follows.

Southern Permian Basin [230]

The Southern Permian Basin gas province corresponds to the location of a large east–west basin in the Variscan foreland. The basin, initiated in early Permian times, stretches from southern England, across the southern North Sea and eastwards to northern

Germany and Poland. Hydrocarbons have been produced commercially onshore from this basin, but since the 1960s most activity and the bulk of the production has been in the offshore area, in the Anglo-Dutch Sub-basin.

The Southern Permian Basin province is a gas producing area, although in The Netherlands sector the rocks of a younger Mesozoic basin have produced and trapped oil, but in relatively small amounts compared to the Mesozoic accumulations farther north. The key factors controlling gas occurrences are:

(1) the presence of gas-prone organic matter in late Carboniferous strata beneath a basal Permian unconformity;

(2) the presence of early Permian (Rotliegende) aeolian dune and wadi sandstones with good porosity and permeability characteristics;

(3) the presence of widely distributed Zechstein salt deposits over the Rotliegende sandstones which provide an excellent seal to any gas accumulation in the sandstones;

(4) sufficient burial of Carboniferous strata to promote heating of the organic matter and consequent release of gas for migration to traps;

(5) deformation of the Permian reservoir strata by faulting to provide structural traps prior to migration of the gas.

Additional reservoir-quality fluvial sandstones occur in the Triassic of the southern North Sea area. However, they receive a gas charge only where the Zechstein salt seal to the Rotliegende has been breached either by faulting or by withdrawal of salt during wholesale salt flow in the subsurface. In the 1980s the improved resolution of deeper geological boundaries on seismic reflection cross-sections has allowed the imaging of the Carboniferous below the basal Permian unconformity. Exploration now has an additional target, namely prospective structures within the Carboniferous.

Notable contributions to the understanding of the geology of the southern North Sea gas province by Geological Survey geologists include the establishment of a structural and stratigraphical framework for the UK and The Netherlands sectors. The publication by BGS in 1985 of the pre-Permian subcrop map, at 1:1 million scale, for the southern UK refined the area of Upper Carboniferous subcrop and collaboration of Survey geologists in the UK and The Netherlands led to publication of regional depth maps for the area. Solid geology maps at a scale of 1:250 000, also published by BGS, provide a useful overview of the area's geology. As part of the Department of Energy/Natural Environment Research Council Regional Mapping Programme, BGS is preparing a report on the geology of the UK southern North Sea.

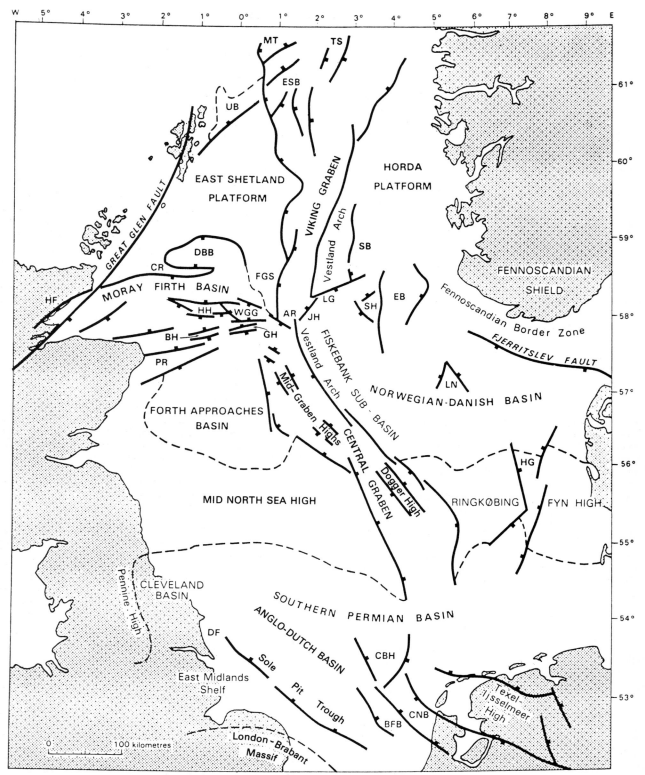

Structural elements of the North Sea Basins

AR	Andrew Ridge	ESB	East Shetland Basin	MT	Magnus Trough
BFB	Broad Fourteens Basin	FGS	Fladen Ground Spur	PR	Peterhead Ridge
BH	Buchan Horst	GH	Glenn Horst	SB	Stord Basin
CBH	Cleaver Bank High	HF	Helmsdale Fault	SH	Sele High
CNB	Central Netherlands Basin	HG	Horn Graben	TS	Tampen Spur
CR	Caithness Ridge	HH	Halibut Horst	UB	Unst Basin
DBB	Dutch Bank Basin	JH	Jaeren High	WGG	Witch Ground Graben
DF	Dowsing Fault	LG	Ling Graben		
EB	Egersund Basin	LN	Lista Nose		

Fig. 3.15 Structural elements of the North Sea Basin.

Central and northern North Sea [230]

The central and northern North Sea area is an oil and gas province, separated from the Southern Permian Basin by the east–west trending Palaeozoic Mid North Sea and Ringkøbing-Fyn highs (Fig. 3.15). Oil and gas are reservoired in rocks ranging in age from Devonian (Old Red Sandstone) to Eocene. However, the main geological factors controlling oil and gas generation and entrapment are related to basin development in the Mesozoic and Tertiary. The area was one of crustal extension (rifting) during much of the Triassic and Jurassic, followed by a phase of thermal subsidence ('sag') which has largely characterized post-Jurassic times.

The key factors controlling oil and gas occurrences are:

(1) the presence of organic shales in the Upper Jurassic over much of the area;

(2) the abundance of rocks with reservoir characteristics in the fill of both the rift and the successor 'sag' basin;

(3) the development of traps, in a range of styles and complexities at different stratigraphical levels before the peak migration of oil and/or gas from the Jurassic source rocks;

(4) sufficient burial (to 3000 m or more) of the Upper Jurassic shales to promote heating of the organic matter leading to the expulsion of hydrocarbons in the late Cretaceous to the mid-Tertiary.

The depths to the principal geological horizons over the North Sea area were published in maps compiled as a result of the collaboration of a number of European Surveys.

A stratigraphical framework for the UK and Norwegian sectors was first established through collaboration between BGS, the Norwegian Petroleum Directorate, and representatives from the petroleum industry. The geology of the Danish North Sea has been well documented, notably through a series of published Danish Geological Survey reports on stratigraphy, structural geology, sedimentary facies analysis, and geochemistry of source rocks. The BGS 1:250 000 solid geology map series, co-funded by the UK Department of Energy and the Natural Environment Research Council, provides a valuable overview of the geology of this area of the UK Continental Shelf. To complement these maps, BGS is preparing a series of regional reports on the three main basins in the UK central and northern North Sea, namely the Viking Graben, the Central Graben, and the Moray Firth Basin.

Traps in the central northern North East basins are predominantly due to the structural configuration of the geological strata, with certain notable exceptions. A common style of structural trap is associated with the crests of tilted, fault-bounded blocks. In the northern North Sea, the fields in the Brent Province (East Shetland Basin), which have reservoirs in Middle Jurassic Deltaic deposits, have traps of this kind, with superjacent Jurassic marine shales, commonly resting unconformably on the reservoir sands, as cap rock. Traps in this area are also developed within horst blocks.

A significant feature of the structural style of the central North Sea, in contrast to the northern area, is the presence of Zechstein salt. Salt pillows cause the arching of overlying strata into anticlinal structures which act locally as traps for hydrocarbons (for example in the Ula Field). In places the process of salt withdrawal causes the formation of complex traps, such as in the Clyde and Fulmar fields.

Stratigraphical entrapment is becoming of increasing interest to the explorationist; as the more obvious, large stuctural traps are drilled, attention is directed to locating smaller, more subtle traps which offer even greater challenges to the explorationist and also to the Survey geologist offering advice to government on undiscovered reserves. In the UK sector the recent interest in the Eocene sand-play involves the search for stratigraphical traps where marine gravity-flow sands, locally at least forming submarine channel-fills and mound-features, have accumulated on basin floors and are enveloped in marine shales. In this play, differential compaction of the shales around the sand has enhanced the convex-up geometry of the reservoir bodies.

Production methods are being introduced to exploit smaller hydrocarbon accumulations more effectively and economically. Sub-sea systems, greater automation, and designs and material which improve system reliability are all areas where development is very active.

Other resources

Whilst the resources in the North Sea account for the vast majority of Western Europe's oil and gas production, the smaller accumulations should not be discounted. The Lower Palaeozoic Baltic Oil Province appears to have its main, though small, accumulations of hydrocarbons in the eastern Baltic regions of Lithuania and Latvia, but very small accumulations have also been tapped in Gotland and north Germany. The potential of the west European parts of the Baltic Basin are considered below, but locally even small fields, like those in Austria, north Germany, or onshore UK, are of value to the economy both directly and indirectly. The following case history illustrates the role that Geological Surveys can play in conjunction with government and industry.

Germany [97]

Following the discovery and exploitation of the Wietze field in north-west Germany in the second half of the nineteenth century, production rose gradually to 214 000 tons in 1932. By that time only four fields had been discovered. The Prussian Geological Survey was given the task of creating the relevant scientific basis and setting up the organizational framework necessary for effective oil exploration. The following important measures were then introduced by government.

In 1934, the Reichs Drilling Programme (*Reichsbohrprogramm*) was introduced allowing a systematic search for oil to be carried out with financial support from the state. The rapid success of these measures was primarily due to the results of the geophysical survey of the German Reich (*Geophysikalische Reichsaufnahme*), which was started in 1932 and continued until 1947, as well as the setting up of an oil–geology exchange service (*erdölgeologischer Austausch*). The latter supplied the companies that were members of the exchange service with the latest exploration results obtained by the other member companies, as well as scientific results obtained by the Geological Survey. This procedure is still in operation.

The companies' exploration activities were carried out within the framework of the law on investigating exploitable deposits within the territory of the German Reich of 4 December 1934 (*Gesetz über die Durchforschung des Reichsgebietes nach nutzbaren Lagerstätten*).

The legislation governing mineral deposits is still in force but has been supplemented by the Federal Mining Law of 13 August 1980. Both laws regulate the activities of the companies, as well as the responsibilities of the mining authorities and the Geological Surveys in controlling the activities of the oil/gas industry.

The Geological Survey of Lower Saxony (NLfB) has been entrusted by the other state Geological Surveys in Germany to act as the central authority on the geology of hydrocarbons. As a result, the NLfB holds a complete collection of the oil and gas industry's data; it not only keeps it up to date but also carries out its own interpretation of the data. The NLfB advises public authorities on issues connected with the protection of energy resources, exploitable hydrocarbon reserves, hydrocarbon exploration prospects, development of oil and gas fields, underground storage of natural gas, the activities of the oil and gas industry, area planning, and regional policy, etc. The NLfB sets up standards, works out definitions, and conducts research programmes. The NLfB, together with the main mining authorities in the various federal states, controls industry's hydrocarbon exploration activities. The mining authorities are responsible for technical, safety, environmental, and legal aspects and allocate exploration and production licences. The NLfB assesses the plans submitted by industry and decides whether they are geologically sound and provide an adequate basis for a straight-forward exploration campaign.

The Geology of Hydrocarbons, Geochemistry subsection of the NLfB is supported in its work on petroleum geology by the central NLfB database, which contains comprehensive data provided by industry. This subsection cooperates closely with the staff of the Federal Institute for Geosciences and Natural Resources (BGR) particularly in the field of organic geochemistry and subsurface exploration. Recently, computer-supported systems have become available for interactive interpretation of 2D and 3D seismics and borehole logs.

The state Geological Surveys are forbidden by law ever to publish or otherwise release companies' data but, in practice, industry is relatively liberal in granting permission to publish geological data on the Quaternary to Rhaetic of north-west Germany and the Quaternary to the base of the Tertiary of southern Germany. In contrast to this, release of data on those parts of the stratigraphical column that are still being explored is highly restricted. If an exploration company that does not belong to the oil–geology exchange service wishes to examine survey data, it only has to obtain permission from the company that carried out the original survey. The Geological Survey of Lower Saxony is obliged to treat all data as confidential.

Onshore hydrocarbon plays in the UK

The principal plays of onshore UK are (1) the Lower Palaeozoic and Devonian, (2) the Carboniferous, (3) the Permo-Triassic, and (4) the Mesozoic.

The Lower Palaeozoic and Devonian play is not being actively pursued in any significant way at present. The Lower Palaeozoic potential is probably restricted to the English Midlands. Devonian prospects may occur in that part of the Orcadian Basin which extends onshore in the north-east of Scotland. This play must be considered to have only very high risk prospects.

The Carboniferous development has been described in two major provinces, one in the north of England and in Scotland, the other in the English Midlands, between the Lower Palaeozoic massif of St George's Land to the south and the Alston and Lake District blocks to the north. Potential for oil and gas occurs in both parts of the northern province, with small-scale commercial gas production having been achieved in the Midland Valley of Scotland at Cousland from 1958 to 1964. Production has been greater in the central province where the oldest commercial oil field in the UK, the Hardstoft Field discovered in 1919, is located. The central province Carboniferous play is second in importance only to the Mesozoic play of southern England in terms of UK onshore hydrocarbon exploration and production.

In the Permo-Triassic basin of eastern England reservoir targets are in the Rotliegende sandstone and more importantly, in Zechstein carbonates. Gas has been discovered in oolitic strata in the area of Malton, in Yorkshire. In the western area of England, the Formby Oilfield, with a Triassic sandstone reservoir, was discovered in 1939.

The Mesozoic strata of southern England have provided the main onshore exploration successes to date, in the Weald and Wessex sub-basins. Most notable is the discovery of the Wytch Farm Oilfield with oil in Triassic and Lower Jurassic sandstones in the Wessex Basin. The principal reservoir in the Weald Basin is the mid-Jurassic Great Oolite which forms the reservoir in the one producing field, at Himbly Grove.

The pace of exploration in onshore UK is currently rather slow. The expectation of finding small accumulations of hydrocarbons and, therefore, future exploration effort will be boosted only by a combination of the following factors: the proving of a new play concept; the maintenance of high oil and gas prices; and an optimal fiscal regime. Environmental issues are and will be increasingly crucial to exploration and production activity onshore.

Shallow gas resources

Gas has been identified in near-surface sediments in many places on- and offshore in Europe and it needs to be treated as a potential hazard for structures, e.g. Abbeystead (England) and west Vanguard (Haltenbanken, offshore Norway). However, just as gas at depth is equally hazardous, shallow gas can also be exploited as a valuable resource, provided adequate precautions are taken.

At shallow depths, it is rare for large gas accumulations to occur; they are limited by low overburden pressures and cap rock integrity (clathrates are a notable exception). Shallow gas often seeps slowly to the surface, sometimes causing cementation or hardening of the surface and/or providing a food input to the local ecosystem. This is most evident when gas seeps occur below sea level. In deeper water, gas seeps locally provide a significant part of the energy inputs into the ecosystem. In the central North Sea, small biotherms develop around gas seeps and this can provide the base line of a food chain leading up to fish and Man.

Where such natural ventings occur on land they have been trapped and utilized, such as in northern Jutland where approximately 20 million cubic metres of hydrocarbon gas was supplied to the Frederikshavn and Strandby area until the 1950s. Man's activities have themselves caused ventings. Gas released by coal workings is often vented to the surface and occasionally exploited as a resource (e.g. Valleyfield Colliery, Scotland). Additionally, man may also create the gas, as well as the conduits, e.g. methane production from waste tips.

The Geological Surveys of Norway, Denmark, Germany (Federal and State Surveys), The Netherlands, Belgium and Britain have been involved in long-term collaborative studies of the Quaternary geology of the North Sea to determine the subsidence and sedimentary history of the basin, with its implications for climate and sea level changes, coastal erosion, and mineral resource potential. *Inter alia* they have mapped areas of the seabed that are characterized by 'pockmarks': surface sediment manifestations of gas venting. No shallow accumulations of gas in Quaternary strata have yet been exploited, but the potential exists in the North Sea for one or more fields where gas occurs less than 1000 m below the sea-bed. The databases of regional shallow seismic and sediment cores, acquired directly by the Surveys, have yielded an understanding of Quaternary stratigraphy and facies which facilitate shallow resource evaluation.

Exploitation and its environmental problems

Appraisal and exploration

Exploration, if successful, is followed by an appraisal stage which aims to evaluate the size and nature of the reservoir, the number of wells necessary to extract the hydrocarbons, and to determine the likely production profile and life of the field. As crude petroleum varies in consistency from very viscous ('heavy') to free-flowing ('light') fluids, this, together with the nature of the reservoir rock, helps to determine the number of wells required. The appraisal stage includes tests to determine whether the oil will flow under the natural pressure, or whether pumping will be necessary.

When and if the viability of a field is proven, more wells may be drilled, though a very small reservoir may be exploited using only the existing exploration and/or appraisal holes. For a number of reasons, including ease of gathering the hydrocarbons at the surface, and where the area at the surface is limited, many deviated wells may be drilled to cover a large volume of the reservoir from one site. Clearly, this is particularly appropriate offshore.

Where two or more well-sites tap a reservoir, a central fluid-gathering station may be installed. This will mostly also separate oil, gas, water, and other wastes. Its size will depend on the nature of the reservoir and how the product is to be transported away.

Any water extracted with the oil or gas may be reinjected, particularly if the reservoir pressures are low. Additional water may need to be brought to the site for injection if pressures are particularly low (see below).

In onshore situations in western Europe, a simple pumping system of 'nodding donkeys' is common. These are generally inconspicuous, being quiet, clean and easily hidden. The gathering stations and associated storage and transport terminals can be much more noticeable.

For fields producing large quantities of oil and gas, either offshore or onshore, pipelines may be the most efficient way of transporting the hydrocarbons. Onshore, these are nearly always placed underground for both environmental and safety reasons.

The volume of oil originally in place in a reservoir always significantly exceeds the reserves. In most cases, and particularly with the heavier crude and in areas with little gas, the natural pressure of a reservoir alone ('primary' extraction) will enable only a proportion of the oil to be extracted. Indeed, the recovery factor for 'primary' producing mechanisms ranges from less than 5 per cent to about 70 per cent.

The proportion recovered can be increased using what are termed 'secondary' and 'tertiary' recovery methods. The injection of water or gas to maintain the reservoir pressure is classed as 'secondary' recovery; this is very common, and is necessary for efficient production in low-pressure fields such as those in onshore Britain. 'Tertiary' methods, the use of carbon dioxide, nitrogen, other chemicals or heat, may be tried as a last resort to achieve the maximum production within the bounds of cost-efficiency.

Environmental aspects

In most onshore situations the wells, once drilled, are rarely obtrusive, though the preceding operations can have local environmental impact. The industry, however, has a good history of collaboration with national and local conservation interests to minimize such effects. More obviously long-lasting is the requirement for storage tanks, major petrochemical plants, and refineries, and the transport infrastructure to move the crude petroleum, e.g. road haulage (onshore), pipelines (on- and offshore) and sea-going tankers. All have problems initially concerned with planning permission (on- and offshore), change of use of land (onshore), and interference with prior use such as fishing and navigation.

Once production is underway, the main potential problems, other than drastic changes in the value of crude petroleum, are pollution and explosion. Whilst rigorous safety precautions are taken to avoid the latter, major disasters such as the destruction of the Piper Alpha platform in 1988 do occur. More frequent, and equally tragic to nature if not so immediately to man, are the incidents of pollution due to spillage. The international nature of the oil trade makes difficult the establishment and enforcement of adequate legislation for the control of such hazards.

Ground subsidence occurring as a result of hydrocarbon extraction from underground strata, whilst more common in shallow relatively unconsolidated fields, has been identified in Europe. The sea-bed subsidence in the Ekofisk field is described more fully in Chapter 5. The reduction in pore fluid pressure brought about by fluid withdrawal leads to a compaction of the reservoir. This 'compaction drive' is an important mechanism of primary recovery. The contraction of the reservoir also generates horizontal stresses which may result in small earthquakes. Such tremors have been identified in the Pau Basin gasfield and at Ekofisk.

Whilst decommissioning of onshore production facilities happens only rarely, the procedures are established and the sites can become sites of special scientific interest or nature reserves. By contrast, the decomissioning of major offshore production platforms will be immeasurably more difficult and costly. Done badly, it could result in severe hazards both to man (e.g. to navigation) and to the environment.

Management policies

The hydrocarbons industry is dominated by a handful of major international companies. All of these have developed, to a greater or lesser degree, a long integrated chain from exploration through refining to product development and sales for a very wide variety of markets, for example, the agricultural, chemical, pharmaceutical, and plastics industries, as well as the more obvious energy and transport interests. Working as they must in many parts of the world, they have learned to accommodate the vagaries of international politics and all display a similar approach to maintaining good relations with national governments. Indeed, it is sometimes said that they act as instruments of government policy.

Whilst the price of petrol at the pump may vary widely from country to country, the international price of crude petroleum is dominated by the spot market and by the coordinated action, when possible, of the majority of the producing nations through OPEC. Within the advanced countries, national policies, related at least in part to taxation, are commonly more influential on the local sales prices than the presence or absence of oil reserves within the country.

The role of the Geological Surveys

Just as the hydrocarbon resources and associated environment differ from country to country within Europe, so do the laws and provisions for controlling its exploration and exploitation, and for safeguarding the environment during these procedures. Given differing legislative frameworks and organizational structures, the varying roles of different Geological Surveys are indicated by the following examples.

Danish Geological Survey [12]

The Geological Survey of Denmark (DGU) has been involved with activities related to hydrocarbon exploration since the first deep test was drilled in 1935. For many years DGU concentrated primarily on the general stratigraphy encountered in the exploration wells, and more specialized disciplines and techniques used in the evaluation of prospectivity. The characterization of reservoir rocks etc. were not covered. This situation altered in the early 1970s, when the Government requested more precise information on the native oil and gas potential; the development was precipitated by the oil supply crisis in 1973. In order to procure an independent assessment, the Government initially hired experienced consultants from abroad; this arrangement continued for several years and provided practical training for DGU personnel. During this period of expansion DGU was granted additional manpower as well as laboratory and computing facilities. Thus by 1982 DGU was in a position to handle most of the geological aspects within the fields of exploration and production.

The Survey now acts as an advisory body for the Danish State authorities, namely the Ministry of Energy and the Danish Energy Agency. Following the introduction in 1981 of new regulations concerning hydrocarbon activities in Denmark, one of the main duties at the Survey has been to map and evaluate prospects in relinquished and open acreage prior to the licensing rounds. After the licensing round closing date, teams of specialists from both the Energy Agency and DGU assess the technical information from the companies and make recommendations to the Ministry concerning the proposed work programmes.

The Geological Survey of Denmark is also involved in the detailed mapping of proved accumulation. Geological reservoir models are developed based on the integration of results from seismic interpretation, sedimentological and biostratigraphical studies, log interpretation, and core analysis. These models are used for the calculation of the volume of hydrocarbons in place, as well as for the Energy Agency's estimation of the production possibilities.

In addition to working for the State authorities, DGU has been engaged in several projects supported financially by a research programme under the Ministry of Energy. This support, together with funding from the EEC and from oil companies operating in Denmark, has resulted in improved insight into the wide spectrum of parameters and processes that are relevant for hydrocarbon exploration and exploitation. The positive effect of this interaction between the work carried out for the authorities and the research project is evident.

Investment in instrumentation for a variety of analytical duties (source rock analysis, core analysis, etc.) and in computer software is considerable. In order to make efficient use of this investment, the laboratories also undertake service jobs for the industry on commercial terms.

British Geological Survey [207]

In the UK, statutory authority governing oil and natural gas exploitation rests with the Department of Energy (DEn). In turn the Department contracts out certain activities to the British Geological Survey (BGS). The contract demands of BGS:

(1) the provision of independent geological advice;

(2) the development and maintenance of a computerized database;

(3) the management of a repository of core samples, which are obtained from exploration and field development activity.

To date the BGS advisory role has been limited to exploration matters, to the provision of information for the calculation of undiscovered reserves, and to the assessment of prospectivity of offshore acreage prior to licensing.

Work under contract to DEn is underpinned by a uniquely comprehensive database of commercially obtained reflection seismic and well information, which is held by BGS acting as the agent of the Secretary of State for Energy. Advice to commercial interests concerning the regional geology of potentially petroliferous areas is available only from staff not involved in the Survey's hydrocarbon contract with DEn.

The Netherlands: 'Drilling Operations Committee' [141]

Introduction

In The Netherlands, the award of permits for onshore exploration and the concern for a number of environmental factors are coordinated through an advisory committee system which also helps to provide the industry with early warning of environmental planning problems.

After an intensive exploration campaign for natural gas in the mid-1960s, the first onshore concessions to be granted after that at Groningen (1963) were issued in late 1968/early 1969. As the ensuing drilling programme went ahead, strong opposition was met in a number of cases from local authorities and action groups. By that time a greater awareness of the environment had arisen, accompanied by the law of 1965 on physical planning, *Wet op de Ruimtelijke Ordening*. The oil companies, having acquired a permit to drill, and potentially to produce hydrocarbons, found that in some places they were unable to pursue their activities. This resulted in disputes which, in some cases, led to the refusal of a drilling site and, in others, to the assignment of a drilling location outside the disputed area with the possibility of reaching the target by deviated drilling.

The Drilling Operations Committee

The Netherlands Minister of Economic Affairs, who grants the Drilling Permits (exclusive permits for exploration drilling in a specified area) and concessions (production licences), took note of the conflicts and concluded that it would be preferable for the companies, when entering an area for exploration, to know in which localities drilling would be forbidden. This led, in mid-1973, to the establishment of an interdepartmental committee with the lengthy name *Commissie van Avies inzake van boringen te vrijwaren kwetsbare gebieden* (Advisory Committee on vulnerable areas to be exempted from drilling operations): in short, *Commissie boorwerken* (Drilling Operations Committee).

The remit of the Committee was given as: 'to advise the Minister on the execution of drilling of areas of high vulnerability and of great importance considering, on the one hand, that the sub-surface mineral resources have to be used optimally and consequently the aim has to be an exploration as active and intensive as possible and an exploitation as rational as possible, and on the other hand that the beauty of nature and landscape needs to be retained as much as possible, and justice has to be done to other planning interests including the protection of the environment'.

The Committee comprises an independent chairman, and representatives from the Ministry of Economic Affairs, the Ministry of Agriculture, Nature Management and Fisheries, the Ministry of Housing, Physical Planning and Environmental Management, the Ministry of Traffic and Public Works, and the Ministry of Defence.

Committee Activity

In preparing advice the Committee first makes an inventory of the vulnerable and sensitive locations in the permit area, with respect to the responsibilities of each member. The opinion of the relevant provincial government(s) is also sought, as provincial authorities have their own responsibilities for planning. To be exempted from drilling, areas must meet set criteria, having either a unique or rare geological environment, or high geomorphological, soil, archaeological, or historical geographical values, or a combination of intertwined scientific and cultural history qualities. The scenic value of the landscape is also a criterion.

Discovered fields (if any), prospective geological structures, and trends for further hydrocarbon exploration form a further inventory. These data are supplied by the Geological Survey of The Netherlands on the basis of company information, supplemented with other available data. Confidential data are treated accordingly by the Committee.

On a map, the areas proposed for exemption from drilling are superimposed on the prospective geological structures. If this shows that certain hydrocarbon targets cannot be reached, special attention is paid to seeking openings or 'windows' in sensitive areas to enable the targets to be achieved by deviated drilling. If no compromise can be found, the environmental conservation and the exploration/production factors have to be 'weighed' which, in some cases, may lead to keeping certain areas closed to drilling activities. In most cases, however, there appears to be reasonable space available for exploration and production drilling.

The Committee reports depict four categories of land:

1. Drilling not allowed.

2. Caution areas. Areas of high scientific and scenic value but with sites where drilling should take place after deliberations in a Planning Committee in which special conditions (such as period of drilling) can be set.

3. Drinking-water supply areas. These areas are not, in principle, exempted from drilling for, given the right controls, the water supply need not be endangered. However, any decision to drill should be dependent on deliberations of the Planning Committee.

4. Other areas. No specific important elements of the natural environment.

This procedure does not cover the non-environmental aspects of planning consent which are built into other regulations. Before an actual drilling location can be set up a series of local consents and licences have to be obtained.

Conclusions

The practice of indicating, at an early stage of licensing, the areas where drilling cannot be allowed, or where special restrictions can be expected, has proved its value. Since the Drilling Operations Committee started its work in 1973, the oil companies know where they can carry out their operations, and how lengthy conflicts and procedures in obtaining a permit for a drilling location within an awarded licence area can be avoided. The input of the Geological Survey, based on its knowledge of the sub-surface geology and hydrocarbon potential, provides one of the important data-sets on which the Committee can make its decisions.

Data and sample archiving and public access

One factor helping to ease or hinder exploration and its coordination, is the accessibility of data and samples from previous surveys and drillings. Like the other policies, the national provisions for archiving and access to such information vary.

Partly under the terms of its contract with the UK Department of Energy, BGS manages a national store for cores and drill cuttings obtained from UK exploration and production wells. Licensing conditions oblige operating companies to provide a slice of all cores taken and a sub-sample of all cuttings collected. These samples have a commercial-in-confidence status for five years after well completion. After this time BGS makes them available for public examination and, with restrictions, for sub-sampling and analysis.

Under present UK legislation, geophysical well logs are also publicly available after five years. These well records are sold on behalf of DEn by the private sector company ERICO. However, commercially acquired seismic data are provided to DEn and BGS, but are held in confidence in perpetuity.

All data covered by Danish subsurface legislation are normally released for sale (paper and magnetic tape copies) or examination (samples) after a five year period of confidentiality. DGU is responsible for the administration of this extensive database.

Input to policy decisions

The economic and strategic importance of energy resources to European governments has resulted increasingly in demands being made on Geological Surveys. Governments, and especially their Departments of Energy, require information and advice specifically on such matters as: exploration potential; identification of potential reserves; field development; suit-

able acreage for licensing; abandonment policies; the monitoring of commercial activity; in summary, advice on all aspects of petroleum geology. Additionally, the geological aspects of possible hydrocarbon accumulations on outer continental shelves, which may give rise to sovereignty disputes, are becoming increasingly important, as industry moves from mature plays to frontier areas in European waters.

Obviously, Geological Surveys must work within their own national legislative frameworks which determine the balance between resource identification and exploitation. This results in the operation of market forces in some European countries and the application of resource-conserving policies in others. In general terms the exploitation of hydrocarbon resources is still largely driven by economic and strategic reasons, rather than by environmental conditions. The environmental problems associated with hydrocarbons, which relate mainly to potential hazards and pollution, are well known onshore and are reflected, for example, in the planning permission policies of the Ministry of Economic Affairs in The Netherlands and the Department of the Environment in the United Kingdom. Environmental concern for developments offshore is increasing, and has been stimulated recently by the Piper Alpha disaster in the North Sea and its attendant loss of life and pollution, together with other spillages in coastal waters from sea-bed installations, pipelines, and tankers. With the abandonment of oilfields and their associated engineering structures now being contemplated, it will be necessary to determine policies which will minimize the future potential for pollution and hazards to both life and navigation.

Conclusions

European society's need for energy resources to sustain its standards of living is likely to remain unchanged, if not to increase, in the coming years. It follows, therefore, that Geological Surveys should have a continuing and, perhaps, increasing role in the identification and classification of hydrocarbon resources, together with the potential for a monitoring and regulatory function which will be dependent on relevant legislative frameworks.

The strategic and positive economic value of hydrocarbon resources has to be set against their potential for damaging the environment by way of direct pollution and hazard, or indirectly by way of the petrochemical industries which they supply. By monitoring the geology of the natural environment and the changes produced by anthropogenic activity, Geological Surveys have another important role.

Arguably, the most crucial role of any Geological Survey may well be the quality of its impartial advice to Governments and others, concerning the balance between an acceptable level of exploitation of resources and the potential damage to the general environment that may result.

Additionally, scientific benefit accrues from the involvement of Geological Surveys in hydrocarbon-related activities by way of: improved earth science databases; regional, three-dimensional geological models; and stratigraphical standards. However, the problem of access to databases, including commercially generated data, remains a problem which is variously treated throughout Europe. For example, borehole cores may be released some five years after completion of drilling in British waters, but remain restricted in the German sectors. Conversely, deep seismic data may be released in Denmark after five years, but not at all in the UK, unless by agreement with the proprietorial owner of each data set; in some countries it is freely available when released, in others it is sold. A consistent policy on the release of data to the public domain would facilitate the work of the Surveys, the earth science community at large, and both Governments and the oil industry.

RECOMMENDED FURTHER READING

Anderson, C. and Doyle, C. (1990). Review of hydrocarbon exploration and production in Denmark. *First Break*. **8**, No. 5.

Bailey, C. C., Price, I. and Spender, A. M. (1981). The Ula Oil Field, block 7/12. Norway. In *Norwegian Symposium on Exploration*. (Norwegian Petroleum Society, Oslo).

British Geological Survey (1990). *World Mineral Statistics 1985–89: production; exports; imports.* (British Geological Survey, Nottingham).

Brooks, J. and Glennie, K. W. (eds) (1987). *Petroleum geology of North West Europe.* (Graham and Trotman, London).

Cameron, T. D. J., Crosby, A., Bulat, J., Lott, G. K., Balson, P. S., Harrison, D. K., and Jeffrey, D. H. (in press). *British Geological Survey, United Kingdom Offshore Regional Report: The geology of the UK southern North Sea.* (HMSO, London).

Damtoft, K., Nielsen, L. H., Johannesen, B. N., Thomsen, E., and Andersen, B. R. (in press). Hydrocarbon plays of the Danish Central Graben. In *Proceedings of the EAPG 2 Conference*.

Glennie, K. W. (1990). *Introduction to the petroleum geology of the North Sea.* (Blackwell, Oxford).

Illing, L. V. and Hobson, G. D. (eds) (1981). *Petroleum geology of the continental shelf of north-west Europe.* (Institute of Petroleum, London).

Johnson, H. O., MacKay, T. A., and Stewart, D. J. (1986). The Fulmar oil-field (Central North Sea): geological aspects of its discovery, appraisal and development. *Marine Petrol Geology.* **3**, 99–125.

Michelsen, O. (ed.) (1982). *Geology of the Danish Central Graben.* Geological Survey of Denmark, Series B, No. 8. Copenhagen.

Nature Conservancy Council (1986). *Nature conservation guidelines for onshore oil and gas development.* (The Nature Conservancy Council, Peterborough).

Schröder, L. and Schöneich, H. (1986). *International map of natural gas fields in Europe: 1:2 500 000. Explanatory Notes*, 2nd edn. (Bundesanstalt für Geowissenschaften und Rohstoffe, Hannover).

Ziegler, P. A. (1980). Northwest European Basin: geology and hydrocarbon provinces. In *Facts and principles of world petroleum occurrence*. Can. Soc. Pet. Geol., Memoir 6, pp. 653–706.

3.3.2. Solid fuels: coal, lignite, peat, oil shale
[104] [65] [36]

Although some coal has been used in the chemical industry and a small amount of peat is still used in agriculture, coal, lignite, and peat are basically fuel for electricity production, or industry or domestic heating. Oil shale is now more a geological reserve in Europe although it was to some extent exploited in the past.

Hard coal has been the dominant energy source in Europe until the middle of the twentieth century. The rise of hydrocarbons, and then of nuclear energy, resulted in a considerable decrease in the use of coal. This evolution went along with increased mining costs, depletion of reserves, and competition from imported coal. Everywhere in Europe now production of hard coal is heavily subsidized. The lignite situation is largely better when the material is of good heating quality and because large-scale opencast mining allows mass production at low cost. This type of mining is concentrated in Germany and to a smaller extent in some other European countries. Peat is a significant energy resource for domestic use mainly in northern Europe. Both opencast lignite and peat mining are land-consuming activities.

Brief description of resources

Introduction
As shown in Chapter 2, the location of solid fuel deposits in Europe is closely linked with basin geology. Hard coal is mostly associated with the last stages of the Hercynian period, and most coal basins correspond to late Hercynian basins and isolated troughs or belts within older terrains. In some cases (France, Austria, Switzerland) the Alpine orogeny has affected coal deposits, and introduced complexities in their structure and disposition which influence their mineability.

Lignite deposits, on the other hand, are much more recent in age and range between Cretaceous and Neogene. The major west German basin is Oligocene to Miocene in age. These deposits are generally flat-lying or gently dipping, and are easier to mine.

Oil shale occurs in sedimentary sequences ranging from late Palaeozoic to Tertiary. Peat has been formed in extensive swampy areas in Quaternary times.

Geology is a key to understanding resource geometry, complexity and extent, and quality relates directly to time of formation. In general, the older the solid fuel, the better its calorific value, as shown in Table 3.3. Within the hard coal category the lower figures correspond to bituminous coals and the higher to anthracite.

These parameters together with mining costs and market conditions determine the feasibility of any mining operation for coal.

Late Hercynian hard coal basins
These basins are present in United Kingdom, Germany, The Netherlands, Belgium, France, and Spain. Their age ranges from Lower Carboniferous to Lower Permian. Coal deposition took place either at the boundry of continental and marine environments (paralic basins on continental margins) or in continental lacustrine environment (limnic basins) within the same time-scale. Contrary to hydrocarbon formation, coal deposition is largely continental and the original organic matter is mainly of continental origin from vegetation grown in a hot climate.

The coal-bearing beds (coal measures) contain several hundred seams of coal, though only a small number are of workable thickness. The sedimentary sequences range from 2000–2500 m in thickness in parts of Great Britain and France to 10 000 m in the Donetz Basin (USSR). The cumulative thickness of coal is 2–4 per cent of the total thickness of sediments.

Due to the number of borings and the availability of underground sections, detailed knowledge of bituminous coal deposits within the mining districts is normally good. Outside the mining districts, however, spatial information of this kind is less complete. It is unlikely that in Western Europe any coal deposits of great economic importance have escaped discovery, but some additional reserves are still to be found.

Figure 3.16 shows the location of the major hard coal deposits in Western Europe and Table 3.4 the resources, reserves, and production situation.

Great Britain is the most important and oldest coal-producing country in Europe and coal deposits are in disconnected basins ranging from north to south. The highest potential is in the central area which includes the Yorkshire, North Derbyshire, and Nottinghamshire deposits. Underground mines produce 85 per cent of the coal and 15 per cent is from opencast pits.

An example of the complex geometry of these late Hercynian deposits is shown in Fig. 3.17 showing the northern France

Table 3.3 Indicative quality characteristics for solid fuels and other fuel sources

Quality parameter	Wood	Peat	Lignite	Hard coal	Crude oil
Carbon[1] %	48–50	50–60	55–77	77–95	83–86
Vitrinite reflectance %	—	<0.25	0.50–0.25	0.5–5	—
Hydrogen[1] %	6.5–6.0	8–5.0	6–5	5.2–2.0	12.5–11.5
Oxygen[1] %	42–38	40–30	30–18	18–2	2.5–1.5
Nitrogen[1] %	2.5–0.5	2.5–1.0	2.0–1.0	1.5–0.5	0.3–0.2
Sulphur[1] %	—	1.0–0.1	1.0–0.2	3.0–0.5	2.8–2.0
Volatile matter[1] %	85–70	80–70	60–50	50–3	—
Moisture[2]	—	95–75	75–10	10–1	—
Calorific value (kcal/kg)[2]	3400–4600	1500–2000	3000–7000	7000–8600	9900–10 000

[1] dry ash-free
[2] ash-free

coal basin which extends into Belgium. The structural complexity and increasing depth of working (down to 1100 m) caused mining to cease here in 1990. The mining activity on the Belgian side ceased in the late 1980s. Only the Campine Basin further east is still mined. The Ruhr Basin is described in more detail below.

Another important coal mining country is Spain with two main extraction areas: the Cantabrian zone to the north with the famous Asturian Basin (Oviedo area) and the Sierra Morena to the south. Many European countries have no or only minor production of hard coal: Portugal, Italy, Switzerland, and Austria. Turkey is producing hard coal of Carboniferous age in the Zonguldak area. Exploration by MTA, the Turkish Geological Survey, which started in 1935 is still continuing. Norway's production is from Spitzbergen and requires special mining techniques because of permafrost.

The Ruhr Basin [100]

The Ruhr Basin (Fig. 3.18) is the most important individual coal basin of Western Europe. It is located on the right bank of the Rhine between the River Ruhr to the south and the River Lippe to the north.

The mineable area is 100 km long and 50 km wide but the coal is present at greater depths farther to the north. The productive coal sequence includes 95 per cent of barren rock (37 per cent sandstone, 58 per cent black pyritic schist). Individual coal seams range between 0.5 and 2.8 metres. There are 75 mineable seams and 94 seams more than 0.3 metres thick. Similar to the Franco-Belgian deposit, the tectonic structure is alternating synclines and anticlines often marked by longitudinal faults.

A total of about 9 billion tonnes of hard coal has been mined in the Ruhr District. According to new calculations the coal reserves in the Ruhr District and its northern extension, for seams more than 120 cm thick, down to a depth of 1500 m and below an overburden of up to 1200 m, are about 36 billion tonnes. Only about 25 per cent of this amount may be exploited under present-day economic and technical conditions.

In Germany, the pressure to keep production costs low by constantly increasing output led to selective exploitation of the coal deposits during the last two decades. This means that only horizontal and undisturbed seams are mined by fully mechanized methods, whereas seams with steep dip or tectonic faulting are left untouched. As a result of selective mining the percentage of abandoned parts of deposits increased from 9 to 27 per cent between 1961 and 1974.

In the 1970s, the energy crisis caused a reassessment of the security of the energy supply in Western Germany. Almost 700 boreholes were drilled between 1973 and 1988. Drilling and comprehensive seismic surveying was concentrated on the northern margin of the Ruhr District to investigate the geology of the areas holding coal reserves. With their plans for shifting mining operations further to the north, the coal industry met public opposition because of fears of industrialization of areas traditionally used for agriculture and recreation. Thus, it was necessary to find a balance between the needs for energy resources and environmental concern.

Plans for the northward extension were developed by the mining companies. They included shafts for personnel access and ventilation, no shafts for removing coal, no waste stockpiles nor processing plants, and no miners' settlements. The coal produced in the northern extension is transported over a distance of more than 20 km underground to the existing shafts in the Ruhr District and processed there. This procedure reduces the negative effects of hard coal mining in the new areas to a technically unavoidable minimum.

From the end of the 1970s to the year 1987, the German Ministry for Research and Technology (BMFT) supported a heavy programme of coal reserve evaluation in terms of quantity and quality. This programme included not only the Ruhr Basin, but also Saarland and Lower Saxony reserves below 1000 m. This programme, which was coordinated by the Geological Survey of Northrhine-Westphalia led to the construction of a data bank which enables the following to be extracted:

(1) tectonics at various scales;

(2) stratigraphic profiles showing all coal seams more than 30 cm thick and correlations between profiles;

(3) data on coal quality;

(4) 237 000 sections of coal seams.

This data bank is now fully operational for any further need.

A specific mention should be made of hard coal basins of Hercynian times which have been modified by the Alpine orogeny. The type example is the Briançonnais Basin which outcrops along an arc extending from Savona in Italy, through Briançon in France to the Valais in Switzerland. The coal-bearing succession has a thickness of 1500–2000 m. It is highly deformed and the seams of anthracite coal have been strongly tectonized. The mining conditions are thus difficult and large mines are not developed. Many small mines have been opened nevertheless during periods of market shortage.

Although the great majority of European coal basins are Carboniferous or Permian in age, there is some coal of Mesozoic times (Liassic) in Sweden. The deposits are small and tectonically disturbed and no longer worked.

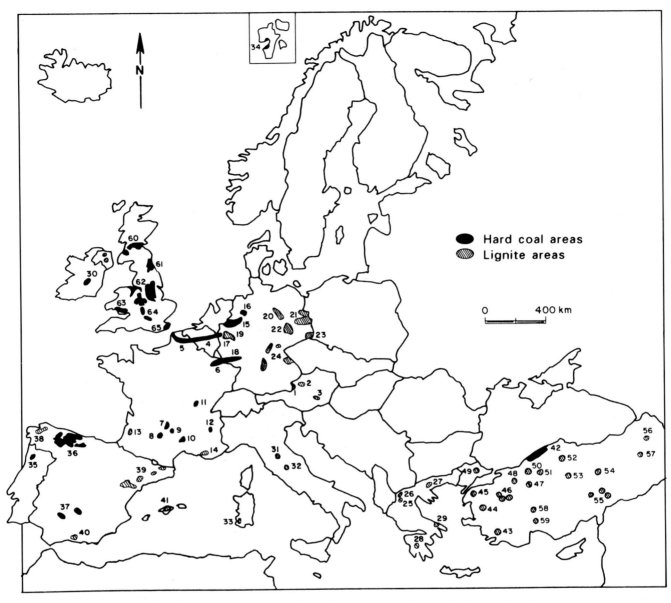

Fig. 3.16   Locations of the major hard coal and lignite deposits of Western Europe.

	Hard coal areas	Lignite areas
Austria		1 Salzach
		2 Wolfsegg—Traunthal
		3 Köflach—Voitsberg
Belgium	4 Campine	
France	5 Franco-Belgian district	13 Arjuzaux
	(closure in 1990)	14 Provence
	6 Lorraine	
	7 Decazeville	
	8 Carmaux	
	9 Cruejouls	
	10 Alès	
	11 Blanzy	
	12 La Mure	
Germany	15 Ruhr	19 Lower Rhine
	16 Ibbenbüren	20 Helmstedt
	17 Aachen	21 Cottbus/Seuftenberg
	18 Saar	(Niederlausitz)
		22 Halle Leipzig
		23 Oberlausitz
		24 Hesse

	Hard coal areas	Lignite areas

Greece

25 Ptolemais
26 Florina
27 Drama
28 Megalopolis
29 Kimi

Ireland   30 Leinster

Italy

31 Valdarno (Castelnuovo dei Sabbioni)
32 Perugia (Pietrafitta)
33 Cagliari (Bacu Abis, Sulcis)

Norway (Spitzbergen)   34 Pyramiden
Longyearsbyen
Barentsburg

Portugal   35 Pejào

Spain   36 Cantabrican zone   38 La Coruna
37 Sierra Morena   39 Catalogne
40 Grenade
41 Mallorca

Turkey   42 Zonguldak   43 Mugla—Yatagan
44 Mamisa—Soma
45 Canakkale
46 Tunçbilek-Serpitömer-Orhonali
47 Beypasari—Andara—Koyunagili
48 Göynüh
49 Saray
50 Mengen
51 Orta
52 Dodurga
53 Sorgun
54 Kangal
55 Elbistan—Tufonbeyli—Gölbasi
56 Hurasan
57 Karliova
58 Elgin
59 Beysehir

United Kingdom   60 Scotland
61 Northumberland
62 Central England
63 South Wales
64 Southern Midlands
65 Kent

Table 3.4 Resources, reserves, and production of hard coal in Western Europe (after BGR)

Country	Proved recoverable reserves (10⁶t)	Proved amount *in situ* reserves (10⁶t)	Total resources (10⁶t)	Production 1988 (10⁶t)	Production 1989 (10⁶t)
Austria	3.0	4.0	10.0	—	—
Belgium	945.0	3 780.0	4 530.0	2.5	1.9
France	303.0	980.0	980.0	12.1	11.5
Germany (West)	25 500.0	43 000.0	229 000.0	79.4	71.0
Ireland	57.0	—	272.0	0.04	0.04
Italy	27.0	60.0	340.0	0.04	0.07
The Netherlands	497.0	1 406.0	1 406.0	—	—
Norway	27.5	32.5	132.5	0.28	0.43
Portugal	18.3	26.9	65.9	0.23	0.26
Spain	531.8	910.5	3 477.3	19.2	18.9
Sweden	1.0	4.0	24.0	—	—
Turkey	59.0	185.1	1 368.2	6.7	6.3
United Kingdom	4 150.0	4 150.0	189 550.0	101.4	98.0

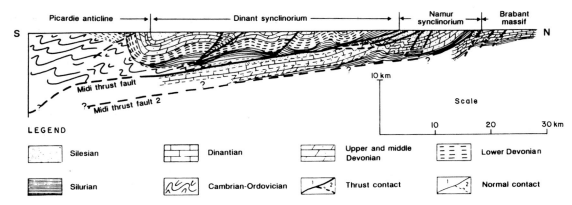

Fig. 3.17 Generalized geological cross-section through Dinant and Namur synclinoria (after Becq-Giraudon, J. F. (1983) *Mem. BRGM No. 123*).

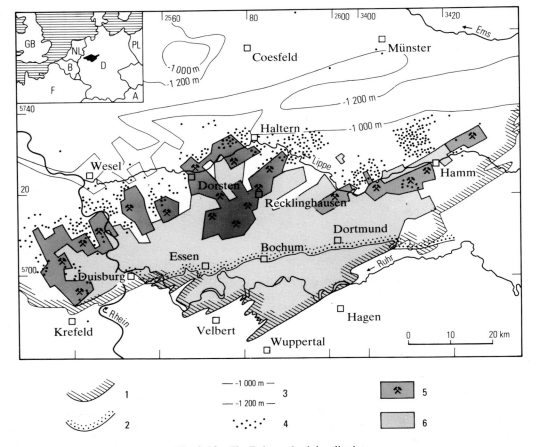

Fig. 3.18 The Ruhr coal-mining district.

1. Southern border of productive coal measures
2. Southern border of overburden
3. Top of Carboniferous below sea-level
4. Exploration boreholes
5. Active mines
6. Abandoned mining area

Lignite resources and mining

Location of the main West European deposits is shown on Fig. 3.16. The resources, reserves and production situation is given in Table 3.5. By far the largest reserves and the highest production levels, are found in Germany, and the Lower Rhine deposit is described in more detail below.

Mining continues in south-east France (Fuveau basin of Cretaceous age) where a unique seam 2.5 m thick is currently exploited. Spain is also an important mining area. The lower Cretaceous lignite of the Teruel province in northern Spain is mined opencast. Other important deposits are in the Coruña, Orense, and Granada provinces. These deposits are of upper Tertiary age.

In Turkey, lignite is a major source of energy, but the sulphur content creates problems. Deposits are scattered over large areas of the country. Lignite is also mined in several

Table 3.5 Resources, reserves, and production of lignite in Western Europe (after BGR)

Country	Proved recoverable reserves (10⁶t)	Proved amount in-place reserves (10⁶t)	Total resources (10⁶t)	Production 1988 (10⁶t)	Production 1989 (10⁶t)
Austria	61.2	64.0	407.6	2.1	2.07
Denmark	80.0	115.0	225.0	—	—
France	2.0	2.0	8.0	1.7	2.19
Germany (West)	17 300.0	34 750.0	56 750.0	105.0	109.9
Greece	2 800.0	5 100.0	5 100.0	50.0	52.0
Italy	6.0	9.0	31.0	1.6	1.53
Portugal	32.8	37.7	40.0	—	—
Spain	154.0	230.0	429.0	17.3	13.0
Turkey	4 575.0	5 103.0	7 944.0	39.0	52.0
United Kingdom	400.0	770.0	1 000.0	—	—

Italian deposits, in three areas of Austria, in Northern Ireland, and in Greece.

Rhenish lignite [101]

The Lower Rhine deposit is the largest lignite deposit in Europe (Fig. 3.19). Annual lignite production averages 110 million tonnes, 85 per cent of which is used to generate about one-fifth of the electric energy consumed in western Germany. The remaining 15 per cent is used to produce briquettes, powdered lignite, coke, and additives for soil improvement.

The lignite is mainly of Miocene age. It is mined entirely by opencast methods. The size of an individual opencast mine may exceed 100 km², its lifespan may be more than 50 years, and its depth may reach 500 m. A section through the lower Rhine lignite deposit is given in Fig. 3.20.

By the end of the Second World War, the thick and near-surface seams of lignite had been extensively exploited. In 1960, the Geological Survey of Northrhine-Westphalia was ordered to interpret all available data and present a detailed synopsis of the Rhenish lignite deposit.

The maps resulting from this work were used by the planning authorities to specify preferred locations for future opencast mines, taking into account the quantity and the quality of the lignite, the thickness of the overburden (ratio overburden:lignite was 1.72:1 in 1962 and reached 4.14:1 in 1988), and land-use conditions: human settlements, major industrial sites, traffic routes, and areas to be protected.

The mapping programme of the Geological Survey made it possible to examine previous estimates of Rhenish lignite reserves, which total 55 billion tonnes, of which 35 billion are mineable. The proven reserves allow for production of 110–120 million tonnes of lignite annually for about 80 years.

The Rhenish lignite reaches a maximum thickness of 100 m in the centre of the basin. The calorific value of the lignite increases from 7000 kJ kg⁻¹ near the surface to about 12 000 kJ kg⁻¹ at a depth of 500 m. The improvement is mainly due to a decrease in water content from about 60 per cent near surface to 45–50 per cent at depth.

The lignite area extends as far as the Limburg Province in The Netherlands where the maximum thickness is 25 m at a depth of more than 300 m. Where lignite occurs near the surface, it has a maximum thickness of only 7 m. Mining was abandoned in 1960.

Peat

The main areas of peat extraction are in northern Europe: Finland, Sweden, Germany, and Ireland. Peat resources are

Fig. 3.19 Location of the Rhenish lignite area. Hatched area, lignite > 3 m. A-B see section in Fig. 3.20.

also found in France and Spain but there is no significant production. Ireland is presented as the type example.

Peat in Ireland [123]

In Ireland about 16 per cent of the land surface is covered by Holocene peat (Fig. 3.21). The Irish peat bogs are subdivided into the raised bogs of the midlands which are up to 15 m thick, and the extensive blanket bogs, generally 1–6 m thick, that occur along the west coast and in mountainous areas throughout the country.

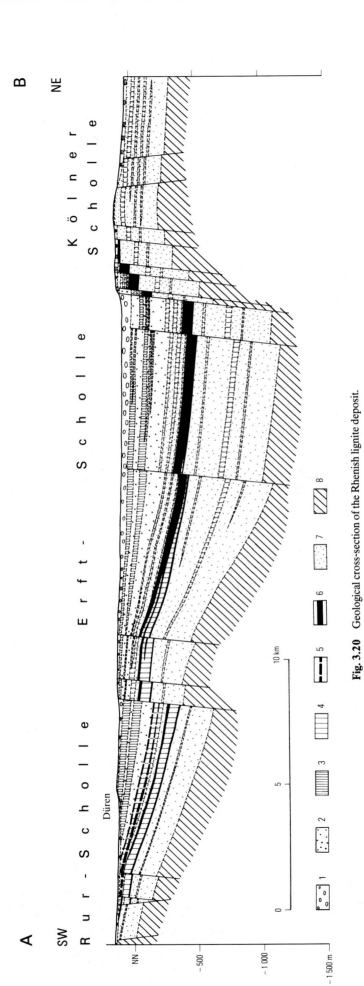

Fig. 3.20 Geological cross-section of the Rhenish lignite deposit.

Quaternary:
1. Gravel with sand

Pliocene:
2. Sand with gravel;
3. Clay, partly sandy, with thin lignite seams.

Oligo-Miocene:
4. Clay, partly sandy, with thin lignite seams
5. Lignite of the upper group of seams;
6. Lignite of main group of seams;
7. Sand, fine-grained, with isolated clay beds and thin lignite seams.

Pre-Tertiary beds: 8.

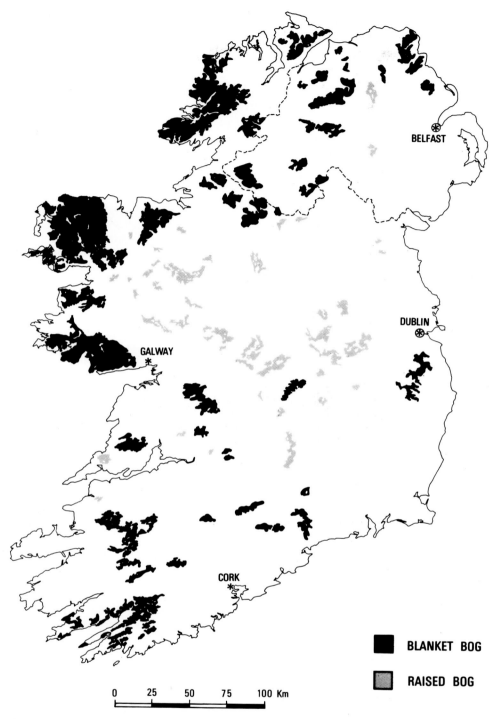

Fig. 3.21 Peatlands of Ireland (after Hammond, R. F. (1979). *The peatlands of Ireland*. Soil Survey Bulletin no. 35. (An Foras Taluntais, Dublin)).

Blanket bogs cover much of the landscape on low and high ground in districts with impervious rocks within the high rainfall western counties (>1200 mm) close to the sea. Raised bogs form up to one-third of the surface area of some of the low-lying poorly drained midland counties.

The use of peat as a source of fuel has a long tradition. In 1946 the Irish government decided to extract peat from the raised bogs on a large scale, both for domestic use and to generate electricity for the rapidly expanding national electricity grid. A special state com-

pany—Bord Na Mona (BNM)—was formed to extract the peat. As a result, planned production by BNM from 1950 to 2030 concerns some 90 000 ha of peat bogs. During the 1970s and 1980s annual BNM production, depending on weather conditions, has fluctuated between 3 and 6 million tonnes/year with private extraction representing less than 20 per cent of BNM production.

This extensive peat extraction has a major impact on the raised bogs which cover some 300 000 ha. What remains will be much fragmented and their overall ecosystem altered. Conservationists

Table 3.6 Characteristics of oil shales in Western Europe (after Schmitz, H. H. (1986). Bituminöse Gesteine. In F. Bender (ed.), *Angewandte Geowissenschaften*, pp. 385–99 (Enke, Stuttgart))

Country	Place	Formation	Oil content		Oil density (g/cm³)	Specific gravity of oil shale (g/cm³)
			g/t	l/t		
Germany (West)	Schandelah	Liassic	14	60	0.950	2.12
	Messel	Eocene	21	89	0.903	1.83
France	Autun	Permian	26	108	0.908	2.03
Sweden	Rockesholm	Cambro-Ordovician	14	58	0.958	2.09
United Kingdom	Lothian (Scotland)	Carboniferous	22	93	0.881	2.22
Spain	Puertollano	Carboniferous/Permian	47	106	0.900	1.80

estimate that up to 95 per cent of raised bogs have been affected, at least to some extent by human activity.

Ireland currently faces a dilemma concerning the use and future status of the bogs. Here is a resource which can be extracted for fuel or for horticultural use. Because of the wide extent of the peatlands, their exploitation in some districts plays a key role in the local economy, providing badly needed employment. On the other hand the bogs can be used for recreation and tourism. At present, Bord Na Mona has set aside for conservation 2000 ha, and has excluded a further 4000 ha from acquisition for ecological/environmental reasons.

The Geological Survey of Ireland produced an initial map of Ireland's peat deposits in 1920. In the current period of controversy about the future of the bogs, the Geological Survey of Ireland, aware both of the ecological and economic aspects, can play a useful role in providing essential factual data to the debate. An example of this is the current Dutch/Irish research project on the hydrogeology of Irish raised bogs. The goal is to develop an understanding of the genesis and evolutionary mechanisms of peatbogs, so that an appropriate management programme can be established.

In Western Germany peatlands cover about 4 per cent of the area. Fens make up about 60 per cent of these peat deposits, and about 40 per cent are raised bogs. Apart from small areas in which peat has been cut for medical purposes, 95 per cent of the low-moor area has for centuries been used for agriculture. Of the raised bogs 67 per cent is used for agriculture and forests, 13 per cent for peat production, and 20 per cent is kept in its natural state.

Finland is one of the world's richest countries in peatlands which cover close to 10 million ha or some 30 per cent of the total land area. According to the present annual consumption of 20–25 TWh, mineable reserves are sufficient for about 200 years. The Geological Survey of Finland is the main body responsible for the inventory of peatlands. An extensive peat data bank has been compiled by the Survey. It contains information on all bogs that cover more than 20 ha and it uses more than 200 parameters concerning the characteristics of both bogs and peat.

Oil shale

Oil shales are fine-grained well-bedded sediments with a high level of organic matter (kerogene). Low-temperature pyrolysis of these sediments produces liquid and gaseous hydrocarbons. Commercial production requires levels of extractable hydrocarbons to exceed at least 40 l t^{-1} (approx. 3.6 wt. per cent).

Oil shales are mined mainly for the production of crude oil or to serve directly as a fuel. Before the American oil deposits were discovered in the middle of the nineteenth century, oil shales served as an important raw material in Western Europe for the production of liquid hydrocarbons. With the increasing import of American oil the low-temperature carbonization of oil-shales became less important during the second half of the nineteenth century.

Following the energy crisis of 1973/74 new interest focused on oil shales as a source rock of so-called unconventional oil. Compared to the huge oil shale deposits in the USA, Canada, China, Brazil and in the USSR, the resources of Western Europe are rather limited. Commercial production took place in Scotland, France, and Germany before the Second World War. There is presently no oil shale production in Western Europe apart from a small operation in Baden-Württemberg, (Dotternhausen) where a cement factory is using oil shale of Liassic age for electricity production. The ash produced is also used for the manufacture of a special cement.

Significant reserves (Table 3.6) still exist in other parts of Germany (BGR inventory post 1975), in Turkey (Tertiary lacustrine sediments), in Sweden (Cambrian alum shales mined until 1965). Minor reserves also exist in the United Kingdom and in Austria. The French situation is given as a type example.

Oil shale in France: an overview [36]

In France, about 50 oil shale occurrences are located in the Permian and Carboniferous (36 per cent), in the Liassic (20 per cent) and in the Cenozoic (44 per cent). The main reserves which have been identified (213 million tonnes of oil shale) are located in the northern edge of the Massif Central, in the east and south-east of France. The most important are located within the Upper Liassic at the eastern border of the Paris Basin (164 million t). Lesser deposits are the Permian shales of the northern Massif Central (45 million t) and some small deposits in the east of the country.

In the area of Autun, south-west of Dijon, oil shale mining and processing prevailed from 1824 (the date of the first pyrolysis experiments) to 1957 when the last mine closed down. In 1865, 21 claims, 16 plants, and 624 retorts were active and the annual oil production reached up to 6700 tonnes.

The Autun Basin is located in the Morvan area and covers about 250 km². It is filled with Carboniferous (Stephanian) and Lower Permian (Autunian) rocks and is more than 1200 m thick in the centre. The whole stratigraphic succession encompasses 20 beds of oil shale interbedded with other Autunian rock types.

The survey carried out by BRGM (Bureau de Recherches Géologique et Minières) in association with DGRST (National Delegation to Scientific and Technical Research) from 1979–82 defined the resources and the quality of the oil shales. Available resources in the unmined areas are estimated at 36.6 million t of oil.

Prospecting, mining, and beneficiation

Prospecting

All solid fuels are of sedimentary origin and represent one component among others of the sedimentary succession within each individual sedimentary basin where they occur. Prospecting methods are directed towards defining:

(1) the geometry of mineral bodies: size, thickness, extension, depth;

(2) the geometry of barren intercalations: sand, sandstone, clay, schist, etc.;

(3) the amount of reserves;

(4) the quality of individual bodies: rank, waste content, calorific value, sulphur content, etc.

Obviously, prospecting methods and strategy will differ with basin structure, type of sediments and mineral fuel, and depth of deposit. We have selected two type examples to illustrate the variety of methods used. The Devay–Lucenay-les-Aix case history relates to the search for a hidden coal basin at the northern border of the Massif Central of France. The methodology for peat prospecting in Scandinavia is the second case chosen. The target here is to delineate potential shallow resources over extensive areas.

Prospecting at Devay–Lucenay-les-Aix [43]

The Devay and Lucenay-les-Aix coal occurrences are part of the same coal deposit separated into two areas by the Loire Valley

about 10 km south-west of Decize (Fig. 3.22). Lucenay-les-Aix is a discovery of the last French coal inventory which was done in the 1980s. Devay has been known since the 1950s. The Lucenay-les-Aix deposit is totally buried below approximately 250 m of Tertiary rocks.

The initial understanding of the geological structure arose from prospecting by geophysical methods: gravimetry and electrical. The gravimetry interpretation showed a slight anomaly which was interpreted as a sedimentary trough. The first drill-hole intersected about 200 m of coal-bearing strata including three groups of coal seams totalling 55 m, below 260 m of overburden (Fig. 3.23).

Fig. 3.22 Location of Lucenay–Devay coal deposit (France) (after Donsimoni, M. (1990). *Doc. BRGM* No. 179).

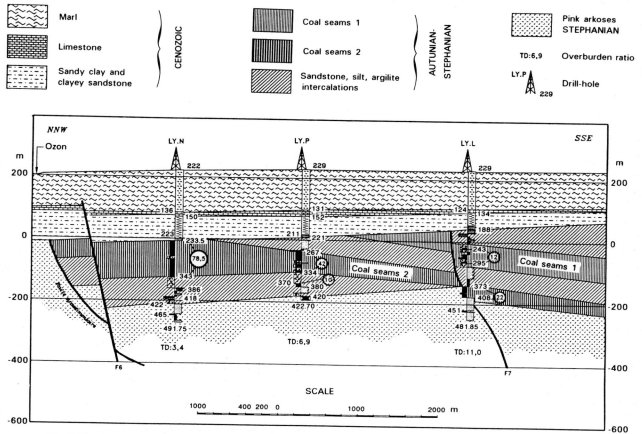

Fig. 3.23 Section through Lucenay-les-Aix coal deposit (France) (after Donsimoni, M. (1990). *Doc. BRGM* No. 179).

Further prospecting work included high resolution seismics (31 km) and systematic drilling (27 boreholes totalling more than 12 800 m). All boreholes were logged to facilitate correlation over an area of 3 × 2 km.

Peat prospecting in Norden [168] [18]

In the two big peat producing countries in the Nordic countries, Sweden and Finland, very similar methodology is used.

The methods used in the investigations originate from decades back, but they have been developed continuously to meet modern demands. These methods can also be applied to conditions which are quite different from those prevailing in the Northern hemisphere. After the first studies of air photos and maps field studies are being made using survey line networks, with study sites at 100 m intervals. At each study site surficial features such as cover type, wetness of mire, amount, size and quality of trees, and amount and size of peat hummocks, are first observed. Notes on valuable berries and fauna may be made, too. After that, the peat is classified according to its botanical composition. The degree of humidification is determined on a scale of ten.

Laboratory samples are taken from selected sites, samples are treated for water content, bulk density (kg of dry peat per m^3 of peat *in situ*), pH-value, heating value of peat, fusion temperature of ash, as well as for some trace elements, sulphur, and radioactivity. The maximum sulphur content tolerated by Swedish law is to be lowered from 0.19 g sulphur per MJ fuel to 0.10 g in 1993. In Sweden there is also a governmental tax of 30 SEK per kg of sulphur emitted from the combustion sites.

The results of the studies are presented in reports containing evaluations of peat resources, their energy content and suggestions for the use of peatlands (industrial, conservation, etc.). Peatland maps, peat profiles, histograms, diagrams on various peat properties, tables of laboratory results, and other necessary data are included.

Peat research units have cooperated with specialists from engineering institutions in developing and testing a subsurface interface radar, with which it is possible to determine the thickness of peat. This radar also gives some idea of the structure of peat deposits. It can well be used to measure the thickness of mineral soil layers and the groundwater levels. The equipment also incudes a device for automatic location.

In peatland inventories the peat research units also utilize gamma radiation maps obtained from the data of the areal measurements made by the Geological Surveys for ore prospecting. This method is based on the fact that gamma radiation from mineral soils indicates areas of deep peat worthy of detailed investigations.

Due to the great diversity of peat deposits, it is not always possible to obtain as many volumetric peat samples as would be needed for the accurate estimation of the dry matter content of a deposit. This is necessary for calculating the energy content. Since the bulk density of peat can be estimated well from its water content, a radio wave probe for determining dry matter and water content in the field has been developed in Finland. The probe is pushed into the peat to measure the dielectric constant automatically. The dielectric constant correlates with the bulk density.

Mining and beneficiation

Mining is carried out underground or in opencast pits at the surface. These exploitation methods are not specific to coal mining.

In recent years underground mining of coal and lignite has progressed by increased mechanization (e.g. long wall system), better recovery of coal, and decreasing the waste to coal ratio. The minimum thickness of seam worked is about 0.8 m. In order to reduce costs and increase output, opencast mining, which is generally the rule for lignite operations, has also been developing for coal with major sites in the UK (15 Mt yr^{-1} produced from about 40 sites), Decazeville in France, and in Spain. But most coal is still mined underground

and it will continue to be so because of the depth of the available seams.

In the large lignite opencast operations, the successive layers of waste and lignite are mined separately using very large equipment (bucket wheel, shovels, conveyor belts). A very sophisticated management system for soil, waste, and lignite has been developed to optimize costs and reclamation (see Section 5.2). Peat mining is in the form of shallow opencast operations, similar to those for sand and gravel.

More information on the various types of mining methods is given in Section 3.5.

Beneficiation of coal and lignite is mainly by crushing, milling, washing, and sizing, and usually leads to three products: cleaned coal (less than 15 per cent ash), mixed (up to 40 per cent ash), and fines. The first step is a heavy medium (water and fine particles of magnetite) separation which aims at separating coal from barren material. Waste is heavier and sinks. Coal is lighter and floats. After separation the coal is washed and sized. A special technique based on mechanical pulsation is used for the fines. Waste fines can be partly used for backfilling into the mines. The second step is the separation into size classes ranging between 0 and 120 mm.

The waste from the washing plant, together with some mining waste, is then stored in gigantic stockpiles which sometimes suffer spontaneous combustion as a result of chemical reactions. Some of the material is used as fill in construction work.

Market and management policy

The mining of hard coal in Western Europe would exist to a very limited extent only if it was not heavily subsidized. Lignite is mostly mined opencast and due to low mining costs large operations are profitable within the present economics of the electricity market. The system is usually fully integrated (mining and power production) either within one company or by the close cooperation of two separate bodies. Peat mining is very much supported by a national policy to reduce the import of energy resources, but the resulting environmental problems must be dealt with by the proper reclamation of large areas, which adds significantly to the overall cost.

In Europe, coal and lignite were the most important energy source until the middle of this century, and hard coal was the commodity which supported the rapid industrial development of Europe during the nineteenth century.

All coal-producing countries of Western Europe have been reducing their hard coal production for some decades. Plans have been set up in every country to reduce the production progressively while improving the operating costs and managing the necessary industrial conversion and manpower reorganization. The problems are dealt with at the national and international (EEC) level. In most cases there is an agreement with the electricity and steel companies which have to buy the coal at a certain price. This policy is long-term and will require many years to stabilize.

Hard coal

Owing to the geology (thin seams, disturbed bedding, and sometimes great depth) and to technical difficulties (deepmining, high pressure and temperatures, mine gas, subsidence damage at the surface) production costs of hard coal in the West European countries are among the highest in the world. As a result it has been steadily replaced in the West European energy market since the 1950s by oil, natural gas, and nuclear power.

Where the demand for coal has remained strong there has been a tendency in some countries to move away from expensive deep-mining to more cost-effective strip-mining methods. Western Europe's supplies are supplemented by about 100 million tonnes per year of low-price imported coal. More than fifty per cent of French requirements (approximately 25 Mt) are imported, mainly from the USA and to a lesser extent from Australia, Germany, and Poland.

Although long-term prospects for coal are quite favourable, many mines in Western European countries remain in operation only with considerable subsidies. To cover future demand it would be wise to preserve as much as possible of the present production capacity. The decision is mainly political, however, as purely market-driven economics would lead to further reduction of production capacities.

In any case mines cannot easily be 'mothballed' as a kind of national energy reserve. Deactivation of a mine generally means giving up all operating facilities and abandoning the part of the deposit being mined. Reactivation of an abandoned mine involves high expense, especially with regard to dewatering and stabilizing the workings, which makes such projects uneconomic except at times of extraordinary crisis.

Several Western European states have attempted alternative mining techniques such as underground gasification, to recover deposits which cannot be recovered by usual mining techniques. This applies to steep, disturbed, or thin seams, and above all to deposits deeper than 1500 m. Although significant results have resulted from research programmes, most of these operations remain at the pilot stage. Such exploitation of coal is likely to remain a marginal contribution to the energy market. The vast majority of coal utilized in future will still have to be exploited and transported using conventional techniques.

Lignite

One of the major problems raised by the use of lignite is the sulphur content, which is much higher than that of hard coal. In some cases, it may reach even 10 per cent. In Western Germany, all lignite-fired power plants that are not due to be shut down will be equipped with flue-gas desulphurization units.

Owing to the decreasing use of lignite for heating, the search for new uses is being intensified. Its chemical and physical composition make lignite a suitable raw material for gasification and liquefaction. Pilot plants in Germany have been producing synthetic methanol and artificial natural gas for several years. Pilot plants also exist for liquefaction of lignite by hydrogenation. Large-scale industrial upgrading of lignite, however, requires higher energy prices than those currently prevailing.

Peat

Peat has a low calorific value and is therefore no longer significant in most countries of Western Europe. It is still a major energy source in Ireland, Finland, and Sweden, however. These countries lack significant quantities of other solid fuels that can be developed economically. Major peat reserves also exist in the United Kingdom and Germany. In these two densely populated countries peat extraction is in conflict with agriculture and forestry and also with the demand of conservationists to protect these often unique landscapes and habitats.

Environmentalists see the Irish bogs as remnants of an ecosystem once spread over much of northern Europe. They consider the bogs to have several functions: conservation of species of specialized flora and fauna as a valuable record of the Earth's recent history (with special emphasis on environmental and climatic changes) for basic research, and for education.

Another environmental problem associated with peat extraction is the increase in peat silt in streams and rivers. The fine silt, which is almost the same density as water, is washed into the drainage channels and from these into the local stream network. Bord Na Mona in Ireland uses settlement ponds which, when carefully sited and regularly maintained, eliminate over 90 per cent of the silt from streams draining midland bogs.

The rehabilitation of the worked-out peatlands is very much a matter of ecological/land-use controversy. In present economic conditions, forestry would be a more attractive option generally than agricultural development.

Oil shale

Oil shales belong to the so-called unconventional energy sources. Worldwide, they represent a huge hydrocarbon potential, and they are regarded as future energy reserves. In Europe, high mining costs, together with environmental problems due to the low-temperature carbonization process (air pollution by dust, sulphur, and nitrogen-oxides, and the contamination of process water by inorganic compounds) and the land requirements for the disposal of spent shale, presently prevent extensive use of oil shale.

RECOMMENDED FURTHER READING

Barrabe, L. and Feys, R. (1965). *Géologie du charbon et des bassins houillers*. (Masson, Paris.)

Becq-Giraudon, J. F. (1983.) *Synthèse structurale et paléogéographique du bassin houiller du Nord et du Pas-de-Calais*. (BRGM, Orléans.)

Bender, F. (ed.) (1986). *Untersuchungsmethoden für Metall-und Nichtmetallrohstoffe, Kernenergierohstoffe, feste Brennstoffe und bituminöse Gesteine*. (Enke, Stuttgart.)

Berkowitz, N. (1979). An introduction to coal technology. (Academic Press, New York.)

Bundesanstalt für Geowissenschaften und Rohstoffe (1989). *Reserven, Ressourcen und Verfügbarkeit von Energierohstoffen*. (Bundesanstalt für Geowissenschaften und Rohstoffe, Hannover.)

Collective Authors (permanently updated). *World coalfields*. Intern. Geol. Correlation Programme 166. (Rijks Geol. Dienst., The Netherlands.)

Collective Authors—International Congresses of Carboniferous Geology and Stratigraphy. Every 4th year. (11th: 1987, Beijing China.) *Abstracts of Papers*. (Nanjing University Press, Nanjing.)

Crickmer, D. F. and Zeegers, D. A. (eds) (1981). *Elements of practical coal mining*. (Soc. Mining Engineers of A.I.M.E., Baltimore.)

Fettweiss, G. B. (1979). *World coal resources. Methods of assessment and results*. (Elsevier, Amsterdam.)

Geologisches Landesamt Nordrhein-Westfalen (1977). *Tagebau Hambach und Umwelt*. (Geologisches Landesamt Nordrhein-Westfalen, Krefeld.)

Grainger, L. and Gibson, J. (1981). *Coal utilization, technology, economics and policy*. (Graham and Trotman, London.)

Instituto Tecnológico Geominero de España (IGME) (1985). *Actualizacion del Inventario de recursos nacionales de carbon*. (IGME, Madrid.)

International Energy Agency (IEA) Coal Research (1983). *Concise guide to world coalfields*. (IEA Coal Research, London.)

James, P. (1984). *The future of coal*. 2nd edn. (MacMillan, London.)

Leonard, J. W. (ed.) (1979). *Coal preparation*. 4th edn (Soc. Mining Engineers of A.I.M.E., New York.)

Lüttig, G. W. (ed.) (1983). Recent technologies in the use of peat. In *Reports of the international Symposium of Deutsche Gesellschaft für Moor-und Torfkunde and section 11 of the International Peat Society*. (E. Schweizerbart, Stuttgart)

Lyons, P. C. and Alpern, B. (eds.) (1989). *Peat and coal: origin, facies, and depositional models*. (Elsevier, Amsterdam.)

Manners, G. (1981). *Coal in Britain: an uncertain future*. (Allen and Unwin, London.)

Moore, P. D. and Bellamy, D. L. (1974). *Peatlands*. (Springer, New York.)

Rühl, W. (1982). *Tar (extra heavy oil) sands and oil shales*. (F. Enke, Stuttgart.)

Stefanko, A. (1983). *Coal mining technology. Theory and practice*. (Soc. Mining Eng. of A.I.M.E., New York.)

Yen, T. F. and Chilingarian, G. V. (eds.) (1976). *Oil shale*. (Elsevier, Amsterdam.)

3.3.3. Geothermal energy

Introduction [120] [31] [209]

Geothermal energy is the heat contained in the Earth's crust which results from the general heat transport from the Earth's hotter interior to its colder surface (by thermal conduction and movement of material from depth). It is generally expressed either in thermal flux (energy dissipated per unit of surface area and time) or geothermal gradient (increase of temperature with depth). Specific, localized conditions exist in many areas, whereby it is possible to exploit such heat economically. Good aquifer permeability and/or high crustal temperatures are the most important parameters giving rise to geothermal fields where the heat can be extracted through boreholes, in the form of hot water or steam. Recent developments in heat pump technology have made it possible to expoit very low-temperature resources economically, and experiments are in progress to develop techniques to exploit the vast heat potential of crustal rocks with insufficient natural aquifer permeability (hot dry rock).

Historically, geothermal effects have had their impact on human society since antiquity. Spas and the domestic use of natural hot water were known to the Etruscans, the Romans, the Maoris, the Turks, and many other civilizations. During the twentieth century this traditional use has been extended to other energy uses, such as district heating and electricity generation, and to industrial applications such as the boron extraction at Larderello in Italy.

The first production of geothermal electricity took place at Larderello in 1904, after which the power plants of Wairakei (New Zealand) and The Geysers (USA) were inaugurated in 1958 and 1960 respectively. The first geothermal district heating system was that of Reykjavik (Iceland), which started in 1930. Since then the use of geothermal energy has been diversified further and the resources have seen an increasingly systematic exploitation that was stimulated by the oil crises of 1973 and 1979. In 1990 the total installed geothermal electric generating capacity worldwide was 5827 MW_e, while the direct heat use corresponded to 11 400 MW_t of thermal power.

Geothermal resources and their evaluation

Geothermal resources are usually classified according to their temperature and to the presence or absence of water and, within the latter, according to their temperature. Low-temperature resources with temperatures below 150°C are fairly common, both in volcanic regions and in sedimentary basins. They can be used in direct heating applications such as district heating, greenhouse horticulture, and some industrial processes. Between 150°C and 200°C, the medium tempera-ture resources can be used both for heating and for the generation of electricity in binary fluid power plants using the Rankin cycle. The high-temperature resources with tempera-tures over 200°C are found almost exclusively in volcanic regions. They can be used for the direct generation of electricity with acceptable efficiency in power plants that operate with single or multiple (flash) evaporation.

The different types of geothermal resources are very unequally distributed throughout the world. Low- and medium-temperature resources are found in the great sedi-mentary basins of France, Hungary, Yugoslavia, and the USA, for instance, and in certain volcanic and tectonically active areas such as Iceland. Most of the sedimentary resources were discovered while exploring for oil and gas. The high-temperature resources occur in volcanic regions and may be located through surface manifestations such as fumaroles, geysers, and mud pools.

The map (Fig. 3.24) of the main geothermal resources of Europe shows that in this region the resources are wide-spread. Low- and medium-temperature resources are found in the large sedimentary basins of, for example, France and Germany, and eastwards in Hungary, Poland, and Yugoslavia. High-temperature resources are confined to Italy, Greece, Iceland, and the Azores. More detailed maps of the geo-thermal resources of the EEC countries have been published in the *Atlas of the geothermal resources in the European Community, Austria and Switzerland*. (CEC, Brussels–Luxembourg, 1988.)

By contrast with coal and hydrocarbons, there are currently no reasonable precise methods for evaluating geothermal resources at the world scale. Estimates of the size of geo-thermal resources are usually based on volume calculations of reservoir heat content together with empirical assumptions regarding the rate at which the energy can be withdrawn from the reservoir. The methodology is still in its infancy, but in certain countries estimates have been carried out.

Exploitation of geothermal resources

In many countries, especially those with warmer climates, the interest in geothermal energy is mainly connected with its use for electric power production. In countries with colder climates the resources can also be of importance for direct heating of such as living quarters and greenhouses. Only the high-temperature resources are economically of interest for electricity production. The low-temperature resources are much more common and have been put to various uses in many countries.

International statistics show that installed geothermal electric power generating capacity worldwide was 5827 MW_e in 1990. Of this, 2770 MW_e are in the USA, 891 in the Philippines, 700 in Mexico, 545 in Italy, 215 in Japan, 283 in New Zealand, 95 in El Salvador, 45 in Kenya, and 45 in Iceland, etc.

Similar statistics indicate that in 1990 the use of geothermal energy for direct heat applications corresponds to an installed thermal capacity of 11 400 MW_t. Of this, 3321 MW_t are in Japan, 2143 in China, 1276 in Hungary, 1133 in the USSR, and 790 in Iceland, etc. The most common uses are for space heating, greenhouse farming, industrial drying, fish farming, medicinal baths, etc.

Geothermal energy is a finite resource in the sense that the energy brought in by natural heat transport processes does not compensate for the energy extracted. An efficient use of the

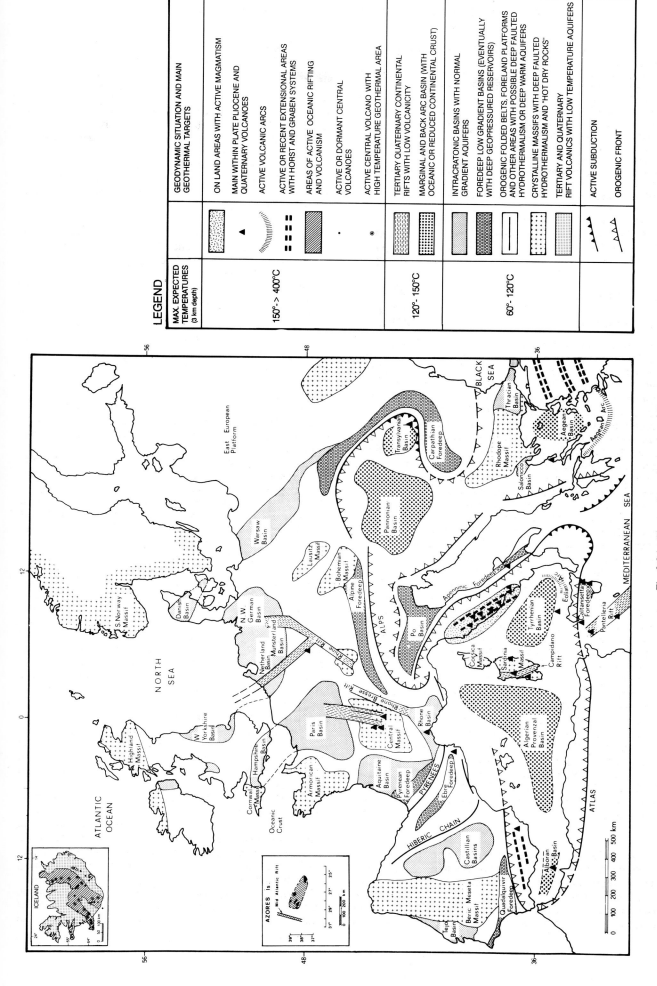

Fig. 3.24 The geothermal resources of Europe.

LEGEND

MAX. EXPECTED TEMPERATURES (3 km depth)		GEODYNAMIC SITUATION AND MAIN GEOTHERMAL TARGETS
150°– > 400°C		ON LAND AREAS WITH ACTIVE MAGMATISM
		MAIN WITHIN PLATE PLIOCENE AND QUATERNARY VOLCANOES
		ACTIVE VOLCANIC ARCS
		ACTIVE OR RECENT EXTENSIONAL AREAS WITH HORST AND GRABEN SYSTEMS
		AREAS OF ACTIVE OCEANIC RIFTING AND VOLCANISM
		ACTIVE OR DORMANT CENTRAL VOLCANOES
		ACTIVE CENTRAL VOLCANO WITH HIGH TEMPERATURE GEOTHERMAL AREA
120°– 150°C		TERTIARY QUATERNARY CONTINENTAL RIFTS WITH LOW VOLCANICITY
		MARGINAL AND BACK ARC BASIN (WITH OCEANIC OR REDUCED CONTINENTAL CRUST)
60°– 120°C		INTRACRATONIC BASINS WITH NORMAL GRADIENT AQUIFERS
		FOREDEEP LOW GRADIENT BASINS (EVENTUALLY WITH DEEP GEOPRESSURED RESERVOIRS)
		OROGENIC FOLDED BELTS, FORELAND PLATFORMS AND OTHER AREAS WITH POSSIBLE DEEP FAULTED HYDROTHERMALISM OR DEEP WARM AQUIFERS
		CRYSTALLINE MASSIFS WITH DEEP FAULTED HYDROTHERMALISM AND "HOT DRY ROCKS"
		TERTIARY AND QUATERNARY RIFT VOLCANICS WITH LOW TEMPERATURE AQUIFERS
		ACTIVE SUBDUCTION
		OROGENIC FRONT

resource calls for strict management policies, such as re-injection of the waste water and optimization of the sustainable fluid flow from boreholes, in order to ensure the longest possible exploitation life at an acceptable operating cost.

Geothermal energy is also a local resource. It can be transported economically for only limited distances. The longest pipeline carrying geothermal water for heating is about 70 km (in Iceland), and steam lines are very much shorter. Thus the resource and the market have to be in practically the same area. This puts constraints in some cases on possible uses of low-temperature geothermal resources.

The analysis of the environmental impact of a geothermal exploitation should consider its nuisance value as well as the possible reduction of pollution in the case of an energy source substitution. This general problem is illustated by the following case history.

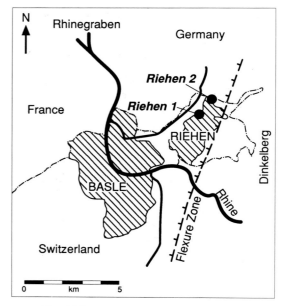

Fig. 3.25 Location of Riehen, Switzerland.

Environment and geothermal energy in Switzerland [179]

The reasons for prospecting for geothermal energy, particularly in a time of low prices, are not always dictated by economic forces alone. In Switzerland, for instance, the reasons for such investigations today are three-fold:

1. The oil crisis in the 1970s showed Switzerland's vulnerability as a result of its total reliance on imports of hydrocarbons. This emphasized the need for diversification of energy sources.

2. The attitude of the Swiss population towards nuclear energy has changed dramatically in the past fifteen years, from supportive to critical. The authorities are thus forced to look at alternative energy sources, one of which is geothermal energy.

3. Geothermal energy may also be regarded as a contribution to environmental protection. This energy requires no combustion, therefore no air pollution. If the cooled down waters—which may be rich in dissolved minerals—are re-injected through a second well (doublet) into the aquifer, there is no water pollution and the water balance of the aquifer is maintained.

In Switzerland, the evaluation of the geothermal energy is done by the cantons and by communities. They have the support of the authorities of the Swiss Confederation on different levels: the Swiss Federal Office of Energy may provide financial support to preliminary studies and a risk insurance for geothermal drilling to promote geothermal energy in general, and to help cover the financial consequences in case of a failure. The risk cover applies to projects with drilling depths greater than 400 m and usually amounts to 50–60 per cent of the total drilling costs, including production as well as re-injection tests but excluding liability.

One of the most outstanding projects recently carried out with the help of the governmental risk cover is that of Riehen (canton Basle-Stadt). The Basle region in north-west Switzerland (Fig. 3.25) extends within the Rhinegraben, a regional rift valley underlain by a relatively thin crust and lithosphere and consequently characterized by a comparatively high geothermal gradient. In view of this geothermal potential, the Canton of Basle-Stadt, jointly with the Community of Riehen, decided to explore the possibility of its practical exploitation for a heating system. A two-well programme for a doublet was set up from the beginning, due to the fact that the Rhinegraben produces highly mineralized waters which need to be re-injected for environmental reasons.

In 1988 the two wells were drilled. In 1989 a short-period re-injection test was carried out with encouraging results, and a long-term test started in 1990. The operation of the pilot project has been highly successful so far. Two problems, however, still remain to be solved: gas content and mineralization require an analysis of pressure and temperature dependence in order to define conditions for doublet operation. The final decision to go into exploitation has been taken in 1991. The necessary heating distribution system is

already under construction and will be operated by classical combustion of oil and gas and with geothermal energy in addition if it becomes available.

The Riehen project has been carried out successfully so far. Experience has shown that this is only feasible where the population is motivated to accept higher expenses for energy to achieve environmental protection, the use of renewable energy, and the diversification of energy supply. A drilling operation demands a well-defined target (aquifer, fracture zone, karst) and a drilling site within reach of the future consumer of the energy, commonly in the neighbourhood of a residential area. In this case, noise protection during the drilling operation is of primary importance. The support of the population is obtained by means of good information on the objectives and progress of the programme.

The environmental impact of geothermal exploitation can be both of a physical and a chemical nature. Fluid withdrawal from the geothermal reservoir results in a pressure decrease which in turn may lead to ground deformation above the reservoir, usually in the form of vertical settlements. Such effects are known from several geothermal fields which have been operated for a long time. An example is given in a case history in Section 5.3 for the Larderello and Travale fields in Italy.

Chemical pollution induced by geothermal exploitation is always much smaller than that caused by fossil fuel thermal power stations of equivalent capacity. The pollution is related to the emission of gases that cannot be condensed (H_2S and CO_2), or to the discharge of brines that can contain certain toxic substances such as heavy metals or arsenic. Today it is normal practice to use specific treatments to restrict polluting emissions. Another form of pollution is noise, which is an inescapable part of any thermodynamic exploitation, but which can be attenuated to a large extent.

In the case of low-temperature geothermal utilization the net result is usually a reduction of the impact on the environment. When geothermal energy replaces a fossil fuel, the polluting emissions (CO, CO_2, NO_x, SO_2, and dust) from boilers are eliminated and the urban impact of road transport for supplying the power plants is greatly reduced. A striking example is that of Reykjavik which is now totally heated with geothermal water: this system has replaced the initial, heavily polluting, coal-fired operation, as well as later oil-fired installations.

For all geothermal exploitations the balancing of resource and requirement is of fundamental importance, which means in particular that the temperature of the identified resource must be higher than that required for a particular application. In the case of very low-temperature resources this is not the case, but the development of high-power heat pumps has made it possible to utilize such shallow heat reservoirs. This is described in the following case history from Sweden.

The heat pump—how to increase the small-scale use of energy [167]

During the 1970s a rapid development of the technology for using various energy sources for heat pumps started in Sweden. The market was mostly concentrated on private use, implying a small-scale utilization of energy resources stored in the near-surface rocks and sediments, and in the groundwater within them.

With heat pump technology (Fig. 3.26), it is possible to provide enough energy for heating houses. There are two main routes for this. The first is to use the heat content in groundwater from an ordinary well with a high capacity. The second is based on the heat flow of the strata. When cooling down a well, a heat flow to the well is initiated. The heat flow energy is transported by a 'heat carrier' to the heat pump. The two methods can also be combined.

The largest and perhaps the most successful heat pump plant in Sweden is based on groundwater from a sandstone aquifer at a depth of about 800 m. This 45 MW low-temperature geothermal heat pump plant came into production within an extremely short time. This was possible thanks to a very successful and accurate geological investigation and hydraulic testing of the aquifer, and the close cooperation of the geologists with the engineers. On the basis of the results a commercial heat plant was built and put into production within two years. Today, this plant supplies about 40 per cent of the heating of the city of Lund (Fig. 3.26).

Essential for planning the utilization of groundwater energy are heat pump technology and geological and hydrogeological knowledge. The role of the Geological Survey of Sweden (SGU) has been to furnish geological and hydrogeological data as a basis for such planning.

The SGU has developed, in connection with its hydrogeological work, a map indicating relative conditions for exploiting heat energy from the ground. The map shows heat conductivity of the heat-producing rocks, tectonic features, some mines, and wells of different capacities.

The total potential extractable energy from groundwater and the heat-flow in the rocks, utilizing heat-pump technology, is very large. One single degree temperature decrease of the annual groundwater usage in Sweden will produce about 200 TWh, which is more than the demand for the total heating of housing in the country. The heat flow in the rocks has at least the same potential as the groundwater.

The small-scale utilization of energy became large-scale utilization in a short time. During the period 1980 to 1986 about 50 000 heat pump plants were installed, converting the energy of groundwater and heat flow of the rocks to a usable form. The exploitation of energy corresponds to a power effect of 500 MW or almost as much as is produced by a block in a nuclear power station.

The most efficient use of geothermal energy is in cascading applications where each step uses a successively lower temperature than the previous one. An example would be to associate, on a single high-temperature resource, a mixed electricity generating plant (a Carnot cycle coupled to a binary fluid power plant), and a heating application that in itself may be designed for a step-like use. In some cases extraction of chemicals from the geothermal fluid or other raw materials may be a part of this process.

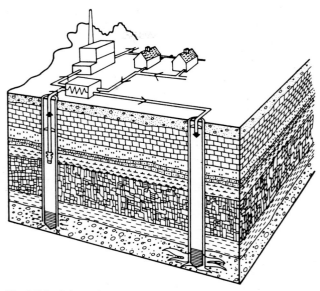

Fig. 3.26 Schematic view of the heat pump plant at the city of Lund, southern Sweden.

Several examples exist of such cascades, even though most are only partial:

- At Svartsengi in Iceland a binary-fluid power-plant was installed below a classical power plant producing both electricity and heat for a district heating system. The condensate is discharged into the 'Blue Lagoon' which is a popular bathing pool.

- In the Paris region, most of the numerous geothermal networks have cascades with radiators, floor heating, and heat pumps.

- At Larderello in Italy the electricity generation is combined with boron extraction.

High-temperature resource applications and techniques
Geothermal resources are used essentially for producing electricity, for heating, and in a more marginal sense for industrial, agricultural, and recreational purposes.

The techniques used for electricity production depend on the degree of saturation of the water vapour and on the thermal power that can be recovered at the well-head. The most favourable conditions are those of dry steam fields. In general the output of several wells is collected by a network that feeds a power plant with one or more generators.

A fraction of the steam produced is often discharged into the atmosphere, whereas the condensate can be either re-injected or discharged at the surface. The non-condensable gases, such as H_2S, must be treated as a function of their toxicity, and eliminated as needed.

In certain geothermal fields the problems include scaling with silica and calcite in the wells and the surface pipelines, in addition to the usual ones of corrosion and erosion caused by humid vapours. To minimize such effects, freshwater deposits are sometimes used to carry the heat from a production loop via heat exchangers to the distribution network.

In unstable areas, volcanic activity may cause disturbances. A situation where both high- and low-temperature fields, as well as freshwater deposits, are exploited under conditions of potential volcanic hazards, as illustrated by the following case history from Iceland.

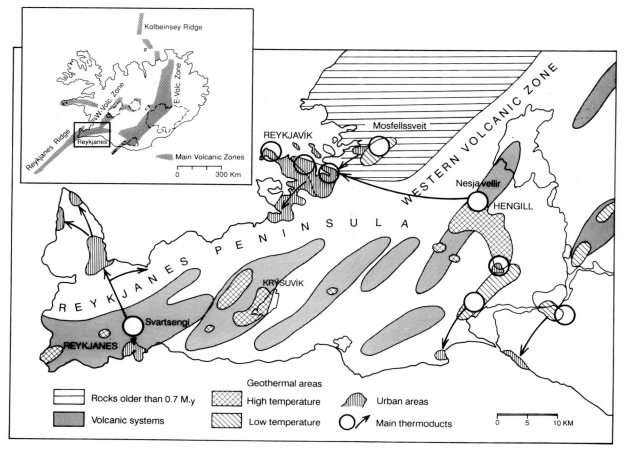

Fig. 3.27   Geothermal fields and volcanic hazards on the Reykjanes Peninsula, Iceland.

Geothermal fields and geological hazards on the Reykjanes Peninsula, Iceland [122]

The exploitation of the geothermal resources of the Reykjanes Peninsula in south-west Iceland is an illustrative example of utilizing such resources at a calculable risk in a volcanically and seismically hazardous area (Fig. 3.27).

The peninsula is part of the Mid-Atlantic Ridge and has thus been very active volcanically in Quaternary and Recent times. Close to 65 per cent of the population of Iceland live on the north coast of the peninsula and on both sides of its neck, of which nearly 40 per cent are in the capital, Reykjavik. The volcanism is indirectly a primary source of vital and very important natural resources (fresh groundwater in recent lava flows and abundant geothermal heat) but also, conversely, the cause of potential geological hazards (volcanic eruptions, earthquakes, and tectonic movements). Reykjavik is located only 20 km away from the nearest site of historical eruptions. Under these circumstances, the role of geology in providing information as a basis for decision-making is clearly very important.

Five distinct volcanic systems run obliquely in a NE–SW direction across the E–W trending volcanic zone of the peninsula. The easternmost one (the Hengill volcanic system) lies inside the western volcanic zone. Each system is characterized by groups of narrowly bundled eruption fissures with variable width, and less active areas in between. One or more shield volcanos, composed of olivine-tholeiitic rocks, is situated on the northern side of the volcanic zone in each system. The rocks from the eruptive fissures are tholeiitic basalts, acid rocks not known to occur on the peninsula, except in the Hengill system.

During the glacial periods, the erupted rocks were turned into pyroclastics and pillow lavas under the subglacial conditions. These rocks were heaped up in mountains, giving rise to the presently high precipitation, 1500–3500 mm yr^{-1}, replenishing the fresh groundwater streams on the peninsula as well as the geothermal fields. During the interglacials and in recent time the eruptive products spread out as lava sheets on the lowlands on both sides of the mountainous volcanic zones.

Each volcanic system corresponds to a tectonic fissure swarm, which also runs NE–SW and extends for tens of kilometres out from the volcanic zone, at least on the north side. In the western part of the peninsula a narrow, seismically very active zone runs just north of the axis of the volcanic zone. Where it crosses the volcanic systems, N–S tectonic fissures are prominent; these are the centres of the volcanic activity and most of the high-temperature geothermal fields.

The young and fresh rocks of the peninsula are of Late Quaternary age and very permeable, the dense fissuring strongly enhancing the permeability and creating at the same time an area of anisotropy, horizontally and vertically. The stratification of the basaltic lava pile creates a vertical, very strong anisotropy in the permeability, increasing the degree of confinement rapidly with depth.

The extraction of naturally clean groundwater is very easy from the highly permeable aquifers. The urban settlements are lavishly supplied with freshwater (near to 1500 l s^{-1} or 900 l d^{-1} per inhabitant), at extremely low costs by European standards.

On the western part of the Reykjanes Peninsula the high permeability leads to freshwater floating as a thin (only 50 m thick) lens on sea water in the bedrock. The fluid in the deeper geothermal systems in this part of the peninsula ranges from two-thirds sea water to salinity noticeably above that of sea water. This indicates concentration of the fluids through evaporation in the geothermal system.

On both sides of the neck of the peninsula, low-temperature geothermal fields are active in older (Early Quarternary) and less permeable rock formations. These fields were the first to be exploited for central heating in the urban areas, the first thermoduct to Reykjavik being put into use in 1930. At present, all municipalities in the region are supplied with hot water from geothermal fields, the energy utilized amounting to 4000 GWh yr^{-1}. Nearly 80 per cent is for space heating, a very important energy supply in a region with mean annual temperature of only $+5\,°C$ and summer temperature seldom exceeding 15°C. In the whole of Iceland there are 29 public geothermal district heating systems supplying about 85 per cent of the space heating needs of the population.

At present, two major geothermal plants are operated on the high-temperature geothermal fields on the Reykjanes Peninsula, together with extraction at numerous places from the low-temperature fields for the Reykjavik Municipal District Heating Service and various small townships.

One of the high-temperature plants is located at Svartsengi in the western part of the peninsula, supplying the adjacent townships with hot water. The geothermal fluid has a chemical composition corresponding to two-thirds sea water mixed with freshwater. It is unsuitable for direct use, so instead the hot steam from the wells (reservoir temperature 230–240°C) is used in a heat exchanger to heat cold freshwater, extracted in the vicinity. The warm and saline waste fluid is said to have excellent balneological qualities, even healing qualities against psoriasis. The Svartsengi plant is a combined heat and power generation plant, producing 125 MW$_t$ of heat and 12 MW$_e$ of electricity in a cascading arrangement.

The other major plant is at Nesjavellir and is intended, in the future, to supply the major parts of the heating needs of the Reykjavik Municipal District Heating Service. The reservoir temperature is 300–400°C and the steam there is also used in heat exchangers to heat fresh water, extracted from wells in the vicinity, drawing on the reservoir of the deep lake Thingvallavatn. The plant is under construction and in its first stage will supply 100 MW$_t$ of heat, later increasing to 400 MW$_t$; generation of electricity is also envisaged.

The highly economical exploitation of these natural resources is not without risk as the sites of production are inside or marginal to the active volcanic zones. The activity of the zone can be gauged from the fact that in post-glacial times more than 200 eruptions are known to have occurred in 10 000 years.

The hazard of volcanic eruptions (especially lava flows, as phreatic activity seems to be insignificant) and earthquakes presents a risk to the exploitation of the most efficient natural resources, and even renders at least some part of the area unsuited for prolonged settlement, a fact which must be taken account of in all regional planning.

For the people living in the area, the natural resources described above are fundamental, and even vital, positive assets of geological origin and of great economic importance. Conversely, the geological hazards are significant restrictions to the exploitation of these resources and to the evolution of urban settlements. Both are major factors in regional and local planning, and it is indeed interesting to notice that urban planning is increasingly being based on geological investigations.

Another problem is that in strongly faulted areas massive exploitation of geothermal resources can entail the invasion of colder water from the surface. In all cases, a particularly careful monitoring of the production parameters is necessary for good management of the resource.

Low- and medium-temperature resource applications

In the case of low- to medium-temperature resources that are used for heating, the fundamental characteristics related to the low heat-level of the resource is the need to develop low-temperature heat networks. Two cases can be distinguished, based on the chemistry of the water: water with little or no salt content, and water with a high content of dissolved salts.

The former case poses few problems. Applications of geothermal energy sources of this type are known from Iceland and France (Aquitaine). In the most favourable cases, such as in Reykjavik, it is even possible to distribute the hot water directly for heating and for sanitary use. The waters are not corrosive and cause almost no scaling. Thus, no particular treatment of the water is necessary, nor is the use of special materials for contact surfaces. Re-injection may, however, be necessary to maintain the reservoir pressure. Other uses for this type of resource are medicinal (spas) and recreational (swimming pools).

Concern for the environment prohibits water with a high content of dissolved substances being discharged at the surface. An original response to this problem has been the use of doublet wells, as in the case of the exploitation of the Dogger resource in the Paris Basin where 50 doublets are operational. The first was put into service in 1971 at Melun, east of Paris. The present state of geothermal exploitation in France and the main problems encountered are related in the following case history.

The history and current state of exploitation of geothermal energy in France [55]

Introduction

In the past twenty years, a rather common but relatively new type of geothermal energy has been developed experimentally in France: the use of heat under normal thermal gradient conditions in sedimentary basins (Fig. 3.28). This development was made possible by favourable reservoir conditions in the two major sedimentary basins of the country, the Paris and Aquitaine basins. New techniques developed, observations made during operations, maintenance of the geothermal systems, and problems encountered in a few projects, have stimulated progress in reservoir engineering and opened up new areas of current research.

As shown in Fig. 3.29, geothermal exploitation started in 1968, with the doublet at Melun, 30 km south of Paris, and reached a large peak in the 1980s.

From 1985 to 1990 the research efforts have been focused on detailed reservoir knowledge, corrosion–scaling processes induced by fluid composition, new methods and techniques for maintenance, and regeneration of wells.

At present, some sixty plants in France use geothermal energy to produce direct heat for district heating. Drilling and connection to networks took place mainly during the years 1980–85.

Two types of reservoir exploitation are in use: single wells without re-injection and doublets (production and injection wells coupled). Development of single well systems is limited to specific, generally shallow reservoirs of low salinity (below 2 g l^{-1}), corresponding to smaller heat loads at the surface.

The doublet technology (Fig. 3.30) has been developed mainly for the Dogger reservoir in the Paris Basin—a highly permeable limestone formation of Jurassic age—allowing optimization of the management of this geothermal resource.

Sedimentary characterization of reservoirs

The Dogger reservoir of the Paris Basin consists of Middle Jurassic deposits, mainly limestone between the marls at the top of the Liassic, and those of the lower Callovian.

While these limestones have been intensely drilled for oil exploration in the Paris Basin in general, information in the area of Paris and its environs, for the obvious reason of intense urbanization, is limited to the data acquired from geothermal boreholes. These holes have been drilled mainly in the northern and south-eastern suburbs.

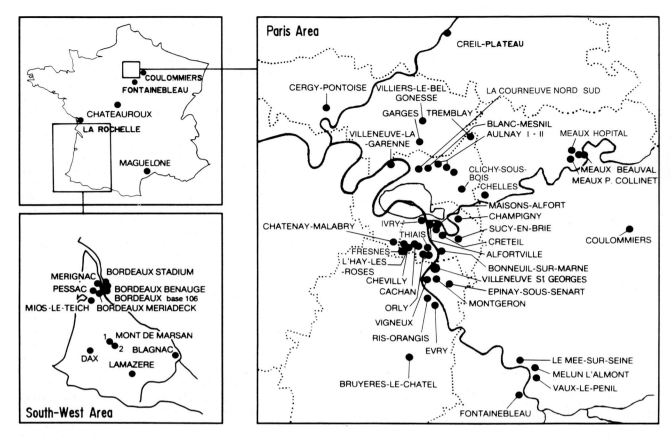

Fig. 3.28   Geothermal exploitation in France: location of wells.

The productive horizons exhibit a very wide range of textural permeability, to which secondary permeabilities may have been added due to fracturing and dissolution.

Analysis of technical problems

Geothermal exploitation of the Dogger has been faced with two main problems, a shortage of reliable pumping systems, and corrosion and scaling.

The pumping problems, which caused considerable exploitation losses, have now been practically resolved thanks to more adequately sized and improved electrical connections in the immersed pumps.

However, the problem of corrosion and scaling is not yet resolved. The action of sulphides in the chloride environment results in corrosion of the metal sections of the geothermal loop, with widespread corrosion of the soft steel casing, corrosive spotting of the sections of stainless steel piping, and the deposition of iron sulphide scale. The scale and suspended particles generated cause pump breakdowns, filter and exchanger clogging, and an overall reduction in hydraulic production, with head losses and clogging of the re-injection reservoir.

New diagnostic and regeneration techniques

Though much of the technology used for geothermal investigations is derived from the petroleum industry, it has been necessary to develop new techniques for the diagnosis and regeneration of geothermal projects in the Dogger. New concepts in doublet design are also being considered now.

To assess the development of the geothermal field and to have available a historical record of the major parameters of doublet projects, l'Agence Française de la Maîtrise de l'Energie (AFME) has set up a remote monitoring system. This unique system, which

has now been in operation for three years, makes it possible, among other things, to know at any given time the operational parameters of a doublet, using a telephonic monitor (Minitel). The centralization, storage, and analysis of the acquired data make it possible to improve the global model of exploitation of the field and to assess the extent of interference between adjacent doublets (see Fig. 3.30).

A new procedure of well descaling by injection of water under high pressure (jetting) has also been tested. Descaling is currently being orientated towards the adaptation of processes linked to the degree of scaling: jetting for slightly fouled wells, and the mechanical cleaning followed by jetting where the fouling is considerable. Chemical descaling processes are also being investigated.

Brief review of the economic context

The period 1985–90 was marked by a fall in the price of the traditional heating fuels—oil, gas, and coal. The price of imported crude oil, for instance, fell on average from FF1842 per tonne in 1985 to FF641 per tonne in 1988. This fall has been directly responsible for the significant reduction of income in new geothermal energy projects.

The majority of projects have been set up by public bodies and consisted of a geothermal doublet and a district heating system. To encourage use of the geothermal network, subscriber contracts generally contain safety clauses stipulating that the cost of geothermal energy shall not exceed the cost of the energy it replaces.

The safety clause puts the financial management of geothermal operations (wells + district heating) at a serious disadvantage, as project economics relate to unpredictable international influences. Nevertheless geothermal energy exploitation is still cheaper than any other classical fossil fuel, but the safety clause makes it difficult for the district heating network to be profitable.

Moreover the viability of certain projects was judged to be independent of the high and rising prices of energy resources, the

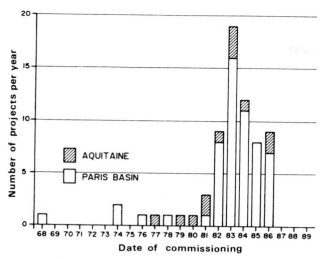

Fig. 3.29 Geothermal activity in France: number of projects per year with date of commissioning.

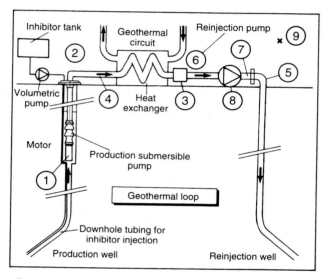

Fig. 3.30 The principle of a geothermal loop.

continuing price rises and interest rates being assumed constant. In fact quite the reverse occurred, incurring considerable financial losses for certain projects.

Conclusions

The geothermal exploitation of the Dogger aquifer in the Paris Basin, which began in 1969, has demonstrated the feasibility of the doublet concept. In common with the Hungarian and Icelandic experiences, where the low enthalpy geothermal resources are exploited mostly by single wells, the re-injection of the total flow is carried out on an industrial scale. This technique has major advantages for reservoir management and environmental protection.

After a period of rapid development where many geothermal heating systems were installed as the result of the two oil crises, new economic conditions, combined with some unexpected corrosion problems, have led to some changes to the initial concept.

The evolution of geothermal energy exploitation in France was marked in 1989 by the actual or expected shutdown of several operations in the Paris region, and no development of new projects. This crisis, brought on by the collapse in the price of fossil fuels (which served as reference for the calculation of the selling price of geothermal heat) has been aggravated by the technical problems related to corrosion and scaling, the cost of overcoming which increases the total cost of exploitation. Research has been given a strong impetus by the search for solutions to the problems related to corrosion and scaling in the wells, and to increase the general knowledge of overall management.

The drilling of a new injection well at Melun (the oldest doublet on the Dogger) after almost twenty years of exploitation, suggests that geothermal energy may be viable in the longer term.

Research trends

Significant research and development work has been carried out in the long term in order to achieve an economically viable development for geothermal energy and to solve the associated technical problems. This work includes the improvement of exploitable methods for known resources, of optimizing techniques for exploration of geothermal deposits, and of the feasibility of extraction from hot dry rock systems.

For high-enthalpy systems one of the present-day research trends is to develop less costly processes for the elimination of hydrogen sulphide. For low-enthalpy highly saline geothermal waters, systematic re-injection methods reduce pollution problems. Re-injection is adequate for carbonate reservoirs, but is still problematical for clay–sand reservoirs.

Some waters are highly corrosive at any temperature or induce scaling in the geothermal exploitation system. Extensive research is carried out to understand corrosion or scaling processes and to improve diagnostic, prevention (inhibitors), and regeneration.

The re-injection of used geothermal water may be necessary to maintain adequate pressure in the geothermal field. Specific research is carried out on the consequential effects on the chemistry in the reservoir. This type of research has many connections with the hot dry rock studies, in particular in the field of hydrodynamic and thermal modelling of fractured reservoirs.

In the high enthalpy field the exploration techniques still have to be improved. These concern in particular microseismicity studies associated with injection–production operations, the study of hydrothermal alterations, the use of small diameter boreholes at shallow depth, and a multidisciplinary approach to decision-taking.

Reduction of exploitation costs is also a major goal. The best compromise must be found between improving performance of geothermal installations and reducing maintenance costs. This leads to the development of expert systems and the appraisal of the economics of the impact on the environment.

A number of national geothermal energy research programmes have been fostered in the last decade or so, and the Commission of the European Communities has taken a major role in stimulation and coordination. The Geological Surveys continue to have an invaluable role.

As previously mentioned, there is a vast potential energy in the heat contained in crustal rocks which have relatively high temperature but low aquifer permeability (hot dry rock). To transfer the heat efficiently to the surface it is necessary to create an artificial permeability by fracturing the rock between two boreholes; this fracture network then serves as a pathway for the circulation of the injected heat carrier (water) between the two holes, the heat being recovered at the surface.

Three projects of this type are known today. In chronological order these are Los Alamos (New Mexico, USA), Camborne (Cornwall, UK), and Soultz-sous-Forêts (Alsace, France). The characteristics of the latter two projects are described in more detail in the following case history.

European hot dry rock programme [206] [46]

In 1976 the UK Department of Energy and the CEC commenced funding R&D programmes on geothermal energy. The agreed programme addressed three topics, principally concerned with the evaluation of the UK geothermal resource potential.

(1) low enthalpy of deep sedimentary basins;

(2) potential for hot dry rock (HDR) development, with particular reference to Caledonian granites;

(3) assessment of the geothermal resources.

Camborne project

The prime aim of the 'demonstration' element of the HDR R&D Programme was to prove the feasibility of creating a commercialized reservoir within an impervious rock. The work was carried out in three phases, mostly by the Camborne School of Mines (CSM) at the Rosemanowes quarry in Cornwall, with the contracted support of the British Geological Survey, Sunderland Polytechnic, Taylor Woodrow Construction, and RTZ Consultants. This work is still under way.

The major work carried out at Rosemanowes, under Phase 1 (1977–80), was a small-scale field trial aimed at linking four boreholes over a horizontal distance of 40 m at a depth of 250–300 m. One objective was to show that the use of controlled explosive stimulation could destroy the stress concentration at the well bore and initiate a number of small fractures which could then be stimulated hydraulically to provide a low impedance link between the boreholes by way of the natural joint system.

Phase 1 was successful in all its aims. The boreholes were linked and explosive stimulation followed by water circulation reduced system impedance by a factor of 50 relative to that achieved (previously) by hydraulic fracturing alone.

Phase 2 (1980–88) was aimed at the creation and management of a commercial-scale HDR reservoir at an intermediate depth (about 2 km) at which the temperature would be about 80 °C.

The principles of the heat extraction using this approach are summarized in Fig. 3.31. The cold water is injected by the first well, circulated and heated as it passes through the network of natural joints after hydraulic stimulation of its permeability. The hot water then returns to the surface by the second well, with the help of pumps if necessary. The heat is extracted from the hot water through a surface heat exchanger and the water is thereafter re-injected. The extracted heat can be exploited for space heating or electrical power production, or combined uses depending upon the flow rates and temperatures at the production well.

The reservoir created in Phase 2 did not behave as expected; water losses were too great for a commercially viable operation and overall pumping pressures for circulation were too high. As a result, it was not possible to get within a factor of ten of the sustainable production flow of $75 \, l \, s^{-1}$.

A review of the programme, started during 1987, identified a number of problems with HDR technology:

1. The size of reservoir then available was considered to be almost two orders of magnitude smaller than required for a commercial system.

2. The thermal behaviour of the reservoir was unsatisfactory (because of excessive temperature drawdown, short-circuiting water losses).

3. The reservoir design process had not been validated.

4. A potentially commercial system was likely to require several reservoir 'segments' to be created in a 'multiple stimulation'.

None the less, HDR was identified as likely to be commercially attractive in the long-term, and a further three years' funding was announced to remove the uncertainties that had been identified and justify a decision to proceed with a prototype-depth reservoir.

In conjunction with all phases of the Camborne programme, other 'generic' studies were conducted. These tackled subjects of general application wherever HDR resources might be appropriate for exploitation. For example, the British Geological Survey participated in geochemical modelling and the monitoring of earthquakes in the Rosemanowes area.

Soultz-sous-Forêts project

Following the invaluable groundwork of the UK programme, a European hot dry rock feasibility project, funded by governmental agencies and the CEC, was set up in 1987 near Soultz-sous-Forêts, in northern Alsace, at the western side of the Upper Rhine Valley (Fig. 3.32). The site is close to the centre of a wide geothermal anomaly (150 km by 20 km)) covering French (Haguenau) and German (Landau) territory. From measurements in the old 'Pechelbronn' field and other available boreholes it was known that the geothermal gradient in the sedimentary cover, close to Soultz, is three times as high as the normal value. From numerous oil wells and extensive seismic surveys the stratigraphy and structure of the sediments, as well as the topography and main tectonic elements of the crystalline basement, are well known.

In this specific Alsacian situation, where both thermal and tectonic conditions seem very favourable, it is hoped that a minimum-cost test programme, carried out jointly by German and French scientists, will demonstrate the effective economic possibilities of this new clean resource at a time of growing costs of conventional energy.

A 2 km deep geothermal investigation hole (GPK1), drilled at the end of 1987, penetrated more than 600 m into the granite basement (Fig. 3.32). During 1988/89, the geology, geothermal, rock mechanic, hydraulic, and geochemical properties and parameters of the crystalline complex were identified; all are necessary for the planning and pre-evaluation of an HDR reservoir. The temperature at the base of the sediments is 124 °C; bottom-hole temperature at 2000 m is 141 °C, whilst geochemical thermometers indicate temperatures above 200 °C in natural fluids along their path before they enter the borehole. These fluids are saline ($100 \, g \, l^{-1}$). The anomalously high geothermal gradient in these sediments provides, within the whole crystalline complex, an advantage in temperature of 75 °C for 2.5 km borehole depth against areas with normal geothermal conditions. During stimulation tests at a depth of 1969–2000 m, more than 500 m^3 of water was injected at a constant flow rate of $3.3 \, l \, s^{-1}$. The stimulated fractures connected to natural hydraulic systems near the borehole. The specific yield of the borehole was improved considerably from 0.06 to $0.18 \, l \, s^{-1} \, bar^{-1}$ and it seems likely that it can be increased much further by massive hydraulic fracturing.

A seismic network was installed in boreholes adjacent to the 2000 m hole. Three sets of three-directional seismometers capable of withstanding temperatures of above 125 °C were cemented in three holes 840, 960, and 1360 m deep, at distances between 200 and 400 m from the geothermal hole. Using this network, the results of the seismic monitoring performed during the stimulation tests have shown the geometrical characteristics of the stimulated joints.

At present work is proceeding to design and construct an HDR heat exchanger with the following characteristics: average rock temperature 175 °C (approx. 3.5 km depth); effective flow rate 25–35 $l \, s^{-1}$; hydraulic impedance 0.3 MPa. $s \, l^{-1}$; exchange surface 0.5–1 km^2; acceptable water losses during circulation. This task is the

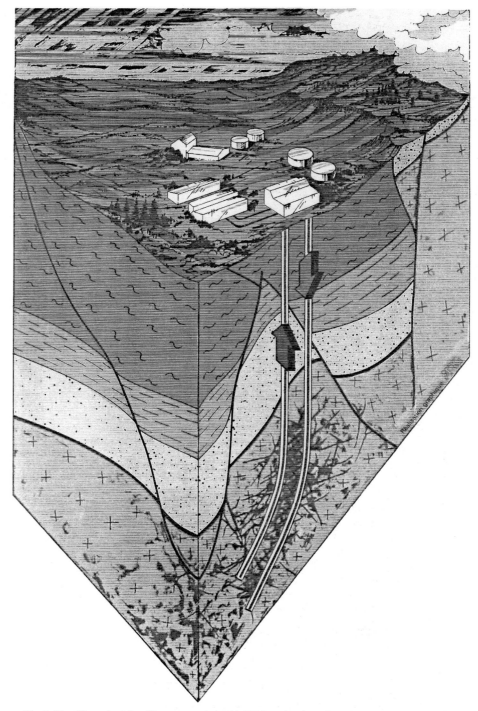

Fig. 3.31   The principle of heat extraction with HDR technology, based on Soultz-sous-Forêts.

first step within a European Project aiming at the construction of a scientific prototype of an HDR demonstration plant.

Consequently, with the complementary support of British partners bringing their experience from the Camborne project in Cornwall, the site operations continued in 1990. They began with the drilling of an exploratory well which will core the granite down to 3200 m and with the recovery/deepening of two abandoned oil wells in order to improve the seismic monitoring of the hydraulic tests.

The first well will be used for better evaluation of new techniques developed for the stimulation of the permeability of fractured deep

rock and the chemistry of water–rock interactions at high temperatures.

In parallel, work will proceed on tests and development of some investigative tools adapted for deep and hot wells.

To conclude this present phase, the details of the scientific prototype will be defined in 1991 to enable construction to begin in 1992.

The scientific concepts and techniques being developed and tested within this programme are at the forefront of international research on deep geothermal resources. The economic viability of further industrial plants will depend strongly on the evolution of

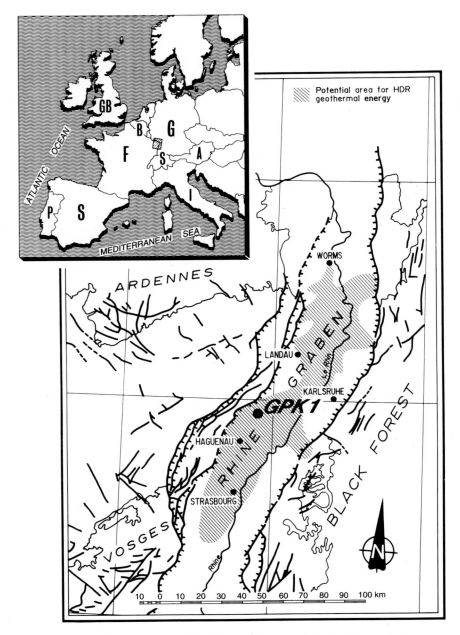

Fig. 3.32   Location of the European HDR project at Soultz-sous-Forêts.

cost-effective drilling and the efficiency of monitoring fluid circulation.

Currently, the tests are undertaken by scientific teams. An industrial consortium involving Siemens, RTZC, BRGM, and several other partners is being established to manage the project during the next phase.

Many of the adverse environmental effects of the exploitation of geothermal energy can be reduced or even eliminated, but at a cost. Pollution can be minimized, for example, by the re-injection of used fluid into the reservoir, but much research remains to be done to perfect such techniques. Geological Surveys have much to contribute by applying their accumulated data and their knowledge of the hydrogeological properties of the reservoir rocks.

RECOMMENDED FURTHER READING

Anon. (1983). *Guide du maître d'ouvrage en géothermie*. (BRGM, Orléans).

Armstead, H. C. H. (1983). *Geothermal energy: its past, present and future contributions to the energy needs of man*. 2nd edn. (Spon, London).

Blair, P. D., Cassel, T. A. V., and Edelstein, R. H. (1982). *Geothermal energy, investment decisions and commercial development*. (Wiley, New York).

Bowen, R. (1989). *Geothermal resources*, 2nd edn. (Elsevier, Amsterdam).

Cermak, V. and Haenel, R. (eds) (1982). Papers from *Symposium on Geothermics and Geothermal Energy. Budapest, 1980*. (Schweizerbart'sche, Stuttgart).

Coudert, J. M. and Jaudin, F. (1989). *La géothermie: du geyser au radiateur. Trésors de la Terre*. (BRGM, Orléans).

Edwards, L. M., Chilingar, G. V., Rieke, H. H. III, and Fertl-Walter, H. (eds) (1982). *Handbook of geothermal energy.* (Gulf, Houston).

Goodman, L. J. and Love, R. N. (1980). *Geothermal energy projects: planning and management.* (Pergamon, New York).

Gupta, H. K. (1980). Geothermal resources: an energy alternative. (Elsevier, Amsterdam).

Harrison, R., Mortimer, N. D., and Smarason, O. B. (1990). Geothermal heating: a handbook of engineering economics. (Pergamon, Oxford).

Lienau, P. J. and Lunis, B. C. (1989). *Geothermal direct use engineering and design guidebook.* (Oregon Institute of Technology, Klamath Falls).

Rybach, L. and Muffler, L. J. P. (eds) (1980). *Geothermal systems: principles and case histories.* (Wiley, Chichester).

Varet, J. (1982). *Géothermie basse énergie: usage direct de la chaleur.* (Masson, Paris).

3.3.4. Nuclear fuels [65]

Introduction

The collective term nuclear fuels is applied to products derived from uranium, the heaviest naturally occurring chemical element known both on Earth and in the universe. The basic particle of uranium, the atom, comprises a nucleus containing 92 positively charged protons and 146 neutrons, 238 nucleons altogether, and 92 negatively charged electrons that orbit the nucleus. Atoms that have the same number of protons but a different number of neutrons are called isotopes. There are 325 natural isotopes, everywhere present in the same proportions within a given element, and almost 1200 have been created artificially.

In nature, there are three uranium isotopes, 234, 235, and 238, each containing 92 protons, and with 142, 143 and 146 neutrons respectively. Natural uranium contains 0.0055 per cent U 234, 0.72 per cent U 235, and 99.2745 per cent U 238. Such uranium disintegrates to produce three types of radiation: α (the helium nucleus), β (the electron), and γ (a photon, a particle with the same properties as light and X-rays). As uranium 238 disintegrates to form thorium, α particles are emitted. The chain continues with the disintegration of thorium to protactinium, with the production of a β particle. A nucleus excited by α or β disintegration finds its equilibrium by emitting a γ particle. The same three types of particle are produced by induced or artificial radioactivity.

Of the three uranium isotopes, uranium 235 is the most easily ruptured by nuclear fission, with the production of considerable energy (82 thousand million Joules from the total destruction of a single gram).

In the course of irradiation in the reactor, the uranium 235 progressively disappears, the uranium 238 generates plutonium by capturing neutrons, and the fission of uranium 235 and plutonium 239 causes products of fission to appear. Prolonged irradiation of uranium and plutonium causes the appearance of elements heavier than uranium, known as transuranians.

Nuclear fission produces a chain reaction that can be mastered to obtain constant and pre-determined production of energy. For the chain reaction to occur, it is necessary to have a sufficient mass of uranium 235, the critical mass. The fission reaction is produced by bombarding uranium 235 with neutrons. Fission is accompanied by the giving-off of energy due to loss of mass. Simultaneously, neutrons are liberated and in turn can cause fission of other nuclei of uranium 235. In a nuclear power plant, it is not the reaction that is directly controlled, but the geometry of the core, using control rods that act as screens by absorbing the neutrons.

The nuclear industry covers all operations ranging from the extraction of uranium ore to the processing of nuclear waste from nuclear power stations, and four major stages are involved:

1. Extraction of uranium ore and its enrichment to produce concentrates containing less than 1 per cent uranium 235.
2. Refining, resulting in the production of tetrafluoride, which is then processed in beneficiation plants to obtain a concentrate enriched in uranium 235 by a factor of about 5 (3–4 per cent). This gaseous concentrate is transformed to oxides used in the manufacture of fuels in nuclear power plants.
3. Combustion in the nuclear power plant, producing heat, subsequently transformed to electrical energy and waste.
4. Processing and storage of nuclear waste from plants.

These four stages are examined successively below, and the problem of the storage of radioactive waste in geological strata is discussed more specifically in Chapter 5.

Uranium deposits: exploration

Two main types of mineral deposit are described in Chapter 2: deposits resulting from endogenic processes related to the emplacement of igneous rocks and accompanying phenomena, and deposits related to exogenic processes in sedimentary formations. Both processes have produced uranium ores.

Endogenic uranium deposits are associated with rocks rich in silica (acidic rocks), principally plutonic and to a lesser extent volcanic. The ore fills fractures or is disseminated in the voids of granite-type rock in which quartz, one of the main constituents, has been dissolved by the natural circulation of hydrothermal fluids. In deposits of this type, the uranium occurs in two chemical forms, a tetravalent state, generally stable in nature, and a hexavalent state, derived from the first by superficial oxidation and passing into solution with or without re-deposition. The second form gives the uranium element its great mobility.

The grade of uranium deposit is usually very low, with 1000 kg of ore producing a maximum of 2–5 kg of uranium concentrate.

In Europe, endogenic uranium deposits are associated with mature granites throughout the Hercynian basement (see Chapter 2), in the Iberian Peninsula, France, Germany, and the United Kingdom.

Exogenic uranium deposits occur in Palaeozoic, Mesozoic, and Cenozoic rocks, the source of the uranium being the basement. The uranium in the alum shales of Sweden results from leaching of the Precambrian basement. In central and southern Europe (the United Kingdom, Germany, France, and the Iberian peninsula), the main deposits occur in sediments immediately overlying rocks affected by the Hercynian orogeny (examples being the Permian of Lodève in France and the small deposits in the Tertiary basins overlying the basement of the Massif Central). They occur also in more distant sediments capable of trapping such mobile elements. This is the case with sediments rich in organic matter, a typical example being the deposits of the Gironde region of France. Elsewhere in the world (in Australia and Canada, for example), very large deposits are associated with rocks of Precambrian age.

The fact that uranium ores are radioactive aids exploration for deposits. The radiation they emit (principally γ) can be detected using Geiger–Müller counters or scintollometers, either on the ground (on foot or using a light vehicle) or airborne (using a plane or helicopter). Instruments for recording radioactivity in boreholes are also used extensively. The problem is to detect anomalies that effectively correspond to concentrations of economic significance. The detection of radon gas (88 protons, 138 neutrons and part of the natural chain of uranium degradation) is also used in exploration (emanometry).

Radiometric exploration is only one of the techniques: others include geochemistry, detailed geological study, and geophysical methods.

Uranium production by western countries was about 28 570 tonnes in 1990 (34 000 tonnes in 1989), of which 11 per cent is from Europe (3236 tonnes), mainly France (2800 tonnes) and to a lesser extent Spain and Portugal (400 tonnes). A third of world production is from Canada (8698 tonnes). Other major producers are the USA (3547 tonnes), the Republic of South Africa (2529 tonnes), Australia (3520 tonnes), Namibia (3210 tonnes), Niger (2830 tonnes), and Gabon (720 tonnes).

Mining, refining, and waste

The mining methods used to exploit uranium deposits are the same as for other metallic ores (see Section 3.5), generally either by selective underground mining or by bulk opencast extraction. The radiation emitted by the ore and the release of radon mean that the atmosphere in underground workings must have powerful ventilation systems and be regularly monitored for radiation, with regular control of the personnel involved.

Uranium minerals are extracted from the ore by acidic leaching (generally using sulphuric acid) or by alkaline leaching, followed by solid/liquid separation. Leaching is carried out on ore reduced to a state of small fragments by crushing and milling. The concentrate resulting from solid/liquid separation is then purified over ion exchange resins and passed through a solvent to produce solid magnesia uranate containing 70 per cent uranium. Another process used is the precipitation of uranium in the form of lime uranate, followed by the production of a uranyl nitrate solution by various processes. The first process is by far the most common.

During the refining stage, the ore concentrates (magnesia uranate or uranyl nitrate) is calcined to prepare uranium trioxide. The raw material, known as uranium tetrafluoride, is then fed into an enriched process to produce uranium 235, or uranium metal.

The separation of uranium 238 is very delicate because the two isotopes have, by definition, identical chemical properties and very similar physical properties. The only difference is in their mass. Gaseous diffusion is one of the main separation processes in industrial use. Another industrial process is ultracentrifugation, and other processes are currently under study, including enrichment by laser beams.

The various types of nuclear fuel and power plant

Almost all nuclear power plants operating industrially use natural uranium fuel, either in the form of metal or oxide (graphite or heavy water reactors) or in the form of oxide from hexafluoride enriched in uranium 235 (ordinary light water reactors, the water being either pressurized—PWR—or boiling—BWR). This light water reactor, and particularly the PWR, is by far the most common today.

Only the graphite–gas reactor uses metal fuel. All the others, and in particular the light water reactor that is the most common today, use ceramic oxide fuel.

The high-temperature reactor (HTR) uses graphite as a moderator and, because of the temperature (750–1000 °C), uses helium as a vector for calorie transport. The fuel consists of oxide or carbon particles of fissile material coated with graphite and agglomerated. This type of reactor is also still at the prototype stage.

The power plants built in France up to 1970 were principally of the graphite–carbon gas type. The heavy water reactor remained at the prototype stage in France, and was developed mainly in Canada. The rapid neutron reactor is known as a breeder reactor. The fuel is plutonium mixed with either natural uranium (essential uranium 238) or impoverished uranium obtained by reprocessing fuels from power plants or isotopic separation plants. Cooling is by liquid sodium. This type of facility is still at the prototype stage, an example being Superphénix at Creys-Malville, in the Isère region of France.

The plutonium recovered during the reprocessing of irradiated fuels is a major source of energy that might have been used in fast breeder reactors. Several countries have envisaged its use in a mixed fuel, consisting of uranium oxides and plutonium, to be used in light water reactors. Such fuels contain about 5 per cent plutonium and are presently used in some countries. In France, tests have been going on for several years in four power stations, and the construction of a plant to manufacture of such fuel was recently approved for the Marcoule site.

Processing of irradiated fuels

After use, the irradiated fuel is extracted from the reactor and plunged into a deactivating bath in which it remains for several months. This bath acts to protect the environment and to cool the fuel. After one month, by normal decay processes, the total radioactivity of the fuel is approximately one-tenth of what it was when it left the reactor.

The next stage involves reprocessing the nuclear waste in a special plant, like the one at La Hague in France. Such reprocessing aims to separate the various components of the used fuels: the uranium impoverished in U235, the plutonium, the transuranians, the products of fission, and other metals alloyed in the fuel or having formed the sheath.

The waste from reprocessing is stored in special containers, and the long-term trend is towards vitrification of the products of fission.

Management of nuclear raw materials and fuels

The greater part of the uranium market is governed by long-term contracts between producers and users. This makes it easier for producers to manage their mining operations and for users to secure their long-term supplies. Additional supplies are obtained on a one-off basis on the spot market, but this market controls between 10 and 20 per cent of the total volume of transactions.

France, for example, which has opted for a strong nuclear industry, secures its fuel supply from domestic production and extensive participation in the exploitation of certain overseas deposits (in Gabon, Niger, and Canada, for example).

The problems posed today are concerned mainly with the management of the oldest nuclear facilities and with the storage of radioactive waste. Investment strategy is now geared towards replacing old equipment, the prime concern being to avoid allowing high-risk equipment to reach a critical age. In the case of storage (see Chapter 5), two separate stages must be distinguished, that of partial decontamination and processing (vitrification, for example), and that of deep storage, still for the most part at the exploratory stage.

RECOMMENDED FURTHER READING

Boyle, R. W. (1982). *Geochemical prospecting for thorium and uranium deposits*. (Elsevier, Amsterdam).

OECD (1990). *Uranium Resources, production and demand 1989*. (OECD, Paris).

Collective Authors (1982). *Uranium exploration case histories*. (IAEA, Vienna).

Collective Authors (1982). *Proceedings of the symposium on uranium exploration methods*. (OECD, Paris).

Collective Authors (1982). Management of wastes from uranium mining and milling. *Proceedings of an International Symposium 1982, Albuquerque, USA*. (IAEA, Vienna).

Collective Authors (1985). Advances in uranium ore processing and recovery from non-conventional resources. *Proceedings of a technical committee meeting, Vienna*. (IAEA, Vienna).

Collective Authors (1985). *Methods for the estimation of uranium ore reserves*. (IAEA, Vienna).

Collective Authors (1988). *Nuclear power and fuel cycle: status and trends*. (IAEA, Vienna).

Collective Authors (1988). Recognition of uranium provinces. *Proceedings of a Technical Committee Meeting, London 1985*. (IAEA, Vienna).

Collective Authors (1989). Isotope techniques in the study of the hydrology of fractured and fissued rocks. *Proceedings of an advisory group meeting, Vienna, 1986*. (IAEA, Vienna).

Collective Authors (1989). Metallogenesis of uranium deposits. *Proceedings of a technical committee meeting Vienna, 1987*. (IAEA, Vienna).

Collective Authors (1989). Uranium deposits in magmatic and metamorphic rocks. *Proceedings of a technical committee meeting held in Salamanca*. (IAEA, Vienna).

3.4. INDUSTRIAL MINERALS [143] [204] [209]

3.4.1. Introduction

Industrial minerals are not the most 'glamorous' of our natural resources and would normally receive little media coverage by comparison with, say, our energy minerals. Yet our reliance on them is just as widespread. Although industrial minerals have been used for several thousand years, consumption has increased dramatically during the last 40 years. Demand is closely related to population size, per capita income, and level of industrialization. Other important factors effective in the shorter term include business cycles, technical changes, substitution, and changing consumer preferences.

With the exception of fertilizer minerals, the major proportion of industrial minerals is consumed by the industrial and construction sector of the economy for use in civil engineering, building materials, the chemical and metallurgical industry, and the production of consumer goods. Because of this, growth rates of mineral consumption are closely associated with structural changes in appropriate sectors of the economy. In typical long-term development patterns, the share of the industrial sector in the total Gross Domestic Product (GDP) ranges from below 20 per cent in the least developed economies to between 30–50 per cent in the most advanced countries.

Whilst both metalliferous and industrial minerals are exploited for a wide range of industrial processes, the metalliferous minerals are used for their metal content only. By contrast, industrial minerals are very sensitive to special niches in the market (bentonite may be sold, for example, for cosmetics, for drilling mud, as a binder for foundry sand, or as pet litter) and sophisticated industrial processes often have special quality requirements for their raw material feed. Thus, in many cases it is necessary to process industrial minerals to meet multiple specifications. The problems for the producers are increased further as there can be considerable substitution of one industrial mineral for another—where the technical advantage of one may be affected by the lower cost of another.

Industrial minerals are many and varied; indeed there are far too many to describe exhaustively here. Instead, examples will be used to indicate the general overall situation. Table 3.7 presents some production statistics for a selection of industrial minerals. The fact that Europe has relatively high production is not an accident of geology. Unless, or until, a country becomes industrialized, or otherwise economically well-developed, its industrial minerals potential will not be fully exploited. In terms of their geographical distribution in economic quantities the industrial minerals may be characterized into 'common' and 'less common'. The following sections of this section will use this categorization as a basis for discussion.

The *common industrial minerals* are those which can be produced in large quantities in most countries. They include aggregates (e.g. sand, gravel and crushed limestone, dolomite, sandstone, and igneous and metamorphic rocks), clay (common brick clay and fireclay), limestone products for cement, lime, and carbonate fillers, salt, gypsum and anhydrite, and industrial sand (for glass making and foundry sand). Being common or geographically widespread, it is difficult for such material to incur the cost of long-distance haulage without becoming uncompetitive compared with local alternatives, although more specialized and higher priced products can usually be moved over greater distances. For example, chalk used in agriculture is seldom moved more than 30 km from the quarry, whereas finely-divided, closely-sized chalk products such as whitings or special fillers are traded internationally.

The *less common industrial minerals* depend on a more select set of geological circumstances for their presence in economic quantities. Their distribution, therefore, is more sporadic, though their use may be widespread. Included in this category are asbestos, bentonite, borates, diatomite, feldspar, graphite, kaolin, magnesia, mica, nepheline syenite, perlite, phosphates, potash, and talc. Whilst Europe has some resources of all of these, its production of some, like phosphates and graphite, amounts to only a small part of the world's output. For these materials transportation costs are a relatively small consideration in their commercial viability.

Considering the large consumption of some industrial minerals, particularly construction materials, it is quite understandable that increasing conflicts arise between mineral workings and other uses of the land, such as housing, roads, recreational areas and water supply. Even where there is no competition for land-use, mineral workings may be a threat to the local environment because of the associated visual impact, noise, traffic, dust, and other effluent.

Stringent rules have to be designed to avoid or minimize conflicts between mineral exploitation and the environment (also see Chapter 6). In this context the Geological Surveys throughout Europe have a major role to play in helping planners by the provision of statistical data on mineral

Table 3.7 Selected industrial minerals; production of the world and Western Europe. 1989 statistics. Based on *World Mineral statistics, 1985–89, 1990* (British Geological Survey)

	World	Western Europe	
	Production (M tonnes)	Production (M tonnes)	Percentage of world
asbestos	4.50	0.167	3.72
barytes	5.40	0.900	16.70
bentonite and Fuller's Earth	13.19	3.300	25.02
diatomite	1.85	0.530	28.65
feldspar	5.50	2.490	45.39
fluorspar	5.80	0.708	12.21
graphite	0.65	0.038	5.86
gypsum and anhydrite	90.90	21.610	23.77
kaolin	23.90	5.250	21.96
perlite	2.34	0.780	33.33
potash	29.50	4.800	16.30
salt	183.30	39.000	21.27
talc	7.70	1.200	15.61
vermiculite	0.54	0.000	0.00

production and uses, and advice on technical and economic aspects of mineral working, substitution, end-use applications, and environment. Mineral inventories are one of the basic tools which can be placed at the disposal of governments and environmental planners for the development of resource and land-use management policies. A very illuminating example of the use of resource inventories in coordinated planning is presented in the case-history 'Inventory of rocky terrains in Finland' on p. 103.

Such inventories must contain sufficient information regarding quality, volume, and location of economically useful minerals and outline possible uses of the raw material, taking account of the more obvious possibilities offered by beneficiation.

The following sections describe the steps from resource evaluation through exploitation and use, together with some comments on resource management policies.

3.4.2. Resource evaluation and description

Resource evaluation for industrial minerals is closely related to their potential application, where physical and engineering properties often play a key role, although in some cases the value lies solely in chemical properties. For a large proportion, however, physical properties are of prime importance; not only is this true for such bulk products as crushed stone, but also for some specialized less common materials such as diatomite, asbestos, and graphite.

Evaluation of mineral resources, particularly industrial minerals, may have to be up-dated quite frequently as the economic exploitation of a given mineral is related to a number of factors which themselves vary with time, namely:

(1) economic market (supply and demand);

(2) changing of raw material specification on the part of the consuming industries;

(3) development of new applications;

(4) availability of substitute minerals;

(5) increasingly elaborate processing techniques;

(6) planning regulations;

(7) increasing concern with health, safety and the environment.

A systematic assessment programme is a necessary part of any mineral resource inventory which should comprise evaluation and estimation of resources. However, in countries with a free market economy, evaluation of industrial mineral deposits is carried out most commonly by industrial enterprises, with the primary aim of securing their own economic activities in the short-term only. Therefore, resource evaluation based solely on information obtained from the mining industry is largely limited to actual reserves of those commodities which have commercial value at a specific time and location; this may well represent only a relatively small part of the total resource.

In developing adequate resource management policy, comprising such topics as self sufficiency, economic and technological development, employment opportunities, and environmental issues, governments can—through their Geological Surveys—implement systematic mineral resource investigations, supplementary to the exploration activities of the mining industry. An adequate appraisal of total mineral resources will be possible only when it is based on a thorough understanding of the Earth's geological processes.

The two categories of industrial minerals, 'common' and 'less common', by this very fact require different approaches in resource evaluation. Perhaps with the exception of the marine resources, the locations of deposits of the common industrial minerals of all geologically well-surveyed countries are known to a large extent, but they require detailed prospecting or exploration programmes for development. What is also required are programmes of evaluation of the capability of the raw material to satisfy, after processing, some of the likely user specifications. On the other hand the less common industrial minerals mostly require initial exploration projects, some of which may be scientifically very comprehensive.

Common industrial minerals

The common mineral most in demand is aggregate. There are two broad types of aggregate, natural sand and gravel, and crushed hard rock. The major uses of both of these are in construction, particularly as concrete aggregate and road-stone. There is considerable substitution between these types according to local availability and cost. Whilst the crushed rock aggregate is normally more expensive to produce, transport costs and material suitability can affect the choice of mineral. In general, natural gravel is preferred for concrete and crushed rock for roadstone. For both cases the material must meet rigorous specifications for particle size (and is normally processed for this purpose by crushing and sizing) (see Table 3.8) and physical quality (hardness, resistance to abrasion, etc.). In the case of sand (<5 mm) used for concrete, for mortar, or for asphalt, it is necessary to reject very fine material (generally less than about 63 microns, depending on end-use).

The following case history illustrates the procedures involved in a regional assessment of natural sand in gravel.

Assessment of sand and gravel resources in the UK [219]

Introduction

Of the industrial minerals that are used in very substantial quantities in the UK (aggregate production in 1990 was about 300 M tonnes), sand and gravel is probably the most widely distributed. The associated land-use planning and mineral supply problems are difficult to resolve. Consequently, it is not surprising that work on the sand and gravel resources of the UK formed a major part of an industrial minerals assessment programme undertaken for the UK Government (mainly the Department of the Environment) by the British Geological Survey. Most of the effort was absorbed on surveys of fluvial and glacio-fluvial sand and gravel. The assessment of resources was concentrated on areas of high demand and high mineral value, rather than on the more remote deposits. Surveys were also directed away from deposits with well-known potential as a result of company prospecting or other detailed geological information.

Assessment procedure

The assessments are statistically based on the findings of regularly-spaced borings (including their laboratory analyses) and the geological relationships available on standard geological maps.

The standard unit of assessment was chosen as 100 km² at the 1:25 000 scale. A map at this scale and size shows large enough tracts on one sheet to provide a regional appreciation, while allowing space for detailed results to be displayed. 'Resource blocks', relatively large areas, arbitrarily fixed at about 10 km², are defined for each sheet. For these the gross volume and chief physical and lithological characteristics of the sand and gravel are assessed, using simple statistical procedures. An open grid of assessment boreholes sited without detailed reference to the geology at roughly 1 km centres, provides the basic data. These are augmented by the examination of exposures, including workings, and the collation of records of wells and boreholes from all available sources.

The boreholes should provide a regular pattern of sample points: otherwise the characteristics of sparsely sampled ground will be under-represented in the calculations and clusters of boreholes may have to be assigned weighting factors, to avoid bias. Such a system of weighting may be applied when data from a group of closely spaced commercial boreholes have to be integrated into the calculations of the resources.

A basic judgement is also involved in the identification of a deposit as a resource, and limiting physical criteria have to be decided by which the day-to-day conduct of field investigations can

be controlled and against which final results can be compared; they are the essential basis for the work, and they can be regarded as the minimum economic criteria by which potentially workable deposits may be recognized. After discussions with the extractive industry and others, the following criteria have been applied in the resource surveys since 1968:

(1) the deposit should average at least 1 m thickness;
(2) the proportion of fines should not exceed 40 per cent by weight;
(3) the ratio of overburden to sand and gravel should be no more than 3:1;
(4) the deposit must lie within 25 m of the surface, this figure representing the likely maximum depth in most circumstances.

It follows from the third and fourth criteria that boreholes are terminated at 19 m if no sand and gravel has been encountered; in general, they are taken only far enough into the bedrock to confirm rock type and establish without doubt the thickness of the deposit.

Large-scale (1:10 000) geological maps, which form the basis of the surveys, provide the area of the deposit, a primary statistic in the calculation of resources. As far as possible, the limits of the resource blocks are based on geological boundaries so that similar deposits are considered together and, provided sufficient data are available, the volume of the resource is calculated with conventional confidence limits at the two-sided 95 per cent probability level; otherwise an inferred assessment is given.

A summary of the gross resources in terms of cubic metres of minerals by resource block is given in the report on each area of survey.

Publication results

Bearing in mind that major uses of the resource surveys are primarily concerned with land-use and mineral planning, the results are presented graphically by overprinting on to a 1:25 000 scale geological map, and the potentially workable deposits are depicted in shades of colour, showing whether they are exposed or carry overburden. Information derived from assessment boreholes is shown by rectangular diagrams the height of which is proportional to the thickness of the mineral, lateral subdivisions showing the proportions of gravel, sand, and fines, calculated from the sieve grading of bulk samples taken continuously from the whole thickness of the deposits. The thickness of overburden and of inter-bedded waste are also shown. Information available from other boreholes, including well records, is presented as far as possible in the same way, although it is rarely so comprehensive. An extract from a sand and gravel resource map is given as Fig. 3.33.

Even amongst the common industrial minerals, the raw material specification and the environmental constraints combine to limit the locally available supply. The following case history provides a good example of the role of geologists in helping to maximize the reserve.

White Chalk and cement production in Denmark [14]

White Chalk is used for cement production in a number of European countries, especially in some of those bordering the southern North Sea, where this particular type of limestone is well exposed.

In Denmark cement production has always been based on the White Chalk of Upper Cretaceous age. Because of the technical properties of the chalk—it is wet, soft, and crumbling—the cement production has been based on a wet process in which the chalk is dispersed in soft water, mixed with clay and/or other components and burned to clinker pebbles in rotary kilns before it is milled ready for use as ordinary Portland cement.

At one time there were seven cement factories in Denmark, all of which used White Chalk, but today, the production is focused on a

Table 3.8 Industrial minerals: processing

Process	Typical application
Crushing etc.	
1. Gyratory crusher	Large machines, some capable of accepting fragments up to 2 m; used for treating blasted hard rock. Smaller machines used for crushing coarse gravel.
2. Jaw crushers	Similar application to gyratories, but preferred for intermittent operation and lower throughputs.
3. Cone crushers	Secondary crushers used to produce coarse aggregate for concrete, roadstone, or feed for industrial processes. A typical product may be in the range of 50–10 mm
4. Impact crushers (hammer mills etc.)	Used as primary and secondary crushers where the rock is not abrasive e.g. limestone, gypsum, etc.
5. Ball mills, rod mills	Used for producing sand-sized and finer materials e.g. agricultural lime, cement, fillers, etc.
Mixers, blenders, and scrubbers	
1. End runner mills	Used for clay in brick making etc. and for pottery etc.
2. Wash mills	Used for preparing chalk slurry for cement making.
3. Scrubbers	Cylindrical mills with a scrubbing action, used to wash gravels and remove any adhering clays.
Sizing	
1. Screens	Vibrating screens used, wet or dry, to produce sized materials for concrete aggregate, roadstone, and feed to industrial processes.
2. Classification	Sizing of sand-sized and finer products, usually carried out in water by exploiting differences in particle settling velocity, using a variety of devices. Applications include removal of quartz and mica from china clay, preparation of fine fillers, preparation of sized material for subsequent processes. Cyclones which use centrifugal forces in a vortex are used for separating clay from sands and for producing sand in specific size ranges.
Mineral separating	
1. Heavy media	Preconcentration and recovery of fluorspar and barite.
2. Jigs, cones, spirals, etc.	Removal of coal from gravel; recovery of zircon, ilmenite, monazite from beach sands.
3. Magnetic separation	Removal of iron-rich minerals from industrial sands; recovery of ilmenite from beach sands.
4. Flotation	Recovery of fluorspar, barite, apatite, pyrite; cleaning of industrial sands and recovery of potash minerals from saturated solution.
5. Acid leaching	Cleaning glass sands.
Chemical and thermal treatment	
1. Calcination and kiln treatment	Lime making, cement making, brick making, and ceramics in general.
2. Roasting	Pyrite burning to produce SO_2.
3. Evaporation (including solar evaporation)	Recovery of salt from brines, potash, etc.
4. Precipitation and filtration	Recovery of MgO from sea water.
5. Chemical activation	Smectite clays.
6. Surface conditioning	Use of surface active agents to enhance adhesion of aggregates to bitumen, and to ensure free flowing of fine fillers or binding with matrix.

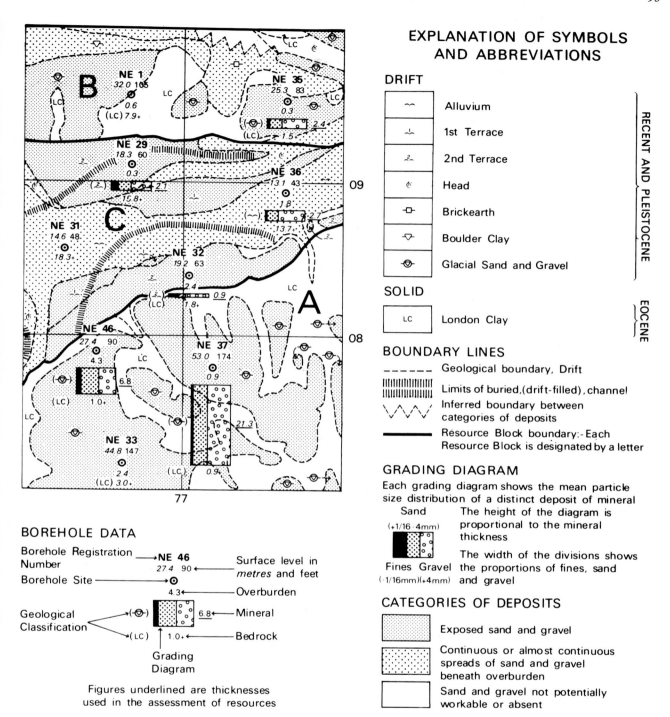

EXPLANATION OF SYMBOLS AND ABBREVIATIONS

DRIFT

~~	Alluvium	
⊥	1st Terrace	
2	2nd Terrace	
ℓ	Head	RECENT AND PLEISTOCENE
⊡	Brickearth	
▽	Boulder Clay	
⊙	Glacial Sand and Gravel	

SOLID

LC	London Clay	EOCENE

BOUNDARY LINES

‑‑‑‑‑‑ Geological boundary, Drift

|||||||||| Limits of buried,(drift‑filled), channel

\/\/\/ Inferred boundary between categories of deposits

▬▬▬ Resource Block boundary:‑ Each Resource Block is designated by a letter

GRADING DIAGRAM

Each grading diagram shows the mean particle size distribution of a distinct deposit of mineral

Sand (+1/16‑4mm) The height of the diagram is proportional to the mineral thickness

The width of the divisions shows the proportions of fines, sand and gravel

Fines (‑1/16mm) Gravel (+4mm)

CATEGORIES OF DEPOSITS

Exposed sand and gravel

Continuous or almost continuous spreads of sand and gravel beneath overburden

Sand and gravel not potentially workable or absent

BOREHOLE DATA

Borehole Registration Number → NE 46

Borehole Site → ⊙

Surface level in metres and feet ← 27.4 90

Overburden ← 4.3

Geological Classification → (‑⊙‑) → (LC)

Mineral ← 6.8

Bedrock ← 1.0+

Grading Diagram

Figures underlined are thicknesses used in the assessment of resources

Fig. 3.33 Example of BGS Resource Assessment Sheet (sand and gravel).

single cement factory producing about 1.65 million t yr^{-1}, approximately equivalent to Danish requirements. Intensive working using special extraction equipment has led to the development of a wet opencast area reaching some 40 m below water level. Before such a development could take place, the Geological Survey investigated the quality of the chalk and the groundwater conditions over a wide area as there was concern that the water supply of the nearby town of Aalborg might be affected.

A special problem related to the quality of the chalk, and a possible obstacle to its exploitation, was the presence of a major system of buried channels (valleys) within the area. The relationship

to the joints of the chalk, and their development by karstic and fluvial processes, were demonstrated. The karstic valleys, sinks, dolines, and cavities may have developed in Eocene times, when tropical climatic conditions prevailed. They were traced some 70 m below the surface of the solid chalk. The cavities and valleys were also eroded by streams, presumably mainly in the Pleistocene. Later they were filled up to the present terrain surface by partly stratified sequences of fine sands, silts, and clays with a varying proportion of chalk ooze. On the basis of the information gained from the geological and hydrogeological investigations, a detailed plan for the mining of White Chalk was developed by the company.

Less common industrial minerals

A major reason for industrial minerals being 'less common' is that their genesis is dependent on unusual and often complex geological circumstances. For example, whilst clay is a common material, some types of clay mineral occur much less widely in economic quantities. The minerals calcium and sodium montmorillonite (Fuller's Earth and bentonite respectively) are thought to have been formed by either the alteration of appropriately constituted volcanic ash or by the direct chemical precipitation of montmorillonite in shallow marine basins. To result in a worthwhile resource this needs to have occurred in an environment virtually free of the sedimentation of other materials. Furthermore, as the montmorillonites are very thixotropic and break down easily to a clay slurry, they are easily dispersed in slumps or washed away. The problem for the geologist, therefore, is finding the location/situation where the montmorillonites may have been deposited in sufficient quantity in a clean quiet environment, and where the deposit has not been removed by subsequent erosion.

The next case history illustrates the complex geological controls affecting the exploration for celestite.

The celestite resources near Bristol, UK [228]

Introduction

Celestite (strontium sulphate) is the main ore of strontium; the other commercial source is strontianite (strontium carbonate). The only celestite deposits of known commercial importance in the UK occur in the area north-east of Bristol. Until 1968 this area had supplied 50–70 per cent of world production, since at least 1875. In the early 1970s the working area was perceived to be under threat from housing development, so the Institute of Geological Sciences (later to become the British Geological Survey) was asked by the Department of the Environment to advise on the origin, mode of occurrence, and distribution of the celestite, to identify an area for potential exploration, and to study appropriate surveying techniques.

The achievement of these objectives depended on establishing the sedimentological and stratigraphical relationships of celestite and determining its connection with surface geochemistry.

Existing data—1:10 560 scale geological maps, chemical analyses of earlier samples, and records of natural sections and boreholes—were evaluated prior to an orientation study to determine, in particular, the value of geochemical sampling techniques. As a result, the field programme included systematic soil sampling and the use of a portable isotope fluorescence analyser (to measure strontium content) to help guide the sampling. Boreholes were sunk to determine the continuity, stratigraphy, and sedimentology of the celestite.

Results

Although celestite may occur at more than one horizon in the Keuper, only that associated with the Severnside Evaporite Bed has been found to be important so far.

The celestite of the Severnside region is thought to be diagenetic after gypsum and/or anhydrite. The primary minerals were formed in a supratidal mudflat environment (sabkha) similar to that found in the Persian Gulf today. The assessment project demonstrated the importance of the proximity of the limestones, and rudaceous and arenaceous rocks of the Carboniferous in providing the right circumstances for the early replacement, during diagenesis, of calcium by strontium in either gypsum or anhydrite.

Three modes of occurrence of the mineral were recognized, each affecting the quality and quantity of mineral present and the manner of working:

Type A: Within the Keuper Marl nodular and disseminated celestite, thin stringers, and veins, forming the Severnside Evaporite bed.

Type B: Nodular and disseminated celestite contemporaneous with the Severnside Evaporite Bed at the unconformity between the Keuper Marl and the underlying Palaeozoic rocks.

Type C: Veins, sheets and 'vuggy' infillings of redistributed celestite in Keuper Marl and in Palaeozoic rocks beneath the present or past cover of Keuper Marl.

The horizontal and vertical relationships of these types is depicted in Fig. 3.34.

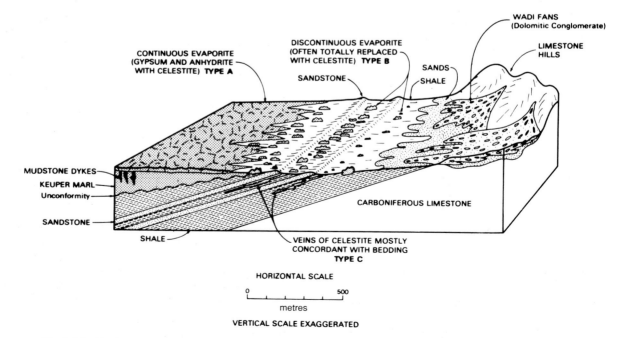

Fig. 3.34   Vertical and horizontal relationships of the three main types of celestite in the Severnside region of the UK.

The resource areas of celestite, in categories defined as indicated, inferred and hypothetical, were presented on a 1:25 000 scale map. The recommended procedures for further elucidating the celestite resources accompanied the report on each area.

The relationship between the strontium content of the flora and the distribution of celestite was investigated. Whilst the method was unable to provide the resolution necessary for the identification of local detail, a geobotanical survey may be useful on a regional scale.

Sometimes a less-common industrial mineral will be discovered during the course of a systematic geological survey: that is, without mounting a specially directed operation to search for that mineral. For example, during a survey of part of the south Midlands of England a deposit of calcium montmorillonite was discovered. Whilst that survey was not intended to produce a commercial assessment, it did define, within reasonable limits, the local lateral extent of the workable horizons. The prospect was then taken up by a commercial company, which proved the deposit's suitability for a range of applications (particularly drilling mud and a binder for foundry sands) and the actual reserves, and exploited the material. As an example of such a discovery, the Couleuvre (France) clay deposit is described below.

Fig. 3.35 Location of Couleuvre clay deposit (France).

Discovery of high quality clay deposits in central France (Couleuvre, Allier) [33] [59] [30]

The Couleuvre clay deposit is locted 35 km north-west of Moulins (Fig. 3.35). It was discovered in the course of the systematic 1:50 000 geological mapping of the French territory. The clay is Oligocene (Tertiary) age.

As defined by systematic auger and core drilling at an average density of 0.8 borehole per hectare over the whole area of the deposit (30 hectares), the clay layer has an average thickness of about 7 m (3–12 m) and is overlain by about 5 m of overburden.

The clay is partly interbedded with limestone. As an average its composition is 40 per cent absorbing clays (sepiolite attapulgite), 40 per cent smectite, and 20 per cent other minerals.

Based on the detailed prospecting work, mineralogical controls, and beneficiation tests, a feasibility study was carried out including a market study which suggested that the clay could be used as cat litter. The study showed that the opencast mining of at least 40 000 t yr^{-1} would be economically viable. At the end of 1990, the new company, Argiles du Bourbonnais, started a pilot test operation.

3.4.3. Exploitation and associated environmental problems

Economics of exploitation

Unless strong political considerations prevail, exploitation of a mineral deposit will only occur when it provides an adequate return on the investment. The most important factors in this context are:

Revenue
(1) the extent of the market into which the mineral is to be sold and the price it will bring;
(2) the required specifications of the mineral products.

Working cost
(3) the extent, geometry, depth, and thickness of the mineral body;
(4) the reserves and grades of the minerals present in the deposit;

(5) processing to achieve product specifications;
(6) the location of the deposit and plant with due regard to:
    (a) freight rates to the market, and
    (b) the environmental constraints placed upon mining and processing of the materials;
(7) the existence of infrastructure close to the mine and/or process site.

Politics
(8) the prevailing taxation and investment climate;
(9) planning conditions.

Of these factors, the first two are of particular importance in the evaluation of industrial mineral deposits. In the absence of a 'London Metal Exchange' or similar open arrangements for the sale of commodities, the same mineral can sell at very different prices from one customer to another depending on individual contracts. In addition, the price of a given mineral product depends upon quality and the amount of processing required. For example, talc is used for cosmetics, paper filler, ceramics, dusting agent, and as a filler in plastics, paints, rubber dusting compounds, and roofing felt, etc., all of which may have different specifications and attract different prices. In contrast, one tonne of a common non-ferrous metal has a more or less agreed world price at any specific time. However, it may be noted that variation of price with time is generally much less with industrial minerals than with metals, and commodity speculation is relatively uncommon.

The market is undoubtedly the most important factor in the evaluation and profitable exploitation of any new non-metallic commodity source, and it is often one of the most complex and indeterminate to evaluate. One has to establish:

1. What proportion of the existing market can be captured and held? What will it be worth, and is it of sufficient size to justify the investment for mining and beneficiation?

2. What are the prospects for opening up new markets for the products either by mineral-for-mineral substitution or by persuading end-users to change their specifications and perhaps technology to accept the new commodity?

In response to the above market questions, there are a number of obvious possibilities which have to be explored, both at customer level (delivered prices, increased technical product merits, improved technical services) and at government level (production and processing in the country, provision of local employment opportunities, and even the concept of self-sufficiency).

The method of mining (Table 3.9) clearly affects the cost of the operation. For example, opencast methods are generally cheaper than underground working. Except in areas of extremely low labour costs, the cheapest operations are those which offer the opportunity of using large-scale rock-moving machines and bulk haulage techniques.

Cost of transport to the market is vitally important and is not necessarily equated with distance. It will completely rule out some deposits, which because of their poor access to cheap transport, cannot compete over long distances. However, this can be mitigated by the provision of a special low-cost bulk transport system, such as a rail connection or a facility for loading large boats at coastal sites.

After location and verification of exploitable mineral resources, mineral development requires capital, labour, and energy to extract the deposit. In addition, a growing awareness of the need for a healthy environment now necessitates inclusion of further burdens into the total costs incurred. The case history 'Phosphate production at Siilinjärvi, Finland' (p. 97) provides an excellent example of the care which can be taken to mitigate the effects of mineral exploitation and processing, and the commensurate costs.

Mining

Mining methods used to exploit industrial minerals are usually significantly different from those used in metal mining. In the case of the more common industrial minerals the relative abundance of deposits and choice of location usually precludes heavy mining costs as there is too much potential competition from local producers. Hence hard rock aggregate producers, cement plants, brickworks, and the like usually exploit conveniently placed resources in large open pits with minimum overburden thickness. However, in recent years, as environmental factors have reduced the availability of working sites, there has been some relaxation of these constraints in areas of high demand where high product prices can be offset against unusually high working costs. Hence there is now considerable interest in long-distance haulage of aggregates by rail or boat and the development of underground limestone mines.

Some of the less common industrial minerals occurring in veins, such as barite or fluorite, may be worked in underground mines by methods similar to those used in metal mining, but more often the geology, mode of occurrence and mechanical properties of industrial mineral deposits are such that each commodity tends to acquire a characteristic mining technology. For example, soluble minerals such as salt, potash, lithium, or sometimes magnesium may be extracted from natural brines, or in the case of salt and potash, from artificial cavities dissolved out of the evaporite bed during controlled solution and pumping operations. China clay workings make great use of hydraulic mining techniques during which the valuable mineral is converted into a slurry and very large piles of waste quartz and mica are accumulated. Building stones are worked by special methods which avoid the use of high explosives that might give rise to defects in the finished article. Sulphur is mined by *in situ* melting.

Some sand and gravel workings below the water table may be worked by dredging, and all marine deposits must be dredged. However, it is much more usual to pump the water out of a gravel pit, even if it is below the water table, and extract the mineral by face shovel.

Where underground mining is resorted to for industrial minerals, great use is made of the room and pillar method (see also 3.5). This is usually because the mineral occurs in horizontal or sub-horizontal beds, often 3 m or more in thickness, and because the value is such that it is economic to leave the pillars in place. Examples include mines for limestone, salt, gypsum, anhydrite, silica sand, and fireclay.

The mining of industrial minerals is an area where large open pits with low overburden, solution mining, the working of sediments by deep underground methods, extraction from sea water and marine dredging comprise some of the more important methods, but many others can be employed depending on the relevant geological and economic conditions. The range of practices is impossible to describe in a work of this type, but some further indications can be gained by reference to Table 3.9.

Beneficiation

As with mining, beneficiation methods applied to industrial minerals are usually significantly different from those encountered in metal mining. In metal extraction it is usually only necessary to reduce an ore to a size range suitable for the application of gravity or flotation separation methods. In contrast, industrial mineral specifications are often very demanding with respect to particle size and shape and hence much effort is expended on the design of optimum crushing, screening, and classification systems. In the case of aggregates, the aim is usually to produce as much material as possible within a narrow size range with a minimum production of fines. In other cases, such as the production of calcium carbonate filters or china clay, very carefully controlled sedimentation procedures are employed to produce closely-sized but very fine products. The production of asbestos fibres requires a special, if not unique, technology.

The removal of deleterious coatings on mineral particles is sometimes necessary in the treatment of glass sands which require leaching in hot acid. Very high-intensity magnetic separation is also used to remove sand particles containing small inclusions of iron-bearing minerals. Both these techniques have no real equivalent in the beneficiation of metal concentrates.

Flotation may be used in industrial mineral treatment, but sulphide recovery is usually of minor importance except where it is necessary to remove small amounts of contaminants such as galena from fluorite. Otherwise, industrial mineral beneficiation tends to make use of non-sulphide flotation. Flotation in saturated solution is used in the recovery of certain evaporite minerals.

Heat treatment is also of great importance with industrial minerals particularly in such areas as the production of cement which may account for a significant part of national industrial energy requirements. Calcination of limestone to produce lime, the making of brick and the conversion of gypsum to plaster are other areas where heat treatment is important.

Industrial mineral operations, even when producing the same commodity, often work slightly different types of deposits at different rates and are likely to serve different product markets. This implies a need to optimize beneficiation methods to a unique set of conditions at each working. Hence, plants working apparently similar material are likely to show considerable differences in processing flowsheet details.

Table 3.9 Industrial minerals: mining methods

Method	Typical application
Dry open pit	
1. With explosives	Most operations in hard rock, particularly in limestone, igneous rock, sandstone.
2. By ripping	Aggregate working in well-jointed rock where explosives are unacceptable.
3. By face shovel or dragline	Unconsolidated sand and gravel, clay, and shale workings for brick making and chalk for agriculture or cement.
4. Multi-bucketed excavators	Large-scale working in clay for brick making or chalk for cement.
Wet open pit	
1. Dragline grab or suction dredge	Unconsolidated sand or gravel deposits (including industrial sands) where pumping to keep the workings dry is impractical.
2. Hydraulic mining	China clay working.
Marine	
1. Sea-going suction dredge	Offshore sand and gravel deposits.
Underground	
1. Pillar and stall working	Most mining in horizontal strata, particularly for gypsum, salt, limestone, fireclay, etc.
2. Stoping	Most mining in steeply dipping deposits, e.g. fluorspar, barite, etc.
Solution	
1. Controlled solution mining and recovery of natural brines	Salt, magnesia, potash.
2. Recovery from sea water	Magnesia, salt, bromine.

A wide range of processing techniques is available to suit individual requirements.

A very relevant description of appropriate beneficiation is included in the next case history.

Phosphate production at Siilinjärvi, Finland [20]

Deposit, mining and beneficiation

The Siilinjärvi apatite deposit in central Finland (Fig. 3.36) is hosted by a carbonatite complex 16 × 1.5 km in area, containing on average 65 per cent phlogopite, 14 per cent calcite, 4 per cent dolomite, 5 per cent amphibole, and 10 per cent apatite. The main orebody covers an area of 4 km^2 with estimated ore reserves of 500 Mt within a depth of 150 m. The average phosphate content is about 4 per cent P_2O_5, making it among the lowest grade phosphate deposits in the world currently in production.

Kemira, a state-owned company, began exploitation of the deposits in 1980 and by 1990 annual output of apatite ore had reached 7 Mt (total mined rock is over 9 Mt). Opencast mining is the method of working at Siilinjärvi and the first pit will ultimately be 1250 m long, 200–600 m wide and 215 m deep; about 100 Mt of apatite ore and 20 Mt waste are to be extracted and some 3.5 million m^3 of overburden removed.

After mining, the broken ore is reduced to <25 mm in a three-stage primary crushing process, then ground and classified. An apatite concentrate containing at least 36 per cent P_2O_5 is recovered by flotation involving four cleaning stages; recovery is 70–80 per cent. As a by-product, calcite is floated from the tailings for agricultural use.

De-watering of the apatite concentrate involves thickening, followed by automatic pressure filtering to reduce the moisture content to 8–9 per cent, after which the concentrate is conveyed to the product store (Fig. 3.37).

The tailings are pumped five kilometres to a storage dam, where the mineral particles settle and the clarified water is recirculated to the concentrator. The tailings basins occupy over 1000 hectares. The apatite concentrate is hauled two kilometres to Kemira's phosphoric acid plant where it is reacted with suphuric acid and ultimately used mainly for fertilizer production. Sulphuric acid is also made at Siilinjärvi, using sulphide concentrate from the Pyhäsalmi mine, some 100 km distant.

Environmental aspects

The environment was taken into careful consideration when planning the Siilinjärvi fertilizer complex. The waste rock is transplanted to storage basins made of removed topsoil and waste rock. Large amounts of gypsum are obtained as a by-product of fertilizer production, but at the moment only a small proportion is utilized as raw material for pigments and plasterboard. The bulk is piled in a waste tip, which, after landscape restoration, would make an excellent ski-slope.

The apatite beneficiation plant uses 1500 m^3 of water per hour. The effluent is channelled to large purification basins and back to the process plant. Some 20 per cent of the water is taken fresh from the nearby lake, and is returned there after careful purification. The process machinery of the whole complex is provided with efficient gas purification equipment and all flue gases released to the air are cleaned. Emissions of sulphur, nitrogen, phosphorous, and fluorine compounds (Fig. 3.38) have decreased through the 1980s: SO_2 by

Fig. 3.36   Location of Siilinjärvi, Finland.

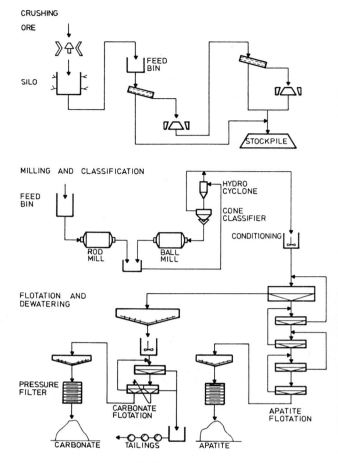

Fig. 3.37   Siilinjärvi processing flowsheet.

90 per cent, NO$_2$ by 58 per cent, P by 69 per cent, N by 80 per cent, and F by 80 per cent. All the values are well below the requirements of the environmental authorities. Even so, Kemira has invested US$20 M in further environmental protection.

Some minerals lend themselves to a multi-product operation. The following example illustrates how a large limestone working can produce several distinct commodities.

Tunstead Quarry, Buxton, UK [204]

Tunstead Quarry, operated by Imperial Chemical Industries (ICI), offers a good example of a large-scale working in Carboniferous Limestone that produces a range of industrial mineral products.

The quarry is located for the most part in a limestone of very high calcium carbonate content in which individual limestone beds are separated by thin (about 0.5 m), horizontal clay layers, or 'wayboards', some 10 m apart. The resulting argillaceous contamination of the coarse limestone is removed by washing.

The operation was originally established to supply stone for calcination at a soda ash plant. This is still the main purpose of the quarry, but part of the washed limestone now goes to a bitumen coating plant for use as roadstone. Lime is also produced on site, in a variety of vertical, rotary, and rotary hearth kilns, for metallurgical and chemical uses. As the processing incorporates a washing stage to remove clay, a significant part of the output consists of a fine slurry of limestone and clay which is utilized by being converted into cement in a small kiln on site.

Environmental effects of exploitation

General

Mineral working is a temporary use of land. Too often in the past the cost of rehabilitation has exceeded the potential value of the site which, in consequence, has become derelict. This is not so much a failure of the economic system as of the regulatory authorities. Mineral development can be detrimental to the environment in a number of ways. For example, most mining operations involve the separation of valuable minerals from the surrounding waste rock, which in turn, requires permanent disposal of the residue. Mine overburden, wastes, tailings, and slimes are often transferred to peripheral areas where they can be unsightly features in the natural landscape.

Environmental degradation or pollution can be viewed as the alteration of the environment by man through the introduction of materials which represent potential or real hazards to human health, disruption to living resources and ecological systems, impairment to structures or amenity, or interference with socially desired uses of the environment. The significance of environmental degradation is reflected in its effects on a range of recipients, including humans as well as the resource and ecological systems upon which we depend.

Environmental degradation caused by mining can be categorized:

1. *Landscape intrusion*: headgear, spoil heaps, tailing ponds, open pits, processing plant.

2. Noise
   (a) blasting in open pits;
   (b) process plant.

3. *Dust* (which can be annoying, expensive, and injurious to health)
   (a) from process plant;
   (b) from open pits;
   (c) from tailing areas, waste dumps, and areas where vegetation has been cleared.

4. *Vibration*: blasting in open pits and sometimes underground.

5. *Chemical pollution*
   (a) water—contamination of local streams by process plant and tailings-pond waters; potential leakage to aquifers.

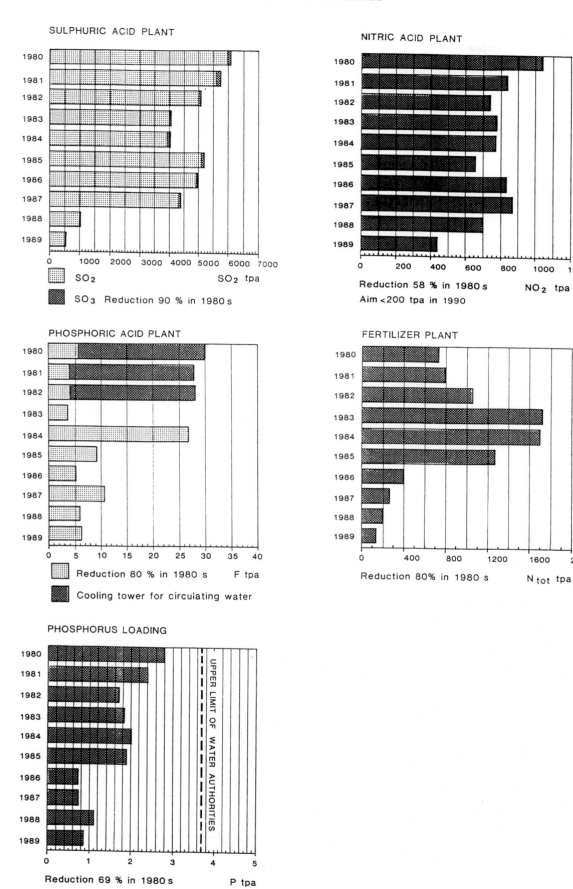

Fig. 3.38 Emissions of sulphur, nitrogen, fluorine, and phosphorus at the Siilinjärvi plant, 1980–89.

(b) air—in addition to the general problem of dust, driers, calciners, bitumen plants, brick kilns, and other processing plant can emit noxious fumes and unpleasant odours.

(c) ground—lubrication oil, fuel oil, processing chemicals and concentrated wastes may all contaminate aquifers, local vegetation and the potential food chain.

6. *Traffic*: Mineral working usually involves extra-heavy traffic on rural road systems, and sometimes through urban areas, which is a secondary source of noise, dust, and vibration.

7. *Subsidence*: All undermined land is suspect to some extent. total subsidence (by long-wall mining or wild-brine pumping in the case of salt) causes permanent alteration to water courses, canals, etc., and damage to property. Shallow underground mining is a permanent blight on future development unless major earthworks are undertaken first. Backfilled opencast sites take a long time to stabilize sufficiently to bear structures. Sides of open pits may be unstable. Old mine shafts are a permanent problem.

8. *Dereliction*: Since mineral workings are temporary they will eventually become idle. Unless remedial action is taken, they will then become derelict, as will the community that serves them.

The environmental problems of bulk mining

The large demand for aggregates, especially gravels, brings with it substantial environmental problems, particularly when those gravels are a source of much of the local water supply. Not only can the exploitation of the mineral pollute the aquifer, but a disused working can be both an eyesore and a hazard to humans and livestock. However, for many years now, companies, mostly as a condition of their land-use planning consent, have restored their old open pits to environmentally acceptable after-use. In some instances this has involved use for recreational purposes, nature reserves, and water parks. Increasingly, as is illustrated in the following case history, this post-exploitation phase is being given as much attention in planning procedures as the extraction phase itself.

Sand and gravel extraction at Renkomaki, Finland [19] [22]

An example of how the environmental concerns can be taken into account properly in sand and gravel extraction occurs at Renkomaki, which is near the town of Lahti. It is one of the biggest gravel extraction sites in Finland (Fig. 3.39) and is well suited to be a test-bed for new approaches to exploitation. Permission has been granted for sand and gravel to be extracted there for 30 years at a rate of 0.5 to 1.0 million $m^3 yr^{-1}$. The Renkomaki Esker, which is fairly large by Finnish standards, will be almost completely exhausted during that time, with extraction extending to below groundwater level in one small area. By concentrating gravel extraction in a coordinated operation in a small area, rather than having a disparate array of small pits over a larger area, it will be possible to save the natural state of the esker landscape at Renkomaki. Although gravel is also being taken from below the groundwater level, the site will continue to serve as a groundwater source in the future. In the course of extraction the area will be landscaped in accordance with carefully made plans and, on completion, the site will be rehabilitated with many facilities (Fig. 3.40).

The Renkomaki case reminds us that the many industrial mineral mining operations, partly because of economics and partly for land-use reasons, are becoming larger and more geographically concentrated. The environmental problems of bulk mining are similar in principle to those of smaller operations but are potentially greater with a more concentrated effect. However, as with smaller mines, which environmental problems occur, and the degree to which they can develop, may depend fundamentally on the geographical location. For example, a large opencast mine situated on a hillside may change the local scenery as well as create noise and dust pollution for the local inhabitants. This would be in contrast to a large open pit excavated deep below the topographic surface, which unless overlooked from a local vantage point might have little scenic impact. Yet the latter, especially if it has steep sides and extends well below groundwater level, would be both a greater threat to the local water quality and a much more difficult site to rehabilitate.

Problems associated with some older mining operations

Many of the existing mining operations commenced well before these more enlightened times when environmental impact assessments are increasingly required for mining and processing proposals. They have brought with them a history that at some time has involved combinations of the foregoing problems. Fortunately, due to a combination of tighter controls by the planning authorities and an increasingly responsible attitude on the part of the minerals industry, many of those inherent problems can be overcome. The next case history is typical of these older mines, but it also serves to illustrate the means by which working practices can be modified even now to bring about a more environmentally acceptable situation.

The asbestos opencast mine, Troodos, Cyprus [8] [10]

The Asbestos Mine of Cyprus is located in the central part of the Troodos Ophiolite Complex, the most impressive topographic feature on the island. The evolution of the ophiolite was associated with important metallogenesis which includes economically valuable asbestos and chromite deposits and abundant cupriferous sulphides. The chrysotile asbestos deposits occur as veins over an area of $23 km^2$ on the east side of the Troodos range. The veins show no regional pattern and vary in width from minute hair-like stringers to veins over 15 mm thick. The Amiandos Mine forms part of a lease owned by Asbestos Mine Ltd.

From 1904–40 the Asbestos Mine operated on a relatively small scale using a large labour force and carrying out selective opencast mining. Mechanization started by the early 1950s, and modern equipment and methods radically changed the scale of operations. The quarry area increased to over 500 000 m^2 with a proportional increase in the area of waste tips.

Environmental impact

The large-scale operations were producing waste of the order of 6 Mm^3 per year and had an immediate and serious impact on the environment. The size of the quarry face increased greatly whilst the traditional method of dumping waste material on the steep slopes below the installations created tips of increasingly doubtful stability, causing much local and national concern. Besides the threat to the Amiantos village just downstream, another serious threat was caused by the fines (including fibres) that were washed from the tips by the rain, creating havoc to the irrigated agricultural land and contributing to the rapid silting up of dams and other water conservation and utilization structures. Initial remedial efforts by Asbestos Mine Ltd only resulted in enhanced problems.

At this point the Geological Survey Department of Cyprus undertook the task of quantifying the potential instability of the tips, in cooperation with the British Geological Survey and the mining

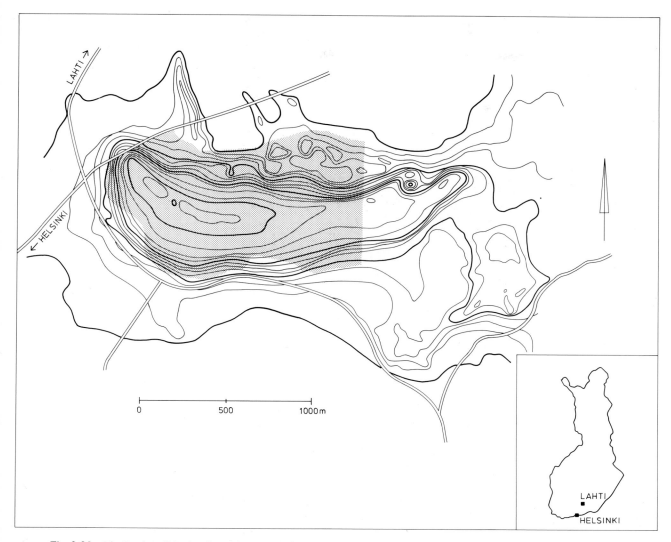

Fig. 3.39   The Renkomäki esker formation, near Lahti, Finland. Contour interval 5 m. The gravel extraction is shown in grey.

company. Detailed surface mapping, sampling, and testing led to preliminary slope stability analyses. Following this study, consultants were hired by Asbestos Mine Ltd to produce detailed design proposals to render the waste tips safe and to provide sufficient room for future wastes. Throughout the study period the consultants worked in close cooperation with the Geological Survey Department, the advice of which was required by government prior to the approval of any proposals for remedial works.

In parallel, the Geological Survey, in cooperation with the Forestry Department, examined the problems of the high pH and low water-retention capacity of the wastes. Experiments to date are encouraging, and involve mixing asbestos tailings with those from local pyrites for pH control, together with various proportions of bentonite for increased water retention.

3.4.4. Management policies

The dominant concern of every commercial company is to satisfy its shareholders by making a profit. Whilst it is by no means inevitable that this concern will run counter to the satisfactory preservation of the environment, or to the wise management of a country's resources, these issues are larger than the interests of any one company. As already indicated, the problems of balancing the often conflicting concerns of agriculture, urbanization, resource exploitation, water quality,

and other aspects of the environment, must eventually be the concern of government at all levels.

These conflicts are difficult enough in any situation; where industrial minerals are involved they can be exceedingly complex.

From the earlier sub-sections of this section it can be seen that such parameters as proximity to market, physical as well as chemical properties, and rarity can be very important. Likewise the environmental effects of their exploitation will vary from place to place. It follows therefore that governments need the best information possible on which to develop their policies for resource management.

The next two case histories illustrate the manner in which Geological Surveys have helped in that endeavour.

Resource assessment surveys in the UK [205] [219]

Introduction

The very rapid increase in demand for industrial minerals in the UK since 1945, and the need to reconcile their production with competing uses of land and with the conservation of the environment, highlighted the need for definite information about the distribution

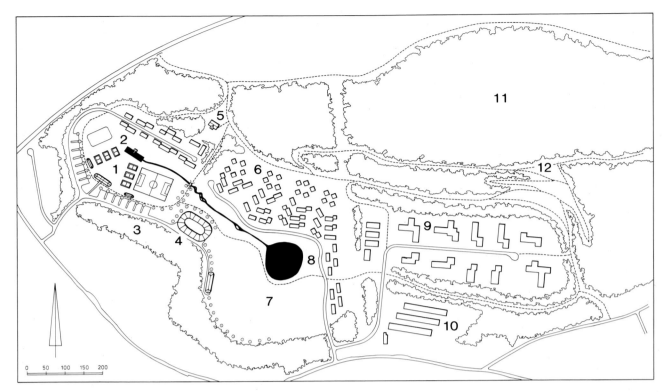

Fig. 3.40 Plan for restoration, including some landscaping, of the gravel extraction area at Renkomäki.

1. Sports ground
2. Outdoor swimming pool
3. Green area
4. Velodrome
5. Skiing cabin
6. Residential area

7. Area for ball games
8. Artificial lake
9. Light industry
10. Industrial area
11. Outdoor recreation area
12. Hiking trail

and main characteristics of the country's mineral resources, particularly the natural aggregates. The UK Government, therefore, initiated a long-term programme of resource assessments, a programme of which the major part was carried out by the British Geological Survey (BGS) for the Department of the Environment.

The primary objective of a resource survey is the establishment of the size and quality of a mineral body from a relatively low frequency of observations. The scale of data acquisition is dictated by the need to achieve good progress and wide coverage at minimum cost. The detail will fall well short of that required for proof of reserves, but the survey will provide more detailed and relevant information than can be derived from existing sources including geological maps, from which the amount and composition of deposits can rarely be inferred.

For some purposes it is also possible to make national or regional resource appraisals based solely on the collation and summation of existing data.

Aggregates surveys

In addition to the sand and gravel resource surveys already described (p. 91), BGS also assessed selected hard rock formations as a source of crushed rock aggregate. The Survey was able to draw on its extensive sub-surface records, and the experience of its field geologists, whose detailed and refined interpretations, embodied in the 1:10 000 scale geological maps, are the essential basis for planning and conducting the assessment work, and for all work on minerals in the United Kingdom.

Despite the lower frequency of sampling and the more generalized and rapid sample appraisal techniques used in a resource survey, the realism and accuracy of the assessments must be

sufficient, of course, to embrace the likely range of variation in the main or index characteristics of the deposits. The availability of commercial data helps to avoid possible duplication of effort and potential waste of funds (for example, from the drilling of boreholes in areas already examined). In terms of the classification of resources, the detail adopted for the BGS resource surveys is at the 'indicated' level.

Other assessment surveys

1. *Limestones*. Limestones occur widely in England and Wales. Their properties differ markedly due to varying amounts of impurities associated with age and location. In volume terms taking limestone, dolomite, and chalk together, the most important usage in the UK is for aggregate (75 per cent), the manufacture of cement (15 per cent), and as a fundamental feedstock for many industrial processes which require raw materials of high chemical grade (7 per cent). Since 1970, the UK Department of the Environment has commissioned the British Geological Survey to carry out resource surveys of major Carboniferous limestone provinces in Derbyshire, North Yorkshire, and Wales, to aid their mineral planning strategies. Specifically, the objectives of these surveys have been the identification of the chemical, physical and mechanical properties of the limestones, their spatial distribution, and their potential suitability for major industrial applications.

2. *Marine aggregates*. UK national guidelines encourage the production of marine dredged aggregate when this does not cause unacceptable damage to sea fisheries nor incur the risk of coastal erosion. Planning authorities, particularly in the south-east of England, are aware of the benefits of marine dredged materials as a means of reducing the demand for land. In 1986, the Crown Estate

and the Department of the Environment commissioned BGS to undertake a programme of resource appraisal for marine aggregates. The first stage of this was a series of desk studies, covering all offshore areas of the UK Continental Shelf, which provided a means of identifying potential resource areas meriting additional surveys to take place in stage two of the programme. It is intended that the results of these surveys will be used by the Crown Estate and the Department of the Environment in forward planning for the provision of aggregates and the assessment of applications for licences to dredge.

Conclusions

The Survey's resource appraisals and assessments are providing new facts, arranged in a clearly understandable form of presentation, against which land-use and mineral planning policies and decisions can be more confidently developed. The results of these outline appraisals augment the guidance available from geological mapping for the preparation of outline mineral plans by local authorities. Figure 3.41 is an index map of the assessment reports published by BGS.

Whilst these reports are available to the industrial minerals industry, they only provide background information on resources. The main concern of the industry must be the detailed evaluation of reserves in areas where they expect to be able to conduct a mining operation.

Inventory of rocky terrains in Finland [19] [22]

Aggregate demand

The total population of the 13 towns, Helsinki included, in the province of Uusimaa is over one million. Since 1645 the aggregate resources of this region have been intensively exploited for building. As about 100 Mm^3 of gravel and sand are used annually in the area, deposits have been exhausted within a 70 km radius of Helsinki. Since 1982, when the Extractable Land Resources Act (ELRA) regulating extraction of soil and rocks came into force, the problems have been further aggravated. According to a study made by the Geological Survey of Finland, gross sand and gravel resources in the province total 1.7 billion m^3. Of the 0.3 billion m^3 of extractable resources only 0.06 billion m^3 are gravel, the material in greatest demand for the building industry. As construction is heavily concentrated in the metropolitan area, the need for sand and gravel is great and transport distances for the material are growing at the rate of 1 km yr^{-1}

Whilst gravel and sand can be extracted from the sea bottom on the coast of Uusimaa province without disfiguring the landscape, the present rate of extraction only matches the demand. Hard rock aggregates, however, have long been used in Uusimaa in places where natural gravel and sand resources are lacking or their use is not permitted.

Apart from replacing gravel, rock aggregates are used for purposes where the durability of the material is crucial, in surfacing roads for example. Studded tyres, which are permitted in Finland in winter, impose special demands on the quality of aggregates in surfacing asphalt. The Precambrian bedrock of Finland is an almost ubiquitous source of rock aggregate. Whilst some lithologies, like schists and gneisses rich in sulphide, micas and possibly graphites, black schists, and the late-Precambrian siltstones and sandstones, have to be rejected for many purposes, the material from the crystalline bedrock is well suited for most applications.

National policy

Finland is a sparsely populated country with a large surface area. To keep transport costs to a reasonable level, a nationwide network of extraction sites for natural aggregate materials is required for the construction and maintenance of roads and railways. Even so, an effort is made to focus the extraction of rock aggregates on fewer

sites, where the exploitation of rocky terrains is determined by the quality of the rock.

To help provide more information on which to base planning decisions, a national programme for appraisal of materials in rocky terrains (1989–98) has commenced. This will assess the resources of high quality rock, thus enabling the authorities to issue permits for extraction, and so ensure that the resources are used sensibly and economically. At the same time the rocky areas will be assessed for their landscape and ecological value. On the basis of the information thus gathered, the future use of the rocky terrains will be planned.

Whilst the earlier case histories have concentrated on local, regional, or national issues, the final case reminds us that the question of resources is international. This brings with it the need for international studies of how to exploit the resources wisely for the good of the community, but without detriment to the local environment at and near the site of the mine and its infrastructure. Associated with the move towards an increased international trade in even the common industrial minerals is the development of the superquarry concept.

Irish coastal quarries for aggregate export [125]

Introduction

The Industrial Revolution passed Ireland by. Consequently, the demand for industrial minerals was, up to the first quarter of this century, domestic and largely for agriculture. The growth in demand for construction materials, therefore, is relatively new and, until recently, the country has been under-explored for industrial minerals. However, as the demand for industrial minerals rises elsewhere in Europe, Ireland is becoming a target for exploration.

The case for the superquarry

The impending problems of aggregate supply were recognized in Britain in the early 1970s. The Verney Committee reported upon the serious shortages likely to affect south-east England and recommended that coastal 'superquarries' be developed to supply that region by sea.

There are cogent environmental reasons in favour of the superquarry. A single huge operation, if environmentally acceptable at the site chosen, is preferable to twenty or more quarries at different locations. The presence of an adequate supply of seaborne aggregate at a competitive price takes pressure off inland resources in sensitive areas. The arguments in favour of superquarries, in terms of benefit to the national economy, hinge on foreign revenue earned from the export of aggregate and a boost for local employment.

The whole concept of the coastal superquarry is based on cheap, high-volume transport, but it also relies on the availability of good unloading facilities in the importing country or region. Such facilities for large bulk cargo carriers are not always close to the areas of highest aggregate demand, and the cost of inland distribution from the unloading points puts a strict limit on the area that can be served.

Ireland's potential for coastal quarry development

There has been considerable interest in the last few years in Ireland's potential as a source of aggregate. The east coast, where quiet waters and proximity to Britain are obvious attractions, has received most attention. A coastal quarry at Arklow Head, south of Dublin, is currently producing dolerite for export to Germany as well as armour stone for coastal defences in Wales. The Tertiary igneous complex at Carlinford, some 80 km north of Dublin, has the advantages of topography (to hide the operation) and proximity to navigable waters; elsewhere on this coast the topography tends to be subdued and the population density relatively high. The south

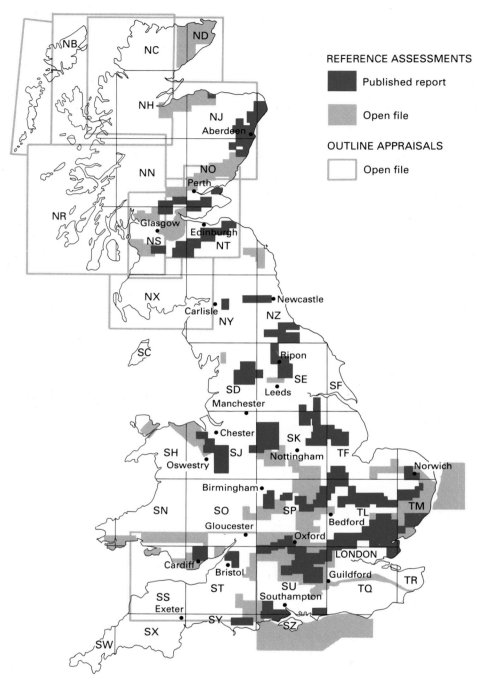

Fig. 3.41   Index map of UK resource-assessment reports by BGS.

coast is considerably more exposed making more difficult the task of matching harbourage and lithology. As the west is approached the scenic potential increases and the problem arises of balancing the benefits of quarrying against the longer-term conservation of unspoilt countryside. However, some areas do not have quite the same scenic rating; one of these is the Shannon Estuary. This natural harbour already receives heavy shipping, delivering coal and bauxite in particular. A serious proposal has been made for an exporting quarry on the Shannon.

Three types of coastal development might be considered:

(1) sites for extraction with only a conveyor-type loading facility;
(2) sites where rock is exported through existing port facilities;

(3) sites where quarrying is integrated with the development of a new harbour facility.

The first is the most probable where a quarry already exists. The size of ship depends almost entirely on the depth of water available and prevailing weather conditions. Export through existing port facilities is impracticable in many cases due to the need to avoid road traffic through urban areas, limited rail facilities, and lack of dockside storage. The recent proposal to develop a coastal quarry on the Shannon was of the third category. The material extracted would have been used initially for the development of a deep-sea port which would have enabled the transhipment of heavy bulk cargoes in ships of up to 200 000 tonnes. Export of aggregates was part of this major scheme. The proposal, however, failed to get planning permission.

Comment

The initial interest in aggregates for export from Ireland has failed to produce a major development. As the pressures for aggregate increase, so will the search for suitable sites. A conflict between economic and environmental considerations must be anticipated. The obligation to contribute to the European resource budget is recognized, as is the attractiveness of increased revenue and employment from quarrying. At the same time, often intangible environmental considerations must be taken into account. These problems are not unique to Ireland and any solution sought should have wider applications. There is a case for integrated studies with the objective of identifying those areas which have potential for coastal quarry development and which, with proper controls, would be environmentally acceptable. In this way the greater interests of Western Europe would be served and the costs and heavy-handed results of confrontation avoided.

3.4.5. Conclusions

Of all the industrial minerals those for use in the construction industry have been the most bouyant over the last few years. Minerals for use in the process industries reflect the general state of the manufacturing industry and the demand for these varies from sector to sector and country to country.

The major assistance that a Geological Survey can provide to the industry, in addition to the availability of good geo-scientific data, maps and advice, is by conducting a 'mineral resources bureau' function. That is to say, by collecting relevant statistics, monitoring the commercial and technical factors affecting the industry, maintaining good contacts with the industry, and being in a position to give impartial advice to government on mineral industry matters.

Indeed, the main role for the Surveys in the field of industrial minerals is in the provision of advice to government to enable it to develop and monitor commercial and environmental policies, and achieve an acceptabe balance between the two.

Looking at the future, it is difficult to forecast details of demand for the less common minerals particularly, for these satisfy the niche markets. However, because they are mostly widely used, and are based on a great range of markets, they will tend to move in harmony with the economy. Notably, the demand for the less common industrial minerals is now coming from an increasing number and diversity of special applications ranging from the 'high-tech' to domestic sectors.

Amongst the common industrial minerals, growth in aggregate consumption is likely to continue over the next decade or so. Whilst estimates of future demand vary, the experience of the last few years suggests that 'official' forecasts can be much lower than those provided by the industry, and much lower than reality. Indeed, until there is a major change in transport patterns—away from roads—the current trends will probably continue. Those trends include an increase in international trade, even in the common industrial minerals like the aggregates, where high-volume cheap transport is available.

The environmental consequences of these longer-term forecasts have not yet been faced up to universally, particularly for the aggregates. For example:

1. How big can inland superquarries be expected to become and how much of the market can they cover?

2. How many sites for coastal superquarries will be tolerated (environment, after-use)?

3. What are the environmental effects of marine dredging and how much material can safely be won from the sea bed?

4. How much environmental degradation may one country suffer for the benefit of European partners, or one part of a country suffer for the benefit of the remainder?

5. How can we improve control of the different impacts on the environment of exploitation and beneficiation of industrial minerals?

Mining is seen by some of the environmental action groups as unnecessarily destructive, but it is clearly unlikely to cease in the next few decades. Rather, the methods of extraction will improve, the technology of processing will become more sophisticated and environmentally cleaner, and the pressure for a high standard of rehabilitation of mining sites will grow worldwide. The end result may leave no hazards, may enhance the landscape and may result in acceptance of the mineral industry by the general public. For that to be achieved, there can be no shirking of responsibility. Commitment and co-operation is required between industry, government, and society, not just nationally but on a wide international scale.

RECOMMENDED FURTHER READING

Advisory Committee on Aggregates (the Verney Committee) (1976). *Aggregates: the way ahead*. (HMSO, London).

Car, D. D. and Herz, N. (1989). *Concise encyclopedia of mineral resources*. (Pergamon, Oxford).

Department of the Environment (1989). Guidelines for aggregate provision in England and Wales. *Mineral Planning Guidance Note MPG 6*. (Department of the Environment, London).

Harben, P. W. and Bates, R. L. (1990). *Industrial minerals, geology and world deposits*. (Metal Bulletin PLC, London).

Harris, P. M., Thurrell, R. G., Healing, R. A., and Archer, A. A. (1974). Aggregates in Britain. *Proceedings of the Royal Society of London*. **339**, 329–53.

Harrison, D. J. and Adlam, K. A. McL. (1985). Limestones of the Peak: a guide to the limestone and dolomite resources of the Peak District of Derbyshire and Staffordshire. *Mineral Assessment Report No. 144* (British Geological Survey, Nottingham).

Industrial Minerals: Published monthly by Metal Bulletin Journal Ltd., London.

Le Fond, S. J. (1985). *Industrial minerals and rocks*, 5th edn. (American Institute of Mining, Metallurgical and Petroleum Engineering, New York).

3.5. METALLIFEROUS MINERALS [194] [15] [65]

3.5.1. Introduction

Europe is a diverse metallogenic province and has been rich in metal mines since prehistoric times. There is evidence for pre-historic metal mining in Turkey and southern Europe. During the Graeco-Roman period the copper, lead, gold, silver, and iron deposits of the Mediterranean area (Cyprus, Greece, Italy, Spain, and later France) were exploited, while the Roman Empire extended organized and technically skilled mining into north-west Europe, including Britain.

The next phase of real progress comes with the renaissance of the fourteenth to sixteenth centuries when central Europe, especially Saxony and Bohemia, became the centre of mining knowledge that was then disseminated throughout northern Europe; the sixteenth century saw Sweden rise to prominence, especially as a producer of iron and copper.

Industrial-scale iron casting was mastered in Europe in the same century, and the beginning of the Industrial Revolution as it is generally understood, can be said to be marked by the first use of coke to produce iron in Britain at the beginning of the eighteenth century. The subsequent explosion of manufacturing technology was greatly assisted by the occurrence of iron ore, coal, limestone, and fireclay in close association in the Carboniferous rocks of Britain, France, Belgium, and Germany.

As the non-ferrous metals, copper, tin, lead, and zinc, also began to be used in heavy machinery, a symbiotic relationship developed whereby the manufacture of machinery demanded a supply of metals that could only be satisfied by using steam-powered machinery to pump water from the mines as the metal ores were followed deeper into the Earth. Until the nineteenth century only the 'old' metals—iron, copper, tin, lead, and zinc—had been deliberately produced and used in metallic form (mercury and the precious metals had also of course been known since antiquity).

The scientific discoveries of the eighteenth and nineteenth centuries gave industry many more metals, notably the light metals (aluminium, magnesium, titanium) and the large group of steel-alloying and refractory metals such as nickel, cobalt, chromium, manganese, vanadium, tungsten, and molybdenum and the minor metals now vital in the electronics industry. These new metals made possible the modern aerospace and electronics industries and the multitude of other technical advances of the twentieth century. At the leading edge of materials technology metals are now being replaced by non-metallic, often composite, materials, but metals will remain a vital component of civilization for the foreseeable future.

Whilst the industrial demands of the continent have outstripped the possible level of indigenous production, metal mining in Europe will continue to make a significant contribution to supply. Indeed, in spite of the present heavy dependence of many European countries on imported metals and metal ores, the sub-continent as a whole still remains a locus of a vigorous metal mining industry (Fig. 3.42). Europe's resources of metals in the geological foundations of the continent have not been exhausted and significant deposits are still being found.

3.5.2. Main types of deposits [194] [15] [65] [49]

As introduced in Chapter 2 a variety of mineral deposit types exist, related to specific types of rocks and two main groups of processes, endogenic ('at depth') and exogenic ('at the surface').

With the exception of iron and the light metals aluminium, magnesium, and titanium, all the metals used by industry are relatively rare in nature, none of them comprising more than 0.1 per cent of the Earth's crust, by weight. To constitute economically viable orebodies they will have been concentrated by natural processes by between 100 and 10 000 times their average abundance in the Earth's crust. This fact partly explains the relative scarcity of viable metal deposits, compared with the non-metallic or 'industrial' minerals, the constituents of which are, in general, the more abundant elements of the lithosphere.

There is, however, no simple relationship between the crustal abundance of a particular metal and the number of viable deposits: vanadium, for example, is ten times as abundant as lead, but while Western Europe has no vanadium mines, it has several major lead mines and is more than 20 per cent self-sufficient in the mined production of lead.

The occurrence of metals in Western Europe is the consequence of the complex tectonic and metallogenic history of the area. The relative scarcity, or absence, of viable deposits of certain metals is related to (1) the relatively small area of exposed pre-Cambrian rocks (older than 570 million years) when compared to the other continents, (2) in particular to the absence of large areas of pre-Cambrian sediments of the type that, elsewhere, host giant deposits of iron, manganese, copper, and gold, and (3) the absence of metal-rich layered basic igneous rocks such as those that occur in southern Africa and Canada and which host huge deposits of chromium, vanadium, nickel, platinum, and other metals.

It must be emphasized that a mineral deposit is not only a geological object but also an economic concept. The evolution of industrial needs and of market prices, together with long-term policies of the mining companies and of governments leads to a flexible approach to deposit types, grades, tonnages, and priority investments in mineral exploration or beneficiation. This explains why so many of the European metal-bearing areas have been re-worked several times over the centuries, in some cases since prehistoric times. Apart from economic considerations, such reworking is based on new geological concepts, new exploration tools, and new beneficiation methods.

Europe's production of iron, once in a leading world position, has been largely replaced by high-grade ore imports. However, high-grade magnetic iron ores are still mined in Sweden, and to a smaller extent in Norway, Finland and Turkey. The French low-grade sedimentary 'minette' ore, once producing 50 million t yr^{-1} but now one-fifth of that tonnage, is still in operation. Also working is a low-grade carbonate ore mine in Austria.

The production of zinc, lead, copper, and associated minor metals (silver, gold, germanium) is still significant from two main types of deposits:

1. Zinc and lead disseminations, lenses, and lodes in sedimentary (mainly carbonate) environments in UK, France, Italy, Austria, Germany, Spain, Ireland, Turkey; silver, and to a lesser extent germanium and gallium, are frequent by-products;

2. Copper, lead, zinc massive sulphide deposits in volcanogenic environments: Huelva-Alentejo extensive belt through southern Spain and Portugal (Neves Corvo); Precambrian Skellefte area of Sweden, Precambrian belts of Finland, Hercynian areas of central France (Chessy) and Brittany, Alpine belts of Turkey and Cyprus.

Gold is frequently associated, as a by-product, with massive sulphide deposits. Pyrite is usually the main mineral component and is either a problematical waste product or a raw material for sulphuric acid manufacture.

Europe has only one 'porphyry copper' type mine, Aitik in Sweden. Deposits of this type usually have copper grades of less than one per cent, and are generally mined opencast with a large daily tonnage of ore raised and processed. Some potential for copper–molybdenum (tungsten) porphyries also exists in other Hercynian and Caledonian areas, but it is undeveloped at present.

Tin and tungsten mining has a very long history in Europe. The ores are oxides, not sulphides as is usually the case of lead, zinc, and copper. Orebodies are mostly lode type or lenses within granite rocks of Hercynian age, or in the surrounding country rock. An exceptional case is Neves Corvo which is a massive copper–zinc sulphide deposit rich in both oxide and sulphide tin minerals. Mining of tin and tungsten has occurred in all Hercynian areas of Europe and in particular the Iberian Peninsula, France, UK (Cornwall), and the Bohemian Massif.

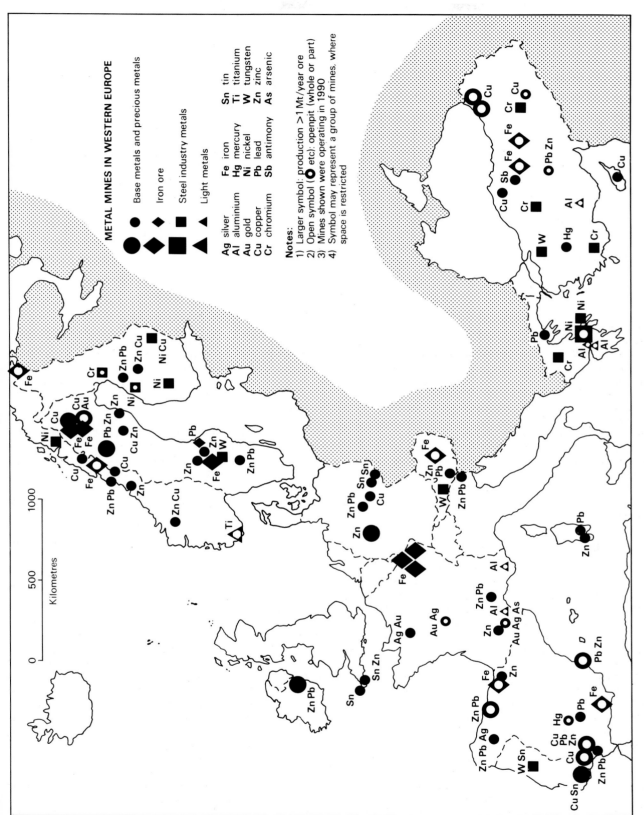

Fig. 3.42 Location of the metal mines of Western Europe in 1990.

Table 3.10 Western European mine production (1989) and reserves of metal (kt = thousand tonnes, t = tonnes). *Sources*: Production: *World Mineral Statistics, 1985–89*, British Geological Survey; Reserves: United States Bureau of Mines (figures are minima, due to incompleteness of data)

	Production		Reserves	
	Amount	Percentage of world	Amount	Percentage of world
Aluminium ore (kt gross)	3 855	3.6	600 000	2.8
Chromium ore (kt gross)	1 480	11.7	21 000	2.0
Iron ore (kt gross)	45 079	4.6	5 200 000	3.5
Titanium ore (kt gross)	900	10.3	32 000[1]	11.2
Antimony (kt metal)	2	2.9	n.a.	n.a.
Copper (kt metal)	274	3.0	n.a.	na.
Lead (kt metal)	263	7.7	6 000	6.3
Nickel (kt metal)	28	3.1	588	1.1
Tin (kt metal)	6	2.7	160	3.7
Tungsten (kt metal)	3	6.8	61	2.4
Zinc (kt metal)	922	12.6	11 000	6.5
Gold (t metal)	25	1.2	n.a.	n.a.
Mercury (t metal)	1 373	25.0	79 000	60.8
Silver (t metal)	1 035	6.9	n.a.	n.a.

[1] TiO_2 content of ilmenite and slag.

Mining of tin has now significantly decreased but still exists in UK, Portugal, Spain, south-eastern Germany, and France (by-product of kaolin production).

For tungsten, there only remain government-assisted operations (Portugal) and/or operations integrated with downstream metal or carbide production (Sweden, Austria). This situation is mainly due to competition from China. The genesis of deposits is similar to that of tin.

Small antimony operations also exist in Hercynian lode-type concentrations (France, Austria) and strata-bound deposits (Turkey).

When the reference price for gold, of 35 US$ per ounce, was abandoned in the 1970s, it created a great incentive for mineral exploration leading to a significant number of small to medium size discoveries. These are mostly lodes, veins, or elongated structures (shear zones) in the old basement areas of UK, France, Spain, Portugal, and the Scandinavian countries. The mine with the highest individual production in Europe is Salsigne in southern France, a sulphide deposit producing more than two tonnes of gold per year, and also the world's leading producer of arsenic. Most of the Spanish production of gold (average: 10 t yr^{-1}) is won as a by-product of massive sulphide mining.

Europe accounts for more than 30 per cent of the world mercury production, mainly from one operation, Almaden, in southern Spain. Some production also comes from Finland (by-product of sulphide mining) and Turkey.

In the category of deposits associated with basic and ultrabasic rocks:

(1) nickel is mined in Finland from sulphide ores and in Greece from laterites. In the latter case, the final product is ferro-nickel. Recently (1989/90) mining of sulphide nickel ore started in Råna in Norway, based largely on supportive data from the Geological Survey (NGU);

(2) chromium is widespread in the Mediterranean basin and closely associated with the ophiolitic rocks emplaced during the Alpine orogeny in Turkey, Greece, Yugoslavia, and Albania. The only European chromium mine associated with Precambrian rocks is the Kemi deposit in Finland;

(3) titanium, in the form of the mineral ilmenite, is mined in Norway;

(4) platinum group elements have been found in UK, Norway, Finland, and Greenland but have not yet proved economically viable.

A significant production of aluminium ore (bauxite) comes from the Mediterranean basin. Most of the deposits are Cretaceous in age and correspond to fossilized laterites, similar to those found at present in tropical regions of the world. The main production comes from Greece with lesser amounts from France, Turkey, Italy, and Spain. Turkey also has deposits of Palaeozoic age which are not mined at present.

Magnesium is produced from dolomite in Norway, France, and Italy. The electrolytic process used in Norway also treats magnesium chloride brines from Germany.

There is no production of rare earths (REE) in Europe, even from the Finnish Siilinjärvi mine (see also 3.4) which produces mainly phosphate (apatite) and various by-products (including carbonate and mica). However, concentrations exist in Palaeozoic rocks of central and southern Europe and also in carbonatites in Greenland and Scandinavia.

Table 3.10 gives an overview of Western Europe mine production and reserves, on a 1989 basis. Table 3.11 provides detailed mine production by country. Table 3.10 shows clearly that for most metals Europe's share is below 10 per cent, except for chromium, mercury, titanium, and zinc. This straightforward statement should not hide the fact that

Table 3.11 Western Europe mine production (1989) by major metal and country (sources: British Geological Survey; *World Mineral Statistics 1985–89*)

	Aluminium		Antimony		Chromium		Copper		Gold		Iron		Lead		Mercury		Nickel		Silver		Tin		Titanium		Tungsten		Zinc	
	(1)	(2)	(3)	(2)	(1)	(2)	(3)	(2)	(4)	(2)	(1)	(2)	(3)	(2)	(4)	(2)	(3)	(2)	(4)	(2)	(3)	(2)	(1)	(2)	(3)	(2)	(3)	(2)
Austria	—	—	0.4	0.6	—	—	—	—	—	—	2 410	0.2	2.3	0.1	—	—	—	—	—	—	—	—	—	—	1.2	2.5	16	0.2
Finland	—	—	—	—	574	4.5	24	0.3	1.8	0.1	35	0	2.6	0.1	202	3.7	9.2	1.0	31	0.2	—	—	—	—	—	—	58	0.8
France	720	0.7	—	—	—	—	1	0	3.5	0.2	9 370	0.9	1.1	0	—	—	—	—	22	0.1	—	—	—	—	—	—	25	0.3
Germany (West)	—	—	—	—	—	—	—	—	0	0	105	0	9.3	0.3	—	—	—	—	—	—	—	—	—	—	—	—	64	0.9
Greece	2 576	2.4	—	—	56	0.4	0	0	0.1	0	—	—	25.1	0.7	—	—	16.1	1.8	9*	0.1	—	—	—	—	—	—	25	0.3
Greenland	—	—	—	—	—	—	—	—	—	—	—	—	24.0	0.7	—	—	—	—	61	0.4	—	—	—	—	—	—	72	1.0
Ireland	—	—	—	—	—	—	—	—	—	—	—	—	32.1	0.9	—	—	—	—	15	0.1	—	—	—	—	—	—	169	2.3
Italy	12	0	—	—	—	—	—	—	—	—	—	—	14.6	0.4	—	—	—	—	7	0	—	—	—	—	—	—	44	0.6
Norway	—	—	—	—	—	—	16	0.2	0.3	0	2 358	0.2	3.2	0.1	—	—	1.3	0.1	91	0.6	—	—	929.8	10	—	—	15	0.2
Portugal	—	—	—	—	—	—	98	1.1	—	—	13	0	0.1	0	—	—	—	—	1	0	0.1	0	0.1	0	1.4	2.9	—	—
Spain	3*	0	—	—	—	—	14	0.2	6.7	0.3	4 610	0.5	74.7	2.2	967	17.6	—	—	530*	3.6	0.1	0	—	—	0.1	0.2	282	3.9
Sweden	—	—	—	—	—	—	71	0.8	5.1	0.3	21 406	2.2	89.0	2.6	—	—	—	—	228	1.5	—	—	—	—	0.1	0.2	174	2.4
Turkey	544	0.5	1.5*	2.3	850*	6.7	45	0.5	—	—	4 043	0.4	14.0	0.4	204	3.7	—	—	22	0.1	—	—	—	—	0.4*	0.8	37	0.5
United Kingdom	—	—	—	—	—	—	1	0	0	0	32	0	2.2	0.1	—	—	—	—	2	0	3.8	1.7	—	—	0	0	6	0.1
Total	3 855	3.6	1.9	2.9	1 480	11.5	270	3.1	17.5	0.9	44 382	4.4	294.3	8.6	1 373	25.0	26.6	2.9	1 019	6.7	4.0	1.7	929.9	10	3.2	6.6	987	13.5

(1) = 1000 t ore; (2) = % of world production; (3) = 1000 t metal; (4) = 1000 kg metal; 0 = less than 0.05 of unit shown; * estimate

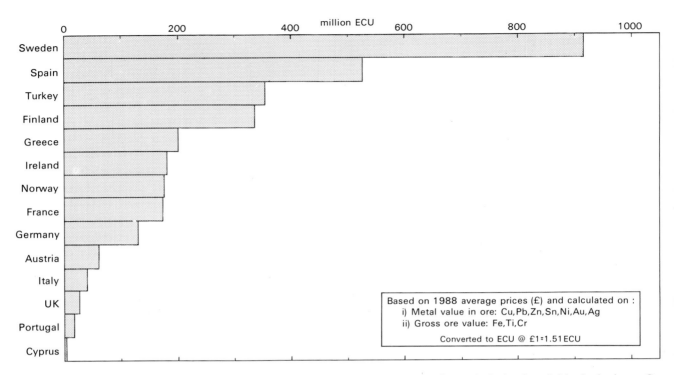

Fig. 3.43 Value of metals mined in Western European countries. Based on 1988 average prices and calculated on: (i) Metal value in ore: Cu, Pb, Zn, Sn, Ni, Au, Ag; (ii) Gross ore value: Fe, Ti, Cr.

Europe keeps a share in the production of ten other metals, thus maintaining an industrial capacity and expertise in the corresponding fields.

It is important to note that 'reserves' is a dynamic concept: reserves tonnages tend to grow with time rather than diminish, due to exploration and development efforts. The ratio of production to reserves has little meaning except as an indicator of the ease of assessment and prediction of ore-bodies: extensive stratiform ores such as bauxite (aluminium ore) and iron ore, will always tend to show high tonnages of reserves, relative to production figures; in contrast base metals tend to show low tonnages due to the high cost of proving reserve tonnages in deep and discontinuous ore bodies.

The histogram (Fig. 3.43) shows the relative value of total metals mined in a number of countries in Western Europe. It is clear that there is a relationship between the land area of the country and its primary metals production, the three countries with the greatest production-by-value being also those that are among the largest in area. The correlation is by no means perfect due to differing histories of industrial development in different countries.

In spite of European production, world supply of many metals is dominated by a relatively small number of foreign countries. Western Europe now has a very heavy dependence on those few countries for primary supplies of many metals. The reason for this heavy dependence on overseas sources is not simply the exhaustion or common uncompetitiveness of European ores but also the enormous increase in demand for metals that has occurred in the past one hundred years; even at the peak production levels of the past, European copper mines, for example, could not have supplied more than a fraction of the present European demand. The histogram (Fig. 3.44) illustrates the extent of European self-sufficiency (or dependence), including the important contribution from post-consumer scrap, for six major metals.

Nevertheless the fact that Parys Mountain (UK) and Chessy (France) may be revived , and that the great mines of Navan (Ireland), Rubiales (Spain), and Neves Corvo (Portugal) have become successful producers shows that metal mining in Europe is not finished. New mines will continue to be discovered: many will probably be 'blind' deposits presently concealed beneath a cover of younger rocks; others will be discoveries of types of mineralization not expected or not hitherto recognized. Their discovery will be brought about by advances in geophysical and geochemical techniques and by refinement of geological theory.

To illustrate these comments short case histories describing Parys Mountain (UK), Chessy (France), and Neves Corvo (Portugal) operations are given below.

Parys Mountain (UK) [194]

The BGS Minerals Reconnaissance Programme has been concerned with geochemical, geophysical, and geological surveys in Anglesey, North Wales. Here the Parys Mountain mine (Fig. 3.45) was a major producer of copper in the late eighteenth and early nineteenth centuries. Following initial interest by companies in the 1960s the Institute of Geological Sciences (now the British Geological Survey) undertook a desk study of the potential reserves of Parys Mountain. Encouraged by high metal prices in the early 1970s, four major companies mounted consecutive exploration campaigns at Parys Mountain, based on diamond drilling and re-interpretation of the geological structure. Part of this work was financially assisted by grants from the Department of Trade and Industry (DTI). A condition of this assistance was that all relevant data were deposited in the archives of the Institute of Geological Sciences, which acted as the Department's inspector on such grant-assisted exploration, and which undertook the curation of borehole cores recovered in the course of the work.

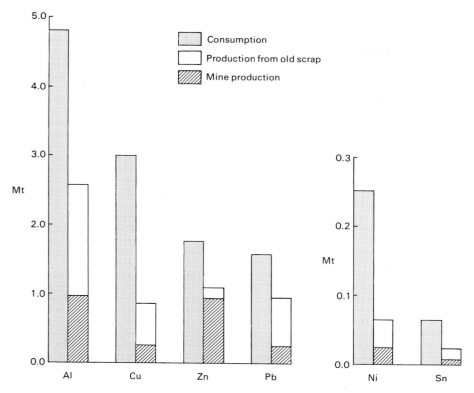

Fig. 3.44   Western European metals budget 1989.

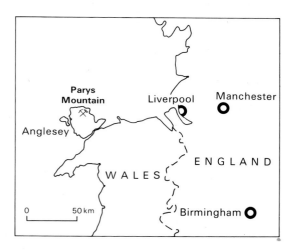

Fig. 3.45   Location of Parys Mountain.

Parys Mountain ore is in the form of submarine volcanic-exhalative sulphide mineralization associated with Ordovician acid to intermediate volcanics, now folded into an eastward-plunging syncline (Fig. 3.46). Minor epigenetic mineralization also occurs, associated with faults and stockworks.

The companies involved abandoned their work at Parys Mountain in favour of other prospects, but recently Anglesey Mining plc took a decision to sink a shaft and develop a mine on the site. In January 1991, mineable reserves were reported to be 4.8 million tonnes, grading 1.49 per cent Cu, 3.03 per cent Pb, 6.04 per cent Zn plus 69 g t^{-1} Ag and 0.4 g t^{-1} Au. A feasibility study envisages milling at a rate of 250 000 t yr^{-1} to produce concentrates containing

13 000 tonnes of zinc, 5700 tonnes of lead, and 2500 tonnes of copper. It has been suggested further that gold could be recovered to the extent of 3000 ounces per year in a lead–gold concentrate. This will be the largest base metal mine ever to operate in the UK.

Chessy (France) [34]

The Chessy deposit is located on the eastern side of the Massif Central of France, 25 km north-west of Lyon (Fig. 3.47). The Chessy mines were possibly worked during Celtic and Roman times but the oldest official record is the fifteenth century. Since that time, production was discontinuous from various occurrences and focused on rich copper ores. The total amount of ore produced was less than 10 000 tonnes. At the end of the nineteenth century, the Chessy mines were estimated to be exhausted and Saint-Gobain operations were concentrated on the nearby Sain-Bel deposit which produced about 40 million tonnes of pyrite until closure in 1972.

During the twentieth century, various phases of prospecting (geophysics in the 1920s, drilling in the 1950s) did not lead to any discovery. Renewed prospecting by BRGM started in the 1960s but was intensified only in the 1980s. The work included:

(1) regional geology, ore mineralogy, geochemical prospecting: 1960–1970s;

(2) helicopter electromagnetic survey: 1978;

(3) ground geophysics (gravity, magnetic, electromagnetic surveys) and 15 drill holes (2415 m): 1980–82;

(4) underground mining development consisting mainly of one main decline, two galleries, and underground drilling: 1983–87.

From top to bottom, the geological sequence includes (Fig. 3.47):

(1) basic volcanic unit IV;

(2) upper acid volcanic and volcano-clastic unit III;

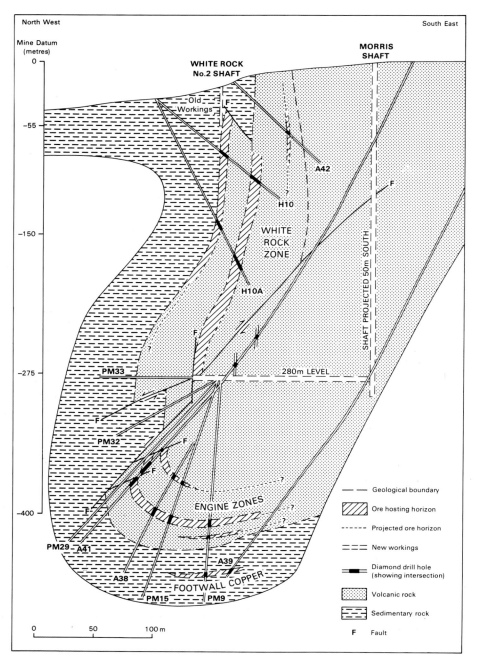

Fig. 3.46 Section across the current development of Parys Mountain.

(3) main mineralized zone II;

(4) lower acid volcanic unit I.

Some mineralization is found in the lower acid volcanic unit. The main ore zone has an average thickness of 14 m and has been traced over more than 700 m along the strike.

Pyrite is the dominant constituent of the ore. The main economic sulphide minerals are chalcopyrite (copper) and sphalerite (zinc). Baryte is also an important component. Reserves have been estimated at 5 million tonnes at 2.5 per cent Cu, 10 per cent Zn and 15 per cent Ba.

Beneficiation studies showed good recovery and adequate separation of four concentrates: copper, zinc, baryte, and pyrite. Planned production level is 1000 tons of ore per day, due to start in 1993.

Neves Corvo (Portugal) [150]

Of recent mineral discoveries, Neves Corvo (Fig. 3.48) is the most outstanding because of the richness of the ores of copper and tin and because the ore bodies are buried by over 250 metres of barren rocks. This has been considered the most important ore discovery in Europe since the Second World War. The Neves Corvo deposits were explored and developed by a Portuguese–French association (Sociedade Mineira de Santiago, Sociedade Mineira e Metallurgique Penarroya Portuguesa, and BRGM). It is important to note that the Neves Corvo target areas were chosen on the basis of studies and the earlier discovery of a 0.5 mgal Bouger anomaly by the Serviço de Fomento Mineiro of Direcçao General de Minas e Serviços Geológicos—the Portuguese Geological and Mining Service.

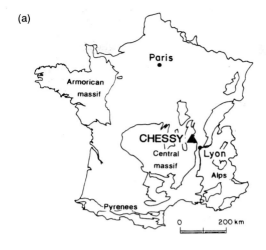

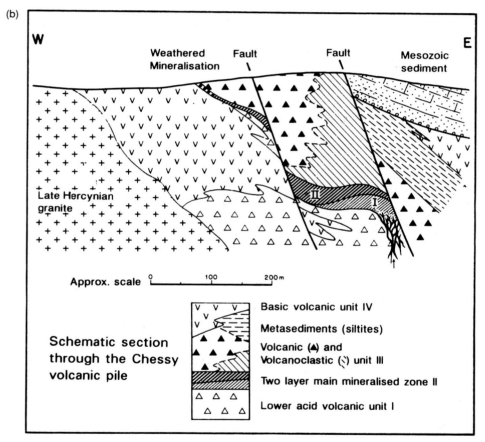

Fig. 3.47 (a) Location of Chessy; (b) section through Chessy. (Adapted from *Chronique de la Recherche Minière* (France) No. 501, Dec. 1990, p. 5).

After allocation of the exploration rights to the Portuguese–French consortium in 1972, early work was concentrated on geophysical and geological data. Re-interpretation of the previous gravimetric surveys led to the choice of Neves as the best potential anomaly.

The first borehole drilled in early 1973 reached a depth of 244 m and was negative. It did not explain the gravimetric anomaly. Re-interpretation of the geological data suggested a significant thickening of the normal geological succession as a result of overthrusting.

Further groundwork between 1973 and 1975 led to the selection of 40 anomalies of which 15 were chosen for drilling, but Neves remained the priority target. Drilling in May 1977 intersected

massive sulphides between 350 and 403 m. Between 1977 and 1978 three other major deposits were found in the vicinity: Corvo, Zambujal, and Graça. A geological cross-section of the Neves Corvo area is given in Figure 3.49.

This major discovery led to the creation of SOMINCOR, a direct continuation of the Portuguese–French exploration association, though for various non-technical reasons, the French withdrew from the operation.

The four deposits of the Neves Corvo area contain more than 150 million tonnes of polymetallic massive sulphides of which 32 million tonnes average 8 per cent Cu; 32 million tonnes average 5.7 per cent Zn, 1.13 per cent Pb, and 40 g t⁻¹ Ag; 2.9 million tonnes average

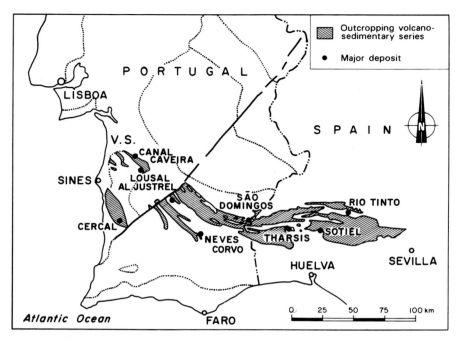

Fig. 3.48   General geology of the Iberian Pyrite Belt showing location of Neves Corvo.

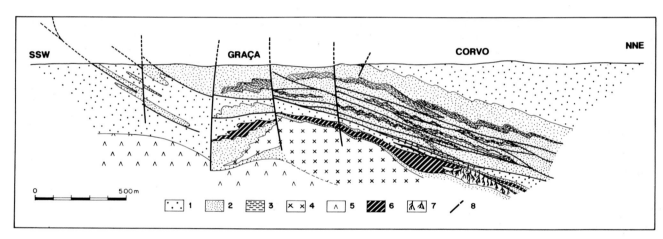

Fig. 3.49   A geological cross-section of the Neves Corvo area.

2.37 per cent Sn, 13.39 per cent Cu, and 1.3 per cent Zn. This is one of the richest copper deposits in the world (Fig. 3.50), and the same can be said for the tin. Before the Neves Corvo discovery, copper ores of 1.5 per cent were considered 'high-grade' in the Iberian Pyrite Belt (IPB). Neves Corvo is the largest copper and tin producer in Western Europe (120 000 t yr^{-1} of Cu and 5000 t yr^{-1} Sn). Total value of economically recovered metals in this deposit exceeds 5.6 billion ECU (January 1991).

The existence of anomalous Cu-rich ore bodies in the Iberian Pyrite Belt (only proved after 3000 years of exploration and mining activity in one of the best known provinces of the world!) revealed by the Neves Corvo discovery demonstrates the feasibility of exploring targets at a depth greater than 300 metres, previously considered to be economically prohibitive. This discovery, along with recent detailed data produced by the Geological Survey of Portugal regarding stratigraphy, structure, and tectonics of the IPB, has not only affected exploration philosophy, it has enlarged the area favourable for exploration in the Portuguese part of the IPB from little more than

3000 km^2 to more than 8000 km^2. This experience may have application in other regions especially in similar metallogenic provinces.

3.5.3. Extraction and processing of ores [32] [27] [65]

Introduction

The decision to exploit a mineral concentration, an orebody, is dependent upon the evaluation of a large number of technical, economic, and political factors. A decision on the style of extraction is obviously an essential part of a mining project. The primary factors influencing the mining methods to be employed are the shape of the deposit, its size and the ground conditions in the vicinity of the future workings. In many cases, especially in the developed countries, environmental considerations also weigh heavily and may be the deciding factor, in particular for opencast or shallow workings.

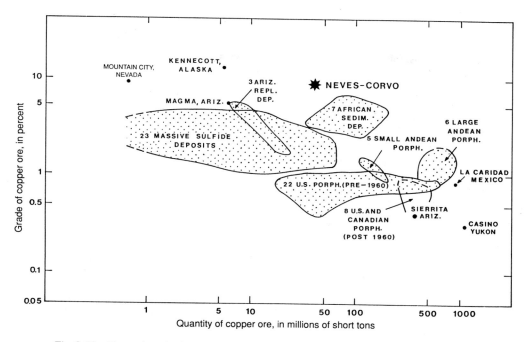

Fig. 3.50 Size and grade characteristics of selected copper deposits throughout the World.

In most cases, waste material forms the greater bulk of the ore. This may attain 80–90 per cent in polymetallic base metal ores (Cu–Pb–Zn), and in the case of precious metals, such as gold, it is even higher, since gold ore values are commonly 5–10 g t^{-1} (1 tonne = 1 000 000 g). This means that it is usually economically out of the question to transport a raw ore as it comes out of the mine. It has to be concentrated on the spot in order to obtain a 'concentrate' from which most of the waste has been eliminated. At this point two other problems come into play: one is the separation of a single ore into concentrates of the different contained metals (copper, lead, and zinc, for example) and the other is what is to be done with the tailings from the processing, which commonly contain significant amounts of metal sulphides or pollutants either from the ore itself or that were introduced during processing.

The last stage is that of metallurgy. This most often takes place on a different site from that of the first two stages, as it is performed upon a concentrated product whose high value permits much greater flexibility than for most ores with regard to long-distance transport. Thus, all combinations of extraction/processing and metallurgy can be found, from metallurgy close to the mining area to metallurgy in or near the areas of consumption in the developed countries operating essentially on imported concentrates.

The main type of mine

There are two main methods of mining, opencast and underground. The former is used for extracting ore that outcrops at the surface or is covered by a relatively small thickness of overburden, but open pits may nevertheless reach several hundred metres in depth. The characteristic feature of this method of mining is that both the ore and any intercalated or overlying barren rock are extracted in bulk.

In underground mining, the quantities to be extracted are smaller, the costs higher, and it is necessary to be much more selective in order to extract as little waste as possible with the ore. Various mining methods are available, suitable for optimal extraction of different shapes, sizes, and three-dimensional attitudes of orebody. These are appropriate to a certain extent to the intended rate of extraction and take into account the existing ground conditions.

A certain number of problems, at different scales, are common to both major types of extraction:

1. It is generally necessary to pump out groundwater present in the rocks hosting the orebody, which would otherwise flood the mine. The amount that has to be pumped may range from a few cubic metres to a few thousand cubic metres per hour.

2. Waste and ore have to be transported between the point of extraction and the point of disposal or concentration.

3. The waste may be disposed of, commonly on the surface. This may involve the preparation of a disposal site. In underground mines, some part of the waste is commonly used as backfill in the mine.

These factors are not confined to metalliferous mining, but may also affect the extraction of industrial minerals and solid fuels (coal, lignite, peat, uranium minerals).

Mineral deposits can also be exploited by dredging beneath the surface of a river or the sea. This is used most often for mineral-bearing sands and gravels, and the dredge may use a bucket chain or, less commonly, a suction tube. In the case of metalliferous deposits, grains of ore minerals are mixed with barren sand and gravel in accumulations found in rivers and in beach sands or other coastal deposits. The economic minerals in this type of deposit are varied, and are mainly oxides, silicates, and phosphates of a variety of elements—titanium, thorium, tin, zirconium—and native gold or diamonds. They are generally heavier than the barren material and therefore separated from it by gravity methods.

Opencast mines, usually referred to by the metal-mining industry as open-pit mines, may be of any size. Extensive pits, covering large areas, are typical of solid fuel minerals, e.g. lignite in Germany, but may also be developed for shallow stratiform metal ores such as nickel laterite, bauxite and certain iron ores. In the developed countries the constraints

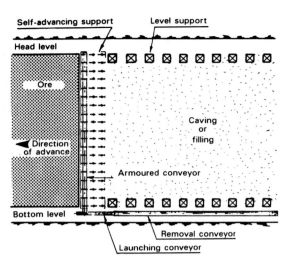

Fig. 3.51   Longwall mining.

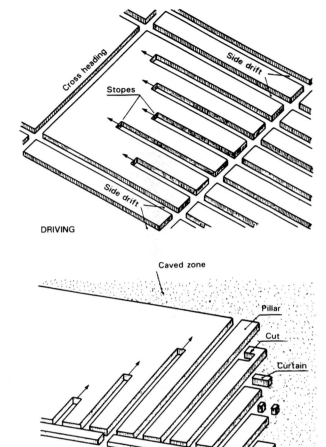

Fig. 3.52   Room and pillar mining.

bearing on the exploited land, and the great volumes of material that have to be transported, require fully-optimized extraction, and re-filling with waste of the abandoned pits with restoration of the land surface to agriculture, or other use. Examples of this type of metal mine that have been worked and restored are sedimentary iron ore mines in the UK and bauxite mines in Western Australia.

More commonly, open-pit metal mines, especially those in base metal ores, are deep in relation to their width, reflecting the usual shapes of the orebodies, which may plunge to great depths while having restricted lateral extent. Although more than 90 per cent of the material raised may be discarded, comprising as it does both waste rock and the tailings produced by the ore beneficiation process, it is normally not feasible to backfill the pit and the waste and tailings have to be disposed of on the surface, with possible environmental problems. When mining ceases the pit remains, requiring safety measures to be undertaken. If dry, the pit may be suitable for domestic or industrial waste disposal: if the pit fills with water as the water table returns to its natural level this may allow its use for water storage or, possibly, for leisure activities.

The form of underground mining is much more variable. When the ground conditions are good, the empty spaces created by the extraction of the ore may be left as they are, but in less favourable conditions they may be filled with waste, a practice that is required in certain mining methods, or they may be left to cave in naturally. These possibilities lead to four main mining methods—long wall, room and pillar, sub-level stoping, and working in horizontal slices: either top slicing (working downward) or cut-and-fill (working upward).

The long wall method (Fig. 3.51) is used when the orebody is in the form of a continuous layer of regular thickness, with little or no offset due to faulting, and dipping gently, at no more than 20°. The ore is extracted by ploughing or by a rotating drum-type shearing machine in a conveyor pass. The hanging wall is supported by self-advancing hydraulic jacks, which move forward as the working face advances, and behind which the roof is allowed to collapse (caving). Few metal mines are suitable for this method, which is used chiefly for coal.

In the room and pillar method (Fig. 3.52), the orebody is extracted from a network of perpendicular drives and cross-

cuts, resulting in a checkerboard arrangement of empty spaces (rooms or stopes), with pillars of ore left in place to support the roof. Before final caving, the pillars are reduced as far as is compatible with stability in order to extract as much as possible of the ore. The caving fills the spaces left by extraction of the ore and reduces the pressure on the neighbouring areas.

Sub-level mining is used to mine steeply inclined orebodies. It is effected in two stages. The first stage consists of driving sub-level passages at regular intervals through a complete level. This is followed by progressively stoping out each sub-level, one after the other upwards, so that the stepped total working face advances regularly as a whole.

In a general way, the ore is withdrawn on the level at the base of the series of sub-levels, in some instances also at each sub-level. The term sub-level stoping (Fig. 3.53) is used when the spaces are left empty, and sub-level caving when the spaces are filled by caving walls.

In the cut-and-fill method (Fig. 3.54), the ore is blasted and removed in horizontal slices working upward, the space left behind being backfilled as the face advances. In top-slicing, the ore is also removed in horizontal slices. As each slice is removed, the next underlying slice is left to form a temporary floor and the slice below that is removed. The space left behind may be abandoned or filled by caving or filling.

Sub-level and slicing methods are used only in orebodies that are steeply inclined (>20°) or of complex form. A large proportion of European base-metal mines are of this type.

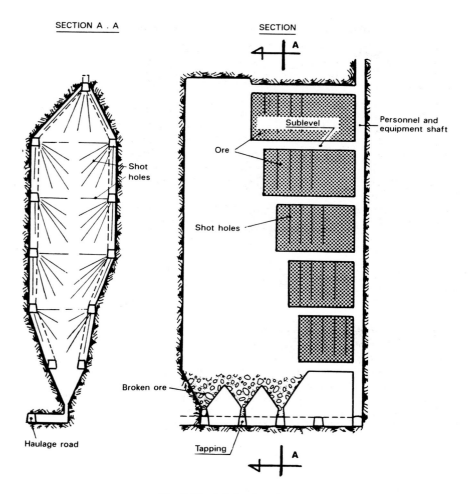

Fig. 3.53 Sub-level stoping.

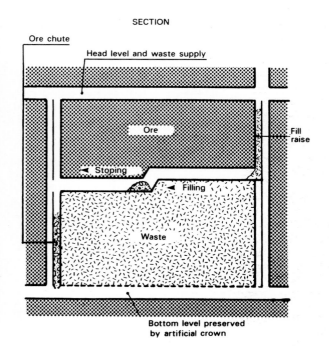

Fig. 3.54 Cut-and-fill mining.

Transporting the ore

An essential part of all mining operations is the transport of the ore from the working face to the site where it is to be beneficiated. The organization of this phase of exploitation varies according to the type of mine (opencast or underground) and the amount of ore to be transported.

In underground mines, subsurface and surface transport are generally quite separate. Underground, specialized vehicles, conveyor belts, and mine cars are used; transport to the surface is by raising through a shaft, by conveyor belt or by dump truck up a system of ramps, and transport on the surface from the mine itself to the processing plant is by truck or conveyor belt.

In opencast mines, transport is by truck, conveyor belt, or in some instances by railway, overhead cableway, or pipeline. The transport stage of mining, with its accompanying dust and noise, in particular, is a source of environmental nuisance, especially in the case of opencast operations.

Processing (concentration) of the ore

It has already been observed that most commonly an ore has to be concentrated in order to be of commercial value, i.e. acceptable for metallurgical processing, for which only concentrates of a certain grade can be used.

The concentration of an ore has three objectives:

(1) separating the different economically-valuable components of an ore, at the same time making sure that each concentrate is as pure as possible;

(2) eliminating the waste material accompanying the ore, such as silicate minerals in base metal sulphide ores;

(3) controlling possible environmental nuisances. Rendering tailings from concentration harmless for the environment has become a major consideration over the years to be dealt with quite separately.

Impurities, metallic ones in particular, remain associated with the valuable metals in the concentrates, decreasing their metallurgical value and consequently reducing their market prices. These are metals, such as arsenic and mercury, chemically related to those for which the ore is mined, or base metals present in uneconomic amounts in the ore, such as lead in a zinc concentrate, for example.

Conversely, it is commonly during metallurgical processing that associated precious or rare metals such as gold, silver, germanium, and indium are recovered.

Environmental nuisances are particularly important in the case of non-ferrous metals, which most commonly are found in the form of sulphides of the valuable metals mixed with iron sulphide (pyrites), which is not of great use. The oxidation of this mineral causes an increase in the acidity of the water passing over it, and the release of pollutants that are usually unacceptable environmentally and must be controlled.

As will be seen later, the use of flotation in concentration involves the addition of chemical reagents which cannot be released at will with the tailings into the environment, without taking certain precautions.

The ore extracted from the mine is in the form of fragments of very variable size, mainly rather large. The ore is subjected to several stages of crushing (which may take place in the mine and/or on surface) and grinding to reduce the particles to a size at which the ore and waste mineral particles can be separated. For economic reasons grinding is kept to a minimum, and each particle-size fraction (0–2 or 2–4 mm, for example) is usually processed separately.

In physical processing, of which grinding and particle-size classification are a part, advantage is taken either of the higher specific gravity of the ore minerals compared with the associated waste minerals (gravity processing), or of their capacity to react with chemical substances, creating a water-repellent surface film on the mineral grains that facilitates their selective entrainment in a stream of bubbles (froth flotation). Flotation is by far the most commonly used process. It is usually carried out in aerated square cells, giving two products, one that sinks and one, consisting of the desired minerals, that is floated off with the bubbles. The minerals are then subjected to several stages of washing. Column flotation, which is a major technological breakthrough in this traditional technique, is currently being developed.

Among the battery of tools for physical processing, mention must also be made of magnetic separation, which takes advantage of the differences in magnetic susceptibility of different minerals, and electrostatic separation, which makes it possible to separate two minerals, both poorly magnetic but having different electrical properties.

The product of physical beneficiation is known as a concentrate; this is the feedstock for pyrometallurgical treatment (smelting) to produce metal. Certain metal ores are smelted directly without significant beneficiation, apart from size grading of mined ore. This is the usual treatment for high-grade iron and manganese ores and laterite nickel ores, the last two being normally smelted to ferro-mangangese and ferro-nickel respectively.

Chemical treatment may be used either instead of, or in addition to, physical beneficiation. 'Oxide' ores (usually metal carbonates, sulphates and silicates) may be dissolved directly in acid or basic solvents and the desired metals are either precipitated selectively from solution as salts, or are recovered as metals by electrolysis ('electro-winning'). A roasting stage, to oxidize and remove sulphur, may precede chemical treatment in the case of sulphide ores. Bioleaching, i.e. using solutions containing bacteria that can digest sulphides, is a promising process for the treatment of refractory gold ores, where the gold is locked into sulphides of iron or arsenic.

It can be seen that the processing of ores covers a very wide range of techniques that can be adapted and modified at will. Optimization of the parameters of the concentration process, of processing cost, metal recovery, and concentrate quality, enabling continuous improvements in the overall control of ore treatment, is fundamental in determining the viability or otherwise of an operation. This has led in recent years to the development of on-line monitoring sensors and of computer models for the simulation, automation, and optimization of ore-processing.

Impact on the environment

Impact on the environment can be felt in several ways:

(1) the utilization of land by the mine itself, especially in the case of opencast mining, and for the housing of personnel;

(2) the utilization of land for waste dumps and tailing dams;

(3) damage caused directly by the mining—subsidence at the surface above underground workings and landslides at the edges of open pits;

(4) the effects on the water table of mine de-watering or of stopping pumping, and any eventual groundwater pollution by the effluents from mine drainage and mineral processing;

(5) vibrations and shocks caused by blasting;

(6) noise from blasting, crushing and grinding, processing, transport;

(7) dust from blasting, transport, and tailings.

Mining practice and legislation have evolved considerably in recent decades towards lessening these impacts and, beyond this, towards restoring former mining sites after exploitation has ceased.

However much progress is made in these respects, it will never be possible to eliminate these impacts completely, but many are temporary, existing only during the life of the mine. Their acceptance, within a framework of environmental regulation, is the price that must be paid by the community if natural resources are to be exploited and new activities created.

3.5.4. National policies [194] [65] [48] [209]

Introduction

The major Western European countries are heavily dependent on overseas sources of metals (see Section 3.5.2), which are essential for the manufacturing, construction, and transportation industries. This fact of overseas dependence has led to fears of both interruption of supply and of the gradual loss of mining and metallurgical expertise in Europe. In turn, such concerns have led to varying degrees of publicly-funded support for mineral exploration, mining, and metallurgy by individual European governments, subject to the socio-economic, technical, and financial considerations in each country. The key role of international scientific and technical communication and collaboration in the field of mineral exploration, together with the increasing tendency towards the internationalization of the minerals and metals industry, has also led the Commission of the European Community to give strong financial support to multi-national research in these fields by members of the European Community.

The main objectives of national policies may include the following:

1. To ensure that indigenous mineral resources are adequately assessed and catalogued. National Geological Surveys play a key role in this field by conducting exploration and compiling systematic inventories and data banks, and keeping them up to date as a long-term policy of availability of information. The EEC supports research and development studies in mineral exploration, beneficiation, mining, and metallurgy. Individual governments may give support to mineral exploration by private industry and to investments in the mining sector either on a national or regional basis.

2. To continue a policy of supply of the main commodities. In certain countries this is effected by support for overseas investment by national companies through cooperation programmes and direct financing. This objective is also favoured by international organizations such as the World Bank which aim at helping investment in developing countries. Diversification of supply may be encouraged, and maintenance of strategic stocks of minerals and metals is national policy in certain countries.

3. To support the rationalization and modernization of appropriate mineral sectors. This process has industrial, commercial, and social components. Iron, tin, and tungsten are such supported commodities in certain countries. Price support is usually necessary during the transition period.

4. To ensure the best use of national energy sources for the metallurgical industry: for example, magnesium and aluminium manufacture in Norway using hydro-electric power.

These four general fields of action are examined in more detail below, with case histories in the first section.

Assessment and promotion of national resources

The fact that a mineral deposit, a reserve, is a concept in permanent evolution with reference to economy and prices, necessitates a long-term approach to inventories and prospecting work. This philosophy is particularly valid in Europe where the existence of political stability and highly-developed infrastructure tends to be offset by the erroneous perception that the chances of exploration success are low.

In addition to the provision of basic geoscientific maps and reports, several major types of geological activity aid the identification and promotion of a country's metalliferous mineral resources. These are regional inventories, mineral databases, and reconnaissance surveys, and research and development which all help to identify targets for potential follow-up by industry. The Geological Surveys play an important role in all these functions.

Regional inventories

Systematic geochemical and mineral deposit inventories are carried out in many countries of Western Europe. The geochemical inventories done by the British Geological Survey (Geochemical Survey Programme) and the Geological Survey of France have used stream sediment sampling.

Geochemical Survey Programme, UK [216]

The primary purpose of the Geochemical Survey Programme (GSP) is to produce a comprehensive database on the concentration of up to 30 elements in stream sediments (or soil samples in the absence of drainage) collected at an average density of 1 per square kilometre throughout Great Britain. Samples of stream waters and heavy mineral concentrates are also taken at the same sites. Sampling and analyses have been completed over Scotland and northern England and are progressing systematically southwards, covering basement areas and sedimentary basins alike.

Data are made available in the form of tapes, discs, listings, and geochemical atlases as well as on an interactive computer system—GISA (Geochemical Interactive Systems Analysis)—which is available for interrogation by industry. The programme is supported by the Department of Trade and Industry.

In France, the general reconnaissance sampling density is 3 per square kilometre and has been organized by 1:50 000 geological sheet area. The coverage of the basement and fringe sedimentary areas is now almost complete. No inventory has been carried out in the sedimentary basin areas. The information is recovered at the scale of 1:50 000, using the geological sheet areas. The whole programme is financed by the Ministry of Industry and the results are available to the public. The present-day tendency is to focus on more restricted targets (1000 km² approximately) and to explore for hidden deposits.

In both cases the work has led to the installation of an extensive data bank accessible to all mining operators. The information collected can also be used for thematic mapping or environment management purposes.

The systematic inventories can be designed to be very specific to one task. The following case history is a good example of work on metalliferous resources.

Nordkalott Project [16]

The Nordkalott Project (NKP) is an integrated operation of the three Nordic countries: Norway, Sweden, and Finland, sponsored by the Council of Ministers of the Nordic Countries. The aim was to carry out a systematic inventory of the least known areas of the three countries, i.e. north of latitude 66°N (Fig. 3.55). The area covered is 250 000 km². The project was divided into five sub-projects dealing with:

(1) hard rock geology;

(2) Quaternary geology;

(3) geophysics (gravity and aeromagnetic);

(4) geochemistry on various types of samples: stream sediments, glacial till, plant roots, stream organic matter, etc.

Full use was made of earlier data, but the existing information base was largely heterogeneous from one country and one region to another.

Most of the results were depicted in the form of 1:1 000 000 scale maps. For geochemistry, the maps were presented per chemical element and sampling media at 1:1 000 000 and 1:4 000 000 scale. Interpretation of the data was also shown cartographically.

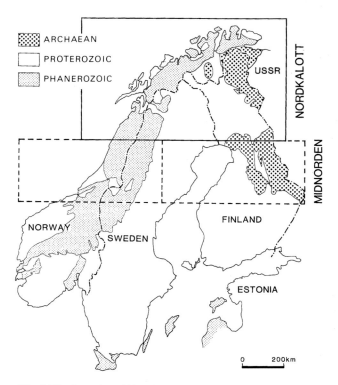

Fig. 3.55 Location of Nordkalott and Midnorden project areas.

Other maps of the NKP programme include:

(1) ice movement directions;

(2) lithology;

(3) regional gravity;

(4) airborne magnetometry (total intensity and interpretation);

(5) mines and mineral occurrences.

The analytical information resulting from the NKP is stored in data banks which can be used according to needs and improved by more detailed information originating from follow-up work. The NKP was carried out between 1980 and 1986 and has been followed by the Midnorden project, its southern continuation.

Many other examples of systematic inventories could be discussed, each involving national organizations: Geological Surveys, Mining Departments, and national companies (such as Minas de Almaden for central Spain). The approach is not universal as to the scale of work, sampling methodology, and presentation of data. Variations are often well founded on differences in geology, climatic conditions, present state of knowledge, and politics (including the level of Government support). All countries, however, identify a need for long-term inventories and data availability, and mining operators use them in planning to invest in mineral exploration—if the financial and legal incentives are encouraging.

Mineral data banks

The mineral data made available to users are not restricted to systematic inventories using geochemical and geophysical methods, but include a spectrum of information related to each individual mineral occurrence investigated, thus providing a background for future development work. Two examples of such mineral data banks, those of Finland and Ireland, are given below, but it is clear that similar data banks exist in every European country.

The Finnish Data bank

The Geological Survey of Finland maintains two computerized databases on metalliferous mineral deposits:

1. The Ore Deposit Database. This includes data on deposits either of economic viability or of significant exploration potential.

2. The Mineral Indication Database is concentrated on small occurrences and indications of mineralization, as well as on the locations of erratic ore-bearing boulders. The metals of interest for both databases are:

    iron metals (Fe, Ti, V, Cr);
    base metals (Cu, Ni, Zn, Pb, Co);
    noble metals (Au, Ag, PGE);
    energy metals (U, Th);
    other metals (Mo, W, Sb, Sn).

The data are collected from company reports, literature, and from interviews with the geologists in charge of the investigations of particular deposits. The aim is to list, as objectively as possible, and in non-genetic terms, the parameters which are collected during the investigation of each ore deposit.

Integrated data analysis has become an important tool in the exploration and assessment of mineral resources. One example is the estimation of the metalliferous mineral resource potential of Finland by using large computerized geodatabases and statistical methods. The deposit model used in the analysis consisted of the values of 2 lithological, 14 geophysical (7 aeromagnetic, 7 gravimetric) and 8 geochemical parameters from cells, each 5 km by 5 km, around the 34 mines, which have been in operation in Finland between 1944 and 1987. The same parameters and their values in cells of the same dimensions (altogether about 18 000) over the whole of Finland, served as the background for comparison. Similarity analysis was used to calculate the mineral resource potential, using the values of the 34 model areas. The result is a map, which can be called the general metal ore potential of Finland. The preliminary results have brought some new ideas to the assessment of the metal potential of Finland.

National mineral resources data bank for Ireland [123] [126]

A national data bank on mineral resources is maintained at the Geological Survey of Ireland and it contains the following major elements:

1. A computerized inventory of localities throughout the country, which contains data on both metallic and industrial minerals. It is also being published on a regional basis.

2. Open-file data submitted to the Survey by exploration companies, which are obliged, under the terms of prospecting licences, to submit regular reports and data on their prospecting activity. Upon surrender of the prospecting licence, such data are placed on open-file and are available for consultation by other parties. It is also intended to publish these data in summary form.

3. Mine plans are available for all the older mines and are available for consultation. In many cases they provide valuable insights into the nature and extent of contiguous mineralization and they also form the basis for understanding the shape of, and controls on, such deposits.

4. Regional magnetic and geochemical data for significant parts of the country are available for consultation. Published versions of many are already available.

Reconnaissance surveys

One of the tools available to governments in stimulating commercial interest in a nation's mineral resources is to support, by one means or another, some of the exploration activity. Various approaches are used; the following is just one example.

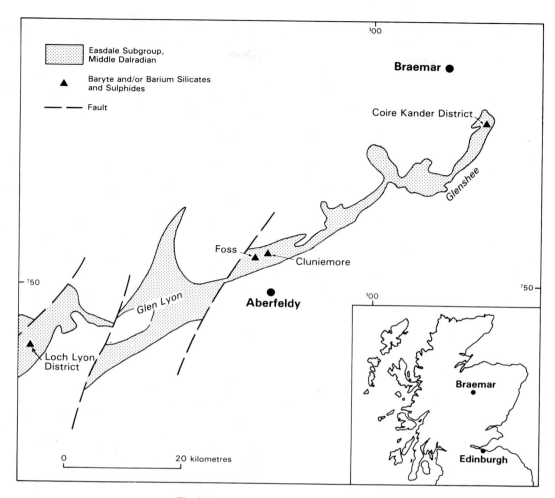

Fig. 3.56 Location of Aberfeldy, Scotland.

Mineral Reconnaissance Programme, UK [201] [194]

The BGS Mineral Reconnaissance Programme (MRP), which begun in 1972, is a multidisciplinary project, funded by the UK Department of Trade and Industry, that draws on a wide range of geochemical, geophysical, and geological expertise. Tried and tested field techniques have been adapted to local conditions and have been complemented by newly developed high-technology methods of chemical and mineralogical analysis and data handling, interpretation, and presentation.

The emphasis was initially on base metals but later became focused on strategic metals such as manganese and tungsten and on other commodities such as barytes, gold, and the platinum group metals. The intention is to keep ahead of the market in the search for resources for which a demand might be anticipated, or for which security of supply might be required.

The results of the MRP are published as a series of reports: by early 1990, 113 reports had been issued. It is estimated that between 30 per cent and 50 per cent of the investigations reported by the MRP have attracted commercial interest. One such case is described in the following paragraphs. It is particularly interesting because it spans the interest of this and the previous section, being a combined industrial and metalliferous mineral deposit.

Aberfeldy barium–lead–zinc deposits, Scotland [227] [229]

Late Proterozoic metamorphic rocks in the Scottish Grampian Highlands were selected in the 1970s for a study as part of the BGS Minerals Reconnaissance Programme. Criteria for selection were a high incidence of historic base metal showings, records of sulphide in hydro-electric tunnels, and metallogenic similarities with Stekenjokk and other Scandinavian deposits.

Stratabound sedimentary-exhalative Ba–Zn–Pb mineralization, extending at intervals over 7 km of the strike length of the Ben Eagach Schist Formation in the mountains north of Aberfeldy (Fig. 3.56), was discovered by drainage geochemical sampling, mapping of sparse outcrops, geophysical definition of the conductive graphite schist host rocks beneath the peat and glacial drift, overburden geochemistry, and by shallow drilling. Following publication of the results, leases were acquired in 1979 by Dresser Industries Mineral Division (now MI Great Britain Ltd) and for the next three years the mining potential of the Foss property, 7 km north of Aberfeldy, was evaluated. Trial stoping commenced in 1983 on a sub-vertical section and production started in 1984. Underground and open-pit operations at Foss produced 47 000 tonnes of barytes in 1989, from a barytes band averaging 4 m in thickness. Sulphides express fine bedding in the massive barytes and the deposit is structurally complex.

Two further deposits were delineated near Aberfeldy: at the Ben Eagach Quarry, barytes production commenced in 1990 from a large outcrop of folded barytes (at the centre of Fig. 3.57), forming the western extension of the barytes bed at Cluniemore. The mineralized zone is dominated by barium-enriched muscovite schist and quartz-celsian rock assaying 8.5 per cent Zn and 3.6 per cent Pb over a thickness of 4.3 m. Coarse-grained galena and sphalerite locally attain concentrations of tens of per cent; at Cluniemore, surface drilling and a decline have revealed a 30 m thick bed of high-grade barytes that is open at depth. The bed is flanked by quartz-celsian rocks enriched in sphalerite and galena

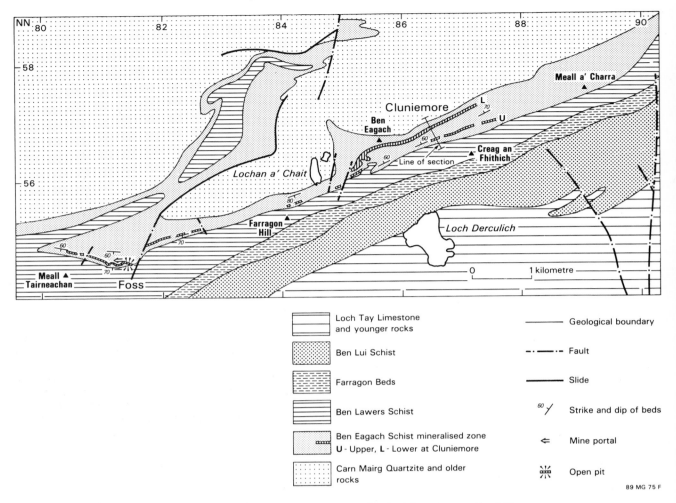

Fig. 3.57   Geological setting of the barium–zinc–lead deposits near Aberfeldy.

and by barium-enriched mica-schists, forming a mineralized zone about 100 m thick.

Further barium/metal sulphide deposits have been identified by BGS for 45 km on either side of Aberfeldy, all in the Ben Eagach Schist Formation, and sulphides unaccompanied by baryte have been found in other formations in the same group.

Research and development

Each European country and many individual companies have developed R&D policies concerning mineral exploitation. Multinational programmes also exist, sponsored by the European Community Commission, and at company level by organizations such as MIRO. In the case of the EEC, programmes correspond to individual sections in the overall R&D programmes.

These aim at improving prospecting tools now that most of the surface occurrences have been found, at investigating new types of ores, and at improving beneficiation and mining technology for ores with which there are problems in raising the quality. Traditionally research bodies in various countries work together in such programmes, and now mining companies are being involved as much as possible. Although there is a trend towards greater emphasis on beneficiation and metallurgical problems, it is essential that R&D programmes

include the whole range of problems concerned with mineral exploration and types of deposit.

National policies

Most major European mining companies have subsidiaries and project partners in foreign countries. For example, RTZ has a network of subsidiary companies around the world: one of this group's latest investments is in the giant Escondida copper mine of Chile.

European mining companies and foreign companies such as INCO and MIM have smelting and refining plant in Europe that treats the raw materials produced at overseas mines. Several European smelting and refining companies, formerly integrated with nearby mines, now depend entirely on the international market for raw material supplies and have no 'captive' or dedicated sources, e.g. Norddeutsche Affinerie in Germany (copper) and Treibacher Chemische Werke AG in Austria (several minor metals).

National governments may or may not support such mining and metallurgical operations depending on the country concerned. They may assist investment in overseas projects, hold equity in mining companies, such as Outokumpu (Finland), and one, at least, has in the past offered financial incentives for overseas exploration, but there are at present wide differences in the policies of different European governments towards such intervention in the market.

Legislation in Europe and most developed countries has established environmental standards that have caused problems for some indigenous mine projects and smelters that use overseas ore. Technology can overcome pollution and potential environmental damage, but in times of weak metal prices companies may not be able to meet the cost involved and may choose to shut down, thus increasing reliance on foreign producers.

Where dependence on overseas supplies is strong and the criticality of the materials concerned is high, governments in some countries (UK, France, Sweden) have set up 'strategic' stockpiles of certain minerals and metals. Producers are traditionally nervous of such stocks but it is clear in the case of the European examples that they exist to guard against actual cut-off of supplies, not to manipulate markets or prices.

In some countries the Geological Survey, or equivalent organization, acts as an adviser to government on the problems of mineral and metal supply, particularly where such considerations are linked to foreign policy.

No European government appears to approach the highly integrated Japanese attitude to raw material supply, but in most countries, even those with the least 'interventionist' policies, governments maintain Geological Surveys that, at the least, monitor the mineral supply situation worldwide, in addition to carrying out mineral-related activities on home territory.

Subsidies

Governments may subsidise metal mining and processing either directly or indirectly.

Indirect subsidies may be in the form of regional incentives, such as grants and 'soft' loans to encourage industry, in general, to invest in underdeveloped or disadvantaged regions of individual countries. Metallic mineral deposits tend to occur in those upland regions most likely to qualify for such incentives, since they are unattractive to manufacturing industry or intensive agriculture (central Ireland is an exception).

Direct subsidies may be made for the purpose of establishing a mining industry where none existed formerly, such as the 'tax holiday' extended to metal mine development in Ireland in the 1960s and 1970s. More usually they exist for the express purpose of modernizing or rationalizing a mine or mining field, where poor profit margins, or other causes, have led to chronic lack of investment, but where it is thought desirable to keep the mines open, either for strategic or socio-economic reasons. Examples are to be found in the cases of the Cornish tin mines in the UK, in Panasquera tungsten mine in Portugal, and in the low-grade iron ore mines of eastern France.

In the last example, the case for continued mining is strengthened by the degree of integration with downstream processing plant (smelters) specifically designed for these ores, and which cannot be used for other types of iron ore. This kind of integration is also found in the cases of government-supported tungsten mines in Austria and Sweden, although in the case of Austria there is also partial supply of concentrates and ammonium paratungstate (APT) from China.

In some countries a depletion allowance system enables companies to deduct exploration and investment costs from profits. This system applies also to investments complementary to the mining activity itself.

Use of national energy resources

Using cheap energy sources to support national industry is often a target for governments when such resources are available. In Norway, for example, there is a long tradition of cheap electricity from hydroelectric plants. This led the Government to support high energy consuming metal manufacturing, such as the magnesium or aluminium industries, or ferro-alloy manufacture. Recently, the state-owned iron and steel complex of Norsk Jernverk stopped its iron smelting activity and concentrated on steel manufacture in electric furnaces using scrap and imported pig iron. In France, the extensive use of electricity of nuclear origin is also significant for the metallurgical sector.

Final remarks

The importance of metal supplies to European industry is so great, the potential for loss of such supplies so obvious, and the consequences of such loss so dire, that few, if any, national governments are content to leave the security of supplies entirely in the hands of market forces. Support for the metals industry and the metals consumers commonly takes the form of publicly-funded activities of Geological Surveys, in undertaking exploration, compiling basic inventories, making data available to mining companies, and advising government on overseas supplies.

The international nature of mineral resource studies and the increasingly multinational nature of the mining industry is leading towards international cooperation especially in the fields of research and technology. In the future it may also include European as well as national strategies for resource development and metals procurement.

RECOMMENDED FURTHER READING

Andrew, C. J., Crowe, R. W. A., Finlay, S., Pennell, W. N., and Pyne, J. F. (1986). *Geology and genesis of mineral deposits in Ireland.* (Irish Assoc. for Economic Geology, Dublin).

Anon (1980). *Guide to underground mining methods.* (Atlas Copco, Stockholm).

Bardosy, G. and Dercourt, J. (1990). Les gisements de bauxites tethysiens (Méditerranée, Proche et Moyen Orient): cadre paléogéographique et contrôles génétiques. *Bull. Soc. Geol. France.* **8**, 869–88.

Carvalho, D. (1991). Case history of the Neves-Corvo massive sulfide deposit, Portugal, and implications for future discoveries. *Econ. Geol. Mon.* 8.

Collective Authors. *Mineral deposits of Europe.* (The Institution of Mining and Metallurgy; The Mineralogical Society, London).
Vol. 1 (Northwest Europe) (1978): eds S. H. U. Bowie, A. Kvalheim, and H. W. Haslam.
Vol. 2 (Southeast Europe) (1982): eds F. W. Dunning, W. Kykaru, and D. Slater.
Vol. 3 (Central Europe) (1986): eds F. W. Dunning and A. M. Evans.
Vol. 4–5 (Southwest and eastern Europe, with Ireland) (1989) eds F. W. Dunning, P. Garrard, H. W. Haslam, and R. A. Iker.

Collective Authors (1979). La valorisation des minerais. *Ann. Mines.* **105**, No. 6, pp. 5–106.

Collective Authors (1984). *Explanatory memoir of the metallogenic map of Europe and neighbouring countries, 1:2 500 000.* (UNESCO, Paris).

Collective Authors. *Mining Journal Annual Review.* (Last issue June 1990).

Crowson, P. (1990). *Minerals handbook 1988–89.* (Stockton Press, New York).

Kelly, E. G. and Spottiswood, D. J. (1982). *Introduction to mineral processing*. (Wiley, New York).

Leca, X. (1990). Discovery of concealed massive-sulphide bodies at Neves-Corvo, Southern Portugal—a case history. *Trans. Inst. Mining and Metallurgy. B.* **99**, 139–52.

Thomas, L. J. (1975). *An introduction to mining*. (Chapman and Hall, New York).

U.S. Bureau of Mines (1991). *Minerals yearbook 1988*. (U.S. Bureau of Mines, Washington DC).

3.6. UNDERGROUND SPACE [165] [176] [174]

3.6.1. Introduction

The use of sub-surface space for storage purposes is by no means new. In China, it has been used for more than 5000 years. It is known that, in the year AD 606, there were more than 3000 underground facilities for the storage of food commodities in the Chinese capital. Some of them are still in use. The underground storage of commodities also has a tradition in Hungary, particularly the storage of liquor, together with mushroom cultivation.

A large variety of storage applications have been brought into practice in recent years. These include storage of water, energy, fuels, sand and gravel, retail goods, cars (car parks), documents in archives, food (cold stores), etc. Underground space is also used for a variety of installations including power stations, military or civil defence installations, transport tunnels, energy links, and so on. Former mines may be used as museums.

The building material in subsurface construction is primarily the rock itself. Rock is not uniform, but consists of many types of material. It has different origins and has been changed over millions of years by temperature and pressure during the Earth's evolution. The geological environment is not static. The continents move, mountains are folded, and tectonic forces affect the rock, creating inhomogeneities in the rock mass in the form of foliations, fractures, and crushed zones. Softer rocks of sedimentary origin, such as limestone and sandstone, are changed through diagenesis, weathering, and erosion (see Chapter 2).

Faults and fractures frequently carry water, which may be useful for supply to the population, but for tunnelling it is a potential problem. Water leakage may lower the groundwater level in the vicinity and dehydrate the rock. For this reason fractures are usually sealed, but in the case of oil storage the water may be utilized to form a barrier against unwanted leakage of the hydrocarbons.

Seismic activity may cause new fractures to appear or reactivate those that already exist, and volcanic activity can pose problems, although this may be turned to advantage in the development of subsurface energy installations.

The suitability of the rock as a construction material varies from one location to another. The hardness, brittleness, and degree of jointing are properties which show different patterns of variation in different kinds of rock. These properties are amongst those which influence the economics of a subsurface project. Crystalline rock, such as granite and gneiss, is usually hard and requires a large percentage of the budget for excavation. This is offset, in most cases, by the low cost of reinforcement and sealing. In softer sedimentary rock the excavations are usually less expensive, but there may be more need for reinforcement work.

Building caverns in cities sometimes can give economic benefits, as the excavated rock can be crushed easily on site and then sold as ballast material. Even after the cavern has been completed the crusher can remain as a central stone-crushing plant utilizing rock and stone from other excavations and providing ballast. This plant can be in operation without people in the vicinity being troubled by noise or dust.

3.6.2. Underground storage of materials

Water

Water can be stored underground either in natural or artificial rock caverns or in permeable rock. The simplest storage is in consolidated rock, such as porous limestone or sandstone, or porous alluvial material. Fractured crystalline rock can also be used in this way.

To achieve the best control, however, water is stored in caverns cut or blasted in the rock. With a suitable topography it is possible to replace conventional water tanks with caverns. Storage of drinking water in rock underground has many advantages. The caverns are safe, invisible and easy to maintain and extend. The water is not exposed to sunlight and pollution from the air. Seasonal variations in water temperature are eliminated. Furthermore, the larger reservoirs cut in reasonably good rock are cheaper to construct than conventional concrete water tanks, and they avoid the sterilization of large areas of the surface environment.

Energy

Energy can be stored in many forms and in many ways. Since the sharp increase in the price of oil in 1973 a number of investigations have been focused on the storage of heat. A number of integrated systems have been built.

At Lyckebo, about 15 km north of Uppsala, Sweden, a solar heating plant with seasonal storage in a cavern has been constructed. Solar collectors supply hot water to the cavern between March and September; water in the cavern reaches the temperature of 90 °C in the autumn. As the energy is consumed during the winter the temperature drops to 40 °C by about April. The heat store is situated in crystalline rock with a rock roof 30 m thick. The outside diameter of the ring-shaped cavern is 75 m, the width is 18 m and the height 30 m (Fig. 3.58). The ring shape was chosen in order to minimize heat loss. The chamber is devoid of artificial insulation as the surrounding rock provides good natural insulation. In the first few years of its operation, some heat was used in raising the temperature of the surrounding rock mass. Since this was achieved, heat losses have been modest, amounting to about 15 per cent of the energy supplied annually.

The storage of compressed air in rock caverns has been utilized for decades in the mining industry. Compressors at the surface generate compressed air which is passed to an underground cavern. The pressure is held constant by a water column in contact with another cavern at a higher level in the rock forming the balancing water reservoir (Fig. 3.59).

Coal

In many countries, coal is one of the main fuels for district heating. In some cases, oil has been replaced by coal for economic reasons. At the Igelsta plant in Södertälje, Sweden, some 30 km south of Stockholm, such a change has been made. Environmental considerations made open-air coal storage in the built-up area unacceptable. All handling of the coal is enclosed, from unloading at the quay right through to ash disposal. The coal arrives at the plant's harbour in vessels of up to 30 000 tonnes and is off-loaded and transported by

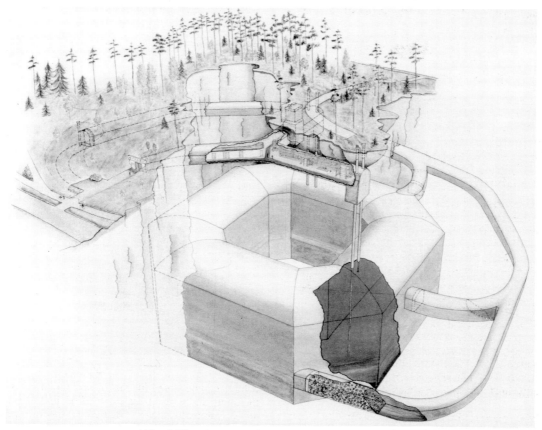

Fig. 3.58 The Lyckebo heat storage cavern, Sweden.

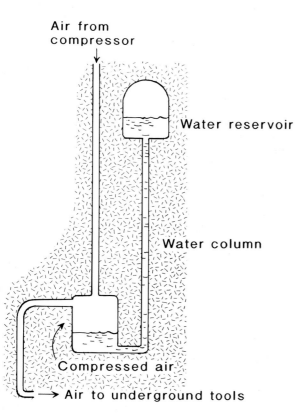

Fig. 3.59 Cavern for storage of compressed air.

an enclosed conveyor belt to pass through sifters, crushers, and metal detectors before being taken directly to either the boilerhouse or to storage.

The coal storage consists of three silos, each 70 m deep and 31 m in diameter with a volume of 50 000 m³ (Fig. 3.60). The coal is poured in from the top and lifted out by a bucket elevator, which runs in a revolving superstructure to ensure uniform lifting, and handling is completely controlled by computers giving the position and level of the coal in the store. The silos can be filled with inert gas or water to prevent fire. Coal from the silos is ground and injected into the boiler for combustion in an overfire-air unit.

Hydrocarbons [207] [230]

Petroleum products have been stored in artificial rock caverns for 50 years. In Sweden the first attempt to store oil, on a bed of water in an abandoned feldspar mine, was made in 1940. Better economy, improved environment and increased security are the main reasons for choosing cavern storage. Storage can be in caverns in hard rock, that is igneous or metamorphic rocks (as in oil storage in Sweden and Finland), or in salt domes (as in gas storage in Denmark).

The storage of oil in unlined caverns in hard rock is now well tested, with hundreds of caverns in existence, some more than a million cubic metres in size. Several methods are in use:

(1) Storage on a fixed water-bed;

(2) Storage on a fluctuating water-bed;

(3) Storage directly on a cavern floor.

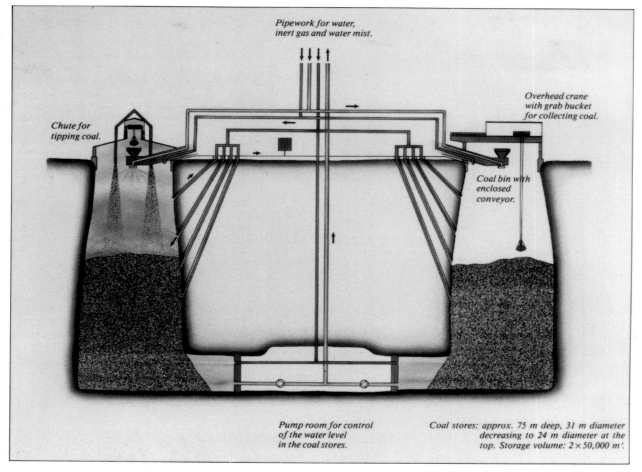

Fig. 3.60   Igelsta coal silos, Södertälje, Sweden.

The second method is used with more volatile products to limit the size of the free-gas volume above stored liquid. The third method is possible with heavy oils only, but these may have to be heated to allow pumping at an economic rate. This is achieved most effectively by heating the water in a fixed water-bed system.

The best medium for cavern storage is homogeneous rock free from fractures. For some systems, the leakage of water into the caverns is essential to provide a natural seal, but the water has to be removed by pumping at high cost.

Research and development work over many years, together with experience gained from the monitoring of existing storage systems, have led to considerable expertise in this subject which now resides in the Western European Geological Surveys and sister organizations, especially those in Sweden, Norway, Finland, and Germany.

The following case history from Finland is a typical example.

Deep storage of hydrocarbons in Finland [25]

Purpose and principle

Finland's lack of its own hydrocarbon resources, and remoteness from other sources, make large stores important for the effective distribution of oil products throughout the country. In particular, large-scale storage of imported crude oil is vital because sea lanes freeze during Finland's severe winters, hindering transport for many months of the year.

More than 20 years of experience shows that storage in unlined rock caverns is by far the best way of storing large quantities of oil.

The caverns operate on the principle that the natural groundwater flows towards the cavern, forced by the pressure gradient, forms a seal and prevents the oil spreading to the surrounding environment and causing pollution. This is practical if the rock caverns are excavated below the groundwater table, and given that the stored material is lighter than water and insoluble in it.

Hydrocarbon storage in Finland

In 1965 the construction of Finland's first unlined storage caverns for crude oil started at the new Porvoo (Borgå) refinery, and it was completed in 1967 at about the same time as similar facilities in Sweden and Norway. Since then, over 40 storage caverns have been built in different parts of the country. At the Porvoo refinery, there are more than 30 individual caverns totalling more than 5 million m³ in capacity; most are for crude oil, but several caverns have been built for the storage of petroleum products.

The potential for the storage of imported natural gas at a depth of over 800 m in a Svecofennian gabbro massif in southern Finland has been under investigation and discussion recently.

Construction of caverns

The relative strength of the bedrock has meant that the recent oil caverns in Finland are unlined and that the construction methods based on drilling and blasting have been the most economical. Little supplementary support has proved to be necessary and careful blasting has saved much of the need for other methods such as rockbolting, shotcreting, and to a lesser extent, grouting.

Geological studies for site selection and planning

The main geological investigations for the Porvoo storage works were concerned with the location of the caverns in zones of the bedrock which are free from discontinuities. The studies were based mainly on the interpretation of aerial photographs and field control, bedrock mapping, and core-drilling. Previous knowledge of rock conditions from the mapping of the Geological Survey was a major factor and reduced the need for these investigations.

In the design phase maximum importance was attached to the geological conditions, including the hydrogeological prerequisites of the area, in relation to the shape, direction, and level of the caverns, as well as to the location of the auxiliary facilities. The stress conditions and the variations in the fracture frequency greatly influenced the decision as to the excavation span; at Porvoo, this varied between 12 m and 16 m.

The geological investigations during and after construction mainly comprised groundwater observations. The drawdown influence of the pumping of leakage water during the excavation itself was generally rather limited and in the maximum case amounted to over 50 m. The groundwater returned almost to its original level after about 3 years of operation.

Environmental effects

There have been certain geologically-related effects during the cavern construction at Porvoo, such as groundwater drawdown, which were directly related to horizontal fractures cutting the shafts. Excluding these temporary effects, the oil storage at Porvoo is causing little harm to the environment. There are no signs of any environmental pollution.

The environmental advantages of the underground storage of hydrocarbons in comparison to surface tank storage, are the following:

(1) the risk of fires and explosions is minimal, involving reduced insurance premiums;

(2) the rock cavern plant is almost invisible; valuable land can be utilized for other purposes;

(3) high pumping capacities can be used, decreasing expensive harbour time for tankers, and reducing pollution risks involved with them;

(4) oil can be stored under pressure, reducing evaporation losses and air pollution;

(5) maintenance and repair needs are low and the lifetime of a cavern is exceptionally long, thus reducing risky surface procedures;

(6) the excavated rock serves as useful material, and it has been produced without creating many ugly and harmful quarries in the densley populated coastal zone;

(7) from a depth of about 10 m below the bedrock surface the rock mass temperature is even all year round. This and the low heat conductivity of rock reduces heating costs for heavy fuel oils thus saving energy in general.

Although some operational problems have occurred, including a fire in one crude oil cavern, it can be stated generally that all the installations are working satisfactorily from an environmental point of view.

Car parks

As land value and traffic levels increase, the environmental benefit of storing cars underground is being recognized. Already subsurface garages have been built in many cities.

In central Stockholm the Vanadis garage has been built in a two-floor rock cavern. It houses the refuse collection vehicles operating in central Stockholm and acts as a car park with space for 200 vehicles, as well as a fuel depot and a workshop for four vehicles (Fig. 3.61). The rubbish unloaded is compacted in a hydraulic press and stored in containers which are taken to an incineration plant by truck, each truck carrying the load from five vehicles. By keeping the vehicles underground in the centre of the city thousands of miles of driving are saved annually, saving fuel and environmental pollution.

Archives

The storage of documents is a problem everywhere. Commonly, archives are located in central positions in cities where space is limited. Expansion is rarely possible, except in distant suburbs or below ground.

The Swedish National Archive, with documents from medieval times up to modern computer records, is situated near the centre of Stockholm. It consists of offices above ground and a storage unit in a rock cavern below, and has been in use since 1968.

This solution provides protection against accidents (such as an aircraft crash), acts of war, and unauthorized access. The control of humidity and temperature is readily achieved and there is no dust, factors which are essential for the safekeeping of stored documents. The overall energy costs are some 30 per cent less than those for comparable surface buildings.

Cold storage

In cold storage it is essential to freeze the goods as quickly as possible to the desired temperature, usually $-25\,°C$. Then it is vital to maintain a steady temperature, as variations cause an increase in ice crystal growth with deterioration in the quality of the goods.

Cold stores in rock caverns are extremely well insulated, the thick rock surround providing favourable thermal inertia. Once that rock mass has reached its new, low, temperature, a high thermal stability is obtained with a minimum input of energy. Usually the energy consumption is 20–30 per cent less than for conventional storerooms. The large rock mass with its high cold capacity also provides a good insurance against interruption in the cooling.

Liquor storage

Storage of wine in barrels and bottles in caverns is common in many countries and has been in use for a long time. There are also modern facilities for liquor storage. One of them is situated in Årstadal, Sweden, where there are five rock caverns with a total area of 40 000 m² giving space for 20 million litres of spirits (mainly vodka and aquavit), more than 20 million bottles of wine and more than 11 million litres of wine in tanks. The rock maintains the appropriate temperature for such storage without any heating or cooling.

Sand

During the winter months large quantities of sand are spread on roads and streets in northern Europe in order to get better friction for vehicles on icy surfaces. The Tantolunden silo in Stockholm was designed as an underground cavern located in the city itself, just where the sand is needed (Fig. 3.62). The underground storage prevents the damp sand from freezing,

Fig. 3.61   The Vanadis garage, Stockholm.

and was built under a city park at low cost as a result of land purchase being minimal.

3.6.3. Underground location of installations

The layout and design of hydroelectric stations depend on local conditions. Flat watercourses and wide river valleys seldom provide any natural sites for high dams and commonly only low dams and moderate heads of water are available. In these circumstances the situation can be much improved by utilizing the subsurface and building power stations and tunnels underground to achieve whatever head is economically feasible.

The fundamental principle in underground construction views the rock mass as a construction material capable of standing with a minimum of additional support, if any. Tunnelling is often so extensive that it determines the construction time of the whole project. It is thus important to have as much information as possible about rock mechanics and other technical aspects of the location. A flexible approach to the design is frequently adopted, however, allowing for changes as the knowledge of the rock increases once tunnelling has started.

Pump generator hydroelectric stations provide a combination of base load and peak load pumped storage power. Such a station has been constructed in a rock cavern at Storjuktan at the end of a 20 km long tunnel between two branches of the River Ume Älv in northern Sweden (Fig. 3.63).

Tunnels take water from Lake Storjuktan to Lake Storuman 60 m below, and from Lake Blaiksjön situated 270 m above Lake Storuman. During nights and weekends, when there is a surplus of electricity, the generator acts as a motor and the turbine as a pump, lifting water from Lake Storjuktan to Lake Blaiksjön.

Both pumping and generating take place at about 85 per cent efficiency, resulting in some energy loss, but at Storkjuktan this loss is compensated for by the gravity flow of the water from Lake Blaiksjön to Lake Storuman. Energy is also gained as water is diverted from a river tributary, which has no power plant, to the Storjuktan power station and three other stations down stream.

Nuclear power plants [74] can be built underground to optimize the protective effect of the rock mantle not only against radioactive contamination of the biosphere, but also against external influences.

Such power plants can be constructed either in pits or caverns. Construction in pits (in unconsolidated sediments) involves the risk of hydraulic fracturing. For this reason, a pit depth of 60 m necessitates considerable lowering of the groundwater table; otherwise, vertical sealing (down to a depth of 120 m) of the sediments by diaphragm walls is necessary. In solid rock, construction will be in a cavern. Assuming that rock behaves elastically and that the primary rock stress is mainly vertical, the diameter and height of the cavern should hardly have any effect on the stability, and consequently the construction of very large caverns should be

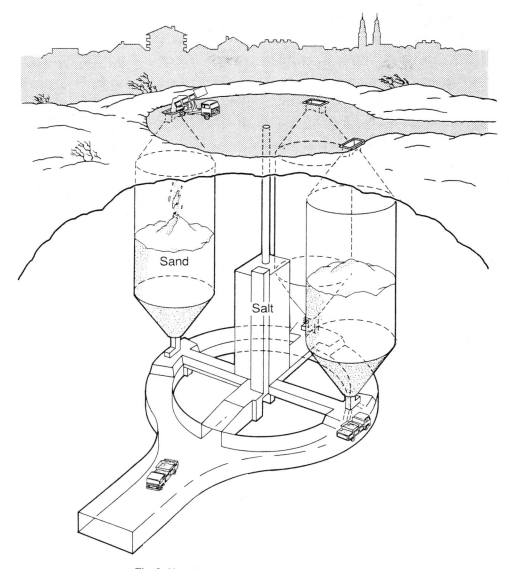

Fig. 3.62 The sand silo of Tantolunden, Stockholm.

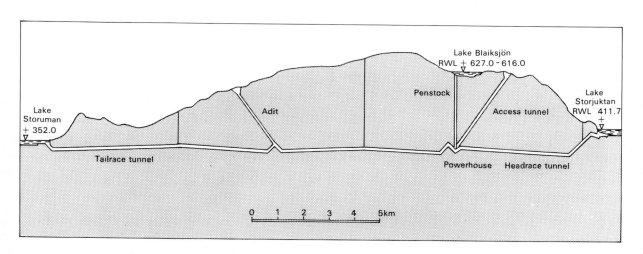

Fig. 3.63 The Storjuktan pump-generator hydroelectric station, Sweden.

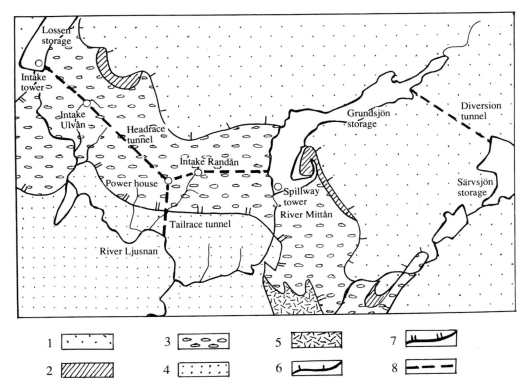

Fig. 3.64 The Långå power station, Sweden. 1. Quartzite—the Särv Nappe; 2. Dolomite—the Särv Nappe; 3. Augen gneiss—the Offerdal Nappe; 4. Sandstone—the Hotagen Nappe; 5. Granite—the Hotagen Nappe; 6. Thrust of the Särv Nappe; 7. Thrust of the Offerdal Nappe; 8. Tunnels for the Långå power station.

feasible under the right conditions. In foliated rock formations, however, the rock may be relatively weak and appropriate measures will be required to guarantee the integrity of the rock mass (such as rock bolts and shotcrete screening). Water in the joints may be a hazard to the construction and must be sealed off.

Earthquakes may be less of a hazard for underground installations than for those built on the surface, but there is no difference in the potential for internal accidents to take place. Preventative measures are essential to ensure that toxic radioactive material does not reach and pollute groundwater and surface water. These may include the installation of special wells downstream from the nuclear power plant to intercept contaminated groundwater.

Two examples of the subsurface installation of a power station are given below.

The Långå power station [165]

The Långå underground power station, situated about 150 km south-west of the town of Östersund in the centre of Sweden, harnesses water from five different watercourses. The power station is situated in the intermediate zone between the Caledonian mountain range and the Proterozoic basement rock area. The construction is made in sedimentary and metamorphic rocks forming nappes caused by tectonic movements from the west.

The geology of the area from west to east is:

- the Särv nappe (sedimentary rocks, sparagmites, quartzites, schists);
- the Offerdal nappe (augen-gneiss, granite, dolerite, schists);
- the Hotagen nappe (red sandstones).

The upper parts of the head-race tunnels were blasted in the Särv nappe. The power station was constructed in the Offerdal nappe and the downstream part of the tail-race tunnel in the Hotagen nappe.

The transition zone between the Särv nappe and the Offerdal nappe, consisting of a series of heavily fractured layers of various types or rock, caused serious problems in the upstream part of the Lossen head-race tunnel. The tunnel had to be completely lined with partly reinforced shotcrete for a length of 1 km (Fig. 3.64).

The augen-gneiss and granite of the Offerdal nappe are homogeneous with a low joint frequency creating no serious problems except one 40 m fracture zone with swelling clay related to a fault. A full concrete lining of the tunnel had to be applied there.

The Hotagen nappe caused no problems in the construction of the 75 m² downstream part of the tail-race tunnel.

Kvarnsveden power station, central Sweden [165]

At Kvarnsveden, an old power station built in a traditional dam was to be replaced by one built in the rock below the dam.

Diamond drilling was performed along the centre line of the proposed power station tunnel. The rock was strong granite except that at the very bottom of some of the drill cores a dolerite was observed. As the dolerite was only at the bottom of the holes it was not considered to be harmful to the project.

Later, when blasting of the tail-race tunnel was started, dolerite was observed in the wall. The possibility of a more extensive dolerite stimulated a further investigation. A dolerite dyke was observed close to the shore of the river, sometimes below the water level of the river (Fig. 3.65).

Detailed geological investigations were then performed on the existing drill cores, the surface and the outlet tunnel as far as it had been driven by that time. This investigation showed that the dolerite exposed on the river bank dips some 45 degrees straight into the tunnel and surge gallery, thus creating a dangerous situation with

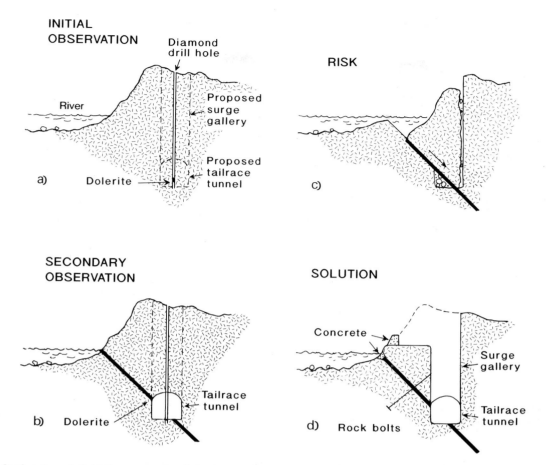

Fig. 3.65 Section through the tail-race tunnel and the surge gallery of the Kvarnsveden power station, Sweden. (a) Diamond drilling was performed along the centre line of the proposed tail-race tunnel and surge gallery. Observations of dolerite at the bottom of the boreholes were made; (b) dolerite was observed in the wall of the tail-race tunnel when the work of blasting the tunnel had started. Detailed geological investigation started; (c) the risk of a collapse of the wall into the surge gallery was apparent as the dolerite dyke could act as a lubricant on the slip-plane; (d) the wall rock was bolted and unnecessary rock at the top was blasted off in order to minimize the weight of the walls.

the risk of the wall caving in. The risk was especially pronounced as the dyke was partly chlorite-rich and thus could act as a lubricant on a fault-slip plane.

The investigation resulted in the walls being bolted and anchored, and unnecessary rock at the top being blasted away in order to minimize the weight of the walls.

Military installations

Installations in rock are protected against most kinds of acts of war and are usually more difficult to discover than conventional installations on the surface. This advantage is utilized in a variety of military installations: hangars at air-bases; naval bases including shipyards with docks, workshops, stores, command centres; and artillery installations.

In Sweden a system of monitoring airspace and providing combat information has been developed, with special attention paid to protecting the central components, notably the radar antenna system. This is constructed in a rock cavern, into which the antenna can be retracted. When retracted the opening is covered by strong shutters. In peacetime, the radar stations are used for civil aviation.

Civil defence installations

Underground shelters have been constructed in several countries for the protection of people against bomb or missile attack. Also, protected control centres for civil defence as well as for telephone exchanges etc, are commonly constructed in rock caverns. The air-raid shelters are used in peacetime also—as garages, storerooms, libraries, classrooms, assembly rooms, and recreational facilities—which provides a better economic utilization of the original investment.

Secondary use

In many cases underground cavities designed for one purpose can be used for another purpose. The value of the original investment is greatly enhanced if the facility is used subsequently in different ways. Cavities originally designed for oil storage can be redesigned for coal storage if the source of energy is changed, and a secondary use can be found for numerous mines. Much research is carried out in order to find new ways for utilizing abandoned mines.

There are many examples of mines which have been entirely or partly turned into museums. In most cases the museums are related to the mining. One example of this kind is part of the Falun mine, Sweden, one of the oldest mines still working. There, in the older part of the mine, a museum has been created with access into the old workings from the surface by stairs.

Mines have also been used as heat accumulators for district heating, as warm water can be stored there during the summer and be utilized for heating during the winter. Mines also function as laboratories for research and development. One example is the Stripa mine in Sweden, which originally was an iron ore mine, but after exhaustion was turned into a research facility for testing methods and theories related mainly to the storage of nuclear waste (p. 256). The headframe and the hoist are used for transportation of men and equipment into the mine where old drifts are utilized, but new cavities have also been created.

In the Kvarntorp mine, not far from the town of Örebro, south central Sweden, a quartzitic sandstone is mined. The method used is room and pillar mining with rooms 11 m wide, 4.8 m high, and 90 m long, separated by 6 m wide pillars. The sandstone is almost horizontal with little variation in thickness, providing good possibilities for rational exploitation. The mine started production in 1968. Underground mining was preferred as an opencast working would have a greater environmental impact and require considerable effort in site rehabilitation after cessation of mining. Furthermore, the sandstone is strong enough to provide safe roofs, although rock bolting is applied for security as blasting might cause minor cracks.

After mining has finished the 'rooms' are used for storage: for local government archives and a variety of goods. There are more than 200 rooms available with an area of 1000 m² each. The round-the-year temperature is 10°C, and the humidity is controlled by dehumidifiers. As the rock is almost white there usually is no need to paint the walls. Access to the rooms is by car, bus, or heavy vehicles

A museum is also planned for this mine; not one related to the mining, but a geological museum mainly focusing on fossils.

3.6.4. Extensive use of tunnelling

Electricity is usually distributed over long distances by surface systems and grids. In areas where conflicts with street traffic, buildings, parks, etc. make it impossible to use surface distribution cables, tunnels can be used. They provide good safety, access for maintenance and upgrading, and the possibility of coordinated use for other purposes. In areas of expensive surface real estate, the sale of the ground used for the old electrical grid can make a notable contribution to financing the tunnels.

District heating systems are common in built-up areas in the countries of northern Europe. They take advantage of the possibilities of the combined generation of electricity and heat, utilize waste heat from industries, or replace many small installations by one large plant, thus providing high efficiency. From an environmental point of view it is easier to get efficient flue-gas purification at big plants. The heat is distributed with hot water as the transport medium, ideally in large underground pipes which do not interfere with surface installations and urban activities generally. The pipes are safer underground and are easy to reach for inspection and maintenance.

Subsurface communication has been adopted to overcome problems of topography and of urban congestion. Examples of the former are the St Gotthard Tunnel in Switzerland overcoming a mountain barrier, and now the Channel Tunnel below the sea barrier between France and England (Fig. 3.69). A number of Western Europe cities, notably London and Paris, have extensive underground railways to overcome the problems of high-volume transport on the surface.

Tunnels are also used for safety reasons; for example, in steep areas where there are risks of landslides, railways are safer in tunnels than on steep hill sides.

Water being so essential for life, it is no wonder that tunnels have been used since ancient times for its transport to urban areas. The Siloam Tunnel in Jerusalem was constructed 2700 years ago and is still in use. It was excavated out of limestone for a length of 533 m. Today, tunnels compete with pipes of steel, concrete, and plastic, but they are most economical when large quantities of water have to be transported over distances of more than 1 km. Tunnels give good protection against sabotage, vandalism, and acts of war and ensure that water is kept cool.

The development of multipurpose tunnels is now proving to be economic in modern urban planning. As well as water supply pipes, these may carry cables for electricity and telecommunication, and district heating systems (Fig. 3.66).

Our use of telecommunications is expanding and increasing the load on existing networks. The network has to be upgraded to raise its capacity and transmission quality. Upgrading is easier in tunnels than in buried cables, and in city centres the work can be carried out without affecting traffic or causing accidents, and without causing subsidence.

3.6.5. Planning and construction

Tunnels and caverns have to be planned very carefully. This involves obtaining a good knowledge of the anticipated difficulties, which in turn affects the design, the number and type of supports, and their probable positions. Such knowledge is obtained through pre-construction geophysical and geological surveys.

Because it is virtually impossible to obtain knowledge of all the important rock variations from pre-construction surveys, the design of the tunnel or cavern tends to evolve as more information becomes available from the early excavations. Indeed, it is sometimes beneficial to commence with a small pilot tunnel to provide that extra information.

As the rock mass itself is the construction material it is obvious that the knowledge of geological factors which can change the general properties is most essential for projects of this kind. Such properties are joints filled with clay, graphite or chlorite, zones with very high initial compressive stress, and bedrock containing water. These will not stop excavation, but they will affect the cost of a project.

Some examples are now given showing the methodology used in planning underground excavations. Two water supply projects of great importance have provided much methodological experience. They are the Päijänne–Helsinki tunnel in Finland and the Bolmen tunnel in Sweden. Also the Channel Tunnel between England and France has been planned for many years and much work was done to get adequate knowledge of the geology, including the use of geostatistical methods.

Päijänne–Helsinki water supply tunnel [25]

Purpose and geology of the tunnel

Since 1982, the Päijänne tunnel has conveyed water from Lake Päijänne to the metropolitan district of Helsinki. It is 120 km in length, making it the world's longest continuous bedrock tunnel. With a cross-section of 15.5 m², the tunnel's flow rate—presently about 10 m³ s⁻¹—is achieved solely by means of a 36 metre water head. The tunnel also forms a large water reservoir where no deterioration of the water quality takes place. It was constructed in 1973–82.

After the selection of Lake Päijänne as the water source with adequate discharge, the route of the tunnel was determined by both

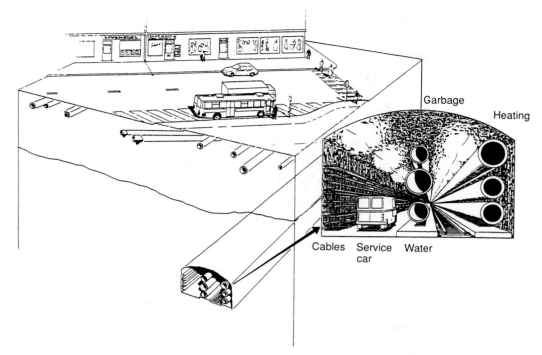

Fig. 3.66 Multi-purpose tunnel.

engineering and environmental factors. Geological features of the bedrock were most decisive for both purposes. Extensive geological field, laboratory, and desk studies were carried out in 1964–70, so that the weathered fracture zones of bedrock, in particular, could be avoided (Fig. 3.67) and the tunnel located as favourably as possible with regard to all the facilities needed at and near the adits.

The Geological Survey of Finland mapped the area, as well as the tunnel after the excavation. The main rocks are granitoid rocks and various gneissose or migmatitic schists. In the valleys, fractures and weathered rocks are common. As the tunnel could not avoid all weak fracture zones, these formed the chief objective of the geological site investigations. Also from the point of view of groundwater flow and its changes, the fracturing (faults, fissures, and joints) was very important.

Route selection and investigation required

For the comparison between many preliminary tunnel routes, the geological conditions had to be classified according to their physical and engineering properties, and directions of anisotropy measured in relation to the alignment of the tunnel. Mainly due to severely weathered fracture zones, it was estimated that lining or grouting the tunnel for about 10 per cent of the length would be necessary. The amount of lining needed was calculated mainly from the seismic velocities in bedrock. The anticipated lining cost remained at 5–6 per cent of the total blasting cost in the most favourable tunnel routes but rose to some 15 per cent in the less favourable preliminary options. Also the estimates for the definitive field studies (seismic sounding and diamond drilling) for each route option varied considerably and thus formed one of the criteria for the route selection.

Effects of changes in groundwater conditions

In addition to the direct effect of groundwater on engineering operations, environmental and economic effects of changed groundwater conditions could be expected.

Owing to rapid grouting, the groundwater changes actually caused little harm. The changes in the groundwater table were carefully monitored during the construction period, and the quality of

the water in nearby wells was examined. The tunnel driving caused some 250 out of 850 wells in the area to dry up until 1980. This problem practically ceased when the tunnel was filled with water.

Role of geology in different phases

In summary, the bedrock quality and water movements, as well as the route selection and the environmental effects of the tunnel, were greatly dependent on the fractures and weathering of the bedrock, on which only sketchy information was available initially. Therefore, in addition to the existing mapping of the Geological Survey and the field studies, it was necessary to determine how certain physiographical criteria reflect the fracture and weathering zones quantitatively and areally. The analysis showed that the variables that best correlated with the *in situ* measured (or 'true') fracturing and weathering were related to the width, depth, and length of the valleys and the toughness/brittleness of the rocks.

The use of such physiographical criteria for the engineering assessment of the fracturing and weathering was a decisive method in the preliminary planning studies for the Päijänne tunnel and the prediction of the associated environmental influences. This procedure was rendered possible by the adequate length of time (1965–70) allowed for the preliminary geological investigations. Consequently this saved an estimated sum of several tens of millions of Finnish Marks that would otherwise have been needed to grout or line the areas most affected by fracturing and weathering.

Due to the extreme length of the tunnel and the great differences between the different route options, the case exemplifies the fundamental environmental and economic importance of careful geological site investigation studies.

The Bolmen tunnel [165]

The Bolmen tunnel forms part of a water supply system for the southern counties of Sweden. It carries untreated water in an unlined 80 km long tunnel in crystalline rock from Lake Bolmen to the vicinity of Perstorp, where it enters a pipeline to the existing water treatment plant at Lake Ringsjön.

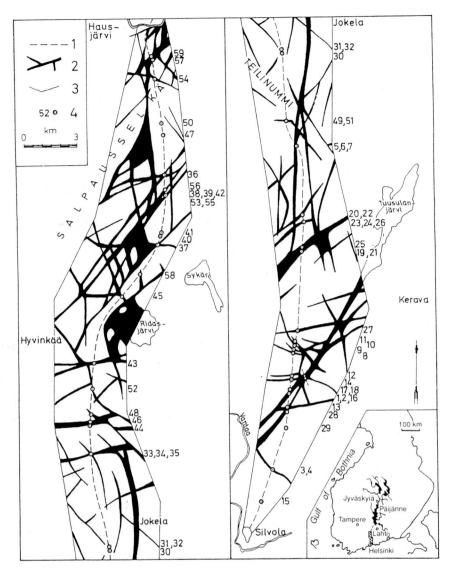

Fig. 3.67 The Päijänne–Helsinki water supply tunnel. 1. Tunnel route; 2. Bedrock valleys representing the weathered and fractured zones; 3. Limits of the zone studied; 4. Diamond drilling point.

Site investigations

The geological investigations for the Bolmen tunnel were carried out between 1967 and 1972. From both economic and environmental points of view it was necessary to base the route of the tunnel on the best possible knowledge of the geology of the area considered. At the start the locations of the intake and the outlet only were fixed. The area between is quite flat with extremely few outcrops of rock. The existing geological maps of the area are old (from the nineteenth century) and on a very small scale 1:200 000. It was decided to perform phased investigations of the geology in order to find the best route for the tunnel (Fig. 3.68).

Stage 1

Geological data collected from older boreholes, construction works in rock and other available information were used. Airborne magnetic measurements and ground-level geophysical surveys (magnetic and seismic) were performed and a basic tectonic assessment was made, resulting in the selection of a narrower zone within which the tunnel route should be located.

Stage 2

Detailed geological mapping and more seismic investigations were included in this stage. Magnetics was used to locate dolerite dykes in the southern part of the area. The investigations were supported by limited diamond drilling. From these investigations a narrow corridor was selected.

Stage 3

The investigations were performed in detail with the objective of finding the most favourable route, especially through zones of fractured and crushed rock.

Seismics was used along the entire length of the tunnel and diamond drilling was carried out at several of the presumed fracture zones. All these investigations provided the basis for a geological–tectonic model of the area.

Tunnelling stage

The model mentioned above was improved during the tunnelling as all the experience gained was used and when necessary, supple-

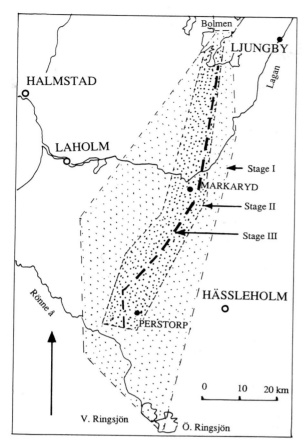

Fig. 3.68 The Bolmen water supply tunnel, Sweden, with the area covered by investigations in Stage I and II. The final tunnel as decided in Stage III.

mented by seismic and electrical measurements, including resistivity. This improved knowledge was utilized for making some ten local changes in the original tunnel route resulting in considerably shorter passages through zones of weakness requiring support work.

The Channel Tunnel [59] [61]

The decision to build a tunnel under the English Channel was taken in January 1985. One year later the French and British Governments announced that there would be a twin-bore tunnel—only for trains, between Folkestone and Coquelles (near Sangatte)—a distance of 50 km.

The first core drillings were carried out between 1958 and 1965, but it had been known for more than a hundred years that the geological formations cropping out on both sides of the Channel are continuous across it. The basic geology comprises generally north-easterly inclined Cretaceous strata, distinguished by the Chalk and underlying Gault Clay, which are separated by a thin transition stratum known as the Glauconitic Marl (Tourtia).

The upper and middle sections of the Chalk comprise the characteristic White Chalk. The lower section is progressively more grey towards the base, reflecting the increasing proportion of clay. At the base of the lower Chalk, the Chalk Marl occurs. The rock, consisting of chalk and clay, is relatively strong, stable, generally free from open discontinuities and hence virtually impermeable, and has long been considered an ideal medium for tunnelling. The underlying Gault Clay is also impermeable, but it is weaker and more plastic and will probably show more tendency to time-dependent deformation as a result of stress changes due to tunnelling. The thickness of

the Gault Clay decreased from some 45 m on the English side to some 15 m on the French side.

It was decided to keep the tunnels within the Chalk Marl (Fig. 3.69), and much effort was devoted to getting as good a knowledge of the geology as possible. Of special interest was the precise location of the top of the Gault Clay.

The 1958–65 drillings resulted in a change of the planned tunnel route on the British side, and supplementary drilling was carried out along the new alignment.

Marine geophysical surveys were performed during the same periods as the drilling. A quantitative interpretation of the seismic survey data and an assessment of its accuracy became available only from the last phase of drilling in 1986–88.

Control of accuracy and validity of interpretation, both with regard to the actual data and the combination thereof, and with regard to the interpolation, is a necessary requirement for a satisfactory analysis of the data. This was achieved by a systematic analysis of the conditions of acquisition and of the scope of variation of each of the parameters included in the final interpretation, and by the use of geostatistical methods at all stages. These methods, which take into account the specific accuracy of each of the data, the distance between data points and the variability of each of the different zones of the project, made it possible to quantify the reliance which can be placed on the prediction of the levels of various geological surfaces.

The evaluation indicates that the level of the base of the Glauconitic Marl over the route corridor can be predicted with a high level of confidence to within ±5 metres. The Chalk Marl has a thickness in the Channel generally of not less than 25 metres and the diameter of the tunnels will be less than 8 metres. The tunnel alignment has been correspondingly positioned some 5 metres above the base of the Chalk Marl wherever possible. The variation of ±5 m considered in the context of a water depth of about 45 to 60 metres and a thickness of soil and rock above the proposed tunnel level of a further 45 to 70 metres, implies that the order of accuracy expected is generally within about ±5 per cent.

3.6.6. Conclusions

Whilst there is now much experience in Europe of utilizing underground space, the potential for its use in a wide variety of environmentally friendly and economic applications is still, in most countries, a curiosity. Yet, excellent examples exist in Western Europe of the wise and efficacious use of this very large resource which lies literally under our feet. The Geological Surveys are in an ideal position to help their governments and industries to plan and develop its potential to the maximum benefit.

RECOMMENDED FURTHER READING

Blanchin, R., Margron, P., and Piraud, J. (1990). Tunnel sous la Manche: les calculs géologiques à l'épreuve de la réalité. *Compte-rendu des journées internationales de l'Association Française des Travaux en souterrain (AFTES) 'Franchissements souterrains pour l'Europe'*. (Balkema, Rotterdam).

Collective Authors (1980). Subsurface use for Energy Savings. *Annex from the Rockstore-80 Symposium*. (BeFo/Swedish State Power Board/SWECO, Stockholm).

Crighton, G. S., Sechet, B., Rankin, W. J., and Margron, P. (1988). The Channel Tunnel; supplementary marine suite investigations 1986–1987. *Proceedings of the 5th International Symposium of Tunnelling, London*. (The Institute of Mining and Metallurgy, London).

Duffaut, P. and Margron, P. (1989). Le tunnel sous la Manche, géologie et géotechnique. *Actes des journées d'études*. (Presses de l'Ecole Nationale des Ponts et Chaussées, Paris).

Rehbinder, G. (ed.) (1988). Hot water storage in rock caverns. (Swedish State Power Board & Swedish Rock Engineering Research Foundation (BeFo), Stockholm).

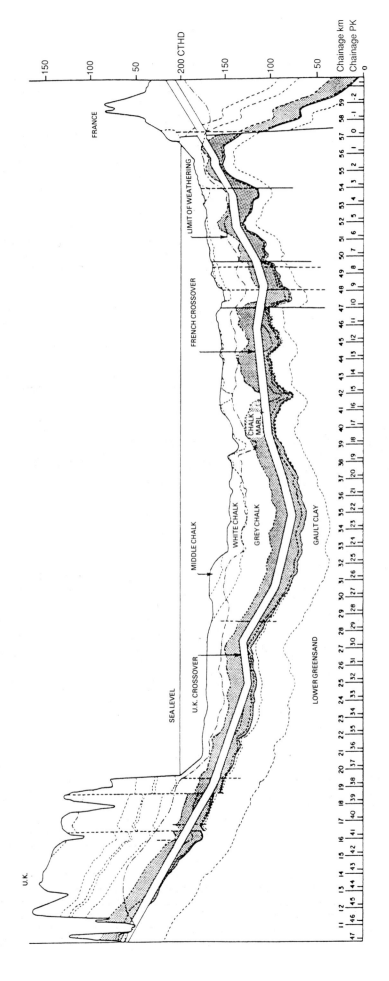

Fig. 3.69 The Channel Tunnel.

Saari, K. (ed.) (1988). *The rock alternative engineering*. (Finnish Tunnelling Association, Espoo).

The Swedish Power Association and the Swedish State Power Board (1981). *Hydro power in Sweden*.

Windqvist, T. and Mellgren, K.-E. (eds) (1988). *Going underground*. (Royal Swedish Academy of Engineering Sciences, Stockholm).

3.7. THE SOIL COVER [70] [65]

3.7.1. A 'non-renewable' resource

Commonly soil is not included in any consideration of the geological environment. Yet it is the material with which man has closest contact, it is a resource of the greatest importance as it supports all our agriculture and forms a fundamental link in the food-chain and it is derived directly from the surface rocks of the Earth.

In geological terms, the soil cover was formed during the 'Recent' period and mostly during the past tens of thousands of years. With respect to that time-scale soil can be considered to be a renewable resource, but in terms of a human generation we are dealing with a non-renewable resource, a 'heritage', the conservation of which is essential. This implies not only that we should try to avoid, or minimize the erosion of soil, but also that we must avoid the degradation of its intrinsic qualities by inappropriate usage, particularly by unsuitable or extreme agricultural practices.

The most important element in the formation of soil is the influence of climate on rock. Second is the effect of vegetation which itself is determined largely by climate. The processes are slow but continuous, and begin with the weathering of rock at the surface. Successive evolutionary stages follow with the progressive alteration of minerals, which is much influenced by any vegetation taking root.

This evolutionary pattern is possible only when erosion is at a low rate, with material being removed less quickly than soil is being formed. This process leads to the characteristic soil profile R-C-B-A. R is the rock or substratum on which the soil is developed; the C-horizon corresponds to weathered rock materials that as yet show no soil structure; the B-horizon is a mineral soil *sensu stricto*; the A-horizon is the soil level at which biological action is greatest. Between the A- and B-horizons a leached horizon E commonly occurs, particularly in the case of 'podzols', 'luvisols', or 'planosols'. In non-cultivated soils, humus consisting of organic debris that has not yet been integrated in the soil-formation process is found on top of the A-horizon. This organic input from above, further aided by infiltration, allows the development of vegetation and animal life on and in the soil, and this in itself is an essential component of soil formation. In many places only part of the complete soil-profile is present.

In cultivated soils, the organic input from the surface is formed by crop residues that are incorporated into the soil surface. In natural soils the equilibrium of the organic component may be modified by artificial practices, such as drainage or the homogenization of the surface layers.

3.7.2. The limitations of soil classifications and the notion of soil aptitude

The existing classifications are still somewhat disparate and arbitrary from one country to another. In Europe, they are based mainly on the difficult interpretation of the formation process for the soil concerned. Because of this, the recent trend has been towards classifying from a better understanding of more objective criteria, such as morphology, physical

and chemical characteristics, and functional properties that are essentially related to water. Table 3.12 presents and equates simplified versions of two modern classifications in use in France.

The main factors of differentiation in the soil cover are climatic variations related to latitude and altitude, types of geological substratum, the geomorphological history, and types of land-use (natural vegetation, cultivated land, habitation, recreation, etc.). In most areas with temperate climate, the geological and geomorphological factors are dominant.

Soils are best considered with respect to their degree of aptitude for a specific use in a given technical and socio-economic context. This degree of aptitude relates to the interplay between several criteria: the characteristics of the soil cover, the local climate, the state of cultural techniques, the needs of plants themselves, and the requirements of man the user. The actual aptitude of an area takes account of the modifications it has undergone already; the potential aptitude varies according to the type and the importance of the improvements that can be made. This notion of potential aptitude is totally flexible, and is of considerable importance when comparing the intended use of soil with the diversity of the resource as a whole and the possibilities for development that arise from that. This does not mean that any soil can be used for any purpose; the optimum position must be found between its intrinsic qualities and socio-economic considerations. For instance, a given agricultural plot, in its function of supporting man, is evaluated quite differently according to whether one wishes:

(1) to grow 100 bushels of wheat per hectare;
(2) to grow vegetables;
(3) to grow animal fodder;
(4) to grow grapes for wine;
(5) to use the land as a tree nursery.

For this reason, care must be taken not to give priority to certain production-oriented uses only. These points underline the fact that, when a 'living' entity such as soil is involved, the evaluation of the aptitude of land must be kept simple, with flexible consideration of choice and use.

3.7.3. Information and management systems

The question that can be posed at this point is whether the available geological information systems are suitable for such a flexible approach to soils. The answer to this question varies for each country, and even for each region in some countries, but, in general, such tools are still mostly unsuitable and there are several reasons for this situation.

First of all, soil classifications are commonly incomplete and based on systems which are insufficiently representative. The scale of data acquisition is often unsuitable (too small) when compared to the scale on which decision-making usually takes place. Nevertheless, systematic work on a regional scale, such as the soil-cover maps of The Netherlands and Great Britain at the scale of 1:50 000, the Bodenkataster system of Lower Saxony at 1:5000, the soil-quality maps of Bavaria, and the computerized soil maps of Northrhine-Westphalia, has been carried out over many years and the results are used for many different thematic purposes, such as establishing the soil potential or the vulnerability of soils.

Other maps at the smaller scales of 1:100 000, 1:200 000 and 1:1 000 000 serve primarily to show a national, or supranational inventory or, in some cases, simply to fill the gaps that still exist at a larger scale. For small scales, remote-

Table 3.12 Example of modern soil classifications used in France

FAO legend (1974) (1)	CPCS, France (1967) (2)	RP, France (1990) (3)	Summary and description
Fluvisol	Young soil of alluvial or colluvial origin	Fluviosol & Colluviosol	Little differentiated alluvium and colluvium soil, not peaty, not hydromorphic, not sandy or salty
Gleysol	Hydromorphic soil, little humus and moderately organic	Reductisol & Redoxisol	Not-peaty soil dominated by excess seasonal or permanent water
Regosol	Raw mineral soil derived from erosion	Regosol	Thin and little differentiated soil formed on soft/unconsolidated rock
Lithosol	Raw mineral soil derived from erosion	Lithosol	Very thin (<20 cm) soil on hard rock
Arenosol	No equivalent	Arenosol	Thick, uniformly sandy, non-hydomorphic soil
Rendzina	Rendzina, brown calcareous or calcic soil	Rendosol, Calcosol & Calcisol	Calcareous or calcic soil with a thickness of less than 50 cm
Ranker	Young humic soil	Ranker	Thin, acid, humic soil developed on hard rock
Andosol	Andosol	Andosol	Humic, fertile soil formed on pyroclastic volcanic material
Vertisol	Vertisol	Litho-Vertisol and Topo-Vertisol	Very clayey, calcic, swelling soil
Solontchak	Saline soil	Salisol and Sodisol	Little differentiated, well-structured, marine alluvium rich in Ca and Na
Cambisol	Brown acidic, calcic and calcareous soil	Brunisol, Calcosol & Calcisol	Little differentiated soil, well aerated, high biological activity, calcareous, calcic, neutral or acid
No suitable term	Soil with surface pseudo-gley	Pelosol	Little differentiated, clayey soil with poor macroporosity
Luvisol	Brown leached soil	Neo-Luvisol and Luvisol	Thick soil, strong textural differentiation, poorly hydromorphic
Podzoluvisol	Glossic leached soil	Degraded or dernic Luvisol	Same as above, but highly hydromorphic and strongly acid (in forests)
Podzol	Podzol, ochre-podzolic soil	Podzosol	Sandy soil, raw humus, very acid, very poor in chemical composition
Planosol	Pseudo-gley soil with perched water	Planosol	Strong textural differentiation and impervious horizontal floor
Histosol	Hydromorphic organic soil	Histosol	Peaty soil formed by accumulation of organic material in water

(1) Food and Agricultural Organization
(2) Commission de Pédologie et de Cartographie des Sols
(3) Référentiel Pédologique Français

sensing has become an important tool of acquisition, particularly where the images are calibrated with ground features.

Traditionally, research and development has been oriented towards understanding the mechanisms of soil formation, rather than towards the search for and promotion of more adequate tools for soil management. Recently, however, the latter area of enquiry has been gaining ground, and there is now a greater awareness of the importance of relevant and flexible Geographical Information Systems.

In agricultural exploitation, economic considerations determine the choices that are made. If the future strategy for the management of soils is to succeed, it will have to be based much more on cost–benefit considerations than on the overall potential of the soil. In the same vein, the organizations that manage agriculture will inevitably be more involved in agri-cultural statistics (surface area of crops, quantities produced, types and amounts of fertilizers used, etc.) than in the value of the ground itself. Nevertheless the latter will continue to be of paramount importance in the wider field of planning for development.

3.7.4. Conclusions

Much R&D work remains to be done on achieving suitable and reliable tools for monitoring and managing one of man's most precious, yet vulnerable resources. The objective, both regional and international, must be to develop tools that are flexible in their use and based on ground-rules common to all countries. Only then will the optimization of relevant decision-making processes become available.

4

Impact of geological hazards to development

P. Masure, with the collaboration of H. M. Kluyver and G. Sustrac

4.1. GENERAL STATEMENT OF THE PROBLEMS INVOLVED [53]

4.1.1. An unstable geological environment

Since the human species first appeared, at least five million years ago, it has had to suffer about twenty glacial periods, each of about 100 000 years, separated by much shorter (about 10 000 years) warmer periods. At first slow and progressive, the cooling-off process that marked each glacial period accelerated after several tens of thousands of years. The general effect was the spreading of ice caps on mountain areas, coupled with a lowering of the sea level that could be as much as 120 m. Our ancestors, faced with polar ice over northern Europe, the disturbance of regional climatic conditions, a strong reduction in floral and faunal diversity, and the intensification of erosion and deepening of stream beds, were certainly not spared their share of natural disasters. These were much worse at the beginning of the interglacial periods with the concomitant fast rise in temperatures (about 10 °C). The resulting melting of the ice, raising sea levels, regression of coastlines, calving of glaciers, flooding from (sub-) glacial meltwater lakes, slope failure and mudflows, drought and desertification in the more southerly latitudes, and increased seismic activity, all created a particularly hostile environment.

Today, we live in an interglacial period that began about 10 000 years ago. It started with a general warming up that may have been as much as 8 °C over about 2000 years. Temperatures 5000 years ago were slightly higher than today's averages and the sea level has risen about 70 m in 8000 years. The climate is now relatively stable with periodic, but slight, fluctuations in the average temperature. The last of these was the Little Ice Age (fifteenth to nineteenth centuries), during which average temperatures in Europe were about 1 °C lower than before.

Geodynamic changes to the Earth's crust, on the other hand, take place over millions of years. In spite of the resulting general stability of the physical environment, our social structures and their developments are often affected—and even disturbed—by geological and/or climatic phenomena that are concentrated in time and space. A true perspective of our environment and its future can be got only from studying all natural phenomena and their interactions.

Earth scientists have much experience of making specific analyses of such phenomena in retrospect, but now they are increasingly called upon to forecast and to be concerned with probability, and with the evolution of instability factors that affect the Earth's surface. The pressure of population growth and the technological development that increasingly allows man to influence his environment, contribute to the degra-

dation of the fragile physical environment. Here, one can speak of a truly 'anthropogenic' risk, where the impact of human activity adds to the natural dangers that manifest themselves on a local, regional, and even planetary scale.

Since the Second World War, the frenetic deforestation in Third World countries has destroyed about 40 per cent of the extant tropical forest. At this rate, in 20 or 30 years' time the forests will have disappeared in many countries, thereby causing flooding, soil erosion, and desertification. The effects, particularly from Africa, will be felt in southern Europe. In many other countries, rural exodus creates a similar effect, with major landslides becoming increasingly common, and with the degradation and erosion of soils having unprecedented economic consequences. The EC policy of 'freezing' the amount of arable land could also lead to an acceleration of this process, as well as the ever-increasing development of major civil engineering and infrastructure works. The deep injection of fluids in the sub-surface for example, such as can happen during the filling of a major reservoir, can cause seismic shocks, some of which can be catastrophic.

In addition to the degradation of the ozone layer in the atmosphere, the increase of carbon dioxide and methane (25 and 240 per cent, respectively, since 1850) is causing concern. Both effects are due to man's activity, the former because of our use of chlorofluorocarbons, the second because of the massive use of fossil fuels. The increase of CO_2 and CH_4 contributes to the so-called greenhouse effect, which will increase the average temperature at ground level by several degrees. According to presently available climatological models, this will cause quite perceptible local changes as the temperature rises. The climatic disturbances resulting from a temperature increase of just a few degrees will cause increased flooding and desertification, increased cyclone activity in tropical regions, and a considerable rise in the average sea level.

Even though these global phenomena merit full analysis, the present and more localized problems of the environment are of sufficient concern to warrant much attention.

4.1.2. The underestimated effects of natural disasters

Violent earthquakes, volcanic eruptions, landslides, tropical cyclones, drought, and floods have killed well over 3 million people over the past quarter century, at the same time adversely affecting the lives of a billion others and causing damage of well over US$1000 billion. Since 1960, the number of victims of natural catastrophes has increased by about 6 per cent per year, which is more than double the rate of increase of the world's population. From 1980 to 1986, for example, 574 million people were affected by the above-mentioned types of disaster.

Tangshan, Mexico City, Medellin, San Salvador, the Sahel, Bangladesh, and Armenia bring to mind the major catastrophes that have affected the five continents in recent times. Europe has known its share as well, and the catastrophes of Skopje, Montenegro, Irpina, and Valtellina are still fresh in our memories. Even so, modern society has difficulty in recognizing the ever-increasing tragedy caused by natural catastrophes and is unwilling to take direct responsibility for improving the situation.

The risks related to dangerous natural phenomena are the product of their probability of occurrence (hazard) and their predicted damage in terms of human life and economic loss. The large increase in natural disasters during the past 30 years is in part due to the degradation of the natural environment, but it is particularly the result of unbridled and unplanned urban expansion. In the year 2000, 50 per cent of the world's population will live in cities and towns, concentrated on only 0.4 per cent of the available land surface. These sites, occupied in many cases for centuries, have seldom been assessed for the potential danger from geodynamic phenomena. Outside northern Europe, most cities had only limited development until the beginning of the twentieth century. Their present rapid and unplanned growth increasingly causes grave natural risks that, until now, have not been recognized. The hyper-concentration of people, goods, infrastructure, industry, and services in urban areas now renders humanity particularly vulnerable to disaster.

The general population increase and political deficiency in many developing countries, by far the most affected by natural catastrophes, further exacerbate their vulnerability. One should not forget, however, that, in the past, Europe was also affected by geodynamic phenomena with catastrophic consequences, and this notwithstanding a much smaller population than today. Earthquakes, for instance (see Table 4.1) have strongly affected Mediterranean countries where, even today, very serious catastrophes could occur.

Generally speaking, natural events have caused 94 per cent of the victims of major catastrophes (see Section 4.3.1), and it can be assumed that the economic damage caused by such natural events is of similar proportions, even though statistics are not clear on this point.

Most natural catastrophes correspond to risks of climatic origin (cyclones, storms, and drought), of geodynamic origin (earthquakes, volcanic eruptions, landslides, and mudflows), or of mixed origin, such as flooding. Placing flooding under climatic risks, one sees that over the period 1960–80 (Table 4.2), with about the same number of climatic and geodynamic events, the latter caused about twice as many victims, even though the material damage caused was hardly more than that caused by climatic events.

As is shown by Table 4.3, the geographic distribution of natural catastrophes over the same period was very unequal.

These figures conform perfectly with the disasters recorded by the Swiss Reinsurance Company for the period 1970–85, which serve further to refine the analysis in relation to the developed countries, based on the figures shown in Table 4.4.

Table 4.4 shows that, with about 2.5 times as many natural catastrophes, North America with only half as many victims as were recorded in Western Europe, and the insurance coverage rates were apparently similar. This very clear difference in the vulnerability of the populations to natural catastrophes, this apparent form of social 'injustice', is not inevitable. On the contrary, it is possible for a modern society to organize itself by implementing an efficient policy for reducing the effects of a natural catastrophe, and the table shows that, in this respect, the European countries are lagging behind North America.

Table 4.1 Loss of human life through earthquakes. This list is certainly incomplete and many of the figures are uncertain, in particular those for the older events. Little is known of the toll taken by the major earthquakes that shook Szechuan (1580), Calabria (1638 and 1693), Tokyo (1703) and Quito (1797). No data are available for the violent earthquakes that occurred in China in 1966 (two), 1970, 1973, 1974, and 1975

Year	From 30 000 to 50 000 casualties	
342	Antioch (Turkey)	40 000
565	Antioch (Turkey)	30 000
856	Corinth (Greece)	45 000
1456	Naples–Brindisi (Italy)	30 000
1641	Tabriz (Iran)	30 000
1755	Kashan (Iran)	40 000
1915	Avezzano (Italy)	30 000
1935	Quetta (Pakistan)	30 000
1939	Chillan (Chile)	30 000
1939	Erzincan (Turkey)	25 000–30 000

Year	From 50 000 to 100 000 casualties	
1268	Seyhan Province (Turkey)	60 000
1667	Shemakha (USSR)	80 000
1727	Tabriz (Iran)	77 000
1755	Lisbon (Portugal)	60 000
1783	Calabria (Italy)	60 000
1908	Messina–Reggio (Italy)	82 000
1927	Tsinghai (China)	several tens of thousands

Year	More than 100 000 casualties	
1202	Aegean Sea	100 000
1290	Chengde (China)	100 000
1556	Shanxi–Henan (China)	800 000–1 000 000
1730	Beijing (China)	100 000
1737	Calcutta (India)	300 000
1920	Gansu–Shaanxi (China)	100 000–180 000
1923	Tokyo–Yokohama (Japan)	143 000
1976	Tangshan (China)	officially 220 000 but probably 600 000–800 000

4.1.3. Effective use of existing means for reducing natural disasters

Beyond the tragic toll paid by the victims of natural catastrophes, the economic and financial foundations of numerous developing countries are seriously affected as well. The Economic Commission for Latin America and the Caribbean has estimated that, for the period 1960–74, the direct effect of natutral catastrophes on the public sector of Central America alone accounted for an average of 2.3 per cent of the GNP. This figure is not exceptional for much of the Third World, for which natural catastrophes form a real brake on development.

Even though the rich countries were capable of putting in place relatively effective measures for the protection of their inhabitants, the material damage recorded from the most exposed countries is considerable. The USA forecast that for

Table 4.2 A comparison between natural disasters and the resulting worldwide loss of life between 1960 and 1980

Origin of catastrophe:	Climatic	Geodynamic	Total
Number of events	21	23	44
Dead	413 000	840 000	1 253 000
Average dead per event	19 700	36 500	28 500
Average dead per year	37 500	76 000	113 500
Origin of catastrophe:	Climatic	Geodynamic	Total
Dead			
Less than 100	11	5	16
Between 100 and 1 000	7	5	12
Between 1 000 and 10 000	2	7	9
Between 10 000 and 50 000	0	4	4
Over 50 000	1	2	3

Table 4.3 Distribution of natural catastrophes as a function of the economic development of countries

GNP of country	Number of catastrophes	Number of victims	Percentage
Low	329	1 090 900	76.0
Medium	451	335 000	23.3
High	79	10 000	0.7
Totals	859	1 435 900	100.0

the coming three decades, the damage caused by natural catastrophes throughout the country will be 0.4 per cent of the GNP and as much as 1.8 per cent of the GNP of the State of California. A similar estimate for Italy is 0.6 per cent of the GNP and in Spain it may be as high as 1 per cent.

As has been shown by some of the most vulnerable countries, such as Japan, most natural catastrophes can be considerably attenuated, if not entirely avoided. Scientific knowledge does not yet allow the accurate prediction of the timing of the paroxysmal phase of threatening phenomena. On the other hand, the means of evaluating risks and of identifying vulnerable areas are increasingly well understood. It is now possible to identify plains prone to flooding, unstable slopes, avalanche paths, areas where seismic or volcanic dangers (direct or induced) are real, and areas where subsidence is likely. Also, constant improvements are being made in the

methodology for determining or forecasting the intensity of the impact of such hazards.

Thus the means for preventing, reducing, and controlling natural catastrophes exist. Their use is still limited in many of the industrialized countries. This is not justified by any cost–benefit analysis of the investment required to reduce natural catastrophies. The cost of the actions necessary to reduce natural risks is usually very much lower than the losses to be avoided. The ratio of preventative cost versus loss reduction in California, for the period 1970–2000, was estimated to be 12 per cent for landslides, 20 per cent for seismic and volcanic hazards, and 60 per cent for flooding. The value of such preventative investments is even greater when they are integrated into urban or regional development plans.

Certain legal and institutional restraints, including those of indemnity, lower the potential for reducing the economic impact of natural disasters. The publication of maps showing the vulnerability to natural hazards, for instance, may affect the individual by reducing land and property values. Furthermore, potential dangers can be ignored and preventative action avoided, as a result of the unrealistic automatic indemnities arranged in insurance against natural catastrophes in the absence of hazard maps and other means of warning.

Faced with an increasing number of natural catastrophes that affect mankind, geologists, together with other scientists such as hydrologists and climatologists, see their role being increasingly identified with the prevention of harmful effects of natural phenomena. To this end, they have developed new techniques and methods in various fields such as seismic

Table 4.4 Distribution of disasters due to natural catastrophes in developed countries

Region	North America		Western Europe		Rest of the world	
	Number	%	Number	%	Number	%
Catastrophes	151	22.5	59	8.8	461	68.7
Victims	3 517	0.2	7 180	0.5	1 411 662	99.2
Insured damages (million dollars)	10 976	61.8	3 773	21.2	3 023	17.0

engineering, geotechnics, and hydrogeology. Supported by their thorough knowledge of the geosphere, they have a pivotal role to fulfil within the multidisciplinary teams concerned with reducing the impact of natural catastrophes. So far, they have not been afforded the means of applying and integrating their expertise to the maximum effect in risk prevention.

4.2. TYPES OF HAZARD AND REGIONAL CHARACTERISTICS

4.2.1. Seismic hazards [57]

The seismicity of Europe

Figure 4.1 and the geological accounts in Sections 2.1 and 2.3 illustrate the distribution of seismicity in Western Europe and the Mediterranean basin, as well as the interpretation of this area in geodynamic terms. This shows that the seismicity of this area and the underlying deformation of the Earth's crust are a function of intra-plate or plate boundary tectonics. In a schematic sense, such deformation is the result of stresses that accumulate in the crust under the effect of two main movements: the opening of the North Atlantic and the northward movement of the African plate against the Eurasian plate. Movements are generally of the order of 2 cm per year.

The area of collision between the continents of Africa and Eurasia corresponds to a zone of major seismic activity that is strong in frequency and intensity (in the sense of liberated energy). It borders the northern part of North Africa, then follows the boundary of the Adriatic microplate from the Apennines, along the Alps to the Dinarides. In the eastern Mediterranean, this boundary becomes even more complex with the existence of the Aegean and Anatolian microplates, around and within which the liberated seismic energy is particularly strong: in Greece it represents 2 per cent of the world's total, compared with a relative surface area of only 0.09 per cent.

On the other hand, in the north-west European platform areas, beyond the great Alpine thrust belts, the progressive deformation of the Earth's crust can be very slow. This is shown by a much lower frequency of earthquakes, in particular major ones with a magnitude of more than 6.0–6.5 on the Richter scale,* with a return period that can be more than 1000 years. Such potential events must be taken into account in the planning of any installations for which the consequences of damage could be catastrophic for the population and the environment.

Historical seismicity, such as that of the Mediterranean basin (see Fig. 2.16 and Table 4.5), shows the major destructive earthquakes that occurred since the year AD 1000. Even though the maps underline once again the intense seismic activity in Italy, Yugoslavia, Greece, Turkey, and North Africa, they also show that seismic activity in other areas is far from negligible. Some examples are:

(1) the western and northern borders of the Alps, with the earthquakes of Basle (1356), the Nice hinterland (1564, 1644), Liguria (1887), and Provence (1909);

(2) the eastern (1428) and western (1660, 1750, 1967, 1980) Pyrenees;

(3) the Rhine Valley area in the wider sense, from the Alps to the North Sea: the Rhine Graben and Vosges (1682), Swabia (1911, 1978), and the Liège/Aachen area (1692, 1983);

(4) the Massif Central (1477, 1490) and the south border of the Armorican Massif (1799);

(5) the North-Artois (1580) and Flanders (1938) transcurrent-fault zone;

(6) the North Sea and western Baltic (1789, 1904, 1927, 1931);

(7) Central England (1903, 1904, 1905), Wales (1906, 1989), and Scotland (1901).

The magnitude of these earthquakes was moderate, but there is at present no way to rule out the possibility that future events may be larger, even though the probability must be low.

Recent or neotectonic movements in Western Europe (p. 000) include all faults with a known or probable Pliocene–Quaternary displacement (deduced from geological observations and seismic data) and all vertical movement (deduced from geomorphological observations and the comparison of high-precision levelling data where available). One of the essential features of western European neotectonics, is that most of the recently active structures are known to have had long and complex geological histories. Examples are provided by the repeated movements along the French 'Sillon Houiller' (Coal Graben) and the Rhine Graben faults.

The comparison of seismicity and recent and actual movements shows the existence of two distinctly active domains, characterized by their specific types of deformation and seismicity:

1. The unit composed of west and central Alps, Jura, Apennines, and the Pyrenean/Provençal chains, characterized by a very recent and still ongoing (measurable!) increase in relief, together with compressive tectonics shown by folds, strike-slip, and reverse faults, and strong seismicity.

2. The continental platform areas in front of the great Alpine thrust faults, where the rate of uplift decreases or even becomes negative (subsidence). The seismicity is much weaker than in the first unit and is correlated particularly with old faults that form the major structural features.

Seismic networks (seismometers and accelerometers)
The first seismometers appeared in Japan and Germany at the end of the nineteenth century. The first seismic network, however, the World Wide Standard Seismic Network (WWSSN), was not created until the 1960s, but has expanded with the growth of electronic and computer sciences since then. The development of networks has benefited from the need to detect underground nuclear explosions, particularly since 1962 in the case of France (the Laboratory for Geophysics and Detection of the French Atomic Energy Commission (CEA)).

At present, even though no working European network exists as yet, most European countries have national networks, and an organization that controls and centralizes the data is located at the University of Strasbourg: the European–Mediterranean Seismologic Centre (EMSC). More than 700 separate stations are operational in the European Community, not counting the special networks that

---

* The Richter scale is open and its most commonly used values fall between 0 and 9.

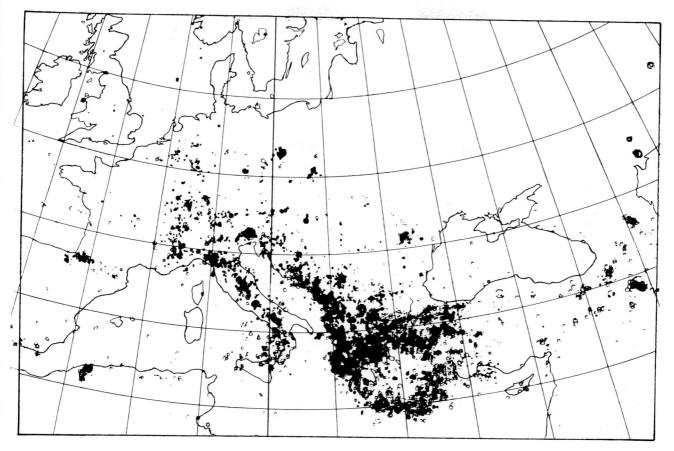

Fig. 4.1 Historical seismicity in Europe during the period 1976–85.

monitor areas of moderate seismic activity, such as in France in the Pyrenees, the Alps, Provence, and Alsace, and in regions of strong earthquakes, such as in Italy, Greece, and Turkey.

These networks provide data on the location of an earthquake and on the energy liberated as seismic waves (magnitude). They usually consist of high-gain *detection* instruments, which means that they are useless near an earthquake, where the magnitude exceeds 3–3.5. For this situation, another type of apparatus is necessary, an accelerometer, which is a low-gain instrument for recording strong motions.

Accelerometer networks are not common in Europe (except in Italy, Greece, and Turkey), for the following reasons:

1. Earthquake engineering in construction is well-developed only in the USA (California) and Japan, where accelerometer stations have existed since the 1930s. These countries have made their data and expertise available generally.

2. Most accelerometers are activated at 0.01 g, g being the acceleration of gravity, which makes it more difficult to obtain recordings rapidly in areas of weak to moderate seismicity, such as much of western Europe.

Highly sensitive accelerometers are now available at moderate cost. This, and the absolute necessity to deploy reference seismic data that are more suitable to the European context, should convince the decision makers of the need for rapidly equipping numerous regions in Western Europe (beginning with the most sensitive areas) with accelerometers of a type from which recordings can be used by both seismologists and engineers. A good example is provided by the following case history.

The UK seismic monitoring network [190]

In the UK, the British Geological Survey (BGS) is officially in charge of seismic monitoring over the whole country. This activity commenced in the latter half of the nineteenth century with the pioneering work of J. Milne on instrument development. However, modern networks of short-period instruments that can accurately locate local earthquakes and determine their parameters were not installed until 1969.

At that time, the network was centred on Edinburgh, which was the collecting point through radio links for instruments located up to 70 km away. From 1976 onwards, the network was steadily enlarged, with stations being installed in various regions of Great Britain. In 1990 the total number of stations was 103 with several more planned, as shown on Fig. 4.2.

Most of the stations were planned to monitor not only natural seismicity, but also the effects of resource development, such as coal mining, abandoned sites, oil and gas development in the North Sea, and the hot dry rock geothermal project in Cornwall. Most of these stations were sponsored by the Department of the Environment (DoE), but recently the EEC has sponsored the extension of the network into SW England.

Table 4.5 Significant earthquakes in Europe since 1965 (after Gerek and Shah 1984)

Year	Location	Magnitude	Deaths	Remarks
1968	western Sicily	6.1	740	17 earthquakes from 4.1 to 6.1 on the Richter scale, from 14 Jan. to 6 Feb.
1970	Turkey Gediz	7.3	1100	Many buildings collapsed
1975	Turkey Lice	6.7	2370	75 per cent of buildings destroyed
1976	Italy Friuli area	6.5	965	Extensive damage to buildings
1976	eastern Turkey	7.3	5000	Many buildings were destroyed in Muradiye and Caldiran
1977	Romania Vrancea distr.	7.2	1570	Many buildings were destroyed in Bucharest
1979	Yugoslavia S. Montenegro	7.0	156	Extensive damage near Adriatic coast
1980	Italy Campania/ Lucania	7.0	3100	Several large shocks; great damage to brick and stone buildings
1981	Greece Corinth	6.7	20	Many villages heavily damaged or destroyed
1983	Turkey Erzurum	6.9	2000	Many villages in east Turkey destroyed
1986	Greece Kalamata	6.2	21	25 per cent of structures destroyed and 22 per cent heavily damaged

Most of the existing recordings are available on analogue tapes; the information from BGS recordings and university research projects has been carefully stored. Bulletins, paper seismograms, and microfilm records from British and worldwide observatories, going back to the beginning of the twentieth century, are preserved in Edinburgh.

In order to develop risk investigations, BGS and the DoE have invited existing and potential customers to form a consortium, which would provide financial support for a 'strategic seismic network' and a pool of equipment for rapid deployment at sites of specific interest. In return, the customers will receive free access to the database, a computer bulletin board of current events, monthly and annual bulletins, plus a faxed alert for any significant event. The need to meet the customers' requirements in terms of data quality and rapid access has induced several developments:

(1) improvement of signal transmission systems (frequency range, transmission speed, etc.);

(2) digitizing on a VAX computer of data stored on tapes, in order to allow various types of calculations to be made, such as Fourier transforms, seismic moments, and modal synthesis;

(3) for greater flexibility a portable cartridge recorder was developed that is based on a microprocessor and has a storage capacity of 14 megabytes, or about 50 typical events.

Seismotectonic maps

The evaluation of seismic hazard at the regional scale is based on the definition of homogeneous seismic and structural areas. This so-called 'seismotectonic analysis' phase is essential, both for determining and forecasting hazards, as is illustrated by the following case history.

The seismotectonic map of France [57]

This project was initiated in the 1970s, mainly to evaluate the safety of nuclear installations. The objective was to discover whether it would be possible, by using all the available knowledge of the structural setting (obtained by such varied means as tectonics, neotectonics, geophysics in its various guises, and remote-sensing studies of lineaments), to resolve the uncertainty concerning the potential seismic activity of homogeneous areas. More specifically, would it be possible to identify those linear features (faults) bordering unstable areas where earthquakes might occur again?

To try to answer these questions, it was necessary to combine on maps all data concerning the distribution of observed seismic events, together with structural, geomorphological, and general geological information. The objective of seismotectonic maps is to provide the basis for a rational synthesis of a larger quantity of data of diverse origin, assembled either during special studies or for the seismotectonic project of an entire country.

In areas of strong seismicity, such as along the Anatolian fault zone in Turkey, it is possible to define the location of those faults that may be the origin of earthquakes. This is not feasible in an area of weak to moderate seismicity, such as France.

For that reason, it was decided to assemble on the map, as objectively as possible, all essential data that needed to be compared. It was left to the user to make correlations that might be significant in evaluating the regional seismic hazard, but updated and more detailed studies might be necessary for greater precision. The resulting seismotectonic map of France, presented at the International Geological Congress of 1980, was based on the following analytical maps:

• historical seismicity (published, see case history on the SIRENE macroseismic data bank, below);

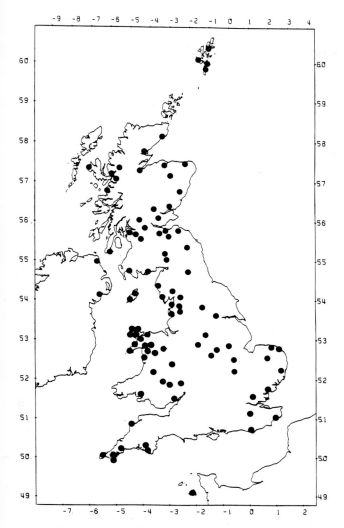

Fig. 4.2 BGS onshore seismograph stations, 1990.

- instrumental seismicity (unpublished);
- tectonic (published);
- neotectonic (unpublished);
- geophysical discontinuities (unpublished);
- satellite image lineaments (unpublished).

At about the same time, similar map projects were launched in Switzerland and, related to reducing seismic risk, in the Balkan countries (Yugoslavia, Albania, Bulgaria, Romania, Greece, and Turkey). Since these first maps appeared, certain of these countries, such as Yugoslavia, Albania, and Turkey, have continued updating; in 1989 Greece published a new detailed seismotectonic map at scale 1:500 000. The position in the other European countries is as follows:

1. In Spain, a seismotectonic map is available for the Granada area and a national seismotectonic project is in progress, at the scale of 1:200 000, the same as that of the forthcoming neotectonic series.

2. In Portugal, only the seismotectonic map of Lisbon (Cabral and Ribeiro) has been published.

3. In Italy, a set of neotectonic maps at scale 1:500 000 is being edited by the CNR (the National Research Centre) in Milan.

4. In Germany, a seismotectonic mapping project has been proposed by the Geophysical Institute at Karlsruhe.

The SIRENE macroseismic database [47]

The SIRENE database of historical seismic information is unique in the following respects:

1. The data stored for a single event are as complete as possible.

2. The coding system takes account of all uncertainties and slight differences of meaning, which are inherent in this type of data of diverse origin.

3. It is possible to express the reliability of the stored data as a function of the precision and quality of the data sources, which leads to the best possible understanding of the seismic events.

For these reasons this system, which was developed by BRGM (France), has attracted many foreign organizations, but, in principle, the use of the SIRENE files is reserved for the funding partners, which are BRGM, CEA (the French Atomic Energy Commission) and EdF (the French Electricity Authority). Even so, as the SIRENE database has become the official reference standard for macroseismicity in France, it is routinely used for all earthquake engineering studies that require data of this quality, in particular for drawing up land-use plans and regulations.

BRGM has installed the database simultaneously in its Marseille office, for continuous updating of the software systems, and in its Orleans headquarters, where the new data are entered, and each year provides its partners with a complete update, which can be stored and used easily on a microcomputer.

At present the graphic restitution capability consists of macroseismic maps of an earthquake event, or epicentre maps. Future development work will comprise the calculation of the focal depth, of the magnitude, and of the intensity–attenuation laws, all directly from the macroseismic maps, as well as the drawing up of maps of maximum observed and probably attained intensities, maps showing the frequency of tremors at different intensity thresholds, and zoning maps incorporating earthquake-engineering building codes.

In the longer term, further development will improve the multi-criteria analysis functions for identifying seismic sources (epicentres, structures, neotectonism), for understanding of the site effects (distribution of the observed intensities, the topography and the lithology, and of the structure in the subsurface), and for an evaluation of the seismic risk (economic simulation of the major historical earthquakes, etc.).

Finally, the computer file is but a reflection of the enormous documentary base that has been assembled by BRGM (corresponding to about 50 000 document pages), which is organized according to geographical and chronological criteria. For reasons of security and ease of handling the data, the technical specifications for hardware, and the software were adapted to the requirements of each user.

In evaluating the seismic risk for constructions requiring a high degree of stability, such as nuclear installations, chemical plants, and large dams, it is essential that numerous seismotectonic analyses are carried out to identify potential seismic movements and to ensure the long-term safety of such structures.

For such analyses, the best input usually comes from neotectonic studies, especially for determining the potential seismicity along faults; seismic data by themselves are generally insufficient for determining the present-day deformation of the Earth's crust. Neotectonic studies, as exhaustive as possible and combining all techniques available today, should be rapidly planned and implemented, in particular in the framework of the International Decade for Natural Disaster Reduction. A recent study of this kind in southern France, around Avignon and Marseille, led to the identification of several faults whose existence was hitherto hardly suspected. Other, similar, studies are underway in Italy (The Benevento Project) and Greece.

The prediction of earthquakes [68]

Why prediction is necessary

Earthquake prediction consists of forecasting where and when a future destructive earthquake will take place. One can speak of *long-term prediction* (several decades) that will serve to define building codes suitable for a given region (and for the possible reinforcing of existing structures); *medium-term predictions* (from a month to a year) which allow scientists to install the most suitable instruments to monitor any menacing faults; and *short-term prediction* (from a few hours to a few days) which is useful for alerting the civil protection authorities, and for preparing aid and, if necessary, evacuation of the population.

Long-term prediction leads to the definition of the seismic hazard of an area, which means the probability that a shock of more than a given magnitude will occur. The analysis of historical seismicity (reconstructing the probable locations and magnitudes of earlier earthquakes and the faults that generate them) and instrumental seismicity, and the identification of active faults in the field, are necessary for defining the hazard. For a given site, once the magnitude and distance of future earthquakes are known, it would be possible to estimate the characteristics of resulting ground movements.

The return time of destructive earthquakes ranges from a few tens of years to a few centuries, or even several millennia, and the seismic cycles of an area are rarely regular. It is thus impossible, using catalogues of historical earthquakes or the geological analysis of past activity along faults, to predict an earthquake at short- or medium term. The seismologists in such a case must base their analysis on precursory effects.

Precursory effects

Precursory effects that can herald the imminent rupture of a fault are: abnormal variation of the local or regional microseismicity, soil deformation, water level variations within wells, variations in the velocity and attenuation of seismic waves, radiation of electromagnetic waves, and subterranean electric currents. Animals are known to be sensitive to certain precursory effects and may give an alert through abnormal behaviour, such as snakes that flee their holes in winter and that are found frozen to death on the roads. In 1974 and 1975 in the Chinese province of Liaoning, many such precursory effects were noted and a public alert was given a few hours before the earthquake of 7.4 magnitude; as a result only a few thousand people died in an area with 3 million inhabitants. In the case of the Loma Prieta earthquake (California, 17 October 1989), precursory effects consisting of two slight but abnormal shocks in a normally calm area (which turned out to be the epicentre of the main event) were noted, and a public warning was given three months before the event. Well-identified precursory effects are not always the rule, however, as was shown by the 1976 earthquake at Tang-Shan, near Beijing, which killed half a million people, and by the 1985 Mexico and 1988 Armenia shocks that each killed tens of thousands.

Such precursory phenomena can be explained by a general fatigue of the environment exposed to tectonic stresses. The rocks that are under such stress around a source zone begin to fracture and rapidly deform before the final rupture. Theoretical models, however, are poorly developed as yet, because of a lack of reliable observational data. In order to advance this infant science of 'precursorology', large amounts of data will have to be gathered and studied.

In the USA, scientists have selected a small segment of only 20 km length of the San Andreas fault system, which is known to rupture regularly, about every 22 years, with earthquakes of a magnitude of around 6. They have installed a dense network of measuring devices for the continuous monitoring of all the precursory phenomena described above. In a similar fashion, a Turkish–German team has selected a 50 km segment of the North Anatolian fault, about 200 km east of Istanbul, and a large number of measurements have been carried out systematically since 1985. In Japan, about 15 institutes carry out similar work in several potentially dangerous areas of the archipelago. Research along the same lines is being developed in China, the USSR, and various other countries. In Greece, a European programme started in 1990. The common denominator of all these efforts is a multidisciplinary approach involving all branches of the Earth Sciences, to arrive at an understanding of the gestation process of an earthquake. For the moment it is necessary to define clearly the appropriate prediction criteria. This is an essential basis for giving public warnings of imminent earthquakes.

The philosophy behind the VAN method for short-term predictions, as proposed by Greek engineers in the early 1980s, is very different. These scientists consider only one precursory phenomenon, subterranean electric currents, and make frequent public alert announcements. This method is still being perfected, however, and a statistical validation of the predictions is difficult because of the very large number of earthquakes in Greece. Rather than prediction, one should speak here of *prediction attempts* and of the fine-tuning of a potential method, as is true for all other research into short-term predictions.

Some promising research into medium-term prediction has been carried out by a team of Russian scientists, who study the variations of the regional seismicity in space and time. This technique, based on annual catalogues of seismicity, was successfully used in southern Italy and in France, in the Alps and the Pyrenees. It makes it possible to locate, within a radius of a hundred kilometres, the place where the major earthquake will strike during the next year or so, and this with a remarkable success rate. These results are of great importance, as it is necessary to know which faults to monitor for the application of very expensive, short-term prediction techniques.

Today and perhaps for several decades, the methods for predicting earthquakes are inadequate. Much more research, using statistical analyses of modern high-quality data, will be necessary.

Seismic hazard assessment

The seismic hazard of a given site is defined by the probability that, over a given period (e.g. an annual probability), a seismic shock of at least a certain intensity will occur. It is one of the components of seismic risk, the others being vulnerability and the value of people and their property.

The evaluation of seismic hazard usually consists of two phases (see Fig. 4.3):

1. *Seismic hazard at the regional scale* (1:200 000 to 1:500 000). In this phase the analysis will be based on:

   • data on historical and instrumental seismicity, contained in a computerized file such as SIRENE, operated by BRGM, the CEA, and EdF, which integrates data on more than 5000 earthquakes that affected France and its neighbouring countries;

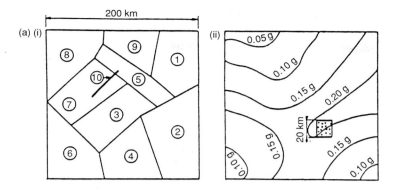

Notes

(i) *Definition of possible earthquake sources in a given area*
• evaluation of historical and instrumental seismicity
• structural geological setting
• seismotectonic analysis
Result: identification of seismotectonic 'provinces' (1 to 9) and an active fault (10)

(ii) *Definition of the activity of a seismic region* (magnitude, frequency, relationship)
Assessment of attenuation laws
Result: assessment and modelling of regional seismic hazard, shown for instance as a map with the probable bicentennial values of ground acceleration; the small square is the surface of the block diagram below.

This result leads to economic studies for, and political choices by decision makers, which in turn lead to:

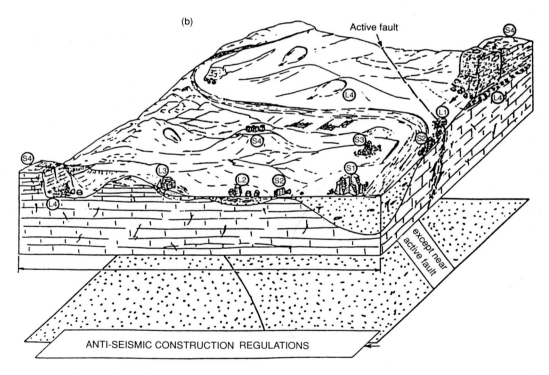

Ground-shaking or 'direct' effects
S1: Thick alluvium: variable amplification but danger to high-rise buildings
S2: Thin alluvium: stronger amplification and danger to low-rise buildings
S3: Rocky subhorizontal surface: no amplification, same danger as S2
S4: Crests and slopes: strong amplification

Such site effects must be taken into account in local building codes, as should:

Ground-failure or 'indirect' effects
L1: Shear along active fault
L2: Liquefaction of thick saturated soil leading to subsidence
L3: Liquefaction of thin soil on slope causing landslide
L4: Rock- or landslides, no liquefaction

Fig. 4.3   The assessment of seismic hazard. (a) Macrozonation of regional seismic hazard; (b) microzonation to counter local seismic hazard.

- structural data, in particular those concerning neotectonic phenomena.

2. *Seismic hazard at the local scale*, significant for locating a structure (in general 1:5000 to 1:20 000). The local hazard concerns the predicted effects of a seismic shock on a particular site, as a function of its specific geological, topographical and geotechnical characteristics. For this phase it will be necessary to observe, record, and analyse:

- *Direct effects* or ground shaking, with reference to the conditions of the site (site effects).

- *Induced effects* in the form of major soil displacements (ground failure) induced by seismic action upon superficial formations, such as:

   liquefaction of loose and saturated sand and silt;
   compaction of such ground, whether saturated or not;
   landslides, rock falls and rock slides;
   displacement caused by movement along a fault.

All these local effects can be analysed and mapped at the scale of a community or township, as is the objective of *seismic microzoning*. Many such studies have been carried out in Europe, in Italy, Greece and several Balkan countries, and in France; in the last country, microzoning forms the purely technical part of the Risk Exposure Plans (PER) that are drawn up for communes. As a result, it is now considered necessary to equip such seismic areas with 'strong motion' accelerometers, both in portable and permanent networks, for the best possible interpretation of a site response and for devising the most appropriate calculation models. Simulation experiments on European sites will be necessary, like those in progess in California, Turkey, and Japan. It will also be necessary to identify the methodology for defining the degree of hazard for induced phenomena, such as liquefaction, landslides, and surface faults.

During the start-up period of the International Decade for Natural Disaster Reduction (IDNDR), the creation of multidisciplinary European teams must be encouraged, in order to improve the knowledge base and compare the various possible approaches. Recently, the following proposals were made (by W. Hays and his co-workers) for a cooperative project for seismic risk reduction in the Mediterranean region:

1. Evaluation of seismic hazards on a regional scale of about 1 200 000 to 1:500 000. This part of a hazard assessment establishes the physical parameters of the region needed to evaluate the earthquake hazards of ground shaking, ground failure, surface fault-rupture, tectonic deformation, and tsunami* run-up. Technical tasks such as the following are required:

   (a) Compilation of a catalogue and map of the prehistorical, historical, and instrumental seismicity, with case histories of the largest events.

   (b) Neotectonic studies (mapping, age determinations, and trenching) to acquire information on earthquake recurrence times during the past several thousand years, not provided by historical or instrumental seismicity.

   (c) Preparation of seismotectonic maps showing the location of active faults and seismogenic structures, and their correlation with seismicity.

   (d) Preparation of maps showing seismogenic zones, the magnitude of a maximum earthquake and the frequency of earthquake occurrence for each zone.

   (e) Preparation of maps showing tsunami-genic zones and their correlation with historical tsunamis.

   (f) Preparation of slope stability and liquefaction potential maps.

   (g) Specification of regional seismic wave attenuation laws and their uncertainty.

   (h) Preparation of ground-shaking hazard maps.

2. Evaluation of seismic hazard on an urban scale of about 1:5000 to 1:25 000. This part of a hazard assessment integrates the seismotectonic and other physical data acquired in the regional study (1 above), with site-specific data acquired in the urban area, to produce ground-shaking and ground vulnerability maps. Technical tasks such as the following are required:

   (a) Acquisition, synthesis, and integration of existing and new geological, geophysical, and geotechnical data, to define the soil/rock columns in terms of their physical properties and their response to various levels of ground-shaking.

   (b) Preparation of ground-shaking hazard maps.

   (c) Preparation of maps showing the potential for surface fault-rupture, tectonic deformation, and flood-wave inundation.

   (d) Preparation of maps showing the potential for liquefaction.

   (e) Preparation of maps showing the potential for seismically induced landslides.

   (f) Preparation of maps that synthesize the ground-shaking and ground failure hazards.

The hazard maps will be used as a basis for undertaking social, scientific, and technical actions such as the following:

(1) using land to the best advantage;

(2) emergency-preparedness and disaster-recovery planning;

(3) reduction of vulnerability;

(4) adoption and enforcement of building and zoning regulations;

(5) avoidance;

(6) coordinated planning, design, and construction practices;

(7) modification of the characteristics of ground-shaking and ground failure;

(8) prediction and warning.

Seismotectonic zonation in Western Provence (France) [57]

Recently, a study was completed in the west of the Provence Region of France, which aimed at defining potential 'seismogenic' structures (faults) in this part of southern France where moderate earthquakes are common. The study first defined the optimal methods for detecting potential earthquake-generating faults and then applied these methods in practical applications.

On the methodological side, various investigation techniques were combined in an area where the detection of neotectonic

---

* Tsunami: Japanese word for a sea wave generated by a sudden movement of the sea bottom, such as an earthquake, the height of which may be devastating near shore.

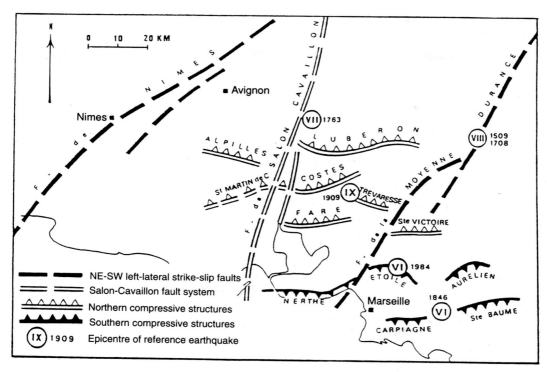

Fig. 4.4 Classification of the main seismogenic structures in Western Provence and associated reference earthquakes.

Table 4.6 Focal characteristics of reference earthquakes retained for earthquake prediction study of Western Provence, France

Date	Epicentral intensity (MSK)	Equivalent magnitude	Focal depth (km)	Associated seismic source		
				Fault system	Orientation	Source mechanism
11/06/1909	IX*	5.9 ± 0.4	3 to 5	"3"*	E–W	thrust fault
14/08/1708	VIII	5.2 ± 0.3	3 to 5	"1"*	NE–SW	left-lateral strike-slip fault
12/07/1763	VII	4.7 ± 0.3	3 to 5	"2"*	N–S	right-lateral strike-slip fault
12/12/1846	VI	4.2 ± 0.2	2	"4"*	E–W	thrust fault
19/02/1984	VI	4.4 ± 0.5	2	"4"*	E–W	thrust fault

* See Fig. 4.4.

features is difficult. Such features indicate where recent movement has taken place in the Earth's crust. The methods used were the following:

(1) microtectonic analysis of terraces of Pliocene to Recent age;

(2) processing and interpretation of SPOT satellite images;

(3) comparison of precision-levelling profiles;

(4) morphostructural analysis.

The study led to the classification of seismogenic faults, based on the physical concept of an earthquake source (geometric and dynamic characteristics). This concept itself is based on several criteria, including the geometry of the faults observed, the fault kinematics and age of recent movements (neotectonics), and the associated seismic activity. The seismogenic and other faults are shown on Fig. 4.4, and are further explained in Table 4.6.

On the practical side, this study led to the following applications:

1. In the context of drawing up Risk Exposure Plans, in particular those for the communes of Salon and Pellissanne, it was possible to locate active seismogenic faults at surface, from which useful information was derived for earthquake building codes.

2. In the framework of the European 'Seveso Directive' for industrial installations posing a risk, it was possible, after defining for each of the identified faults the maximum historical earthquake, to establish for each fault a corresponding maximum seismic movement.

Seismic zonation in an area of strong seismicity near Bolu (Turkey) for earthquake hazard minimization and planning [183] [184] [185] [186]

Although Turkey lies along an important earthquake belt, research activities concerning the minimization of earthquake hazards are not carried out at adequate levels.

The objective of the Bolu study is to construct a seismic zonation map of the Bolu region, showing the distribution of intensity and expected acceleration rates of possible future earthquakes, and to help with the implementation of the various aspects of minimizing the earthquake hazard.

The expected magnitude of the possible large earthquakes along this section of the North Anatolian fault zone, was computed to be

7.3 on the Richter scale, or higher. It is assumed that the epicentre of such an earthquake can be located on any part of the fault trace.

It is suggested that the zones defined as being liable to very violent or violent hazards should be reserved for forestry and agriculture.

For planning purposes other studies in the Bolu area included the evaluation for potential flooding areas, unsuitable zones for urbanization, and pollution problems. This led to the establishment of a land potential and usage map of the area.

The Benevento seismic risk project [57]

In 1986–87, the 'Climatology and natural hazard research programme', issued by the Commission of the European Communities (CEC), established a test site in southern Italy to develop a microzonation and hazard methodology relevant to similar situations in southern Europe. The Italian Gruppo Nazionale per la Difesa dai Terremoti (GNDT) joined the project and extended the objectives to include seismic risk analysis. The Sannio-Matese area, located in the southern Apennines mountains, was chosen as a hazard test site, and Benevento is the town in this region that is to be investigated to achieve microzonation and risk analysis.

Benevento, which consists of an old historical centre with ancient monuments and a section recently developed with reinforced concrete buildings, can be considered to be a typical southern European town because of its size, its population, and its types of building.

Sannio-Matese is one of the most earthquake-prone areas in southern Italy, and Benevento, in particular, has been repeatedly struck by earthquakes. The strong seismicity of the area is shown by several historical earthquakes; the active seismic region is characterized by earthquakes with magnitudes as large as 7, occurring within a narrow belt that is 1 km wide. The moderately strong earthquake (6.9) of 23 November 1990 occurred in an area that had been quiescent since 1694, thus filling a gap of seismic activity. Another seismic gap seems to be present in the Apennines adjacent to the northern boundary of the apparent rupture zone of the 1980 event. This region extends for about 100 km along the mountain range and has been quiescent since 1688. Benevento is located between the two above-mentioned regions.

The study consists of an exhaustive evaluation of the seismic risk. Its main objectives are to make proposals for application to sites in similar settings in southern Europe. It will comprise an evaluation of the regional seismic hazard by means of classical methods and the advanced results of recent research. Fieldwork will include:

(1) the installation of a local digital seismic network for recording micro-earthquakes;

(2) the installation of a 3-level down-hole array, with the deepest of the three seismometers at a depth of several hundreds of metres;

(3) a seismic refraction survey.

The study of the local seismic hazard (seizmic microzoning) will be carried out for the town of Benevento itself, which is located at the confluence of two rivers with part of the town built on the alluvial plain; the historical centre of Benevento lies along a NW–SE ridge. The international project, which began in 1989, will continue until the end of 1991. The participating European institutions are:

Italy:
CNR-IGL (Istituto per la Geofisica della Litosfera, Milano)
Dipartimento di Scienze della Terra, Università di Napoli .
Dipartimento di Geofisica e Vulcanologia, Università di Napoli
Dipartimento di Fisica, Università de l'Aquila
Istituto di Tecnica delle Fondazioni e Costruzoni in Terra, Università di Napoli
Istituto di Scienze della Terra, Università di Catania

France:
BRGM (Bureau de Recherches Géologiques et Minières, Marseille and Orléans)

LCPC (Laboratoire Central des Ponts et Chaussées, Paris and Nice)
IRIGM (Institut de Recherches Interdisciplinaires en Géologie et en Mécanique, Grenoble)
Géotecsis and the Laboratoire de Géologie Structurale de l'Université des Sciences et Techniques du Languedoc at Montpellier

Spain:
Instituto Geográfico Nacional, Servicio de Geofísica, Madrid
Departament de Política Territorial i Obres Publiques, Servei Geologic, Barcelona

Portugal:
Instituto Nacional de Investigaçao Cientifica, Lisboa

RECOMMENDED FURTHER READING

Ambraseys, N . N. (1985). Magnitude assessment of northwestern European earthquake. *Earthquakes Engineering and Structural Dynamics*. **13**, 307–20.

Bard, P. Y., Durville, J. L., Meneroud, J. P., and Mouroux, P. (1989). Seismic microzonation as applied to risk exposure plans (P.E.R.) in the French context. In *Recent advances in earthquake-engineering and structural dynamics* (ed. V. Davidovici). (Presses de l'Ecole Nationale des Ponts et Chaussées, Paris).

Bonnin, J., Cara, M., Cisternas, A., and Fantechi, R. (eds) (1988). *Seismic hazard in Mediterranean regions*. (Commission of the European Communities, Kluwer, Dordrecht).

Browitt, C. W. A. and Newmark, R. (1981). Seismic activity in the North Sea to April 1981. *Inst. Geol. Sci. Glob. Seism. Unit report no. 150*. (Institute of Geological Science, Nottingham).

Combes, Ph., Godefroy, P., Goula, X., Levret, P., Sauret, B., and Terrier, M. (1991). The contribution of neotectonics to the deterministic evaluation of seismic hazard in the Western Provence (France). *Proceedings of IVth International Conference on Seismic Zonation, Stanford, California (USA), Aug. 25–29 1991, vol. II, 541–8*.

Gerek, J. M. and Shah, H. C. (1984). *Terra non firma—understanding and preparing for earthquakes*. (Stanford Alumni Association, Stanford CA).

Godefroy, P., Thirion, S., Lambert, J., and Cadiot, B. (1980). Informatisation du partrimoine de séismicité historique de la France. *Bull. du B.R.G.M., 2e série, section IV, No. 2*, pp. 139–45.

Hays, W. W., Van Essche, L., and Maranzana, F. (1990). SEISMED and IDNDR: opportunities to reduce the risk from earthquakes and other natural hazards. *SEISMED workshop, Genova, Italy, May 1990*, UNDRO.

Lopez-Arroyo, A., Iannacone, G., Mendes Viclor, L., Mouroux, P., Scarpa, R., Marcellini, A., *et al.* (1988). Benevento seismic risk project. *Seminar on the prediction of earthquakes, Lisbon, Portugal, 14–18 November 1988*. United Nations Economic Commission for Europe.

Philip, H. (1988). Recent and present tectonics in the Mediterranean region. In *Seismic hazard in Mediterranean regions*. (Commission of the European Communities, Kluwer, Dordrecht).

4.2.2. Volcanic hazards [66]

Introduction

Several recent and well-publicized volcanic eruptions have served to underline the awesome destructive power of volcanoes: Mount St Helens (USA, 1980), El Chichón (Mexico) in 1982, Nevado del Ruiz (Colombia, 1985), and Nyos (Cameroon) in 1986. These four eruptions caused the death of tens of thousands of people, as well as large-scale destruction and economic losses.

Europe has not experienced volcanic disasters on such a scale in recent times; the last one that affected European lives and property, though geographically not located in Europe itself, was the eruption of Mont Pelée on 8 May 1902 on the island of Martinique, French Antilles, which destroyed the town of St Pierre and killed over 28 000 people. This is fortunate because, even though several European countries have active volcanoes, their eruptions during the past century caused relatively little damage. The reasons for this are manifold: most of the eruptions were limited in extent, most of the areas directly affected were thinly populated, and there was only minor economic development. Because of this, most people have no first-hand experience of the problem and the public perception of volcanic hazard in Western Europe is generally low.

Even so, the past shows that European countries are not sheltered from volcanic hazards. Two great eruptions have affected the Mediterranean civilizations: those of Santorini in about 1500 BC and Vesuvius in AD 79. There is not the slightest doubt that, were an eruption of such magnitude to occur today, it would be an unprecedented disaster.

On 24 August AD 79, Vesuvius suddenly exploded and destroyed the Roman cities of Pompeii and Herculaneum. Although Vesuvius had shown stirrings of life when a succession of earthquakes in AD 63 caused some damage, it had been quiet for hundreds of years and was considered extinct; its surface and crater were covered in vegetation, so the eruption was totally unexpected. It took but a few hours for hot volcanic ash and dust to bury the two cities. Vesuvius has continued its activity intermittently ever since, with numerous minor and several major eruptions. The last, highly explosive, Plinian eruption occurred in 1944. The high population density around Vesuvius (Naples has 3 500 000 inhabitants) now poses a real problem if Vesuvius erupts again.

The other famous eruption that affected the Mediterranean area during historic times, was the explosion of the Santorini volcano, located on the Greek island with the same name (formerly known as Thera) in the Aegean sea. The pyroclastic activity (tephra falls, base surge, flow deposits) and tsunamis that accompanied the collapse of the caldera are thought to have affected a major part of the Mediterranean region. This cataclysmic eruption has been widely blamed for the destruction of the Minoan civilization in Crete. Plato's mythical tale of Atlantis has been hypothetically identified with this eruption.

Ironically, volcanoes have always attracted, and will continue to attract, people to live on their flanks. Thus, as the population increases in density, the potential problems also become increasingly complex and hard to solve. Table 4.7 presents an estimate of the population that could be affected today by a volcanic eruption in a European country. People living in volcanic areas therefore must be made aware of the risks they run and learn to live with volcanoes, between and during their periodic, but almost always violent, outbursts.

Distribution of volcanoes in Western Europe

The distribution of volcanoes is not fortuitous. The current theory of plate tectonics provides a unifying framework, which explains the distribution and diversity of volcanoes on the Earth's surface. Most volcanoes lie on or near two main types of boundary between moving crustal plates: (1) spreading boundaries, where plates move apart; and (2) compressive boundaries where plates collide or ride over each other. A third type of plate boundary is formed by transform faults, where plates move laterally past each other; volcanism, however, is rarely associated with this type. Some 'intraplate' volcanoes are distant from plate boundaries; most of these are explained by the plate moving over a magma-generating spot

Table 4.7 Estimate of the population in several European countries and their overseas territories that could be affected by volcanic activity (figures in millions; Peterson 1988)

Country	Total population	Population near volcano
Greece	9	0.4
Italy	54	22
Canary Islands, Azores	1.3	1
Iceland	0.3	0.3
France (Réunion, French Antilles)	1.3	0.5

beneath the crust: a 'hot spot'. Also, some volcanoes have been proved to be related to intraplate structural lineaments.

The distribution of the (active or dormant) volcanoes in Europe and its contiguous areas shows examples of all types (Fig. 4.5):

1. The Icelandic volcanism (5), which occurs on an emergent portion of the Mid-Atlantic ridge, and thus on a spreading boundary, combined with a hot-spot occurrence.
2. The Mediterranean volcanism of Greece (2) and Italy (1), which is related to the rather complicated collision and subduction of the African plate against Europe (a compressive boundary).
3. The Atlantic volcanism of the Azores (6) and the Canary Islands (7), which is related to the existence of hot-spots.
4. Intraplate volcanism along continental rift structures is found in several places in Europe, in the Eifel, western Germany (4), and the Chaîne des Puys of the French Massif Central (3).

Table 4.8 lists the active volcanoes found in the various Western European countries. More than 30 historic volcanic eruptions are known from this area alone, which clearly shows the importance of taking volcanism into account as a major natural risk. The term 'historic' has a flexible meaning: in the Mediterranean basin this is reckoned in millennia, whereas on the Azores this can mean only a few centuries.

This raises the question of whether a volcano is active, dormant, or extinct. We consider that any volcano that has erupted in historic times, must be seen as *active* and deemed capable of erupting again. However, there are also volcanoes that have not erupted during recorded history, but that are liable to erupt at some time in the future. These are termed *dormant*, either because they have a long period of repose, or because the historical record is short. Geological investigations can tell if activity has occurred within the past (tens of) thousands of years, and conclude whether the volcano must be seen as dormant or as *extinct*, and what the probability is of renewed activity in the future.

Products of volcanoes and related hazards

The morphology of volcanoes and the type of eruptions are closely related to the physical and chemical characteristics of the emitted material. Among these, the *viscosity* of the magmas is one of the most important parameters that control the eruptive activity. It depends on several factors, such as chemical composition of the magma, its temperature and the

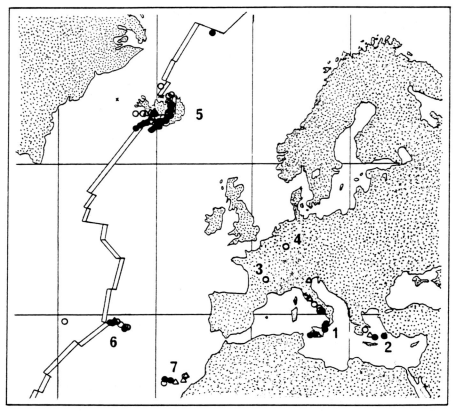

1. Italy   2. Greece   3. France   4. Germany   5. Iceland   6. Azores   7. Canary Islands

● Historically dated eruptions   ○ Eruptions dated by radiocarbon, anthropology, etc.

△ Undated, but almost certain, eruptions

Fig. 4.5   Location of volcanic sites in Europe (from: Simkin *et al.* 1981).

volatile content. Low-viscosity basaltic magmas are character-istically fluid and thus spread easily. They give rise to extensive lava flows that build broad, gently dipping shield volcanoes as in the spreading zones of Iceland. Acid magmas are generally very viscous and produce explosively fragmen-ted pyroclastic material. They form lava domes or thick lava flows of limited extent. Magmas with intermediate composi-tion display a wide range of eruptive activity, from mild explosions to cataclysmic eruptions, and from the quiet over-flowing of lava to the extrusion of viscous domes. They build steep-sided, composite volcanoes (both lava flows and pyro-clastic material), often capped with a caldera, such as at Vesu-vius and Mont Pelée.

The types of eruptions were generally named after volca-noes or volcanic areas where the activity was first observed, or of which the style of eruption was thought to be characteristic. Some widely used types are: Strombolian, Hawaiian, Vesu-vian, Peléan, Vulcanian, Surtseyan, and Plinian.

Eruptions cover a wide spectrum of size, violence and travel of the volcanic material. Hazards vary accordingly, with the assumption that the main hazards arise from explosive eruptions. Volcanic products assume a variety of forms, including lava flows, pyroclastic material (flows, surges, ash-falls, etc.), lahars and mudflows, debris slides and avalanches, and volcanic gas emissions.

Lava flows originate from a vent or a fissure on the flank of the volcano. They are topographically controlled and may travel long distances. Speeds are around a few metres to a few

hundreds of metres per hour. Lava flows commonly destroy all the structures and lands, which are subsequently unusable for a long time, but few have taken human lives (see the Heimaey case history, p. 154).

Pyroclastic material, produced by explosive fragmentation of the magma, can be emitted as flows, surges, or ash-falls. They probably represent the most dangerous volcanic pheno-mena. Flows and surges travel on the surface as hot gas/solid dispersion, partly fluidized by exsolution of magmatic gases and the expansion of trapped, heated air. They are highly mobile and destructive: during the eruption on 8 May 1902 of Mont Pelée, Martinique Island, combined pyroclastic flows and surge effects completely destroyed the town of Saint Pierre, killing 28 000 of its inhabitants.

Flows are topographically controlled, filling valleys and depressions, such as pumice flows and 'nuées ardentes' produced by many volcanoes. Voluminous pyroclastic flows can also produce sheet-like deposits, hundreds of metres thick and covering large areas (ignimbrites). In the Mediterranean area, the prehistoric Campanian ignimbrite (Italy) and the Kos ignimbrite (Greece) are examples; if they were to occur today, they would certainly cause major disasters.

Surges commonly travel at high speed (more than 10 m s^{-1}) and over long distances (more than 10 km) from the source. They occur alone or associated with flows, but their deposits, especially in older volcanic rocks, are difficult to recognize.

Pyroclastic falls can be formed directly from eruptive columns or from clouds developed above a pyroclastic flow.

Table 4.8 Non-exhaustive list of the main active and dormant volcanic sites in European countries, which may involve the problems posed by volcanic hazards (from *The catalogue of active volcanoes of the world*, Simkin *et al.* 1981)

Country	Volcano	Eruptions
Greece	Methana	(250 BC)
	Milos	(AD 80–205)
	Nysiros	1888, 1873, 1871, 1830, 1422
	Santorini	1950, 1939–41, 1928, 1925–6, 1866–7
Italy	Etna	very frequent eruptions
	Ischia	(1301, 150, 350 BC)
	Phlegrean Fields	1538, 1198, (700 BC), (2150 BC)
	Stromboli	very frequent eruptions
	Vesuvius	1944, 1942, 1941, 1929, 1913, etc.
	Vulcano	1886–90, 1873–9, 1812–31, 1771–86
France	Chaîne des Puys	(1500 BC)
Germany	Eifel district	(8300 BC)
Canary Islands	Hierro	1692, 1677
	Lanzarote	1824, 1733
	La Palma	1949, 1712, 1677, 1646, 1585
	Tenerife	1909, 1708, 1706, 1704, 1605
Azores	Faial	1957, 1672
	Pico	1963, 1720, 1718, 1562
	São Jorge	1964, 1808, 1580
	São Miguel	1811, 1682, 1652, 1638, 1630
	Terceira	1761, 1400, 1200
Iceland	Reykjanes Trölladyngja Brennisteinsfjöll	several eruptions AD 900–1200
	Langjökull	around AD 1000
	Snaefellsness	around AD 900
	Vestmannaeyjar (Surtsey and Heimaey)	1963–73
	Katla	the last of many in 1918
	Eyjafjallajökull	1823, 1612
	Torfajökull	1477, around AD 900
	Hekla (this century)	1980–1, 1970, 1947–8, 1913
	Grimsvötn (this century)	1982, 1934, 1922, 1910, 1902–3
	Öraefajökull	1727, 1372
	Askja	1874–5
	Krafla	1974–84, 1724–9

Ash-falls probably constitute the most common of all volcanic hazards, together with mudflows. Thin deposits may cause anything from a slight nuisance to a severe disturbance for humans and animals, and for economic and industrial activities. Air-fall deposits several metres thick are known to have buried large tracts of land and entire cities. Vestmannaeyjar on Heimaey (Iceland) was partly buried by ash-falls during the 1963–1973 eruptive activity of Surtsey; during the AD 79 eruption of Vesuvius, pumice air-falls buried Pompeii.

Mudflows and lahars (an Indonesian term for volcanic mudflows) are common and most destructive volcanic hazards. They are generated during an eruption, when unconsolidated pyroclastic material is mobilized by any type of water (river, crater lake, rainfall, melting of ice or snow, groundwater). They can also be generated without an eruption, when heavy rain, for example, mobilizes unconsolidated deposits on the flanks of a volcano (air-fall ash, alluvial material, landslide material, etc.). Lahars flow in valleys and depressions at high speed, can vary from a small volume to very large masses, and can travel well over 100 km from their source, menacing all life and settlements in the valleys they scour. In 1985, a small eruption at the top of the Nevado del Ruiz volcano (Colombia) melted an icecap and generated lahars that destroyed the city of Armero, 50 km from the volcano, killing more than 20 000 people.

Debris avalanches are even more catastrophic than lahars. They are generated by the collapse of the upper part of a volcano following the emplacement of a subterranean magmatic intrusion (e.g. the 1980 eruption of Mount St Helens, USA). They spread an enormous volume of material around the volcano and in neighbouring valleys, and can thereby generate catastrophic lahars over tens of kilometres. These large landslides may be accompanied by a cataclysmic directed explosion ('blast'), generated by the depressurization of the underlying hydrothermal system. These blasts are highly destructive for tens of kilometres from their source.

The most abundant *volcanic gases* are: H_2O, CO_2, SO_2, SO_3, H_2S, HCl, and HF. They can pose a serious hazard for humans, animals and plants if large volumes are released during an eruption (e.g. Dieng, Indonesia, where a small eruption in 1979 killed 150 people). The poisonous effects of volcanic gases may also persist after the eruption has ended, if gas-laden ash deposits cover land and crops: crops poisoned by sulphur and fluorine resulted in a famine that led to well over 10 000 deaths during the 1783 eruption of Laki, Iceland. The sudden expulsion of gas dissolved in the Nyos crater lake in Cameroon (1986) killed 1700 people and a large amount of livestock.

Human and economic aspects of volcanic hazards

The following case histories show quite clearly that for many European countries a volcanic risk exists and must be taken into account. Table 4.7 presents an estimate of the population that could be affected by a volcanic eruption. In the case of oceanic islands like Iceland and the Azores, most of the population might be affected to varying degrees by the consequences of an eruption. In countries like Italy, where the population density around volcanoes is amongst the highest in Europe, an eruption covering a small part of the territory would affect many people.

The impact of a volcanic eruption on the economic activity of an area or a country is difficult to evaluate entirely. Volcanic eruptions can have numerous direct and indirect consequences on the environment and on human activity, such as the destruction of infrastructure elements, industry and population centres, the pollution of soil, air and water, the destruction of crops and the complete or temporary sterilization of arable land, and even the interruption of air traffic. Other effects can be the destruction of harbour installations through tsunami action generated by distant volcanic eruptions, and the (partial) destruction of tourist facilities.

An example of the cost of a volcanic catastrophe, is provided by the event at Mount St Helens (USA) on 18 May 1980, which caused significant loss of life (62 dead or missing) and property damage in excess of US$ 1.2 billion, in a lowland area with low densities of population and industry.

Evacuation in response to a volcanic crisis creates an economic loss in terms of the cost of such evacuation and the interruption of normal activities. The longer people are away from their homes, the higher will be such losses. A major problem is posed by the fact that evacuation too late may result in loss of life, but that unnecessary evacuation (when no eruption takes place after all) still costs the same. During the phreatic eruption of La Soufrière on Guadeloupe in 1976, the economic and social cost of evacuating 72 000 people for 3 months was very high indeed.

A case history in Iceland: the Heimaey eruption [121]

Site of eruption

Vestmannaeyjar is a group of small volcanic islands off Iceland's south coast, of which Heimaey is the largest and only one inhabited (5000 inhabitants before 1973). The group of islands form the southernmost volcanic system of Iceland's Eastern Volcanic Zone. The core area of volcanic activity within the Vestmannaeyjar volcanic system is Heimaey, which has coalesced from several eruption foci into an island of about 13 km^2.

Surtsey eruption—heralding an eruptive episode

Vestmannaeyjar went through a volcanic episode from 1963 to 1973. The first chapter of this episode was the Surtsey eruption that lasted intermittently from November 1963 to June 1967. The eruptions occurred on short right-stepping en-echelon fissures with NE–SW trend on the sea bottom 20 km SW of Heimaey. Small islands were built up, but all were quickly eroded except Surtsey, where the eruption proceeded from steam to lava production, without causing damage. The total volume of erupted material amounted to about 1.2 km^3 of which one-third was lava.

Heimaey eruption precursors

In the next (and the last so far) chapter of this volcanic episode, the volcanic activity shifted from the distal SW-outlier of this volcanic system to its core area on Heimaey. The eruption there started on 23 January 1973 and lasted until late June 1973. An earthquake swarm started two days earlier and lasted for 14 hours. Four hours before the eruption an earthquake swarm started again. The largest was of magnitude 3 and this time a few were felt on Heimaey. It was discovered later that the foci of the latter swarm became gradually shallower.

The eruption

The eruptive fissure opened at about 1:55 a.m. It was 1500 m long on land with a NNE–SSW trend and passed about 500 m east of the town. After two days the activity had become restricted to a short fissure segment where a 200 m high scoria cone was built up gradually. The cone was open to the north and the lava spread in a fan, 1–1.5 km wide, towards the north and east. The total volume of lava and scoria was 0.25 km^3 of which the lava constituted 90 per cent. The main part of the lava flow is up to 100 m thick. Only the last erupted lava was of lower viscosity and hence thinner. The explosiveness of the eruption decreased with time as the melt became more basaltic and the rate of eruption decreased.

Damage caused by eruption

Damage was of two kinds: (1) to property, caused by falling ejecta and the flowing lava, and (2) economic, due to temporary evacuation of the residents of Heimaey together with the transport of movable property and industrial machinery from the Island. The total cost has been estimated as several tens of millions of dollars, a number that should be compared to Iceland's Gross National Product of 500 million dollars in 1971.

Damage by bombs and ash-fall occurred during two storms within five days after the eruption began. Later ash-falls were less heavy but still added 40 per cent to the total of ash that fell on the town. Houses in the east of the Island were completely buried, and many of them burnt or collapsed. The lava flow buried 200 houses, more than half in late March when a tongue of lava, 300 m wide, advanced swiftly into the town.

The important fishing port of Vestmannaeyjar did not suffer directly from the eruption. However, there were troubles later due to the drifting of sand and gravel eroded from the edge of the flow. In the spring and summer of 1973 people began to move back to Vestmannaeyjar to clean the town and get the fishing plants working again. By November 1973, 2000 residents had returned and in summer 1974 they numbered 4300, some 80 per cent of the pre-eruption population.

Preventative measures

Preventative measures taken against the damaging effects of the eruption included:

(1) clearing the roofs of accumulated ash;
(2) tacking corrugated iron over the windows (to prevent fires from glowing lava bombs);

(3) piling up of diversion barriers;

(4) spraying sea water on the lava to speed its cooling.

Large-scale cooling by sea water did not start until the end of March, after the eastern part of the town had been engulfed by lava. At its peak, up to 1 t s^{-1} of sea water was pumped on the lava margin and large areas behind it, in total some 50 hectares. The effect was to create internal lava barriers which caused the flow to thicken and ride over itself. The estimated cost for the lava-cooling operations (labour, equipment, transportation, fuel, etc.) was US$ 1.5 million.

Gas flux

From February to April, poisonous gas (98 per cent CO_2) accumulated in low areas within the eastern part of the town. In May the gas flux diminished and finally stopped in September. Breathing of the gas caused the only fatality reported as a result of the eruption. An ash wall and a long trench were made to divert and permit the escape of the gas, but neither was completely effective.

Monitoring

Vestmannaeyjar is being monitored seismically and by repeated measurements of horizontal distance.

Not all eruptions are catastrophic. Some are considered by volcanologists to be minor, but even these can have major repercussions on human and economic activity. A simple phreatic eruption can cause ash-falls that are without real danger for the population, but which can severely hinder circulation and wear out machinery because of the abrasive character of such ashes. Heavy rains can mobilize ash deposits, causing mud-flows that can cut roads and sweep away bridges. Such destruction, apparently minor, can have grave implications. If, furthermore, such a phreatic eruption rapidly evolves into a paroxysmal magmatic phase, the resulting evacuation of the population can pose very serious problems.

On top of the destructive effects of the volcanic activity itself, it should be stressed that such areas are commonly subject to other natural hazards during the periods of volcanic quiescence, such as mudflows and landslides. This is particularly true for volcanoes with explosive activity, which deposit unconsolidated material on steep slopes.

Assessment and zonation of volcanic hazards

The forecasting of volcanic hazards is necessary for civil protection, but it is also necessary to plan the land-use of a volcanic area properly in order to limit the economic damage, especially in densely inhabited industrial or agricultural areas. In places where volcanic activity is frequent, such as Japan or Iceland, people generally learn to co-exist with volcanoes, carrying out mitigating measures as necessary. Each new eruption also leads to better knowledge of the volcanic processes and, usually, to improvement in surveillance techniques. For many volcanoes, however, the rest period is very long and there has been no historic experience from which to benefit. It is clear that people who live close to a volcano have little or no understanding of the risks they run.

The catastrophic eruption of Mont Pelée, Martinique in 1902, which destroyed Saint Pierre and killed 28 000 people, is an illustration of the underestimated danger of a volcano that had not seriously erupted in historic times. Before 1902, only two small phreatic eruptions had occurred in 1851 and 1792, when ash-falls did little damage. Numerous precursory signs (sulphurous gas emissions, weak ejection of ash, abnormal river floods, etc.) lasted several months before the disaster on 8 May 1902. At the time, the commission, appointed by the Governor to investigate the renewed volcanic activity, concluded that it was taking the same course as that of 1851, and that there was no immediate danger for Saint Pierre. Because of this sincere, but wrong diagnosis, based on an incomplete understanding of the history of the volcano, 28 000 people died.

The recollection of this catastrophe played an important role during the volcanic crisis at La Soufrière on Guadeloupe Island in 1976, when a series of alarming, but small-scale, eruptions occurred. Deficient knowledge of the past eruptive history and a lack of volcanic hazard appraisal made it difficult to interpret these volcanic events, and to determine what actions should be taken to prevent loss of life. Poor communication between officials and scientists rendered the management of this crisis even more difficult. About 72 000 people were evacuated for 3 months and, though no eruption took place in the end, the social and economic price of this non-event was high.

Both examples clearly show that, whenever a volcano becomes active again, its eruptive history and processes are often virtually unknown. Combined with insufficient monitoring and, above all, the strong variability in behaviour of all volcanoes, it is very difficult for the responsible volcanologists, as well as for the civil authorities that are entrusted with the protection of persons and property, to manage the crisis effectively.

Hazard assessment

It is useful to distinguish 'hazard' and 'risk', two widely used but sometimes ambiguous terms. Following Peterson (1988), *Hazard* can be defined as the possibility that a dangerous phenomenon might occur. Whereas *risk* is a measure of the loss to society that would result from the occurrence of the phenomenon. Loss might include human life, property, or productive capacity. *Acceptable risk* is the level of risk people are willing to assume when confronted by a specific hazard. Determination of the acceptable risk is important in long-term land-use planning or development, or for access to or evacuation of particular areas during crises.

The assessment of volcanic hazards has as its objective the forecasting of the types of eruptions that can be expected, the anticipation of volcanic events and their resulting deposits (which might endanger human lives and property), the location of areas that could be affected by the events, and the probability of a future eruption. Potential volcanic hazards can be forecast by assuming that, for a given volcano, future volcanism will, generally, be of the same type and magnitude as that of past events. The type and magnitude of such past eruptions, and their recurrence, as deduced from historical records and geological observation, are thus critical data for the assessment of the future behaviour of a volcano.

For a volcano with an abundant historical activity, the observations and documentation of frequency, type, magnitude, duration, destructive effects, precursory effects, and extent of the deposits, will provide the basic data for future volcanic hazard assessment. For a volcano with little or no historical activity, the information on future behaviour must come from the extrapolation of geological observations. Reconstructing the history of a volcano is done from geological mapping, ash-fall stratigraphy, chronology, and the evaluation of the genesis of the various deposits found (see the previous section for the description of the various types of volcanogenic deposits). The type of volcanic event is deduced from the type(s) of deposit and the mode of transportation (flow, fall, or surge) that formed them. The age of the rock can be determined by dating certain minerals in the rock or by radio-carbon dating of organic material trapped in the deposits. Mapping the distribution of the products shows what areas were affected by which type of event, and how frequently. The chemical and mineralogical composition of the rock, and thus of the parent magma, is another major clue; any change in such composition can indicate a potential change in eruptive behaviour.

The recognition of cycles or patterns of volcanic activity may be of value in prediction. When such cycles are well established and the volcano enters a period of new activity, then it should be possible to predict the probable sequence of events and prepare for any hazards that might threaten the population. For example, the regular activity of Vesuvius is apparently interrupted every 1000–2000 years by a major Plinian (pumice) eruption.

Mont Pelée on Martinique Island also shows a cyclic pattern of activity, with periods of frequent and voluminous eruptions followed by periods of infrequent and minor eruptions (Fig. 4.6). In fact, geological study showed that about 20 eruptions occurred during the past 5000 years. The figure shows two types of eruption and their approximate magnitude; their distribution through time indicated that no simple law exists that can explain the occurrence of either a pumice or a nuée ardente eruption. On the other hand, there seems to be a certain periodicity in the frequency and magnitude of the eruptions: at present the volcano is apparently in a major phase of activity.

Even so, the study of volcanic deposits that record past activity is not always easy and the origin of some deposits is obscure. Highly destructive phenomena, such as the lateral blast of Mount St Helens, leave thin and relatively small deposits that are difficult to discern in the geological record. Without doubt, the frequency of such 'minor' eruptions is commonly underestimated when working from the geological record alone, even though they can cause economic havoc in an area around a volcano. In addition, there is always the possibility that an unexpected type of eruption or one of exceptional magnitude will occur, as was the case with Mount St Helens in 1980.

Mapping volcanic hazards and risk areas
Volcanic hazard mapping, although a delicate exercise, provides the basic data for long-term land-use planning around (and on!) volcanoes. It should also help to anticipate such problems as may arise in the case of a new eruption and contribute towards the preparation of emergency plans.

Hazard zonation is first of all based on the determination, by geological means, of those areas that were affected by past eruptions, and of which effects were involved. This allows the prediction of the possible extent of similar phenomena in the future. The resulting information is presented as volcanic hazard maps that show which areas will be particularly susceptible to what type of volcanic hazard (e.g. lava flows, ash-fall, toxic gases, mudflows), and that outline the areas of greatest risk around a volcano. Figure 4.7 presents a hazard map of the Plinian (pumice) eruptions for Vesuvius, which has been under observation for a long time (the following case history).

The Vesuvius observatory and volcano monitoring [136] [137]

The 'Osservatorio Vesuviano' (OV), or Vesuvius Observatory, was built in 1841 to 1845, when it was decided to build a laboratory for monitoring and making measurements. Since then, the OV has gradually grown into an organization of international repute in volcanology that now employs 45 people. Its activities cover not only Vesuvius, but all of the volcanically active regions in southern Italy and in the Tyrrhenian and Mediterranean seas (Fig. 4.8).

The monitoring is done by means of fixed networks and temporary control measurements. The networks are installed on Vesuvius, the Phlegrean Fields to the west of Naples, Etna on Sicily, the Eolian Islands (Vulcano, Stromboli, Lipari, etc.), as well as the islands of Ischia and Pantelleria (Fig. 4.8). In addition to seismographs, the networks comprise tidal recorder stations, gravimeters and altimeters, telemeters, and inclinometers for measuring, respectively, horizontal and angular movements. The monitoring further includes temperature and geochemical measurements of the liquids and gases sampled in fumaroles and soils, including, according to the case, their contents in water, hydrogen sulphide, sulphur dioxide, carbonic gas, the rare gases, and helium and carbon isotopes.

Table 4.9 shows a summary of all stations and their activities.

In addition to its surveillance work, the OV carries out specific work in the framework of major research programmes. Examples of these are the microgravimetric measurements on all volcanoes for the Geodynamics Project, and the joint research programme by the Italian Consiglio Nazionale delle Ricerche and the French Centre National de la Recherche Scientifique on the causes and predictions of earthquakes in Calabria. Other work covers engineering activities, such as in the Ipinia area after the earthquake of November 1980 or the integrated programme on the Phlegrean Fields that started in 1982.

Because of its importance, the monitoring and research work on the Phlegrean Fields merits a special comment. This area of about

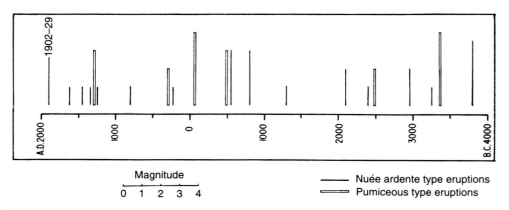

Fig. 4.6 Absolute radio-carbon chronology of the Mont Pelée eruptions during the past 5000 years. The periodicity in the frequency and magnitude of the eruptions is prominent.

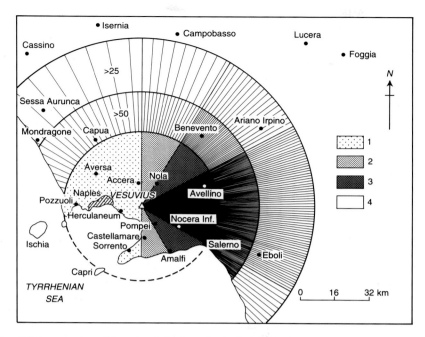

Fig. 4.7 Hazard map for a Plinian eruption of Vesuvius, Italy (from Barberi *et al.* 1983). The circular risk zones around the vent are based on observations of the maximum thickness (contours of 25, 50, and 100 cm) of pumice—fall deposits at various distances from the vent. Also, the dominant westerly winds create areas with increasing hazard, shown by increasing darkness.

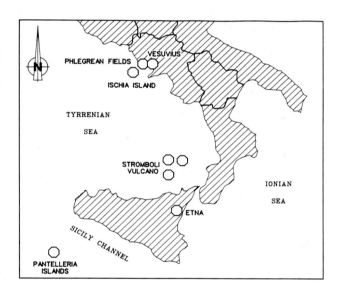

Fig. 4.8 Map showing volcanic activity in southern Italy.

80 km² lies just west of Naples and has a population of about 400 000. In 1982, an abnormal seismic activity combined with a measurable bulging of the ground was observed. This alarming situation culminated in October 1983 with the hurried evacuation of 40 000 inhabitants from the centre of Pozzuoli.

The Fields correspond to a complex structure around a large, ancient, basin-shaped volcanic depression, or caldera, of about 12 km diameter and an age of at least 35 000 years, which comprises several phreato-magmatic (extruding both gases and water) craters, including the Solfatare that was created about 5000 years ago. All through historical time, this area was characterized by strong vertical ground movements, which are very well illustrated by the presence of Roman buildings that are now 14 m below sea level. After successive periods of uplift and sinking, a strong uplift of 3.1 m was measured between 1969 and late 1984. Thanks to the measures made with the monitoring network, it was noticed that the area affected by the strongest bulging between 1982 and 1984 had a radius of about 6 km and was centred on Pozzuoli, the whole more-or-less coinciding with the known outline of the ancient caldera. The analysis of the measurements, particularly from gas geochemistry, led to the conclusion that a phreatic eruption of low energy was imminent, as none of the measurements seemed to indicate that new magma had been injected in the chamber below.

Since 1985, the geochemical measurements have shown a sudden drop in the contents of hydrogen sulphide, methane, and chloride that could be correlated with the end of the period of uplift. At the same time, the measurements of carbon dioxide and helium, which are indicative of magmatic activity, showed only little variation over time; in particular, no significant changes in the helium isotopes were observed.

Over the years, the OV has developed many lines of international collaboration, particularly with the French Institut du Globe (Paris) and with the BRGM, which comprises the French Geological Survey. The joint work with the last two organizations concerns the geochemistry of soil gases and those found in fumaroles.

Another type of hazard mapping uses predictive models that were developed for mudflows, lahars, and lava flows. By knowing or estimating the various relevant parameters, such as flow rate, average slope and volume of the material that may be mobilized, it is possible to predict which area is likely to be affected and what will be the extent of the phenomena. In the same manner, one can use models of the mechanical stability of the volcano itself, in order to predict which parts of the cone will be susceptible to collapse.

Thus, several zones of relative hazard and risk can be defined. The assignment of the degree of risk is based not only on the history of past events affecting the area, but also on other circumstantial factors, such as local meteorological conditions, or the present-day morphology of the volcano. One example is shown in Table 4.10, in which are defined four hazard zones found around Mont Pelée (Fig. 4.9).

Table 4.9 Summary of the number of stations of the Vesuvius Observatory in the various regions of southern Italy and its surrounding seas

Area	Seismographs	Tidal recorders	Topography	Gravimetry	Geochemistry
Phlegrean Fields	8	2	A: 240 (72 km) T: 11 I: 4	14	Pozzuoli Solfatares, submarine fumaroles
Vesuvius	3 permanent 9 mobile	2	A: 24 (17 km) T: 21 I: 4	14	Crater fumaroles
Ischia	1		A: 164	11	Rione Bocca fumaroles
Campanian plain			A: 246 (180 km)	13	
Gulf of Naples				800 (4 cruises 1988–90)	
Stromboli	mobile stations				
Vulcano	70 (the 1986 campaign on Vulcano, Lipari and Salina)		A: 73 (21 km)	16	Fumaroles of crater and coast, soil gases over entire island
Eolian islands				9	
Etna	4		A: 100 (60 km)		
Pantelleria			A: 78 (45 km)	13	

A = Altimetric station; T = Telemetric station; I = Inclinometric station

Table 4.10 Classification of hazard zones around Mont Pelée volcano, Martinique Island, deduced from past eruptive history and hazard assessment (from Westercamp 1986)

1. *Very high hazard zone* (I, II, III, Ia, and IIa on Fig. 4.9)
Region that will be totally destroyed by any volcanic event, either through burying under abundant cold or glowing material, or through suffering the effects of thermal or blast phenomena.

2. *High hazard zone* (IIa and IIb on Fig. 4.9)
Region in which the infrastructure will be partially destroyed. Well-protected inhabitants will escape. Thermal and blast effects are major risks, as are thick ash-fall deposits and pumice-block (up to 1 m) falls.

3. *Moderate hazard zone* (not shown on Fig. 4.9)
Region threatened by fall of cold ash and small pumice blocks, less than 1 m thick, or located downwind of toxic-ash plumes. The boundary between *2* and *3* is that between real danger and serious nuisance.

4. *Low hazard zone* (outside Fig. 4.9)
Region that will be covered by a fine ash layer, without real danger to the inhabitants but likely to disturb social and economic activity and to cause substantial damage to crops and livestock.

Emergency plans

The response to a volcanic crisis is difficult. Both the population under threat and the authorities want to know if the volcano will erupt, and when, but the volcanologists can rarely answer with certainty. Particularly after a long period of repose, the type and magnitude of the next eruption will be highly uncertain.

On the other hand, it is possible to draw up emergency plans in advance, determining what course of action should be taken to restrict or prevent loss of life and property after precursory effects have taken place and during the initial stages of an eruption. Risk-zonation maps are an essential tool for deciding if, when, and how people are to be evacuated, for designating the boundaries of an area to be evacuated, and for determining evacuation routes and modes of transportation.

The drawing up of possible eruption scenarios based on past experience can also serve to anticipate the effects of the eruption and to assist with taking adequate protective measures. Moreover, it is highly important to educate the public in threatened areas about the volcanic processes that can be expected and the effects they may have, and on the emergency plans available to counter such risks.

Figure 4.10 presents a framework of relationship between a volcano and the population. To avoid misunderstanding or even panic, and for maximum efficiency during a crisis, the relationships and lines of communication within the population, that is to say between scientists, authorities, the media, and the general public, must be unambiguous.

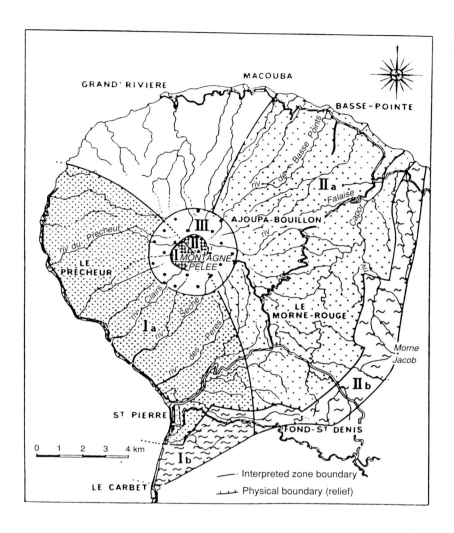

Notes

1 Influence of the location of the eruptive vent

If the eruptive vent opens into Zone I, the eruption will preferentially develop towards the west and south (Zones Ia and Ib), but if it opens into Zone II the entire volcano will be threatened (Zones I and II), with the exception of the north flank. In either case, a 2 km wide area around the vent (Zone III) will be destroyed.

2 Areas threatened by very-high-intensity hazards (Ia and IIa)

These zones will be affected by blast and thermal effects of ash surges, and by pyroclastic flows and associated clouds. Total destruction is likely.

3 Areas threatened by high-intensity hazards (Ib and IIb)

These areas will be variably affected by the thermal effects that are associated with flows and surges. Destruction may occur. The boundary between these 'b' areas and the outer zone of moderate risk coincides more or less with a 10 km radius circle, but locally the role played by relief is important.

Probability of destruction	Intensity			
	Very high		High	
Certain	Zone III			
Very high	Zone Ia		Zone Ib	
Less high	Zone IIa		Zone IIb	

Fig. 4.9 Zonation map of hazard related to 1902-type incandescent cloud eruptions and associated ash blasts at Mont Pelée, Martinique Island.

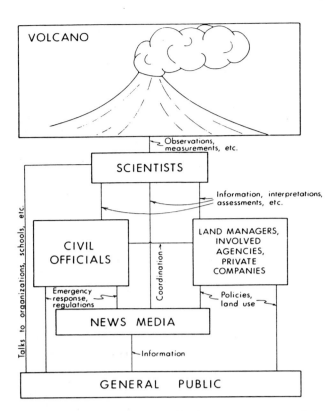

Fig. 4.10 The possible relationship between a volcano and the population living around it, as well as the lines of communication within the population. This schematic framework was drawn up as a result of the May 1980 eruption of Mount St Helens, USA (from Peterson, 1988).

Volcano monitoring

Many active volcanoes located in developed countries have permanent volcanological observatories equipped with a full array of monitoring equipment. Continuous monitoring through complete volcanic cycles, including the rest period, commonly makes it possible to forecast an eruption. At times, however, even these well-monitored volcanoes depart from their predicted behaviour. Most volcanoes with only slight activity have no permanent installations on them, but, when signs of awakening become apparent, mobile monitoring equipment is available for installation. In this case, however, the lack of background data usually makes it difficult to interpret the measurements, even after monitoring has started.

Several instrumental methods are used for the monitoring of volcanoes, both for forecasting and for improving the understanding of the volcanic processes. Seismic methods are amongst the most widely used; they detect earthquakes and tremors generated by ascending magma. Rarely, if ever, does an eruption start without having been preceded by an increase in seismic activity. For this reason, seismic monitoring is a reliable method for giving warning of an impending eruption.

Another important surveillance technique is the monitoring of ground deformation. Horizontal and vertical displacement, such as bulges in slopes caused by the intrusion of magma into the volcanic cone, can be measured with various types of instruments. Systematic measurements can indicate when and where dangerous phenomena could occur. Furthermore, a rising magma body causes changes in the gravity and magnetic fields, as well as in the electrical and electromagnetic proper-

ties of the rock. Measuring such changes can further improve the accuracy of forecasting a volcanic eruption. Another predictive tool is the monitoring of gas emissions.

Conclusions

Several European countries are concerned about the possible occurrence of volcanic hazards, either on their home territory or in their overseas territories. Historic eruptions have highlighted the destructive potential of volcanoes such as Vesuvius, Mont Pelée, or Heimaey. As the population density will certainly continue to increase around most volcanoes, more and more lives, property and infrastructure will be threatened in future. It is certain that, to avoid incalculable loss of life and economic cost, it is necessary to develop coherent programmes for the prediction and mitigation of volcanic risk. Earth scientists are in the position to be aware of the dangers presented by the various active and dormant volcanoes in Western Europe and its overseas territories. Geological and historical studies of the volcanic past are essential in arriving at reliable predictions for the future behaviour of such volcanoes. Volcanic hazard assessment based on such studies will help the authorities to draw up the necessary long-term plans for land-use, and the general public to understand the reasons for concern. Furthermore, such assessment constitutes an essential basis for preparing emergency plans, and for laying out the necessary lines of communication, the warning systems to be used, and the evacuation routes.

The Western European Geological Surveys (WEGS) have a key role to play in the development of volcanic hazard management. They are in the best position to create the necessary liaison between earth scientists, the various government organizations concerned with protecting the population and its property, and the population involved.

There is a clear mandate for WEGS to spearhead the development of the tools and methods of hazard management by being deeply involved in computer modelling, prediction studies, warning and evacuation studies, and land-use planning. This means they must be concerned with volcanological research, environmental sciences, agricultural planning, and economics. Volcanic processes are complex, as is the field of investigations required to manage the risks they provoke.

RECOMMENDED FURTHER READING

Barberi, F., Corrado, G., Innocenti, F., and Luongo, G. (1984). Phlegrean Fields 1982–1984: brief chronicle of a volcano emergency in a densely populated area. *Bulletin of Volcanology.* **47**, 175–85.

Barberi, R., Rosi, M., Santacroce, R., and Sheridan, M. F. (1983). Volcanic hazard zonation. Mt Vesuvius. In *Forecasting volcanic events* (eds H. Tazieff and J. C. Sabroux). (Elsevier, Amsterdam).

Crandell, D. R. and Mullineaux, D. R. (1978). Potential hazards from future eruptions of Mount St Helens volcano. Washington. *U.S. Geological Survey Bulletin 1383–C.*

Crandell, D. R., Booth, B., Kazumadinata, K., Shimozuru, D., Walker, G. P. L. and Westercamp, D. (1984). *Source book for volcanic-hazard zonation.* (UNESCO, Paris).

Luongo, G., Cubellis, E., Obrizzo, F., and Petrazzouli, S. M. (1991). The mechanics of the Campi Flegrei (Italy) resurgent caldera—a model. *Journal of Volcanological and Geothermal Research*, **45**, 161–72.

Osservatorio Vesuviano (1990). *Bolletino Trimestriale d'Informazione.* Anno I. September 1990.

Peterson, D. W. (1988). Volcanic hazards and public response. *Journal of Geophysical Research.* **93**, 4161–70.

Rosi, M. and Santacroce, R. (1984). Volcanic hazard assessment in the Phlegrean Fields. A contribution based on stratigraphic and historical data. *Bulletin of Volcanology*, **47**, 359–70.

Simkin, T. L., Siebert, L., McClelland, L., Bridge, D., Newhall, C. G., and Latter, J. H. (1981). Volcanoes of the world: a regional directory, gazetteer and chronology of volcanism during the last 10,000 years. (Hutchinson & Ross, Stroudsburg, PA).

Tazieff, H. and Sabroux, J. C. (eds) (1983). *Forecasting volcanic events*. (Elsevier, Amsterdam).

Tedesco, D., Pece, R., and Sabroux, J. C. (1988). No evidence of a new magmatic gas contribution to the Solfatara volcanic gases, during the bradyseismic crisis at Campi Flegrei (Italy). *Geophysical Research Letters*, **15**, 1441–4.

UNDRO and UNESCO (1985). *Volcanic emergency management*. (United Nations, New York).

Westercamp, D. (1987). Zonation of volcanic hazards at Mont Pelée, Martinique, French West Indies. In *The role of geology in urban development. Geological Society of Hong Kong Bull.*, **3**, 413–41.

4.2.3. Ground movement hazards [26]

Diversity of ground movement phenomena

The equilibrium of superficial materials in the Earth's crust is governed by the general laws of mechanics, especially gravity, by limiting conditions (internal stresses, external forces) and by laws governing the behaviour of these materials. Man observes how this equilibrium changes almost continuously, mainly due to natural causes, but also because of his own actions. His attempts to slow down such changes count for very little as compared with the natural forces at play, at least as far as major ground movements are concerned.

The main natural phenomena that shaped Western Europe's morphology are, from north to south:

(1) erosion and deposition during the last Ice Age;

(2) the melting of the ice and the subsequent isostatic readjustment of Northern Europe;

(3) sedimentation and variation in erosion base levels in the central and western lower regions;

(4) present-day external hydro-climatological conditions that can vary enormously, aggravating and accelerating the process of imbalance;

(5) glaciation in the Alps;

(6) a vigorous neotectonic activity in southern Europe.

Europe, even if it displays an external geodynamic evolution that is less spectacular than that of certain other areas in the world, is much affected, both in mountainous and coastal regions, by numerous ground movements that generate very high risk for human installations.

The various types of deformation are schematically shown on Fig. 4.11. Ground movement is always demonstrated by external deformation which can be vertical in the case of subsidence, parallel to the slope in the case of landslides, or almost horizontal for certain types of flow or collapse in quick-clays. It can be very slow (subsidence or creep) or extremely fast (rock falls, sink-holes and flows). It can be moderate (settlement) or strong, in the latter case involving a complete break and redistribution of materials (debris-flows, falling blocks).

Ground instability results from two major groups of factors:

(1) Long-term natural topographic features and the mechanisms that govern their equilibrium;

(2) Short-term temporary factors, which may be of natural origin (rain, earthquakes, etc.) or induced by man (earth-

moving, negligence in maintenance, water-proofing of the surface, overloading, etc.).

The former are characterized by their permanence and extensive spatial distribution, both at the surface and at depth, and the latter by their irregular and transient nature. In all of these, water plays a major role, either through physical and chemical reactions with the strata, or as a result of the short-term effects such as pressure, hydraulic thrust, and freezing and thawing.

In earthquake-prone areas, seismic activity can be a major triggering factor, particularly in the case of superficial landslides (see Section 4.2.1.).

Vertical movements

Vertical movements are usually localized in nature: they can be the result of rebound, swelling or settlement, or the formation of sink-holes above subterranean cavities of natural or man-made origin. Only rarely, such as in the case of subsidence, are such movements areally extensive.

Mechanically, these movements relate mainly to the characteristics of the materials involved, such as swelling clays or organic material, and to the effective stresses they undergo. They are the result of readjustment in these stresses, through the lowering of the groundwater table (Venice, Po basin) for example, or because of the natural or artificial formation of an underground cavity.

Except in the case of sink-holes and rock falls, vertical deformation takes place slowly and is only rarely instantaneous. In fact, it is the differential character of such deformation that causes damage. In the case of generalized subsidence, damage in the early stages is hardly noticeable, as a result of the slow rate and the low level of the deformation. When these increase, high levels of risk are induced affecting infrastructure networks first of all. Much more serious consequences can arise at a later stage, such as progressive flooding around rivers and coasts.

Movements on slopes, with limited displacements

Such movements have variable rhythm and speed of deformation, and can result in the redistribution of materials. Mechanically, these are the most varied phenomena, but they invariably include a localized break due to lack of resistance to compression, such as toppling, or to traction and shearing, such as landslides or rock-falls. This break then develops with the appearance of deformation of varying size and speed.

Traditionally, a distinction is made between movements that affect unconsolidated material and those that affect rocks. Unconsolidated material can be the deposits of earlier transport or it can be material derived from *in situ* weathering of rocks. In this material the movements are usually major and rapid, and the dislocation surface is commonly deep-seated. When movement is slow it is generally known as 'creep' and usually affects only a thin layer of material. Nevertheless, it causes permanent or repetitive damage to property, but the low deformation speed normally limits the amount of the damage.

Landslides represent much greater and more rapid deformation. The dislocation surface can be shallow or deep and flat or curved. They usually move very fast and displace much ground, thus causing substantial damage to the property at the surface, or in their path.

In a rocky environment, the rupture surface usually follows a pre-existing plane of weakness. A single rock or a whole cliff-face may be involved and 'sagging' takes place when an extensive slope is involved. Deformation is strongly variable in

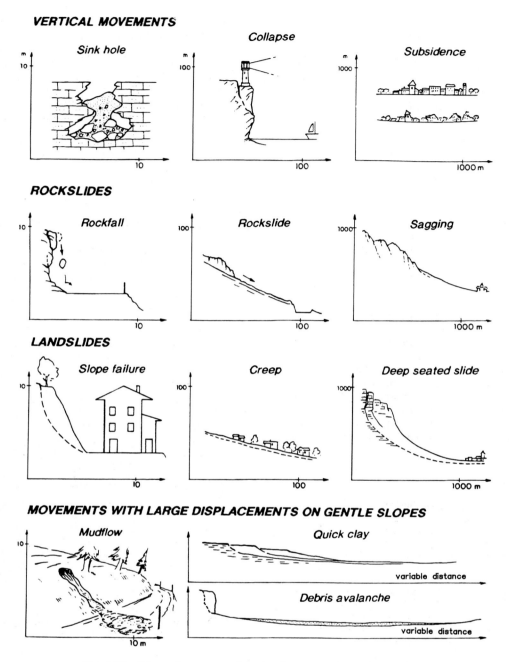

Fig. 4.11 Schematic representation of various types of ground movement.

a discontinuous medium, and certainly more complex than that found in unconsolidated material. The damage is almost invariably great, regardless of the volume involved: a rock fall on a mountain road can cause extensive indirect damage, while the collapse of only a few cubic metres of rock can destroy a house.

Movements with large displacements on gentle slopes
Major landslides involving earth and debris flows result mainly from accumulations of water, the origin of which can be:

1. Exceptional rainfall over a catchment area where the fine sediments are relatively easily moved.

2. Rapid accumulation behind, and sudden release from, a temporary dam created by an earlier landslide, which occurs periodically and especially in mountainous areas (debris flow). They are difficult to predict and cause widespread damage. Such damage is often considerable because of the speed with which the phenomenon occurs and the distance material may be transported. Rock avalanches are always catastrophic, whether due to their direct effects, such as burial, or because of their indirect effects, such as the damming of valleys, followed by catastrophic flooding downstream.

Landslides of quick clays are characteristic and devastating phenomena in northern Europe, related to the existence of marine clays, which were weakened by the leaching out of salts during progressive uplift. A state of critical equilibrium is reached over vast areas of flat-lying deposits and even a minor break can trigger a sudden slide of the material overlying the clay layer.

Socio-economic importance of ground movements

Historical sources

Historial records of ground movements are too imprecise and dispersed to be of much value. Only the more spectacular events are remembered, such as those as Gauldalen in Sweden in 1345, Mont Granier in France in 1248, or Helike and Voura in Greece in 373 BC.

Nevertheless, ground movements leave tell-tale traces in the landscape that the geologist can easily recognise. Linking these to recent progress in understanding paleo-climatology and to observations of similar present-day phenomena in different parts of the world, provides a sound basis from which to tackle future problems. The Western European Geological Surveys could lead the integration of fundamental research in this field.

Already, databases are being assembled by several European Surveys, such as the British Geological Survey, the Geological Survey of the Piemonte (Italy), the French Geological Survey (BRGM), and several Surveys in Germany. Furthermore, a joint task force of the three international committees on soil mechanics, rock mechanics, and engineering geology is in the process of drawing up an international inventory of the great catastrophic events.

Something of the European efforts is illustrated by the following accounts of the geologial data bank of the Piemonte, and of the methodology tests carried out by BRGM in the Isère.

The geological data bank of the Piemonte [135]

The GEOS data bank was set up by the Geological Survey of the Piemonte region and by the Institute for the Hydrologic Protection of the Po Basin (IRPI), with the help of the Regional Computer Centre of the Piemonte (CSI). GEOS combines geological data with a documentary database; it is not only a geological database, but is especially related to the risks from ground movements and floods.

Historical data that cover the period 1801–1980 have shown that the greatest risks are related to meteorological events of high density and long duration. For that reason it was considered essential to include as much information as possible on the effects induced by the observed phenomena.

As regards the level of risk, the following main parameters were used:

(1) the type of process involved, with particular attention to their immediate or delayed response to the meteorological effects, and with emphasis on their probable rate of movement;

(2) The spatial distribution of the processes and potential for their correct identification and location;

(3) the recurrence interval of the processes, obtained by a frequency analysis of the hydrological events identified as the primary triggering cause.

The data thus collected were processed into the following data files:

(1) *Lithological units* (lithological and structural characteristics, qualitative geomechanical rock mass behaviour, the main soil properties, degree of instability and fluvial processes);

(2) *Ancient and recent landslides* (7 different types, mention of particularly active areas, information on rates of movement);

(3) *Damage to man-made structures* (type of geological event, gravity of the damage, and recurrence during the period 1830–1981);

(4) *Areas most susceptible to oversaturation and sliding of the soil* (statistical probability of a landslide occurring, as a function of critical rainfall values that will cause the landslide; shown in red colour on maps);

(5) *Potential stream processes and their activity* (debris and mud-flows on alluvial fans, the frequency of the main events during the period 1830–1981. Thematic maps then show in colour the types of the past and expected processes).

(6) *Characterization of the river channels and description of the main fluvial processes.*

The documentary data bank completes the process data bank. It contains all the sources of historical and scientific information, and serves to add new or updated information to the process data bank. This double approach makes it possible to process and compare highly diverse data, and to use them according to one of the selected themes, which will be mainly the study of a selected geographical area, and/or the probability of the occurrence of a certain risk.

The general organization of the data bank is shown on Fig. 4.12. In this way seven different thematic maps can be generated systematically at scales 1:100 000 or 1:250 000:

(1) maps showing lithological units;

(2) maps showing the vulnerability of the surface to soil oversaturation and slippage as a result of the rainstorms (Fig. 4.13);

(3) maps showing the type of landslide and rock fall: graphic representation of a file;

(4) maps showing areas prone to flooding in valley bottoms and plains;

(5) maps showing portions of stream beds that are distinguished as a function of their plani-altimetric stability, with an indication of the measured rates of flow;

(6) maps showing potentially active alluvial fans;

(7) maps showing all damage to man-made structures in the past century.

The scale adopted for these maps is that of the 1:100 000 geological map of Italy. However, it is obvious that for studying all local problems, further processing of the GEOS data is necessary at a larger scale. For that reason, the topographic database of the GEOS data bank needs upgrading just as much as the geological database.

The French national inventory of unstable slopes (INVI): a test case in the Isère Département [26]

In view of the diversity in type and socio-economic impact of ground movements, it is quite difficult to collect and compare the (often disparate) data that derive from the analysis of each event. The BRGM, with the assistance of the Major Risks Delegation (DRM), has assembled a prototype INVI database on unstable slopes with a triple objective:

1. To facilitate the decentralized collection of available data, both old and scattered over the files of the various organizations that collected them, and newly collected in real-time by the service that is responsible for the property that is exposed to risk.

2. To make such data available to decision makers and the persons that are responsible for crisis management.

3. To assemble and process the necessary data for scientific and technical R&D work, to improve preventative methods.

The prototype is designed to be portable with respect to the data it contains and to the specific types of data collection of each relevant organization. The government restricts its function to centralization of the files and making the results known to the

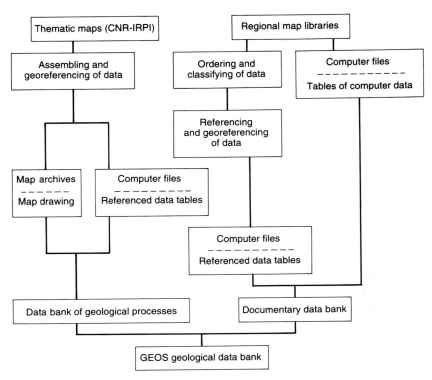

Fig. 4.12   Organization of the GEOS databank.

interested parties. Keeping in mind the wish for completely open access to the system and the existing data, INVI was developed on several levels, from providing basic information that requires no scientific knowledge to the most technical level that can be used by specialists only. The various menus that are successively proposed to the user, are:

(1) the location in space and time;

(2) an inventory of victims and aid;

(3) damage to property, whether recorded or potential;

(4) description and definition of the movement;

(5) intervening parties;

(6) physical quantification of the phenomenon.

In order to make the procedure as simple as possible and avoid errors in the answers, the user can select an answer by means of a window. The answers were selected on the basis of field experience and it is clear that the corresponding keyword files are evolving documents that require periodic updating.

The data, once they are processed with software based on dBASE III, can be interfaced with ORACLE. This allows quite powerful development work with geographic information systems (GIS) such as SYNERGIS and with expert systems like SISYPHE, a slope-instability expert system that was developed by BRGM.

The prototype was tested in the Isère Department in southeastern France, where data on over 1000 events were recorded by the various relevant organizations, which range from government departments and the BRGM to university institutes.

The system is open and can be interfaced with other databases on the same subject, developed elsewhere in Europe. It can be used by government authorities, civil protection organizations, researchers, and engineering firms.

Much work still needs to be done to harmonize these databases and to develop their structures, so that, rather than being simple repositories of data, they can contribute to furthering our knowledge of these phenomena.

Scattered events

If the disasters that occurred throughout history were few and far between, it was because population density was low. Today, with much higher densities of population, similar phenomena, whether natural or man-made, are likely to cause greater damage and loss of life. Although the public in general shows little reaction to or awareness of such events, the authorities must recognize the significance of their cumulative economic impact and include them in local, regional, and national long-term planning.

Increased vulnerability from land-use development

Proper awareness of the risks involved becomes essential with the increased development of human activity in areas that for a long time have remained undisturbed, particularly in mountainous and coastal areas.

In the mountains, the increase in the opening of remote areas, in the number of winter sports resorts, in the development of hydroelectric and water storage schemes, and in the creation of international road links, continuously confronts man with a natural environment in delicate equilibrium.

The problems are no less in coastal areas, where erosion and flooding threatens tourist and industrial development planned for the more vulnerable locations.

Flat land is at a premium in parts of Western Europe, and more and more development takes place on slopes, with consequent impact on the natural equilibrium. Paving, changes in run-off patterns, and disturbance of vegetation can all lead to major ground movements.

Economic impact of ground movements

The socio-economic impact of numerous, varied, and dispersed events is extremely difficult to evaluate with any precision. Nevertheless, some countries are attempting to

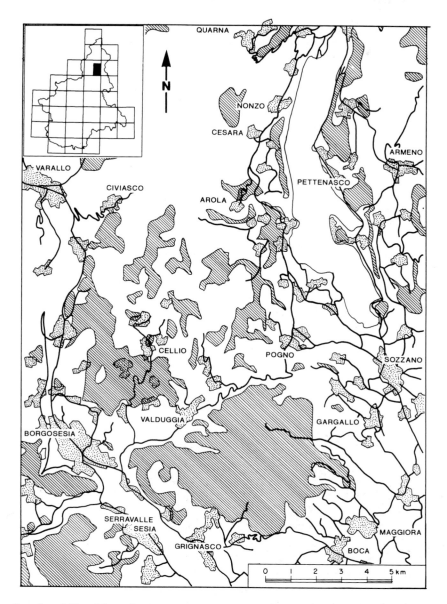

Fig. 4.13 Map showing the vulnerability of the surface to ground movements. The black areas are slopes vulnerable as a result of fluidization of the unconsolidated cover material.

identify the scale of the problem and to get some impression of the cost to the community.

The first initiative of this kind was in California in 1973, where an estimate of projected costs of the damage caused by each major category of natural hazard was assembled for the period 1970–2000. In Spain a similar projection has been made up to the year 2010. Studies in France and Italy suggest that the national average annual cost is in the range of 100 to 200 million ECUs. So far there is little information from other countries, but enough work has been done to recognize that the cost to mankind is very great. Already it can be estimated that the average cumulative cost of damage related to ground movements for all countries in Western Europe is of the order of one thousand million ECUs per year. To get this estimate in perspective, it should be noted that the cost of the rain catastrophe at Nîmes, France, in 1989, was itself of the same order.

Geographical distribution of typical ground movements

Ground movements vary greatly in space and time, but they do relate broadly to the underlying geology, to the topography, and to the past and present climate. In the following sections they are described in relation to three principal geological environments: basement areas, sedimentary basins, and Alpine regions.

Basement areas
Basement areas correspond to very old mountain ranges worn down by several erosion cycles and rejuvenated only locally by the younger Alpine orogenesis. The rocks are mainly granite and metamorphic rocks. There is a noticeable difference between ground movements in granite and metamorphic rocks:

1. Ground movements are quite rare in granite. Most of those that occur are of rock, and relate to the development of decompression or thermal fractures that are more or less parallel to the surface, particularly on steep slopes and on scarps along valley sides. Residual soils are usually thin and instability phenomena are rare except in crush zones along faults or in hydrothermally altered rock.

2. Metamorphic rock commonly shows intense deformation and shattering in the surface zone. Slides are more numerous, and may in part develop in the superficial detrital deposits, and in part along faults in the substratum. Moreover, due to the very ancient history of these materials, fossilized forms of landslide exist, and these can be reactivated by earth-moving or by exceptional precipitation. 'Planar' slides, or creep, with slow deformation, are quite common. Terminal bending or downward curvature of layers can give rise to ground movements affecting both surface deposits and areas of decompressed bedrock.

In the mountainous coastal zone of Norway, debris slides, rock falls and rock slides are common. Rock falls commonly occur in spring and autumn as a result of weather factors such as frost shattering. Rock slides, on the other hand, with volumes of millions of cubic metres, occur along deep-seated dislocation planes. Many such planes are joints lying almost parallel to the surface, resulting from changes in the stress that accompanied glacial erosion. Variations of pore pressure in the joints are likely to cause the initial failure of the rock slope. Great disasters occur when large rock slides fall into relatively narrow and deep lakes or fjords, and generate waves that travel over long distances. This has occurred three times recently in Norway, in 1905 (3 000 000 m³), 1934, and 1936 (1 000 000 m³). So far, landslides into reservoir lakes in Norway have not caused disaster.

Landslides in high mountain areas of Norway [145] [147a]

Landslides, as well as debris flows and avalanches, are typical phenomena in the glaciated valleys and fjords of Norway, where they can cause major problems. These processes involve a mixture of water, debris and organic material (trees, etc.), which can move at speeds as high as 20 m s⁻¹, over long distances and cause great damage. The volume of such landslides varies from a few cubic metres to 20 000 m³, or more, depending on the local topography, drainage, and the presence of surficial deposits. Debris flows are usually triggered during rain storms and snow-melt, and are commonly related to gullying, creep, and slumping (Fig. 4.14).

It is assumed that the debris flows have return intervals of 50–500 years, with the larger ones possibly having even longer intervals. The short history of human settlement as compared to the long return period of destructive mass movements makes it difficult to evaluate the potential hazards properly. Land-use planning based only on present landslide information can lead to major mistakes being made. The complexity of landforms and surficial deposits, and the great regional differences, make it difficult to predict the types of landslides likely to occur on the basis of data on gradients, catchment areas, drainage patterns, or debris type. For these reasons, geological or thematic maps should include data on past and present landslide deposits and rock falls, and should also include information on other rapid mass-movements such as snow avalanches. These maps could then serve as a basis for the construction of natural hazard maps.

Because of the exaggerated topography found in much of Norway, debris slides and flows occur on steep hillsides covered with glacial till and colluvium. The structure of the soil profile is strongly influenced by the effects of frost penetration. Debris slides glide over one or several superficial slip planes at relatively moderate velocities. Debris flows have a higher velocity due to high

water content and commonly follow the natural drainage, from which water is added causing the material to flow even faster. Catastrophic events have been rare, but in July 1789 heavy rain and rapid melting of the snow caused one that killed 68 people.

Sedimentary basins

Sedimentary basins usually have little relief. Their deposits are the result of erosion in the long term and they vary greatly in age and in the degree of induraton. In general, such basins are affected only slightly by tectonic deformation. Nevertheless, in northern Europe the effects of isostatic readjustment after the melting of the ice caps are noticeable, and in southern Europe the neotectonic movements that form the last spasms of the Alpine orogeny, have affected some of the youngest sediments in Europe.

Because of the great variety of material involved, ground movements have many different forms. However, due to the generally low relief they are only rarely catastrophic, and they relate much more to the physical and chemical properties of the rocks, to their structure, to hydrogeological factors and to climate.

Three types of movement are fairly common: *Landslides* are commonly triggered by erosion at the foot of a slope resulting from wave action. *Sink-holes* are the result of solution in carbonate or gypsum rocks. *Coastal subsidence*, especially around the North Sea Basin or in the Po Basin, can be of the order of several millimetres per year, part of which is attributable to eustatic rises in sea level.

In coastal areas, in particular along the English Channel in both England and France, the crumbling of chalk cliffs is quite common. Property is at risk during stormy periods in particular, and in France, especially, resources of building material are depleted as a result of being transported along the Normandy coast. Locally, where the cliffs are composed of clayey or marly material, crumbling may develop into major rotational sliding.

The case history of the Biarritz cliffs clearly shows the amount of work that has to be undertaken for the conservation of some coastal sites.

The Biarritz cliffs [26] (Fig. 4.15)

The Basque Cliff at Biarritz, bordering the Atlantic Ocean, is 50 m high and extends for 1200 m from north to south. Alluvial deposits, some 10 m thick and consisting of sand, gravel, shingle, and shale, rest on marls with a bed of weathered plastic clay at the junction. The alluvial deposits contain groundwater that percolates into the marls through cracks and joints, and dissolves the calcareous fraction. Along the entire cliff, progressive ground movements have brought about the progressive collapse of cliff-top buildings, as well as of the protective cladding built in the 1930s.

Five major destructive elements have been recognized:

1. Dissolution by, and pressure from, groundwater, mixed with sand and gravel, caused the collapse of buttresses that were not anchored.

2. Groundwater flow has caused debris slides, initially in the alluvium and later in the marl, in areas not protected by buttresses.

3. The seepage of surface- and groundwater through the marl is further affected at the base of the cliff by variations in temperature and humidity.

4. Waterlogging at the top of the cliff coincides with the presence of fractures and joints.

5. Marine erosion affects the base of the cliff in the absence of a protective embankment.

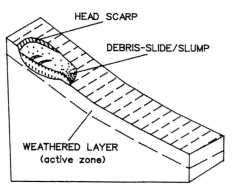

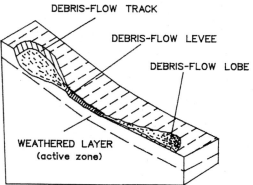

Fig. 4.14 Debris flows.

Rock falls, landslides, and general erosion have caused serious risks to both people and property and, as a result, a major programme of consolidation and rehabilitation has been undertaken. A geological and geotechnical study has identified the problems and the following remedial actions are carried out in successive phases:

(1) drainage of the groundwater, particularly in the alluvium;

(2) reduction of the average gradient of the cliff face to achieve stability of the various materials involved;

(3) building of an embankment at the base of the cliff to act as a protection against marine erosion;

(4) drainage and landscaping of the whole cliff to stabilize the slope, to re-establish access from the top to the beach below, and to render this part of the coast aesthetically pleasing once more.

Along valleys cut in clayey, marly, or calcereous slopes, localized landslides can be triggered by stream erosion. Such slides are quite widespread in the London, Paris, and Aquitaine basins. Similar phenomena occur in the Mesozoic and Tertiary basins of Spain and Portugal. Examples are the Condeixa, Monsanto, and Santarem landslides in Portugal, and the Inza

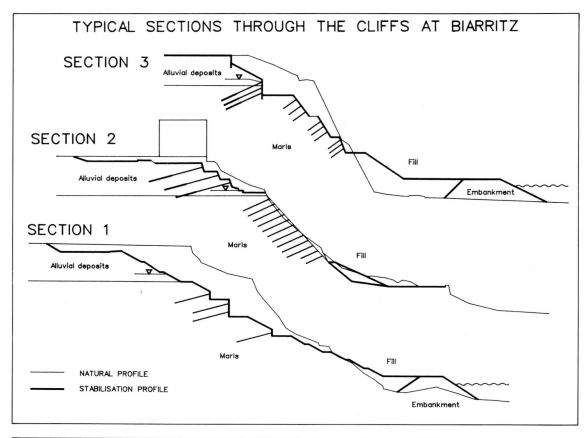

Fig. 4.15 Coastal erosion at Biarritz.

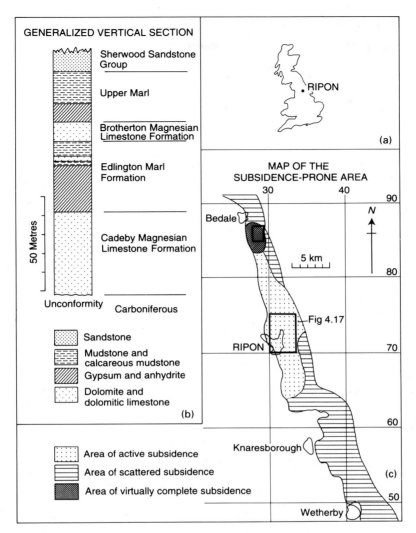

Fig. 4.16 Location map, generalized vertical section and map of the area prone to subsidence caused by gypsum dissolution (from Cooper 1986).

and Olivans slides in Spain, and those that affect a large part of southern Italy.

The formation of sink-holes related to underground cavities is common in calcareous or gypsiferous rocks. These phenomena, very widespread in France, also occur in Great Britain. The case history of Ripon in North Yorkshire shows how catastrophic subsidence can occur because of natural dissolution of gypsum underground.

Subsidence caused by the dissolution of gypsum in North Yorkshire [12]

The Nosterfield–Ripon–Bishop Monkton area (Fig. 4.16(a)) has suffered about 40 episodes of subsidence in the past 150 years. Underground dissolution of thick gypsum beds in the Edlington and Roxby formations of the Zechstein sequence (Fig. 4.16(b)), has resulted in a 3 km wide and 100 km long belt of ground that is susceptible to foundering (Fig. 4.16(c)). Mapping of the area resulted in the recognition of several hundreds of subsidence hollows. The presence of competent rocks result in the formation of steep-sided cylindrical shaft-like depressions, whereas the areas underlain by drift deposits show conical depressions. The most recent subsidence had catastrophic consequences.

Figure 4.17 gives an overview of the distribution of the hollows in the Ripon area, together with the recorded date of collapse. The subsidence is controlled by differences in the solubility of gypsum and limestone, by the gypsum–anhydrite transition, by the orientation of bedding- and joint-planes, and by the collapse of earlier caverns and breccia pipes.

Predicting future subsidence is very difficult, except in areas where the hollows are aligned in regular patterns. These indicate the presence of zones of weakness that are probably caused by underlying faults or joints. Here, the density of hollows gives an indication of the likelihood of further collapse, which can be calculated by statistical means. Land adjoining a subsidence structure is also at increased risk, because most hollows occur in groups of two or three.

Remedial action is, in part, the infilling of the hollows with inert fill. This may prevent the sides from further slipping, but it is far from foolproof as several cases are known of further subsidence of the in-fill. Obviously, in view of the porous substrate, no toxic waste or water should be poured into these hollows. Furthermore, pumping groundwater from the gypsum may increase dissolution and thus cause further subsidence.

Detailed site investigations by means of drilling, and down-hole and surface geophysics, are essential for ascertaining the suitability of a site within this area for development.

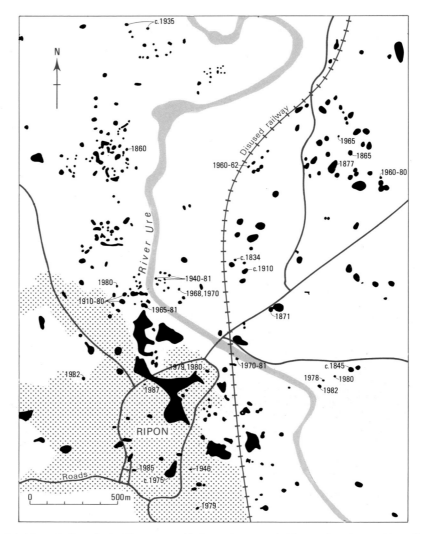

Fig. 4.17   Subsidence hollows in the Ripon area with dates of recent subsidence where known (from Cooper 1986).

A special type of ground movement, major quick-clay landslides, is common in northern Europe, in Norway, Sweden, and southern Finland. They occur in marine sediments deposited after the last glaciation, about 10 000 years ago. They are due to two parallel processes: isostatic uplift that exposes marine sediments to surface erosion, and leaching of salt water from the soil matrix. Two mechanisms seem to predominate in the phenomenon:

(1) a retrogressive process, initiated by a local slide;

(2) monolithic flake-type slides, involving large areas, the overall stability of which is permanently low.

The largest quick-clay slide known from Fennoscandia occurred in Verdal, Norway, in 1893 when the movement of 55 million m³ of clay over an area of 3 million m² killed 112 people. Another very large slide occured in 1345 in Gualdalen when 500 people were killed. The most famous recent quick-clay slide was at Rissa in 1978 (300 000 m², 5–6 million m³). In Sweden, one very large clay landslide (more than 10 ha) is thought to occur every 10 years and one larger than 1 hectare every two years.

Quick-clay slides in Norway [144]

Quick-clay slides, relating to the mass movement of marine clays on land, are a common hazard in Norway. Late Weichselian and Holocene marine clays are widespread in the lowlands of southeastern and central Norway. As a result of isostatic rebound, these sediments are now found at 150–200 m above sea level. The shift in shoreline, which was a 5–10 m per century during the early Holocene, led to intense fluvial erosion of the marine clays, which was accompanied by the leaching out of the salt water from the clay pores and its replacement by freshwater. This process strongly increased the sensitivity and the potential mobility of such clays (Fig. 4.18).

The decrease in shear strength and the reduced ability of many slopes because of erosion has resulted in a large number of quick-clay slides in Norway during the Holocene. The greatest slide known to have taken place in historic times occurred in Verdal in 1893; this displaced 55 million m³ of liquid clay, which flowed down the valley. Sixteen farms disappeared altogether, a further thirty farms were destroyed, and 112 people were killed (Fig. 4.18(b)).

More recently, in 1978, a quick-clay slide occurred at Rissa, near Trondheim (Fig. 4.19). In Norway alone such slides involve well over 100 000 m³ each year. Because of this, a systematic mapping

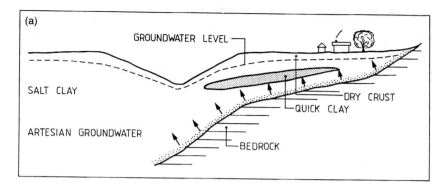

Fig. 4.18 (a) Sketch profile showing the leaching of marine clay by groundwater flow; (b) the Verdal quick-clay landslide in 1893, when 55 million m³ of liquid clay floated out, 16 farms disappeared, a further 30 farms were destroyed, and 112 people were killed.

project of quick-clays was carried out by the Geological Survey of Norway (NGU) and the Norwegian Geotechnical Institute (NGI). The NGU was responsible for the mapping of superficial deposits in areas with marine clays, whereas the NGI evaluated the risk for quick-clay slides. The work resulted in the publication of a series of thematic hazard maps.

One type of movement, not well-known in Europe until the droughts of 1989 and 1990, is the disturbance of the foundations of buildings by the contraction and swelling associated with changes in the water content of clays. Individually, these phenomena may have quite a modest socio-economic impact, but their collective effect is considerable.

In Greece, particularly in the Gulf of Corinth, the coincidence of earthquakes along coastal faults with seasonal coastal sedimentation forming littoral alluvial cones, has caused sliding from time to time with catastrophic results. In 373 BC, for example, the cities of Helike and Voura disappeared under the sea as a result of such phenomena.

Alpine regions
The Alps, be they French, Italian, Swiss, Bavarian, or Austrian, are the scene of numerous and varied ground movements resulting from many factors.

1. Tectonic factors such as joints, fractures, and faults, together with permeability, resistance to shearing and traction, and physico-chemical weathering, generate the states of recent (neotectonic) and current (seismogenic) stresses.

2. The 'glacial history' factor is exemplified by two main modes of action:

   (i) an intense *mechanical* action during glaciation, with considerable steepening of the slope gradients;

   (ii) a destabilization *feedback* action by elimination of stresses during the melting of the ice, with several associated actions, such as the thermodynamic stresses synchronous with the increase of thermal contrasts, hydraulic stresses that occur with increasing water content and pore pressure, and the absence of vegetation.

3. The deepening of valleys and corresponding base-level erosion.

4. Localized dissolution phenomena, in particular in Triassic gypsum.

5. Seismic activity as in northern Italy.

Numerous authors have attempted to relate the various types of ground-movement to major litho-structural units in the Alps. Although encouraging, the results are not yet satisfactory for at least three reasons:

Fig. 4.19 Aerial view of the quick-clay landslide at Rissa, Trondheim, Norway, in 1978, involving 5–6 million m³ of clay, destroying 12 farms and homes, and killing one person.

1. Inventories of areas affected by ground movements are in their infancy as yet.

2. The distinction between present and old (historically recorded) or very old movements is made too infrequently.

3. Finally, and especially, only very little is known about the internal mechanisms of such movements.

In spite of these reservations, correlations between types of landslide and lithological and structural units seem to be appropriate.

Crystalline units These are largely made up of ancient metamorphic rocks (crystalline schists and gneisses) with granite intrusions. In granite, rock falls are the main form of ground movement. In metamorphic rocks, where foliation and other discontinuities are well developed, major slides are common. Where heavily fractured and strongly affected by weathering, these rocks may be covered by thick soils, in which clays (smectite) cause instability. Steep slopes at high elevations can be affected by gelifraction, producing material that is prone to ground movements, particularly when triggered by seismic events.

One of the most important cases of ground movement within a crystalline rock unit occurred recently at Valtellina in Italy.

The Valtellina floods and the Val Pola landslide, 1987
[132] [133] [134]

During the summer of 1987, heavy rainstorms fell on large parts of the central Alps, with the most intense rainfall being recorded on the southern slopes that include the Valtellina. The immediate effect of this inflow and downflow can be summarized as follows:

1. Widespread flooding of the Adda river in the middle and lower parts of the Valtellina, with the transportation and deposition of vast amounts of sandy–silty material, which caused great damage to farmland in the valley.

2. Erosion along the streams at the foot of the slopes, with local destabilization initiating landslides. The most important of these occurred in the Val Tartano, the Val Torreggio, and the Val Pola.

The Val Pola landslide occurred in the upper part of the Valtellina, about 10 km south of Bormio. On 28 July, at 7.23 a.m. following the opening of a crack observed during previous days, a huge mass of rock and debris fell down from Monte Zandila. The villages of San Antonio Morignone, Morignone, Poz, Tirindre, and San Martino, and part of Aquilone, were buried; 27 people died, most of them residents of Aquilone.

The extent of this phenomenon is illustrated by Fig. 4.20. The slide scar volume was 32–33 million m³, but the volume of the slide mass was evaluated at 45 million m³. It travelled about 2.5 km down the valley and up to 300 m up the opposite valley side.

Remedial action

The landslide blocked the valley, creating a lake in the main valley. The remedial action here consisted of the installation of three 15 m³ s⁻¹ pumping units and the digging of two bypass tunnels of 6 m and 4.5 m diameter. The slide mass was consolidated by grouting and a spillway was dug across its crest to divert further floods. Finally, an embankment was constructed at the base of the slope where the slide had started, as protection against further block-falls and debris flows.

After the landslide, numerous other areas of instability were identified, in particular in the area of the slide scar. The work

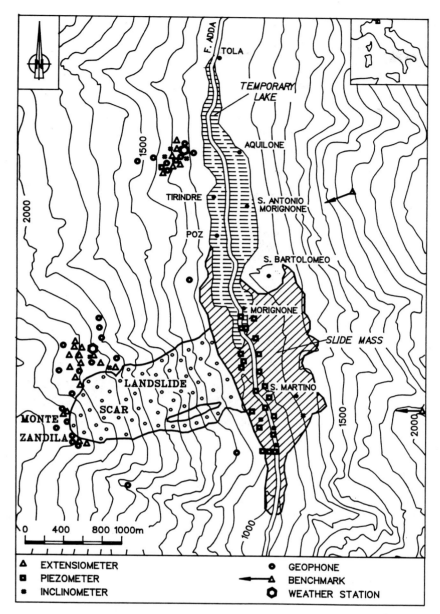

Fig. 4.20 The area affected by the Val Pola landslide, 1987. The slide-scar (dotted) can be seen on the left. The oblique shading is the slide mass, damming the main valley and creating the temporary lake (horizontal dashes) that stretched north as far as Tola. The valley was equipped with an instrumentation network after the landslide to monitor further ground movements.

consisted of a geological assessment of the various materials present and of modelling of the possible trajectories of further falls. Unstable areas were removed, where possible, and a monitoring and alarm system was installed, controlled by the Regional Geological Survey.

It was also necessary to reconstruct a suitable infrastructure of roads, power lines and built-up areas. In the long term, it is planned to bypass the whole of the unstable area.

The monitoring and alarm system as installed is also shown on Fig. 4.20. It consists of geotechnical instruments, and of three control networks for microseismicity, for hydrometeorology, and for topographical monitoring, and functions almost entirely automatically.

Schistose rocks The rocks show considerable variation in content, related to the original sediments that were subjected to a low degree of metamorphism only. Certain rocks weather into talc, asbestos, or swelling clays, particularly along faults. Rock settlements or falls of considerable size occur, as do bed-over-bed movements.

Flysch units The flysch zone consists of great thicknesses of alternating hard (conglomerate, limestone, and sandstone), medium (schist, coal) and soft (marl, shale, and pelite), materials. Large landslides occur with dislocation surfaces several tens or a hundred metres below the surface. Due to decompression, the resulting material is essentially loose rubble. Such movements constitute large rock settlements and complex slides affecting whole mountain sides (sagging). It is difficult to define the age of these movements. Many started long ago and, by progressive re-working, continue to produce powerful flows even today.

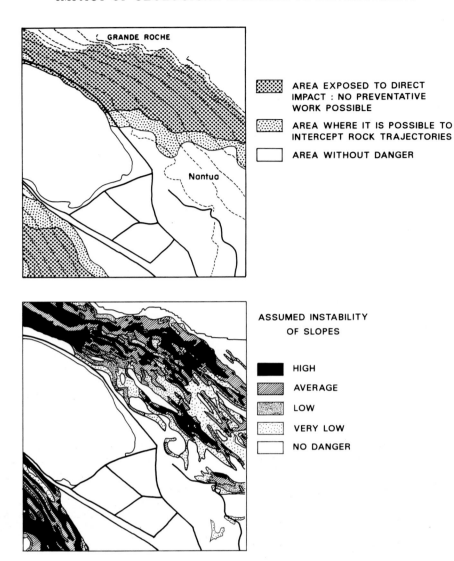

Fig. 4.21 Location of the Nantua site in south-east France.

In the south, in the Apennine range, these units consist mostly of poorly consolidated marine deposits of Oligocene and Miocene age, formed after the paroxysmal mountain-building phase of the Alps and the Apennines. Such deposits vary greatly in lithology, grain size, and cohesion. The frequent landslides are closely connected with the distribution of rocks of variable strength, such as the alterations of sandy and clayey layers. Such terrain is usually covered by a thick layer of weathered material readily affected by ground movements.

Marly–calcareous rocks Marly and calcareous rocks occur in highly heterogeneous and strongly deformed areas. Because of contrasts in permeability and structural discontinuities, the hydrogeological conditions can vary considerably. The most widespread movements are:

(1) large-scale movements, such as that at Nantua, in the French Jura (see below);

(2) bed-over-bed landslides;

(3) slides of cover rocks over marl, which may develop into flows.

Limestone and dolomites are commonly affected by closely spaced fractures, which may be widened by karst, gelifraction, and the normal weathering processes. Massive limestones can be interspersed with thinly bedded marl. The softer rock types are usually covered by a weathered layer up to several metres thick. This layer commonly shows traces of having been mobilized in the past by solifluction processes that were active during colder periods.

The most common processes that affect these formations are solifluction, with its slow deformation, and the toppling and collapse of ledges of more resistant material projecting over weak strata.

Monitoring of the Fècles ridge, Nantua, France: methodology for hazard evaluation [26]

The following case history illustrates two distinct operations: the monitoring of the Fècles ridge near Nantua (Fig. 4.21) and the development of a method for multiple hazard evaluation.

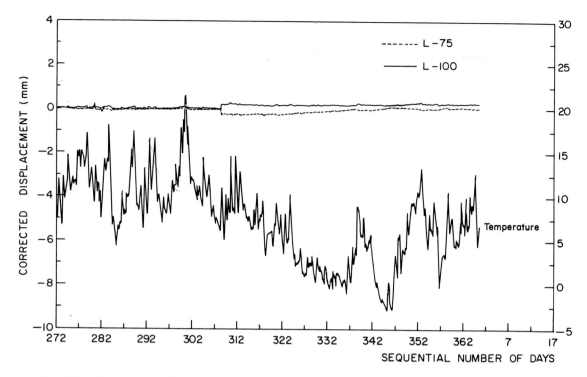

Fig. 4.22 Example of monitoring ground movements at the Fècles Ridge site near Nantua, south-east France.

The town and the lake of Nantua, situated at an altitude of about 400 m in south-eastern France, are dominated by a limestone plateau that reaches a height of about 1000 m. Rockfalls have threatened the town since time immemorial. In 1973, a rock column of about 60 000 m³, which had become unstable and could have toppled, was destroyed in a controlled fashion.

The Fècles ridge is an isolated part of this plateau. Topographical measurements carried out since 1985 by the French topographical service, IGN, showed that this ridge, which is about 450 m long and whose volume is estimated at around 20 000 000 m³, is moving slowly downwards with a velocity of 3 to 4 cm yr⁻¹, a process that is accompanied by fragmentation that loosens numerous small blocks. Even though the sudden collapse of this ridge is not imminent, many isolated rocks and columns no longer forming part of the ridge are heavily cracked and could lead to falls of several hundreds of cubic metres.

BRGM, in the framework of its ongoing R&D work on the prevention of terrain movements, has selected this mountain site, with its rigorous climatic conditions, for the installation of an experimental rock movement monitoring system. Two systems have been completed (Fig. 4.22).

1. The installation on the eastern extremity of the ridge of two Blum clinometers, consisting of a quartz pendulum of very high resolution (10^{-6} rd) and two Schaevitz accelerometers with a resolution of 5×10^{-4} rd, all of which are linked to a MADO data logger that was designed by the BRGM Instruments department.

2. The installation in September 1987, again on the eastern extremity of the ridge, of a unit transmitting data via the ARGOS satellite. The unit consists of:

   (i) on site: four displacement sensors of the resistive type with a resolution of 1/100 mm; two temperature sensors with a resolution of 1/100°C; two MADO modules; a MADO/BALISE interface for sending the data to the ARGOS emission beacon and antenna;

   (ii) in the office: in Lyons and Orléans, two duplicate systems consisting of a Minitel (KORTEX card) and a micro-

computer, for receiving, sorting, and using the measurements that are transmitted continuously by the ARGOS beacon.

In this manner, two major fissures in the Fècles ridge are remotely monitored by a totally autonomous field unit, which, in the office, requires action by an operator only every three to four days on the microcomputer. This experimental unit has proved itself to be entirely reliable and operational. The data received since September 1987 show that the fissures have moved only very slightly.

Hazard evaluation

Three types of phenomena that might generate risks were studied: subsidence of the valley bottom, rock falls, and landslides.

The major geodynamic phenomenon of the wholesale sliding of the Fècles ridge itself was deliberately ignored as it is considered that this is a long-term hazard that is properly monitored, but associated minor rock falls were taken into consideration.

The evaluation of subsidence and compaction of the valley bottom is based on simple criteria, such as the distribution and thickness of compressible formations, which were obtained from drilling. A simple algorithm then allows the evaluation of possible compaction below a given type of foundation, selected as being representative of the existing buildings.

For the potential rock falls, the evaluation procedure was more elaborate, illustrating the potential benefits that can be expected from using a tool such as a Geographic Information System (GIS) and from well-defined expert evaluation rules. The latter were supported by the use of software for the determination of the probable trajectory envelope from a given point of departure (the top of the cliff) and for a given rock mass.

Special software was developed for determining the size of the areas exposed to rebounding blocks and of those where the blocks will roll only. This defined those areas where it will be impossible to defend against the hazard of bouncing blocks, and those where it will be possible to construct defences against rolling blocks.

The evaluation of potential localized landslides is more complex and calls upon greater expertise. It begins with the mapping of all instability parameters, the spatial extension of which can be determined. All altitude, slope-angle, and exposure data are digitized from a 1:5000 scale topographical map, which in itself is obtained by photometric restitution. These data are then enlarged by a factor of ten, in order to obtain a data density that relates to the small scale of the expected phenomena.

Especially in the case of unconsolidated formations, geological data concerning the boundaries between individual units, their thicknesses, and their mechanical properties are approximate only. Their evaluation must then be justified, in order to assign specific values of a stability index to individual pixels. This justification will, typically, be an expert rule, based on the regional experience of the person working on the file. Such experience is further based on regional data banks. Once the stability index has been established, it is possible to evaluate its sensitivity to the infiltration of water into the soil. At this point, two different configurations of ground saturation are introduced and evaluated for each pixel: the variation of the stability index as a function of the water effect, which, in turn, provides the index of sensitivity to the water effect. Finally, the criteria are defined for an evaluation of run-off and infiltration, as a function of vegetation and of the position of the pixel in the catchment basin.

This work has led to a new way of determining the 'instability presumption' or 'hazard' in its commonly accepted usage, but further development and refinement are necessary. It has been demonstrated that it works as a method, but a greater interest lies in the fact that it illustrates the benefits of simultaneous use of:

(1) traditional data, as contained in databases;
(2) calculation algorithms; and
(3) expert rules.

The work carried out on the Nantua site does not constitute a Risk Exposure Plan, as parameters such as the type and vulnerability of the property and the soil exposed to the ground movements, were not included. The last aspect is currently being studied.

Clay and sandy clay materials Poorly consolidated clay and sandy clay forming low, rolling hills, is commonly affected by ground movements, especially in the form of large flows.

Movements of the surface layers on unstable slopes cover very large areas slowly. They are commonly located in the superficial deposits that mask the palaeo-topography and may be triggered off by the presence of features such as concealed and abandoned drainage systems.

Other materials In the Alps, ground movements occur to a lesser extent in:

1. Quarternary glacial-lacustrine clays, which are locally present in most of the Alpine glacial valleys. In places, such as in the Trièves and the Drac valley near Grenoble (France), such deposits are ubiquitous. Locally, more than 10 per cent of surface area is affected by ground movements, either as superficial viscous flows or as slides of up to 70 m deep.

2. Gypsum formations along major faults, which provoke collapse and destabilization of the slope by dissolution of the basal thrust plane.

All these units are usually covered by superficial deposits such as scree, moraine, and colluvium, which are mainly loose in nature, and which are commonly involved in all types of movement.

The need for research and development programmes
Socio-economic losses resulting from ground movements are very high in Europe and the need for risk mitigation is of primary importance.

The Geological Surveys have a key role to play in the prevention of disasters. In the short-term this means helping with the management of crisis situations, and in the long term with the planning of sustainable development.

In order to prepare for this role, long-term R&D programmes are essential on a European, rather than national, scale, to enlarge and consolidate the empirical database that is needed for such work, and also to coordinate the administrative, legal, and technological approaches.

With regard to the scientific objectives, four main fields require attention:

(1) the detection of movement-prone areas;
(2) the comprehension and analysis of the phenomena leading to ground movements;
(3) the forecasting of the occurrence and extent of such movements;
(4) construction design, and the formulation of recommendations and regulations.

The detection of movement-prone areas This type of work is currently carried out in three steps:

(1) the identification of known phenomena;
(2) the mapping of permanent disequilibrium factors that have been identified;
(3) the identification of the main transient aggravating factors.

The construction of databases is a priority for this purpose, both for past events and for transient factors. Geographical Information Systems are the key for open-ended mapping of movement-prone areas, and as a tool for decision makers. Such documents allow quick revision when new data become available. The example of the work done in the Emilia Romagna Region in Italy, is very instructive in this context.

Thematic mapping for the identification of landslide risks in the Emilia-Romagna Region, Italy [130] [131]

Three basic features determine the types of thematic maps produced in the Emilia-Romagna Region: (1) the introduction of a 'Basic Methodology for Establishing District Plans' in 1973–74; (2) the promulgation of the regional law 'Formation of a Regional Cartography' in 1975 and modified in 1977; (3) the creation of a Cartographic Board, conceived as an organization capable of acquiring, updating, and promoting the knowledge of the environmental aspects of the Region. The Region lies in north-eastern Italy, incorporating the Apennines in its southern half and the Po valley in the north.

In 1978, using earlier work carried out to provide a better knowledge of the Region, the production of a detailed 1:10 000 geological map was started. The need for such work lies in the peculiar geological characteristics of the northern Apennines, which is a mountain chain subject to present-day uplift and thus seismically active. This situation is complicated by the fact that the Region has a high density of population and infrastructure.

The substratum of the Region consists almost entirely of sedimentary rocks, such as sandy and marly flysch-type rocks and clayey chaotic complexes. Such 'rock' is particularly prone to landslides, because of its structural weakness, and the hazard to which people and their possessions are exposed is thus great. Moreover, human development on this environment commonly accelerates the natural degradation processes.

As of mid-1990, out of the 12 000 km² included in this project, 10 500 have been surveyed and the work should have been completed by the end of 1991. Since 1986, the first 70 sheets of a total of 350 at the scale of 1:10 000 have been produced. The production of 1:50 000 scale geological maps, which began experimentally in 1989, should be accomplished by 1994.

Because of the remarkable structural and stratigraphic complexity of the northern Apennines, it was decided to take lithology as the main mapping criterion. Areas with and without outcropping rock were carefully distinguished, and various geomorphological criteria and data concerning the type and age of landslides are also shown on the maps. In order to guarantee the homogeneity of the maps, the Region adopted a set of standards and set up a coordinating committee.

The fieldwork is carried out with the assistance of the geological faculties of the universities of Bologna, Modena, Padua, Parma, Pavia, and Pisa.

The final goal is to arrive at a diversified thematic cartography for the whole Region, for which older and thematic maps are difficult and time-consuming to update. The new maps should offer suitable information on the various technical problems as they arise. The basis of the mapping is a modern and computerized topographical map at the 1:5000 scale, which is a fundamental working tool for the mapping and monitoring of ground movements anywhere.

Analysis of the phenomena causing ground movements The methods employed here are traditional geological observations, soil mechanics, and rock mechanics. They have to be reviewed and adapted to the complexity of the phenomena observed, and integrated with input from geomorphology, palaeo-climatology, and pedology. New tools are being developed for modelling such as the transit and accumulation of water in unsaturated layers. Particularly useful is quantitative morphology as applied to recent ground movements, when it is possible to get aerial photographs before and after the event. It is impossible to apply these new methods to all the common events as yet, but with continued study of suitable examples, it will be possible to develop new and more effective methods.

An interesting example is provided by the Geological Survey of Austria in the following case history.

Relationships between slope instability, slope-water balance and mass movement in Saalbach-Hinterglemm, Austria [2]

Each year, mud-flows, floods, and landslides affect large parts of the Austrian territory, in particular the mountainous areas. As is shown by Table 4.11, the damage caused is important and the cost is high. However, the events of the summer of 1987 provided the straw that broke the camel's back, and the Geological Survey of Austria decided to start an R&D programme on the Saalbach-Hinterglemm area (about 200 km²) near Salzburg (Fig. 4.23), the main objective of which was to establish the relationship between slope instability, slope-water balance and mass movement.

The analysis of disasters involving floods and mud-flows increasingly indicates complex correlations between changes in the ecosystem, which have occurred in the long-term. For that reason, any investigation has to begin with an assessment of historical factors.

After the melting of the glaciers of the last glacial period, processes of erosion, sliding, and settling established relative 'stability' on rock slopes and in unconsolidated sediments. The simultaneous development of vegetation (especially forests) further contributed to this stability. Nevertheless, numerous slow 'creeping' processes continue even today. Changes in vegetation, such changes in the timberline due to climatic variations, as well as recent cultural and technical effects, have brought about many reactions in the vulnerable ecosystem of the Alps. The project has been concerned with many of these, one of the most important being changes in the water regime (correlation between surface run-off, infiltration, and the storage and retention capacity in vegetation, soil, and rock), which are caused by:

(1) increased surface run-off in the area of ski-runs, ski-lifts and access roads;

(2) soil compaction and the impairment of young growth by forest development;

(3) reduced grazing on Alpine meadows;

(4) impairment of young growth by excessive game population;

(5) impaired forests or reafforestation of unsuitable areas.

The water cycle in Alpine slopes and valley sides is part of several larger, standard cycles. The importance and relativity of the natural element as part of the standard cycle had to be evaluated, and the construction of a matrix clearly shows the cumulative effect. This type of representation demonstrates known correlations and corroborates inter-relationships previously only assumed.

The system 'forest–forest soil' is the central standard cycle, so the factors related to it, such as forest culture and density of the game population, have to be given priority consideration.

The work on the Saalbach-Hinterglemm area led to the following results:

1. Detailed geological, geomorphological, hydrogeological, and geotechnical maps of different types of mass movement.

2. Description of the development of mass movements, slope instability, and changes of the water balance of the ecosystem.

3. Detailed presentation of the interdependence between landscape development, economic use, geological subsurface conditions, the hydrological situation, and soil forest condition, which can cause slope instabilities and induce mass movements.

4. Contributions to the modelling of hydrology of small watersheds.

5. Contributions to the protection of the quality of drinking water (high aluminium contents indicate forest-soil degradation!).

6. Efficiency tests of an interdisciplinary diagnostic system to evaluate:

(i) the accuracy of the selected elements and indicators (describing the feedback relationship between geo- and bio-cycles);

(ii) trends of stability or instability of slopes (inducing or damping mass movements) as a dynamic factor of time and the resilience of ecosystems.

7. Contributions to systems for assessing geological hazards.

In addition, it was proved that detailed geological and geotechnical mapping for risk evaluation in complex Alpine valleys can be done economically and quickly. There is a clear demand for this work and it was demonstrated that the organizations involved must collaborate in an interdisciplinary fashion.

The downstream applications of this work by the Geological Survey of Austria are three-fold:

For the production of (applied) geological maps, it contributes an adequate presentation of morphological phenomena, including Pleistocene and post-Pleistocene dynamics.

For regional planning and the expert services for each province, it offers an assessment of geological–geotechnical data, in studies on land-use and environmental compatibility, and expert opinions on development projects in new areas, on disasters, and on local and regional planning.

For the necessities of stream and avalanche regulation, it offers the assessment of catchment areas (geological–geotechnical or comprehensive ecological assessment), as well as the assessment of the effects of development on a catchment area, or the planning of administrative measures to be taken and shown on maps of areas of potential hazard.

Forecasting the occurrence and extent of ground movements Forecasting involves the understanding of both the extent and the time of any event. With respect to time, significant progress has been made in recent years in the development of portable equipment for the entry and storage of data obtained

Table 4.11 Estimated damage in two valleys of Tyrol, Austria, 1987

Nature of the damage	Ötz valley	Stubai valley
Cut banks	about 19 km	about 6 km
Destroyed roads and tracks	about 5 km of State roads	about 4 km provincial and private roads
Destroyed bridges	4	3
Destroyed gangways	—	2
Destroyed dwellings	2	1
Destroyed farm buildings	1	—
Houses flooded and covered with mudflows (partly damaged)	45	24
Destroyed camp sites	1	1
Fatalities	14	—
Cattle killed	—	22
Destroyed motor cars and agricultural vehicles	6	35
Destroyed trailers	—	15
Flooded fields	about 150 ha	about 80 ha
Rubble from breached escarpments and terraces	about 1.2 M m^3	about 1.0 M m^3
Removed by streams	about 0.3 M m^3	about 0.2 M m^3
Expenditure to date on immediate relief measures	about 70 million S*	about 25 million S
Further damage as evaluated by Project	altogether about 300 million S	
Total Damage	altogether about 400 million S	

* all costs in Austrian schillings

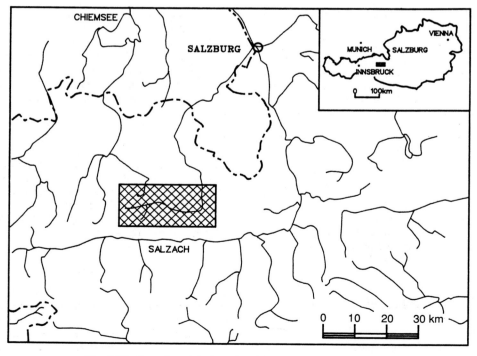

Fig. 4.23 Location of the Saalbach–Hinterglemm area in Austria.

in situ, with independent power supply and low energy consumption. The development of methods for the remote transmission of data (see the Nantua case history, p. 174) and the development of new data-processing software, allow real-time information on the development of ground movements to be available. Adequate decisions can be made, however, only when appropriate methods of interpretation become available.

In the case of the extent of ground movement phenomena, interesting tools have been developed to be applied towards understanding the propagation of rocky material, mud flows and flood waves. Some progress has been made on forecasting slow movements, by relating the amount of spatial variation in the degree of saturation of the soil, after persistent heavy rainfall. Several sites are being monitored in Europe, and interesting results are starting to appear. For major landslides, photogrammetic monitoring and methods for evaluating the three-dimensional equilibrium show significant progress.

RECOMMENDED FURTHER READING

Anagnosti, P. (1988). Instability phenomena in the zone of the alpine arc in Yugoslavia. Special lecture in *Proceedings of the Vth International Symposium on Landslides, Lausanne, July 1988*, Vol. 2, ed. C. Bonnard. A. A. Balkema, Rotterdam–Brookfield.

Aste, J. P. (1989). Some reflections on methodologies of landslide prevention in France. In *Workshop on natural disasters in European Mediterranean countries. Perugia, Italy—27 June–1 July 1988*. US National Science Foundation, CNR Italy. F. Siccardi and R. L. Barnes (eds).

Brabb, E. and Harold, E. L. (eds) (1989). Landslides. Extent and Economic significance. *Proceedings of the 28th International Geological Congress; Symposium on Landslides*, A. A. Balkema, Rotterdam–Brookfield.

BRGM (1984). Ground movements. *Colloquium of Caen. 22–24 mars 1984*. (Editions du BRGM, No. 83).

Canuti, P., Cascini, L., Dramis, F., Pellegrino, A., and Picarelli, L. (1989). Landslides in Italy: occurrence, analysis and control. In *Workshop on natural disasters in European Mediterranean countries, Perugia, Italy—27 June–1 July 1988*. US National Science Foundation, CNR Italy. F. Siccardi and R. L. Barnes (eds).

CETE/BRGM/ADRGT/IRIGM working group (1988). *Recherche régionale sur les risques naturels en montagne: les mouvements de terrain dans les Alpes du Nord*. (BRGM Report October 1988).

Collective Authors (1990). *Banca Dati Geologica*. Consorzio per il sistema informativo settore territorio (CSI, Piemonte).

Cooper, A. H. (1986). Subsidence and foundering of strata caused by the dissolution of Permian gypsum in the Ripon and Bedale areas, North Yorkshire. In *The English Zechstein and related topics* (eds G. M. Harwood and D. B. Smith). (Geol. Soc. Spec. Publ., No. 22, pp. 127–39).

Govi, M. (1984). The instability processes induced by meteorological events: an approach for hazard evaluation in the Piedmont Region (NW Italy). *CNR-PAN meeting on progress in mass movement and sediment transport studies, 5 to 7 December 1984*, Torino, Italy.

Pilot, G. and Durville J. L. (1990). Slope instabilities in the French Alps. Special lecture in *Proceedings of the Vth International Symposium on Landslides, Lausanne, July 1988*, Vol. 3, ed. C. Bonnard. A. A. Balkema, Rotterdam–Brookfield.

Schindler, C. (1988). Unstable zones in Switzerland. Special lecture in *Proceedings of the Vth International Symposium on Landslides, Lausanne, July 1988*, Vol. 2, ed. C. Bonnard. A. A. Balkema, Rotterdam–Brookfield.

4.2.4. Coastal and marine hazards [58]

Introduction

Life in the coastal areas of Europe, particularly around the great ports, has centred on commerce and fishing. Recently, socio-economic changes have led to a particularly strong diversification of industrial activities and services in these areas. Littoral areas have certainly suffered the most drastic modifications during the past 30 years: new industry has developed on land, tourism has grown enormously, and off-shore exploitation of hydrocarbons and sand and gravel has mushroomed. In France, for instance, tourism is the primary economic activity of the 850 communes along the seaboard, with an annual turnover of 50 billion francs, which is twelve times the turnover of the commercial ports and eight times that of offshore fishing.

The effect of these changes has been a large increase in the population, and an explosive development of the littoral infrastructure, comprising industry, harbour installations and general urbanization, and even heavy pressure on the sea itself. The following figures demonstrate the situation:

1. Each summer, almost 14 million Frenchmen and about 6 million foreigners invade well over 2000 km of French coastline and beaches.

2. 51 per cent of the French coastline is already urbanized and each year sees the building of a further 13 000 holiday homes.

3. In the past three decades, 180 marinas have been constructed on the French coast, and plans for 80 more are well advanced.

In addition to his traditional concern with protection against the sea, man is now engaged in environmental protection (risk and pollution management), and the creation of parks and reserves to safeguard the natural environment. Coastal areas are of such economic importance that decision makers are progressively integrating environmental elements with coastal dynamics, in creating master plans for development and construction.

Changes in the configuration of the coastline, which are constant and readily observed, are mostly the result of sea water action on beaches, dunes, and cliffs. In addition, wind action and the influences of streams, lagoons, and marshes are important, as are the impact of man's activities in such coastal developments as harbour construction and beach conservation.

Marine erosion affects most European coasts by destroying fabric, flooding, and attacking dunes and cliffs. At the same time, tidal inlets, harbours, and their access channels are silted up. All these phenomena represent a considerable financial burden on the community and the loss to the natural heritage is usually irreversible.

The impact of man on the coast may be great, but the influence of natural phenomena in the long-term is even greater. The scale of the latter is well-illustrated by the well-studied example of parts of the coastline of south-western France (Fig. 4.24). But short-term erosion is also important, and a cause of considerable damage and concern for the authorities involved. The problem is well shown by Table 4.12, in which estimates of the rates of erosion along the European coasts are given.

Natural causes of coastal erosion

The coastline, which is the mobile boundary between the lithosphere and the hydrosphere, is subjected to influences variable both in space and time, such as the wind, tides, currents, swell, and breakers. These agents attack coasts, which show great variation as they can consist of unconsolidated material, soft rocks, and hard rocks. They also transport

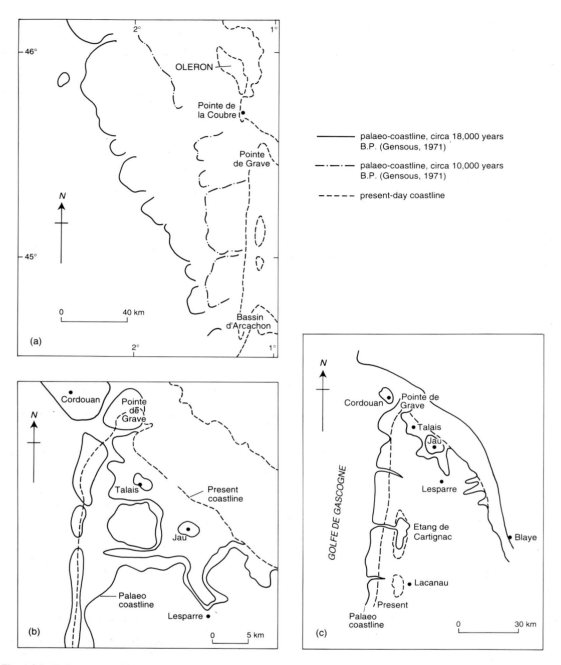

Fig. 4.24 Palaeogeographical evolution of parts of the coastline of south-west France.
(a) General view.
(b) The mouth of the Gironde in Roman times.
(c) The Gironde and the Medoc peninsula in the Middle Ages.
(after Howa, H. (1987) *Le littoral du Nord-médoc (Gironde). Evolution d'une côte sableuse en érosion. Thesis*, Univ. Bordeaux).

sediment along the coast. Such sediment can be derived from river erosion, coastal erosion, or the re-working of sediments.

During heavy storms, especially in conjunction with spring tides, the attack on the coast becomes critical, when abnormally high sea levels affect the most vulnerable parts of the coast, which are the tops of beaches and the bases of cliffs. During the catastrophic storm of 31 January 1953, which flooded great parts of The Netherlands and eastern England, a surge level of 3 m was recorded, while in November 1982, the Camargue was swept by a storm during which surge levels rose to 2 m above mean sea level.

Furthermore, the analysis of observations made between 1880 and 1960 has led certain authors to postulate a general increase of wave height and energy over that period. Others who have analysed meteorological observations have the impression that the frequency and strength of storms has noticeably increased over the past decade.

The intensity of coastal erosion relates to variations in the supply of marine and continental sediment. During the Flanders transgression, for example, the sea level rise led to a mobilization of cobbles and pebbles, which were deposited on certain shores. By contrast, there seems to be a scarcity of

Table 4.12 Coastal erosion rates in the European Community (from L'Etat de l'Environnement (CEC 1988)

Member state	Location	Geology	Erosion rate (m/year)
United Kingdom	Humberside	Glacial deposit	0.3–3.3
	Norfolk	Glacial deposit	0.2–5.7
	Suffolk	Glacial deposit	0.6–5.1
	Kent	London Clay	0.7–3.4
	South Coast	Chalk	0.05–1.0
	Wales	Liassic claystone	0.008–0.1
	Yorkshire	Liassic claystone	0.02–0.04
	Northern Ireland	Glacial deposit	0.2–0.8
France	Somme	Chalk	0.08–0.37
	Seine Maritime	Chalk	0–0.4
	Camargue	Alluvium	1.5–4.0
	Landes	—	1–3
	Manche	—	1
	Aquitaine	—	5
	Charente Maritime	—	0.5–2.0
Germany	Helgoland	Sandstone	1
	Baltic Coast	Glacial clay	0.6–2.0
Italy	Romagna	Sand dunes	<9
	Latium	Sand dunes and marl	2.5
	Calabria	Alluvium	1.2–3.3
	Calabria	Sand and gravel	8–11
	Gulf of Tarento	—	4

sediment today, which would explain the almost universal tendency towards erosion of the coasts.

Subsidence is another factor of importance. Where it occurs, it can vary from a few to several tens of millimetres each year, and low, sandy, coasts are particularly vulnerable. For instance, the coastline around Ravenna, Italy, which forms part of the Po plain, is subject to strong subsidence. Between 1949 and 1977 the surface level fell by 1–6 cm each year and in some places subsidence of as much as 45 cm was recorded. As a result the shoreline retreated by as much as 126 m in places.

Any rise in sea level as a result of the so-called greenhouse effect, a combination of natural causes and the effects of industrial activity, will be of great concern in the future. Much research into its potential effects is already ongoing in Europe. Figures for the possible rises in mean sea level (MSL) for the next century vary greatly, but it could be as much as 60 cm and the pessimistic scenario suggests even 85 cm.

The consequences of such an MSL rise on the Dutch coastline have been analysed by the Rijkswaterstaat. The effect of the shoreline retreat on the safety level of the dune coast can be described as follows. Since sand is lost from the dune front, the amount of sand available for defence of the hinterland against the sea decreases, and the coastal defence weakens. Each part of the coast has its own critical width, depending on the geometry of the coastal profile. When, because of the retreating shoreline, the actual dune width drops beneath the critical figure, the required safety level can no longer be guaranteed.

For the various scenarios of sea level rise, the length of the coast where this level of safety is no longer reached, is shown on Fig. 4.25(a) for the next 100 years. By the year 2000, 20 km of coastline will be unsafe with the present-day amount of MSL rise, and this figure will increase to 40 km by the year 2090. For the unfavourable scenario one must reckon with a doubling of these figures.

The effects of shoreline retreat on the dune area and its functions are also evaluated. Figure 4.25(b) shows the length of the coast where the loss of very valuable neutral habitats can be expected. For present-day values of MSL rise, such losses will be along 40 km of coastline by the year 2000, increasing to about 60 km by the year 2090, which will be about 40 per cent of such areas. In the pessimistic scenario, such loss will climb to 50 per cent.

Human influence on coastal erosion

Human factors that influence the erosion of coastlines are multiple and add to a natural situation that in itself is a cause for concern.

Marine dredging, and the abstraction of beach and dune sands, are important factors in causing the progressive reduction of littoral deposits. The regression of natural marine vegetation, caused by pollution or dredging, weakens the coastal sedimentary environment through a reduction in the elements stabilizing marine sediments. As an example, the regression of the Posidonia vegetation in the Mediterranean has already reached alarming proportions.

Coastal dunes degrade as a result of the wanton removal of vegetation, human overpopulation, and the off-road use of vehicles. This degradation eventually leads to destruction of the dunes under the onslaught of water and wind. The urbanization of the shore and the construction of ports, jetties, dykes, and the like, further contribute to the overall disturbance of the fragile natural equilibrium.

Urban development of the littoral area, more often than not authorized without prior relevant technical studies, leads to a

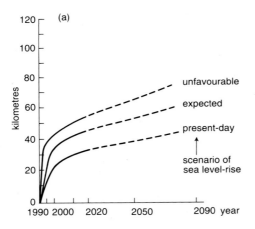

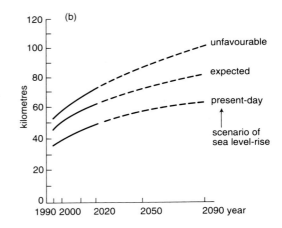

Fig. 4.25 (a) Length of coastline of The Netherlands where safety requirements are not met; (b) length of coastline where loss of valuable natural habitat must be expected (from Louisse and van der Meulen 1990).

snowball effect of various undesirable events that inexorably follow each other:

(1) increased occupancy leading to overpopulation;

(2) far-reaching ecological modifications;

(3) development of near-shore and onshore leisure areas;

(4) (artificial) increase of the beach surfaces open to bathers;

(5) extension of marinas.

All these actions disturb the natural dynamics to such a degree that defensive measures have to be taken to protect structures. Such reaction is born from the justified fear that considerable investments may be lost as a result of marine erosion. The decision makers and developers thus find themselves caught up in a vicious circle, as they resort increasingly to the use of concrete and riprap.

The impact of man's inland development on coastal processes

Man's actions inland, such as building dams, deforestation, and the abandonment of farmland, commonly have a measurable impact on coastal erosion and sedimentation processes. With respect to their ecological and economic impact, sedimentation problems such as silting up or sanding up commonly lead to as much difficulty for mankind as does erosion. Here again, history provides many examples. The problem of the silting up of the Mont St Michel bay in France is well known, as is the gigantic scale of the processes that have led to the development of deltas, such as those of the Po and the Rhone. Another example is provided by the considerable sums that are spent each year in dredging shipping channels within or leading to many harbours.

The following case history, published by the Instituto Tecnológico Geominero de España in the November 1987 issue of *Riesgos Geológicos*, clearly illustrates the impact of degrading catchment areas upon littoral processes.

The silting up of the Vélez Málaga estuary [153] [157]

Studies were carried out in the Río Vélez estuary near Málaga, Spain, on a coast with minimal tides. It was possible to reconstruct the palaeogeography of the area with precision, thanks to the presence of archaeological and historical remains, which led to the dating of the various phases of infilling of the estuary (Fig. 4.26).

The post-glacial transgression transformed the Río Vélez valley into an estuary that reached about 7 km inland. Infilling began immediately, with a sedimentation rate of 1–2 m per 1000 years, which lasted until the end of the Middle Ages. Phoenician settlements of 750 BC have been found 7 m below the present-day valley level. At the end of the fifteenth century, the fleet of the Catholic Kings had to sail 5 km up the estuary in order to lay siege to Vélez Málaga that was held by the Moors. After their expulsion, massive deforestation caused extensive soil erosion and an enormous increase in the sedimentation of the estuary. Deposits which accumulated during the sixteenth and seventeenth centuries are as much as 20 m thick. During this period, the estuary filled up completely and the Río Vélez started forming a marine delta, on which the modern town of Torre del Mar has now been built. It is recognized that any type of soil conservation measure in the catchment area, such as reforestation, would lead to a smaller sediment load in the Río Vélez and to possible erosion of the delta. In that case it would be necessary to construct coastal defences for the protection of the town and the beaches of Torre del Mar.

The management of rivers, which usually means the building of dams, has a profound impact on the sedimentary balance of estuaries and deltas. Dams trap a major part of the sediment load transported by a river, and the regular flushing of dams only partly, but forcefully, restores the original sediment load.

A good example of this is provided by the River Rhone and its tributaries, mainly the Durance, in France. Dams have been built since the 1950s which contribute to the trapping of several millions of cubic metres of sediment each year (Fig. 4.27):

1. The French–Swiss flushing sluices installed on the Upper Rhone in 1987 caused the removal of 1.5 million tonnes of material.

2. The hydroelectrical development of the central third of the Lower Rhone in France, between the confluence of the Isère and the Ardèche rivers, trap about 300 000 m³ sediment each year.

3. On the Durance, two dams have trapped about 17 million m³ sediment in 20 years, whereas a third dam is silting up at the rate of 1 million m³ yr⁻¹.

The impact of sediment trapping on such a scale is easily understood, not only on fluvial dynamics but also on the delta dynamics at the mouth of any river system.

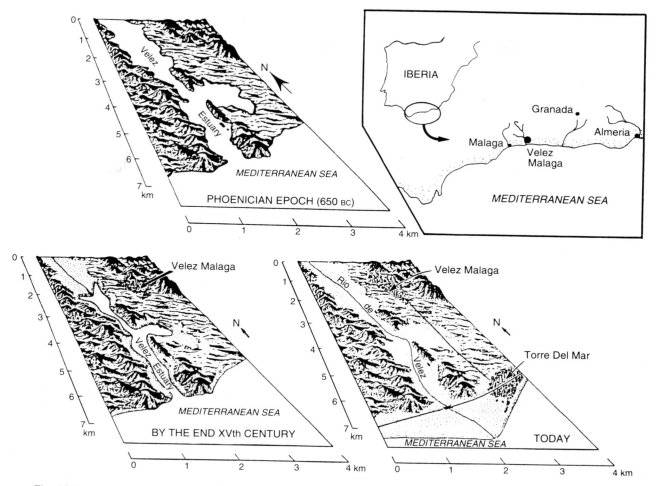

Fig. 4.26 Historical evolution of the Vélez Málaga coast, southern Spain (Hoffmann and Schulz (1987) in *Riesgos Geologicos*).

Complex influences on the coastal zone: the case of Venice

The example of the lagoon of Venice is a particularly interesting case. Here, natural phenomena such as subsidence and eustatics, together with man's activities (the battle against silting up, construction activity, exploitation of aquifers, and so on) destabilize the coastal region. In fact, this region has been studied in depth for well-nigh 40 years, and many recommendations have been made for reversing the trend.

The following case-history presents some of the results obtained by the Italian Istituto per Lo Studio della Dinamica delle Grandi Massi [ISDGM].

Dynamics of the Venice lagoon [58]

Four disciplines were covered by the studies carried out by the ISDGM: stratigraphy, hydrology, geodetics, and geotechnics. The work itself was carried out in three phases:

(1) collecting, processing, compiling, and interpreting historical and geological data;

(2) field observations and the creation of relevant data banks;

(3) hydrogeological and hydrological modelling, and the simulation of subsidence.

The lagoon of Venice is located between the mouth of the River Sile in the north and that of the River Brenta in the south. It covers a surface area of about 550 km², with a length of 50 km and a width of 8 to 14 km; three channels form the connection with the Adriatic Sea.

The lagoon in its present form has existed since about 700 BC. Human intervention in the environment began around AD 1400. Its objective was to avoid the complete filling up of the lagoon by sediment from the major rivers that flow into the northern Adriatic. Carried out in successive phases, the work was finished in the eighteenth century, and caused enormous disruption of the entire eco-system of the lagoon, resulting in coastal erosion, natural subsidence not being compensated by sediment inflow, increasing salinity of the water, and development of marine features for the lagoon as a whole.

In the end, the surface of the lagoon itself had been reduced to 187 km² (only 32 per cent of the original area) through filling in with rubble, the development of areas for fish conservation, and land improvement. Moreover, the abstraction of groundwater for industrial uses since the 1930s had lowered the groundwater level by more than 20 m, mostly between 1950 and 1970. This lowering of the water table caused increasing subsidence, which led to the decision in 1970 to halt the pumps. Since 1975, an artesian groundwater regime has been restored.

The hydraulics and ecology of the basin as a whole were also radically changed, not only by the diversion of several streams, but also by the closing of five of the eight openings to the sea, and by the dredging of internal shipping channels. In addition, the coastline

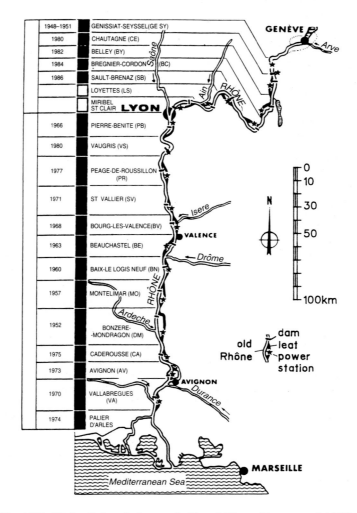

Fig. 4.27   Hydraulic installations on the French Rhone (Tormos *et al*. 1989).

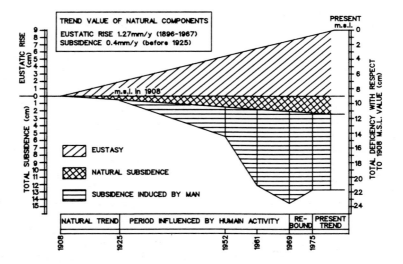

Fig. 4.28   Eustasy and subsidence in the Venice lagoon (after Gatto, P. and Carbognin, L. (1981). The lagoon of Venice: natural environmental trend and man-induced modifications. *Hydro. Sci.*).

was under attack from wave action, amplified by secondary circulation caused by the construction of breakwaters outside the harbour installations.

Man's influence on the littoral erosion processes has been fundamental, not only for the coastal current patterns, but also for the differential erosion of the coastline. Generally speaking, the subsidence of the lagoon is due to a combination of three factors:

1. The pumping from aquifers.
2. The natural subsidence of 1.3 mm yr^{-1} on average over the past 40 000 years, which reduced to 0.4 mm yr^{-1} after work carried out during the period 1900–30 (the 'Caranto' palaeosol that occurs in outcrops on the mainland is covered by about 13 m of sediment along the coastline).
3. The eustatic rise of the mean sea-level recorded at Trieste between 1900 and 1960, was at times as much as 1.27 mm yr^{-1}. Since then this rise has become less.

The combined effect of the three factors is shown on Fig. 4.28. The total subsidence of 22.4 cm since the beginning of the twentieth century can be broken down as follows: 9.1 cm caused by eustatic effects, 2.9 cm due to natural subsidence, and 10.4 cm due to man-made subsidence.

The negative effects of isostatic uplift [172]

During the latest ice age, called the Weichselian, much of Fennoscandia was covered under an ice sheet that reached a thickness of 2 to 3 km, depressing the Earth's crust as much as 500 m. When the ice sheet began to contract, about 20 000 years ago, the ground surface started to revert gradually to its earlier position of equilibrium by epeirogenic movements in the crust. This geological process, called isostatic uplift, is still taking place in some areas. On the whole, the effect of this process has gradually decreased, but it still affects the landscape and the human activity in both positive and negative ways.

Recent measurements have shown that the present apparent uplift, which is the combined effect of isostatic uplift of land and eustatic sea level changes, is nearly 1 cm yr^{-1} in the northern part of the Bothnian Gulf and almost zero in southern Sweden and Norway. In fact, as the sea level itself shows a general rise, the uplift of the Earth's crust is a few millimetres more.

A positive effect of the apparent uplift is the formation of new arable areas along the shores. However, there are some negative effects as well. An example relates to groundwater in Sweden, which is the main source of potable water; its level, mainly located at a few metres below the surface, usually follows the contours of the land surface. Thus, as the ground level rises because of the uplift, the water level rises as well. This causes an increase in the main groundwater gradient towards the sea and thus in its flow velocity. The net result is a lowering of the water table in relation to the ground surface. This may cause the drying up of wells and springs, and can also lead to quality problems due to chemical processes in the new unsaturated zone.

Because of the uplift, wooden piles used as the foundations of many older houses, such as in the medieval centre of Stockholm, are no longer protected by oxygen-poor groundwater. They have started to rot, resulting in serious subsidence problems.

Furthermore, the uplift is known to cause or to reactivate tectonic movements, including large-scale faulting. Such neotectonic movements affect the flow patterns of groundwater, causing problems for underground storage sites for hazardous waste. The uplift and the frequency of earthquakes in northern Fennoscandia may also be related.

Geotechnical problems are caused by the lowering of the base level, which increases river erosion and coastal instability that can lead to landslides. Several such slides are known to have occurred during historical times, usually causing costly damage. Another, related, problem is the gradual reduction in the depth of shipping channels, either because of uplift of the bottom or excessive silting-up. The two major ports in Sweden, Stockholm, and Gothenburg, were both located in different places until they could no longer be reached by sea-going ships. Finally, several of the large lakes in central Sweden experience differential uplift; Lake Vättern, 150 km long, is lifted more in the north where its outlet is located, causing major flooding problems at the urbanized south end.

Sea bed instability [207]

Large movements of sediment take place on the sea bottom itself. These are recognized by the presence of typical stacked mass flow deposits.

The north-west European continental margin has been the scene of a wide range of slope failures throughout the Quaternary, varying greatly in style, size, and mobility. They include large slides that generated mass flows, and small, shallow, rotational, and planar slides affecting only a few metres of sediment. Such slides may have a significant surface expression that is easily recognized by bathymetric surveys, or they may be almost invisible, either because of scale or through subsequent burial. A good example of such a slide is the Miller Slide on the West Shetland Slope (north of Scotland) that involved more than 50 cubic kilometres of sediment.

A thick succession of Pliocene–Pleistocene sediments is present over most slope areas, suggesting high sedimentation rates in the order of several hundreds of metres per million years. On the West Shetland Slope, these sediments comprise glaciofluvial mass flow deposits that can be 500 m thick. Such sediments are likely to be non-consolidated, with a high interstitial water content that makes the sediment susceptible to liquefaction. Excess pore pressures, sufficient to overcome gravitational and cohesive forces, can be induced by natural events, such as seismic loading, the build-up of shallow gas, the cyclic loading by storm waves in shallow water (less than 100 m depth), high sedimentation rates, or the rapid lowering of the sea level during a glacial period.

The largest events have occurred along the Norwegian continental margin, including the Storegga slides that displaced about 5580 cubic kilometres of poorly consolidated glaciofluvial sediment during three main events. The resulting turbiditic mass flows were highly mobile, travelling up to 800 km into the Norwegian Sea basin. The sudden displacement of such large volumes of sediment can generate a tsunami. One of these is thought to have occurred during the second Storegga event some 7000 years ago, flooding the coasts of Scotland and the Shetlands up to a height of 19 m above the then sea level.

The spatial and temporal distribution of submarine slides is not uniform, being dependent on several geological and climatic factors. The late Weichselian to early Holocene was a period of high susceptibility for the occurrence of such phenomena, because of rapid changes in the regional stress-fields (the melting of a heavy ice-cap, high sedimentation rates, and rapid fluctuations in temperatures and sea levels). Similar conditions must have prevailed during other transition times between glacial and interglacial periods, with similar results. Today, the chances of slope failure through natural causes are low, due to a high sea level and low sedimentation rates along the continental margin. Even though the likelihood of a large event is thus small, small slides can easily happen because of the high water content and low yield-strength of the slope sediments. Such slides can have a disastrous impact locally on sea bed installations, such as submarine cables.

Coastal erosion

The CORINE coastal erosion inventory: an example of international collaboration [58]

A recent study carried out for the Commission of the European Communities has led to a much better understanding of the type, location, and magnitude of the problems concerning littoral erosion. This study was part of the CORINE programme (1985–89), which had as its objective the collection, coordination, and standardization of all information on the condition of the environment and the natural resources of the EC.

This study was carried out by a multinational (11 European countries) and multidisciplinary (geologists, geographers, geomorphologists, civil engineers, and hydraulic engineers) team, in which BRGM (Bureau de Recherches Géologiques et Minières—the French Geological Survey) played the role of coordinator.

In this study, the littoral zone is described as a sequence of coastal segments according to morpho-sedimentological descriptors, and defined as belonging to a given class within one of the following three categories:

(1) the morpho-sedimentological environment (beach, cliff, dune, etc.);

(2) the presence or absence of artificial structures and/or coastal defences;

(3) the present evolutionary tendency: sedimentation, erosion, stability, absence of data.

The identification and definition of the coastal segments was based on available documents, and completed by enquiries and field visits. The scale used in the investigations ranged from 1:25 000 to 1:100 000, depending on the available base maps and other compilation documents of each member country. The smallest coastal segment identified varies from 100 to 500 m, depending on the scale of the base map.

In this manner, well over 57 000 km of coastline were analysed and divided into more than 17 000 'homogeneous' segments, whose distribution by country is shown in Table 4.13.

These figures also take estuaries and harbour areas into account, as well as all islands more than 1 km long; excluded are the Greek Islands (except Euboea) and the interior seas of Denmark.

Each segment is defined by its geometry and geographic position (for digitizing), as well as by three groups of attributes:

(1) the morpho-sedimentological facies (19 codes);

(2) the evolutionary tendency (9 codes);

(3) the presence or absence of coastal defence structures.

When considering the main features of the morpho-sedimentological facies, four main classes can be distinguished:

(1) rocky coasts and cliffs;

(2) beaches;

(3) coastal marshes, wadden, and polders;

(4) artificial coastline.

Excluding the estuary and port segments, which explains the difference from the total kilometres of coastline shown in the table above, the distribution per class, expressed in percentage of the linear total, is shown in Table 4.14.

The evolutionary tendencies of the coastlines can be assessed as belonging to four groups:

(1) unknown tendency;

(2) stable coast;

Table 4.13 Coastlines of the EEC countries (as defined in the CORINE project)

	Total length of coastline (km)	Number of coastal segments
Belgium	98	40
Denmark	4 486	1 094
France	7 201	3 997
Germany (West)	1 863	456
Greece	5 335	525
Ireland	5 148	574
Italy	7 406	2 541
Netherlands	860	95
Portugal	923	207
Spain	6 568	1 706
United Kingdom	15 901	5 800

(3) coast being eroded (certain or probable);

(4) growing coast (certain or probable).

Of the eleven countries investigated, the evolutionary tendency is shown in Table 4.15.

One can thus say that, for all EC countries, 10 800 km coastline or 21 per cent of the linear total are affected by erosion, of which:

(1) almost 6000 km are beach (29 per cent of the linear 'beach' total);

(2) 4500 km are rocky coast/cliffs (17 per cent of the linear 'rocky' total).

These figures can be further explained. With the exception of Denmark, Greece, Ireland, and the UK, where between 12 and 24 per cent of the linear beach kilometres are being eroded, the other EC countries have beach erosion rates of between 35 and 48 per cent of their linear total. Furthermore, 3400 km of the rocky coasts in the United Kingdom, of a total of 9900 km of this type, are subject to erosion as well.

Notwithstanding these interesting results, the CORINE inventory has its limitations. For instance, the seriousness of littoral erosion phenomena cannot be related to the portion of the total length of the coastline that is subjected to erosion. In fact, only two major criteria for the evaluation of the magnitude of the phenomena are relevant:

(1) the intensity of the phenomenon (retreat of the coastline in metres per year, or whether or not the erosion is a general phenomenon over the entire segment);

(2) the risks incurred to housing, industry, and the ecology.

A test was carried out by the working group on coastline segments that straddle international boundaries, concerning the acquisition of certain data related to these critiera. This showed the need to have access to a large database, but it also showed how difficult it is to use such a database in an effective way for the whole of the EC. The reasons for such difficulties were found to be:

(1) the varying quality and incompleteness of such studies;

(2) the fact that such studies are commonly very localized, both in space and in time;

(3) the confidential nature of some studies, which rendered them inaccessible;

(4) the absence of measurement and observation networks.

For these reasons, the working group could only base its evaluation of the criteria for 'Evolutionary tendencies' on (possibly subjective) assumptions in certain cases. This is one of the major weaknesses of the present state of the inventory, which has been attenuated by the introduction of specific codes, such as: erosion (or silting up) 'probable but not documented'.

Table 4.14 Distribution of types of coastline in countries of the European Community

	Class 1 (%)	Class 2 (%)	Class 3 (%)	Class 4 (%)	Total (km)
Belgium	0.0	100.0	0.0	0.0	65
Denmark	10.0	80.3	8.7	1.0	4 071
France	46.5	39.9	11.7	1.9	6 337
Germany (West)	7.5	35.9	35.2	21.4	1 678
Greece	60.6	39.1	0.1	0.2	5 230
Ireland	56.5	41.1	2.4	0.0	4 669
Italy	45.3	52.4	0.0	2.3	6 899
Netherlands	0.0	51.7	32.9	15.4	788
Portugal	43.2	55.9	0.0	0.9	871
Spain	67.1	29.2	1.3	2.4	6 034
UK	65.0	24.1	10.3	0.6	15 221
Average/total	51.5	39.3	7.2	2.0	51 863

Table 4.15 Tendency towards change in the beaches of the European Community countries

	Group1 (%)	Group 2 (%)	Group 3 (%)	Group 4 (%)
Belgium	0.0	40.1	40.1	19.8
Denmark	2.0	68.6	14.8	14.6
France	10.8	52.7	25.0	11.5
Germany (West)	1.3	26.5	20.9	51.3
Greece	32.5	55.2	10.1	2.2
Ireland	37.9	50.4	11.2	0.5
Italy	17.7	52.1	26.2	4.0
Netherlands	0.0	51.9	23.1	25.0
Portugal	2.3	60.9	34.3	2.5
Spain	1.4	83.3	12.8	2.5
United Kingdom	0.8	60.7	27.3	11.2
Average	11.0	59.1	20.9	9.0

RECOMMENDED FURTHER READING

Carbognin, L., Gatto, P., Narabini, F., Nozzi, G., and Zanbon, G. (1982). La tendance evolutive du littoral émilien-romagnol (Italie), *Oceanologica Acta*. (Special issue with proceedings of the 1981 Bordeaux colloquium: *Symposium International sur les lagunes côtières*.)

Carbognin, L., Gatto, P., and Marabini, F. (1984). *Guide book of the eastern Po plain (Italy)*. A short illustration about environment and land subsidence. Instituto per lo Studio della Dinamica delle Grandi Masse (ISDGM)-CNR Venezia. Proc. III Convegno sulla Subsidenza (Venezia 1984).

Cravero, J. M. and Guichon, P. (1989). Exploitation des retenues et transport des sédiments. *La Houille Blanche*. No. 3/4, 292–5.

Hoffman, G. and Schulz, H. D. (1987). Holocene stratigraphy and changing coastlines at the Mediterranean coast of Andalucia (S.E. Spain). In *Late Quaternary sea-level changes in Spain* (ed. C. Zazo). (*Trab. Neog.-Cuat. Museo Nal. C. Nat. C.S.I.C., Madrid.* **10**, 227–47.)

ISDGM and Carbognin, L. (in press). Evoluzione naturale e antropica della laguna di Venezia. In Proceedings of the Symposium 'Geological Research related to the Environment', (ed. G. Giardini). Rome–Viareggio, September 1988. *Mem. Deser. Conta Geologica d'Italia*, Vol. XLII.

Louisse, C. J. and van der Meulen, F. (1990). Future coastal defence in The Netherlands; strategies for protection and sustainable development. *Proceedings International Symposium of the European Association 'Eurocoast—Littoral 1990'*, ed. R. E. Quelennec. Eurocast, Marseille.

Ricard, C. and Levasseur, L. (1989). Les chasses franco-suisses sur le Haut-Rhone. Reflexions sur les opérations de 1987. *La Houille Blanche*. No. 3/4, 306–10.

Riesgos Geológicos (1987). Volume containing the papers by the organizing members of the 1st Workshop on Geological Risks, Madrid, September 1987. Published by the Instituto Tecnológico GeoMinero de España (ITGE) in the series *Geologia Ambiental*.

Stive, M. J. F., Roelvink, J. A., and de Vriend, H. J. (1990). Large-scale coastal evolution concept. To be published in the *Proceedings of the International Conference on Coastal Engineering*, Session on the Dutch Coast. Paper No. 9, Delft, 2–6 July 1990.

Tormos, E., Cornier, P., and Scotti, M. (1989). Influence d'une chaine d'aménagements hydroelectriques du Bas-Rhone français. *La Houille Blanche*, No. 3/4, 215–20.

4.2.5. Flood hazard [181] [180] [182]

Introduction

Hazards like landslides, rock falls, debris flows and mud-flows commonly occur during periods of heavy rainfall and/or snow-melt, which may also produce floods. In these cases,

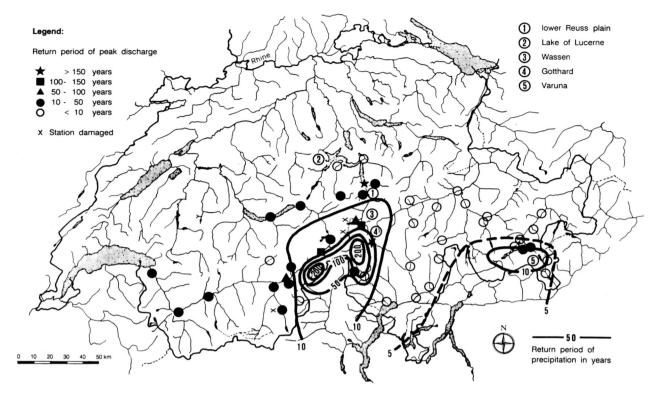

Fig. 4.29 Switzerland, showing the areas affected by the unusual precipitation conditions of the summer of 1987. Comparison of the return periods of measured precipitations (sum of two days) and peak discharge.

water, in association with relevant geological conditions, is the triggering mechanism for natural disasters. Because of the strong links between the geological environment and water, it is essential that geologists and hydrologists colaborate. Most countries in Western Europe have separate Geological and Hydrological Surveys, and it is in Switzerland only that these services are combined in a single Swiss National Hydrological and Geological Survey. The following pages clearly illustrate the interdependence between geological and hydrological hazards.

Meteorological and rainfall conditions

Swiss Alpine regions are regularly struck by severe natural disasters, mainly snow avalanches, debris flows, and flooding. Damage in 1987 was the worst for a long time. During the whole year, many extraordinary storms were recorded in various regions of the Alps. Along with many local storms with high precipitation intensities and flooding in smaller areas, two disastrous episodes were recorded on 18/19 July and 24/25 August 1987.

The meteorological conditions were similar for both events. An active depression supplied cold air towards the Alps, where it met warm humid air from the south. The high air temperature differences of about 12 °C (on the 850 hPa level) caused the formation of a secondary low over the Alps, which intensified the inflow of warm air from the south. These factors were the main reason for the long duration of heavy showers that fell mainly on the southern part of the Alps. There, extreme values of precipitation were recorded, with 55 mm within 1 hour, 174 mm in 4 hours, 270 mm in 10 hours and 392 mm in 24 hours. In many regions 120–275 mm in 24 hours and 150–440 mm in 48 hours were measured.

These amounts correspond to a probability of occurrence of once every 50 to 200 years (Fig. 4.29).

The period May–June was very wet (150–300 per cent of mean amount) with frequent rainfall (about 40 rainy days). Therefore, the soil and the underlying strata were well saturated in July and quite well saturated in August. In addition, due to high temperatures, liquid precipitation occurred up to 3500–4000 m above sea level, an unusual situation in heavy storms. Moreover, in July the contribution of melting water from the snowpack above 2500 m was considerable.

In spite of the absence of very high rainfall intensities in many parts of the country, the combination of all these factors caused the formation of many extreme or even catastrophic peak discharges.

Observation and assessment of flood discharge

For more than 125 years, the Swiss National Hydrological and Geological Survey has been responsible for the observation of river stages and discharge. With the large database collected within that time, flood frequency analysis has recently been established for about 135 stations with 30 to 115 years of observation, and for about 110 stations with 10 to 30 years. These analyses provide an important basis for decision-making in hydraulic planning and for the assessment of probability of extreme events such as those of the summer of 1987.

The network of about 250 recording discharge stations and of about 90 flood-crest-stage gauges supplies data on flood events. For flood warning purposes the Survey operates a network of teletransmitting stations that automatically transmit alarm messages to the responsible authorities when predefined water levels are exceeded. Furthermore, the flood forecast centre of the National Hydrological and Geological Survey provides short-term discharge forecasts for the River Rhine system.

During the most important flood events in July and August 1987, extreme peak discharges and very large volumes were observed. Three stations, built as concrete structures, were destroyed and disappeared almost completely. Many of the stations were damaged, but most of the data were saved. By means of the above-mentioned flood frequency analysis, return periods were estimated for all stations with remarkable floods. For 18 stations, the return periods of 50 years, and for 4 stations those of 200 years, were exceeded (Fig. 4.29), but the variations are quite important.

Early in 1988, the National Hydrological and Geological Survey published the results of the discharge measurements together with a statistical analysis, so that basic data needed by other offices and scientific institutes for the analysis of the causes of these events could be available as soon as possible.

Flood damage

Damage in 1987 has been the worst since systematic flood-damage recording started in 1972. Eight persons lost their lives. Total damage for the year was estimated to be about SFr 1200 million, which is about 10 times the annual average and twice the second largest value (Fig. 4.30). About 50 per cent of the damage was caused to the transport infrastructure, such as roads and railways, and 25 per cent to flood protection works, while only 9 per cent was to forest and agriculture. About 110 years earlier, in 1868, a similar flood ravaged the same area. Damage then, related to income, was similar or even higher than in 1987, but its distribution was very different. At that time, more than 50 per cent of damage was related to agriculture and only 8 per cent to roads and bridges. These variations reflect the changes in the national economy, but it appears that the mute acceptance of risk with respect to very rare floods has remained unchanged.

The Federal government charged the Federal Office for Water Resources Management, together with the Swiss National Hydrological and Geological Survey, with a study of the causes of these extraordinary floods. One of the largest problems was to collect reliable data, even shortly after the flood. During the event local authorities are preoccupied by rescuing people and saving property, and no attention is paid to collecting data or making observations. This underlines the necessity for an independent service, such as the National Hydrological and Geological Survey, which would not be involved in rescue operations, but would be responsible only for observation and data acquisition. This service must be well-equipped to ensure rapid mobilization and to allow it to operate independently in the case of major events. Since any new flood protection work must depend on past experience, the economic importance of the analysis of data related to extraordinary floods must not be underestimated.

In respect of the type of damage, the following basic processes have been distinguished:

Flooding by water: Although this is the classical type of damage, it was of minor importance during the 1987 flood. There was flooding in the lower Reuss plain, just upstream from the Vierwaldstättersee (Lake of Lucerne), since discharges in the Reuss exceeded the capacity of the river bed. Dams were eroded as a result of their capacity being exceeded.

Erosion of banks: Erosion is more dangerous than flooding, since it can lead to a collapse of a structure situated well above the water level at a location that might be regarded as safe if

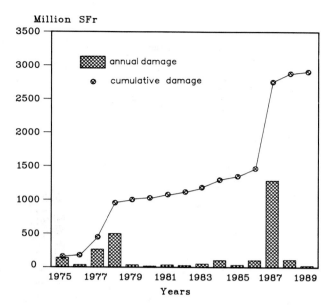

Fig. 4.30 Flood damage in Switzerland over the period 1975–89.

flooding is considered to be the only danger. At least four houses were destroyed because of the undermining of their foundations. The most severe local damage occurred at Wassen, where the foundation of a pillar of a national highway bridge was washed away, causing a settling of the roadway of more than one metre. Since the railway was eroded nearby as well, all traffic crossing the Alps through the Gotthard was interrupted.

River bed aggradation: As a result of excessive bed-load transport, caused either by bank erosion or the input of debris flows, the level of the river bed may rise, forcing the river to divert to a new bed. This is a particularly dangerous process, since it causes flooding at unexpected places.

Debris flows: Debris flows were the outstanding characteristic of the 1987 flood events, with respect to both number and size of the events (the debris flow of the Varuna deposited about 200 000 m³). In establishing the causes of the 1987 floods (see above), the Hydraulic Laboratories and the Geological Institute of the Swiss Federal Institute of Technology in Zürich have been charged with investigating the reasons for debris flows. The following observations have been made:

1. Debris flows occurred in well-defined areas only. Reasons for the sharp separation between areas with numerous debris flows and unaffected areas are not obvious, since the neighbouring areas usually have the same geological conditions and the levels of precipitation must have been similar.

2. For the initiation of debris flows, loose material with little or no silt and clay content and slopes at the limit of stability must be present. The amount of rain necessary to start the debris flow varies widely. Small debris flows are common, which indicates that the basic conditions to start the process do not require extreme amounts of water. The same is indicated by the fact that many debris flows started at very high elevations within small catchment areas.

3. Traces of debris flows show characteristic levées. Some flows travel only short distances; some start with large volumes, and others increase in volume on their way to the valley bottom.

Despite the fact that the mechanism of debris flows is only partially understood, it seems possible to delimit the areas of risk. Knowledge of the existence of a risk is always the basis for taking protective measures.

Conclusions

The amount of material moved either by debris flows or by bank erosion is a feature of the 1987 floods, and the present topography was formed by such extraordinary events. This natural process is well recognized but, because of the intensive use of land, we are no longer willing to accept these changes. In view of the increasing population, the consequences for man also become more and more severe. However, comparable events will occur again.

Events of this magnitude are rare, but investigations into the causes and the processes are of the utmost importance. The outlining of areas at risk and the design of protective works depend to a large extent on the knowledge of past extreme events. Frequent lesser events cause damage, but the rare, extreme floods endanger life.

Although damage was great, numerous flood protection works resisted the flows and proved to be well-designed, but protection can never be complete and further improvements are necessary. The events showed again that floods are only partly a hydrological and hydraulic problem. The design of protective installations must take into account the possible large inputs of bed-load or debris flows, and the instability of banks with respect to erosion.

A sound design and an effective protection can be based only on deep knowledge of the processes, and on good hydrological and geological databases. Services that systematically collect the necessary high-quality data on a long term basis are of vital importance.

Furthermore, good cooperation between these services and those who plan and implement protective measures, is essential for the safety of man and his property.

4.3. RISK MANAGEMENT AND MITIGATION [53]

4.3.1. The cost of geological hazards and the necessity for risk management

The socio-economic impact of geological hazards

In order to limit the impact of natural disasters, the following parameters should be determined as rapidly as possible: type, frequency, and characteristics of the dangerous phenomena, and their effects on people and their property, and on the socio-economic fabric and institutions of the affected communities. The assessment of risk and the definitions of the necessary means for its reduction, pose a dilemma for the decision makers: should they invest in a future reduction of the potential risks, or should their investments aim at the immediate improvement of the economy and living standards? For this reason, the analysis and evaluation of risks forms an invaluable aid for the decision-making process of politicians.

In its current meaning, the word 'risk' covers two concepts: that of a peril, a danger, a potential for damage, as well as that of a hazard, a possibility, a probability. This convergence of two different concepts can also be found in the mathematical expression of risk, which is defined as the mathematical expectation of a cost function that corresponds to an abstract evaluation of damage.

Faced with a given hazardous phenomena (e.g. landslides, flooding, earthquakes, etc.), the risk notion R_x links two complementary concepts:

(1) the hazard concept P_x, corresponding to the probability of reaching a level x of the dangerous phenomena during a given period;

(2) the concept of vulnerability D_x, corresponding to the probability of a given damage when level x (see above) is reached

By definition, the risk R_x is the product of hazard P_x with the vulnerability D_x:

$$R_x = P_x \times D_x.$$

As a town grows, its vulnerability ΣD_x increases with the increase in dwellings, in infrastructure, in individual or collective property, in production capacity, and in the number of inhabitants. In summary, the risk R_x grows geometrically in such a fashion that a policy of prevention and rehabilitation becomes increasingly profitable in order to ensure the continuous economic development of the town.

Creating a risk-control programme is as profitable in an economic as in a social sense; this concept is schematically shown on Fig. 4.31, as well as on Table 4.16. Faced with a given hazard x, one can compare for each situation the amount of predicted damages C_d with the amount of necessary investment C_p to limit such damages to C'_d or to eliminate them. The benefit/cost ratio of such investments is:

$$B/C = (C_d - C'_d)/C_p.$$

The curves of Fig. 4.31 must be defined for each type of hazard. For the seismic risk for instance, which affects large areas, the damage and cost curves justify a general control policy. When risks of more limited extent threaten, such as landslides or floods, the curves are more sectorial and must be defined for each type of hazard. In such a case, a classical cost/benefit evaluation is entirely suitable. An example for seismic risk is presented in Fig. 4.32.

The evaluation of the socio-economic impact of dangerous natural phenomena is extremely difficult; for this reason it has been done in only a few countries, the first attempt dating from 1973 for the *Master Plan of California*. Such a venture is commonly hindered primarily by a lack of sufficiently precise data concerning both the direct and indirect effects of disasters ('indirect' meaning a reduction in production capacity or trade, for instance). The most practical reference unit, that of finance, stumbles over the evaluation of the socio-economic cost resulting from the death of and damages to the

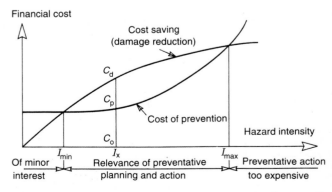

Fig. 4.31 Economic effect of a preventative policy for natural disasters.

Table 4.16 Projected losses due to geological hazards in California, for the period 1970–2000 (figures in millions of dollars, 1970 value). (from Alfors, J. T. *et al.* (1973). Urban geology master plan for California: the nature, magnitude and cost of geologic hazards in California and recommendations for their mitigation. *California Division of Mines and Geology, Bulletin No. 198*, USA)

Geological hazard	Projected total losses, without preventative policy	Possible total loss reduction with prevention		Estimated total cost of prevention		Benefit/ cost ratio
		% total loss	Amount (US$M)	% total loss	Amount (US$M)	
Earthquake shaking	21 035	50	10 517.5	10	2 103.5	5
Landsliding	9 850	90	8 865	10.3	1 018	8.7
Flooding	6 532	52.5	3 432	41.4	2 703	1.3
Erosion activity	565	66	377	45.7	250	1.5
Expanding soils	150	99	145.5	5	7.5	20
Fault displacement	76	17	12.6	10	7.5	1.7
Volcanic hazard	49.38	16.5	8.135	3.5	1.655	4.9
Tsunami hazard	40.8	95	37.76	63	25.7	1.5
Subsidence	26.4	50	13.2	65.1	8.79	1.5
Total	38 324.58	61	23 411.695	16.0	6 125.695	3.8

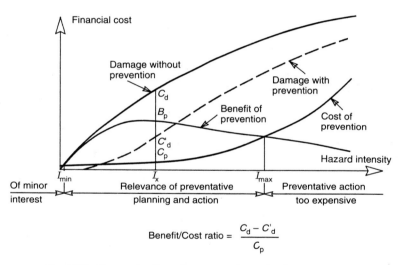

Fig. 4.32 Economic effect of a preventative policy for earthquakes.

affected population. In comparison, it is clear that the number of road accident victims is much higher for the European countries, but it is spread out over time and space, and does not have the same socio-economic impact as natural disasters, the effects of which are amplified by their spatial concentration and by the degree of unpreparedness of the population.

Notwithstanding such difficulties, further research on this subject is very necessary and should be developed as quickly as possible if a coherent preventative policy is to be installed in the threatened countries. The Instituto Tecnologico Geominero de España (IGME) has estimated the socio-economic impact of natural hazards in Spain over the period 1986–2016, basing itself on the methods developed in California, which are schematically shown on Fig. 4.33. The losses related to a destructive event x of a given intensity, are calculated on the basis of an evaluation of the direct economic damages, multiplied by a catastrophe factor K that is based on the number of victims ($K = 1.1$ for 1–10 victims, 1.5 for 11–100 victims, 2 for 101–1000 victims, etc.), which integrates

therefore in a still rather empirical fashion the indirect socio-economic effects of the disaster.

$$D_x = d_x \times K$$

The risk corresponding to R_x is the product of D_x with the probability P_x of the event x of a corresponding intensity occurring over the time span under consideration (30 years).

$$R_x = P_x \times D_x$$

The results obtained by adding up all risks R_x, correspond to the various hazards that might affect the Spanish territory and are presented on Fig. 4.34 and 4.35, based on two different hypotheses: [152] [157]

(1) *Maximum hazard* corresponds to the most serious historical event for each of the dangers taken into account.

(2) *Average hazard* corresponds to the most probable event occurring in the time span under consideration.

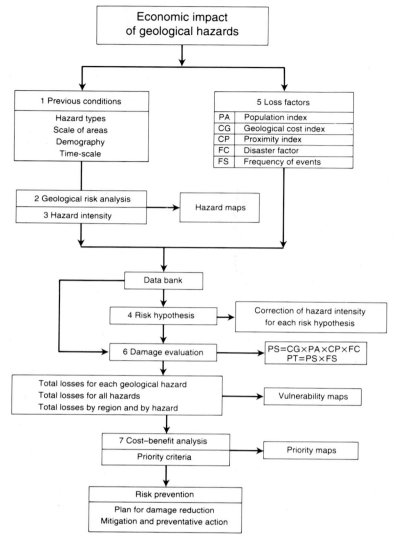

Fig. 4.33   The economic impact of geological hazards.

Table 4.17 shows the results obtained. With a risk that falls between 39 and 64 billion ECUs for the period 1986–2016, the socio-economic impact of natural hazards has been evaluated at between 0.68 and 1.13 per cent of Spain's GNP. These figures are higher than those available from similar estimates for Italy, which are 0.33–0.57 per cent of the GNP, even though in absolute values they are virtually the same, with a 'high' of 62 ECUs/inhabitant/year for both countries (Fig. 4.36). The figure for damages as evaluated for the USA (at 0.4 per cent of the GNP), is much higher in absolute value at 82 ECUs/inhabitant/year.

All these figures require further refinement in the future, as do the concepts and methods that were used. Even so, they show quite clearly that natural catastrophes have a major socio-economic impact on most countries, whether rich or poor. The means for limiting such impact are available and their implementation gives excellent results, as is shown by the examples of Japan and California in particular.

Risk-management concepts

Management and control of natural risks must be based on a series of four complementary actions: development of risk analysis, selection of solutions for reducing or eliminating their harmful effects, creation of mechanisms for risk management and prevention, and provision of suitable information and training to the community.

Risk analysis combines the identification and evaluation of hazards with the estimation of the vulnerability of the environment and any human presence. It allows the selection of the most cost-effective solutions and of the priority actions to be taken, in order to reduce the occurrence and the effects of predictable natural disasters. Such selection is based for a major part on cost–benefit analyses.

Risk analyses form a major decision-making aid for politicians and funding organizations, as they form an integral part of a fabric of socio-economic considerations that govern sustainable development.

Means for the prevention or reduction of the effects of natural disasters are legion; they include rational land-use and, in particular, the guidance of urban development towards the least exposed and least fragile areas, and the modification of man's actions that might cause risks. Other means are: suitable building codes (earthquakes, construction on slopes, protection of the vulnerable natural environment), the development of corrective measures (slope stabilization,

MAXIMUM HAZARD

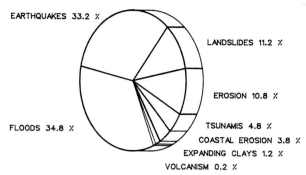

AVERAGE HAZARD

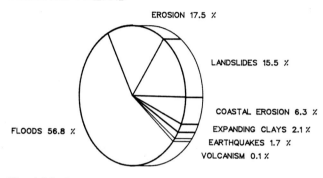

Fig. 4.34 Representation of maximum and average hazards in Spain.

MAXIMUM HAZARD

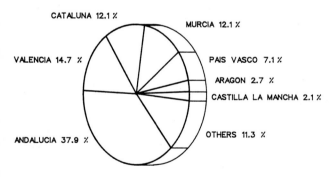

AVERAGE HAZARD

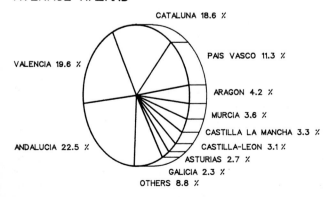

Fig. 4.35 Regional distribution of maximum and average hazards in Spain.

regulation and channeling of streams, protection against erosion, etc.), the installation of monitoring networks covering potentially dangerous areas, the organization of alarm and rescue systems, the training of technical and operational services in charge of protection and rescue actions, and the implementation of rehabilitation measures.

Such means will only then be truly operational if the selected solutions take into account the sociological conditions, the technical means, and the construction and agricultural practices appropriate for each country. Risk analysis and technical recommendations for the reduction of natural hazards represent the basis of *preventative planning*.

Mechanisms for the management of risks and their prevention constitute the indispensable complement for preventative planning in areas exposed to natural hazards. The installation of such risk-management systems is dependent upon coordinated political, legal, administrative, and institutional action.

The political area decides the type of risk to be taken into account, the level of protection and coverage required, and the designation of the responsible authorities and necessary funding. The legal, administrative, and institutional areas ensure the structuring and coordination of planning, financing, and execution mechanisms, as well as the control over preventative and rescue actions, based on laws, regulations, and decrees, as well as on standards that are adapted to each project. At this level, the basis for insurance and indemnification for natural catastrophes is fixed as well.

Information, training, education, and community organization, which are necessary for reducing the identified risks, form a fundamental part of the success of preventative planning programmes. Only when made aware and fully informed, as well as integrated into the decision and planning process, will the population and the local authorities be able and willing to ensure the proper implementation of such preventative plans. By 'authorities' are meant all technical and security people, such as contractors, architects, town-planners, fire brigades, etc., as well as legislative bodies.

The completion of these actions for the control of natural hazards is essential in any sustainable urban and regional development.

4.3.2. Risk analysis

The analysis of natural hazards should include the following:

1. *Development of geological hazards database*, combining all known events (historical and actual) or all types of natural hazard with an evolution of the recorded damage.

2. *Phenomenology and evaluation of natural hazards*, for which factual and forecasting methods are developed; modelling is used for harmonizing and, if necessary, completing existing studies.

3. *Zoning and mapping of natural hazards*, to cover the entire area of interest in a first stage, after which particularly vulnerable areas are studied in detail, in relation to land-use planning.

4. *Assessment of the vulnerability of areas exposed to natural hazards*, including not only the victims, but also the socio-economic and institutional trauma caused by the natural disaster.

5. *Elaboration of potential crisis scenarios* for each type of hazard identified, such as earthquakes or volcanic eruptions, to serve as the basis for defensive programmes for the efficient tackling of and management of crisis periods.

Table 4.17 Total losses for maximum and average hazards, as forecast by the IGME for Spain, over the period 1986–2016 (the figures include an annual economic growth rate of 2 per cent over the 30 year period)

Hazard	Total losses (10^9 ECUs)	
	Average hazard	Maximum hazard
Flooding	22.22	22.22
Earthquakes	0.67	21.10
Landslides	6.04	7.06
Soil erosion	6.86	6.86
Tsunamis	0.00	3.08
Coastal erosion	2.45	2.45
Expanding soils	0.82	0.82
Volcanic activity	0.02	0.13
Total	39.08	63.72

The aim, in terms of the management of existing property or the planning of sustainable development schemes, is the definition, on a drawing, of those parts of the territory that are in danger, and of the global value of foreseeable damage. The decision makers need to be able to chose which action to take, whether it be technical, administrative, or legal in nature, by appraising the relative advantage that it will offer to the community in danger. This is based on the assumption that the analysis of the expected phenomenon and, beyond that, the estimation of the risk, have been properly assessed. As an example, we present below an application of this method to landslides, this geological hazard having been the subject of quite interesting, but as yet too scattered, analyses in several European countries.

Natural hazard analysis

Such analysis aims at defining the type of phenomena likely to occur in the observed zone. Three basic steps need to be taken for this:

1. *Refer to events that have already occurred*, provided that one has access to exhaustive knowledge of past events on the reference territory, and on other territories that underwent the same forms of geomorphological evolution.

2. *Identify predisposing factors* linked to the initiation or acceleration of movements.

3. *Interpret such movements in the light of knowledge* acquired in analysing the factors, by taking into account detected mechanisms and rules resulting from accumulated experience.

With regard to past events, knowledge is generally available only from oral witnesses, who become less precise with time. Moreover, management of the relevant files is poorly organized, and spread over the various filing systems of the organizations that collect them. The arrival of relational database systems provides the opportunity to reorganize and make the best use of all available data. Care must be taken that such data are not distorted when being transferred into a database management sytem. A DBMS must thus be provided that is appropriate for the general hazard prevention issue, i.e. it must allow input of data that may evolve over time, making

way for new knowledge on type of mechanisms, and it must inform not only on technical aspects of the recorded event, but also on administrative, legal, and economic issues. For the Isère region in France, a prototype of such a DBMS (INVI) has been designed by BRGM (the French Geological Survey).

It forms a necessary complement to the zoning of ground movement dangers, as presented in the so-called ZERMOS maps, a French acronym for zones exposed to movements of the surface and subsurface, which were drawn up in France during the 1970s. Similar programmes were developed in other European Geological Surveys, such as the GEORISK database for landslides in the Bavarian Geological Survey; such databases usually go together with slope instability maps at scales varying from 1:5000 to 1:50 000. In Italy, the Geological Survey (Piemonte) has set up the BDPG database that covers many geological processes in the Piemonte region, as well as the BDD documentary database that manages information on geological and meteorological hazards.

Obviously, at the same time, considerable effort has to be put into collecting and inputting data. Such an effort can be made when regional inventories are drawn up and published, or sometimes within the context of university work. The classifications that are established usually have a strong regional flavour.

As far as *factors determining equilibrium* are concerned, these vary from one type of movement to another, but there are two main types: localized permanent factors, and transitory aggravating factors.

Examples of *localized permanent factors* are slope, exposure, position inside a catchment area, lithology, thickness of superficial layers, etc. They may be important, depending upon the type of movement involved; the way they are geographically distributed, even if fixed, is but little known. Even if the slope can be appraised in a continuous way, and if traditional geological maps give a fairly good account of the geological nature of the substratum, this is often not so for the type and thickness of superficial layers. For these factors some degree of uncertainty must be expressed and controlled.

Spatial variation in these localized factors can be expressed with maps, provided that such maps present the possibility of accounting for uncertainty in the factors. This implies that such maps should not be considered as set and traditional documents, but rather as evolving documents, bases for successive layers of data, available for combinatory analysis processing or for expert analysis based on rules.

Today, the necessary tools exist for creating such documents, which act as bases for complex data. Horizontal databases are geographical information systems (GIS), which enable a given x, y parameter to be associated with a certain number of information elements. These can be altitude, slope, lithology, thickness, and type of first ground layer, but also the uncertainty (presumed value and possible deviations) of such thickness, or of another element.

The localized or permanent factors are those that will lead to definition of the type of phenomenon and evaluation of its spatial distribution. They will enable definition of the dangerous zones, but not of the time-scale, frequency, or intensity (only the volumetric component of intensity).

The *transitory or aggravating factors* can be of natural origin, such as variation in hydro-meteorological conditions or seismic events, or of human origin such as civil engineering earthworks or changes in agricultural practices. The role of the latter can be evaluated in an empirical manner, provided that the mechanisms of their impact on stability are well known. For the former, analysis of the conditions of their repetitiousness provides the basis for forecasting the moment of occurrence for the expected phenomenon. This analysis

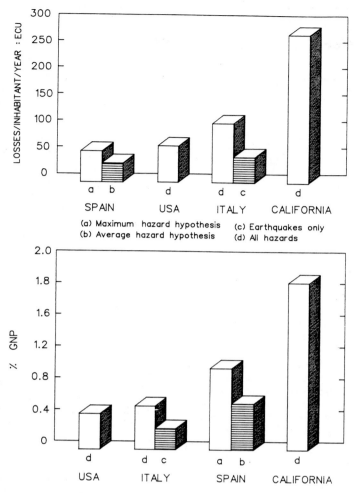

Fig. 4.36 Comparison of losses as a result of natural hazards, in Spain, Italy, California, and the USA as a whole, expressed in Ecus per person per year (upper diagram) and as a percentage of GNP (lower diagram).

presupposes the ready availability of sets of past recorded data on the evolution of natural conditions for the site in question.

Here, again, the need to organize databases arises. Generally speaking, such databases exist, but with quite varying density of data.

Whether it be a question of permanent or transitory factors, and thus of building up GIS or DBMS systems, it needs to be made clear that the bringing together of these elements, which are necessary for a rational evaluation of the hazard, is a heavy task requiring medium-term planning and good organization. This calls more for an 'observatory' type of long-term continuous operation, rather than for occasional or isolated studies. The people responsible for such work thus require scientific competency, but must also be ready to carry out a great deal of fieldwork, making use of the best quality of data and efficiency of the tools that are available.

Evaluation of landslide damage

Risk exists when exposed people, property, or activities may suffer injury, damage, or interruption, respectively, should the phenomenon occur.

For each damageable element that exists on the territory involved, the probable damage level should be evaluated.

Thus, for example:

1. Will the people be psychologically disturbed, bothered, slightly injured, seriously injured, or killed?

2. Will property be slightly damaged, or partially or completely destroyed?

3. Will activities be slowed down, stopped briefly, or shut down for long periods?

4. Will the social and institutional organization be affected?

Each analysis raises its own specific issues. For instance, the primary problem for communities or activities is that of their evacuation or interruption. To provide the answers to such a problem one must be able to evaluate the impact of the economic, environmental, and socio-political consequences.

For building or structures, the problem posed is rather in terms of relevant life-span, and interaction with the phenomenon. Moreover, it is not enough to evaluate direct damage alone, but also the indirect damage.

Given the complexity of the phenomena involved and the diversity of the parameters (type, intensity, frequency, etc.) that intervene in the production of damage, the evaluation of the level of foreseeable damage remains highly uncertain. Hopefully, damage will be progressive, based on the mechanical interaction between the structure and the phenomenon, but in the short-term it seems much more realistic to work on a retro-analysis of damage that has already occurred, as is the case for seismic pathology. The considerable investigation and data processing work required for this should make use of the same computing tools as previously described for hazard analysis (p. 194).

Risk zoning and mapping

Once the hazard has been analysed, taking into account the foreseeable damage level for a given type of property and integrating all elements exposed to the danger, it will be possible to evaluate the economic cost of the phenomenon and its consequences to society. For a given portion of territory, this cost will depend on the type of phenomenon and the full set of elements exposed.

The difficulty faced in applying the notion of risk mapping, which was introduced several years ago, can thus be appreciated. The most commonly used decision-making element for development schemes at the time was the drawing or the map. This was primarily because a map is a document upon which the extent of the affected area and the occupied land surface can be superimposed. In this sense, the drawing is thus the information document for a given period. *However, risk is a concept that is much less well defined, and which depends upon far too many factors to allow it to be 'fixed' on a map.* The objective of evaluating risk is to manage and to decide upon the most appropriate type of preventative action. This will vary, depending upon the short- or long-term perspective, or on the legal or socio-political framework.

It would thus seem desirable to single out those elements that are geographically well-defined, such as hazard or land-use modes, from those elements that depend, for purposes of interpretation, on factors which may change, depending upon the requirements of the analysis.

Development perspectives

For the reasons explained above, work needs to be done in two directions:

1. Facilitating input, updating, and use of all data (geographical data in the GIS, and symbolic data or world knowledge in the DBMS). In this way maps can be generated that group all data or factors necessary for defining:

   • zones likely to be affected by slope movements,

   • present or future land-use.

   Once these data have been properly organized and periodically re-evaluated, they can be manipulated, converted, and interpreted from a hazard analysis point of view, based on official guidelines and the experience of specialists and field-workers.

2. Initiate the creation of institutional guidelines and regulations, possibly of a regional nature, for both hazard analysis and risk analysis. These may either refer to empirical work, or else allow for simple calculation or modelling procedures based on data grouped together in geographical documents or in more traditional databases, such as those used for hydro-meteorology.

To recapitulate, the means are available today for creating expert systems or systems to aid decision-making that help in the interpretation of data from highly different sources, so as to identify zones likely to be affected by ground movements, and to analyse the resulting risks.

This is by no means an easy task, but it is the price to pay if we hope to rid hazard- and risk-evaluation procedures of the subjectiveness that makes them subject to criticism today.

4.3.3. Mitigation or elimination of risks and preventative planning

Preventative planning is based on three principal actions as shown in Fig. 4.37: prevention, defence, and site rehabilitation. However, only a few European countries have the organizational structure on a national level capable of ensuring the efficient management of natural hazards. In these countries, the national Geological Surveys have a vital role to play in the evaluation and drawing up of the basic guidelines for preventative development planning. The experience acquired by France in this domain is described in Section 4.3.4.

Natural risk prevention programmes (Fig. 4.38)

All programmes that aim at preventing natural hazards must be developed on three levels: regulation, operation, and education.

Regulation and planning

All preventative regulation is based on a political evaluation of the accepted level of risk, or of the desired level of protection. The *technical* evaluation of the vulnerability of persons and goods to identified hazards, is an indispensable basis for the *political* choice concerning a rational use and development of the land. Further choices to be made concern the adoption of suitable building regulations (in particular in areas exposed to seismic risk), and the correction of human actions that might cause risks. Drawing up decrees, regulations, and technical codes, as well as their enforcement, requires a level of technical counselling that can be provided only by organizations that are highly specialized in all aspects of the earth sciences. However, under no circumstances should the role of such an organization go beyond that of a consultant.

Technical operations

The reduction or elimination of natural disasters requires the use of preventative or corrective works. These can include the regulation and channelling of water courses, the stabilization or strengthening of slopes, or measures to counter erosion. Such operations must be carefully studied and planned in the framework of rational programming that is based on the urgency of the preventative actions and on their socioeconomic interest. The latter point requires an evaluation of the investment costs versus the expected benefits, which would lead to a reduction of the vulnerability and thus a saving on the expected damage, as was expounded in Section 4.3.1. The consultant's role at this operational level is obvious.

Education and training

The effectiveness of all preventative policy depends on making the people aware of its necessity and having them accept it. Communication and educational action, using the media and schools, elicits a positive response from the population. Similarly, technical training courses for urban planners, architects, engineers, and technicians, are essential in gaining acceptance of and respect for regulations and standards that govern planning, development, and construction. Sociological polls have shown that scientists and technical specialists have much 'credibility' in the eyes of the

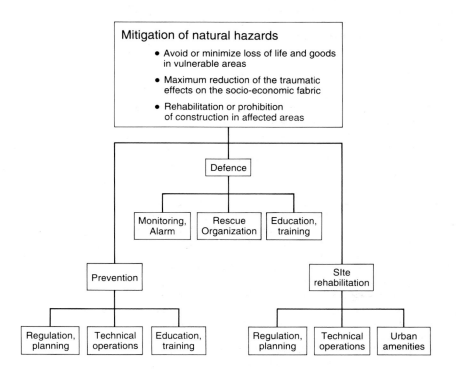

Fig. 4.37 Planning against hazards.

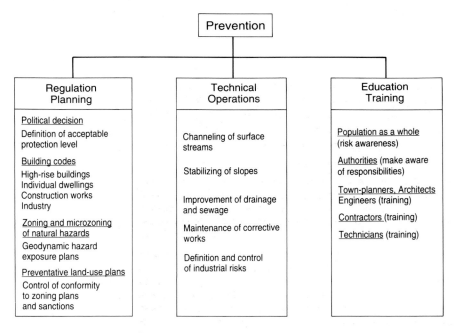

Fig. 4.38 The organization of programmes preventing natural risks.

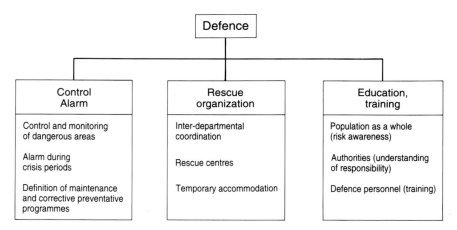

Fig. 4.39 Organizing the defence against natural disasters.

population. Unfortunately, the national Geological Surveys, which could play a key role in these matters, have seldom if ever been given the chance to do so.

The defence against natural disasters (Fig. 4.39)

It would be illusory to try to eliminate all risks. For this, the technical means are not available and the cost would be prohibitive. Even an efficient preventative policy will only limit, but not eliminate, the occurrence of disastrous or dangerous phenomena. For defences against such events to be effective, four types of action are necessary: the setting up of monitoring and alarm systems in danger zones; the organization of crisis management cells; the organization of aid; and the initiation of education and training.

Concerning regulation and planning (see above), the role of technical consultants must not go beyond the level of the design and preliminary organization of these four types of defensive programme.

Recourse to the assistance of the Geological Surveys in the defence against geological hazards has been spasmodic and limited in scope. This is quite harmful to the community as a whole, as the competence and means for developing relevant action usually exist in these organizations. The main limiting factor appears to be political will.

Hazard monitoring and alarm systems for danger zones

The construction of risk maps leads to the identification of the most vulnerable areas requiring control and continuous monitoring, which includes topographic monitoring, hydro-climatological measurements, seismographic and accelerographic recordings, clinometer readings, etc., with the objective of setting up effective preventative management of persistent hazards. For this, specialized teams must be created, whose task is not only the control and monitoring of critical areas, but also the management of the data obtained, and the maintenance of the networks.

Each team would have access to an alarm system that is triggered when a crisis level is reached. In this manner it is possible to be forewarned of a dangerous event that would require the mobilization of a crisis management cell, the installation of a close-observation network, and the possible evacuation of the population. The alarm criteria for these actions must be established beforehand, based on the potential crisis scenarios mentioned above.

Organization of crisis management cells

During a period of alert, a crisis management cell must be mobilized. The monitoring team will then act under the orders of the crisis cell, the composition and functions of which must have been established beforehand by the competent authorities. The setting up of an efficient system for aiding in decision-making is of fundamental importance. Improvisation is not appropriate, as detailed and deliberate preparation is essential. The construction of scenarios for potential crises is a useful foundation for appropriate action in the event.

Organization of aid

The organization of aid, before or after the event, requires an inter-institutional coordination in direct liaison with the crisis management cell. The zoning of natural hazards and practice with scenarios form the indispensable base for locating aid and temporary accommodation for the menaced or stricken population.

Education and training

Training and educating the population by creating local or regional defence committees, is an important part of preparation. Such action must complement that by specialized government organizations. Practice alarm drills provide useful training for the people involved, and will identify weaknesses in equipment and organization.

The training of the personnel making up the defence committee is a further fundamental objective for improving the efficiency of their work. Technical and scientific organizations, such as Geological Surveys, can contribute significantly to such educational programmes.

Protecting and rehabilitating fragile sites (Fig. 4.40)

The high growth-rate of the numbers of people affected by natural catastrophes is due mostly to population increase. Two problems are involved:

1. Under the impact of man's actions, the natural environment undergoes strong degradation, which leads to an increase in the number of destructive natural phenomena such as landslides, floods, erosion, and drought.

2. The uncontrolled urban growth in some countries extends to sites exposed to natural hazards, increasing the vulnerability of the population in such areas.

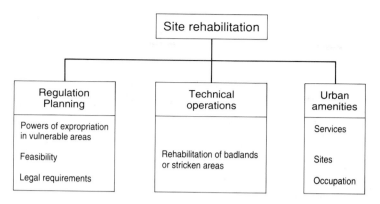

Fig. 4.40 Organizing the rehabilitation of a site.

Faced with such problems, all policy makers concerned with land-use development must ensure that such projects are integrated into their physical environment so as to preserve fragile sites, and also that effective stabilization is carried out on all sites affected by dangerous natural phenomena.

The Geological Surveys of Western Europe have much expertise in both respects, which, except in rare cases, has seldom been used by the authorities concerned.

The technical organization for risk-management

The effectiveness of a plan for the identification of natural hazards and the management of natural risks depends on the capacity for mobilizing a suitable inter-disciplinary and inter-institutional cooperation. Furthermore, perfect agreement must exist between the responsible authorities and the specialists who will carry out the technical work. It should go without saying that each case will require a unique solution, based on the specific technical needs, and on the geological, geographic, and social frameworks.

The role and the importance of the Geological Surveys in such an undertaking would be evident.

4.3.4. Institutional and legal mechanisms for the prevention and management of risks

In France, for ten years or so now, a policy of prevention of risks generated by natural hazards has been operational; this covers landslides, earthquakes, avalanches, floods, forest fires, cyclones, etc. This policy was launched to complement the law concerning insurance, passed on 13 July 1982, which is aimed at developing national solidarity for the compensation of victims of natural phenomena. This law enabled an original document to be drawn up, the PER (Plan d'exposition aux risques) or risk exposure plan. A PER not only makes it possible to prepare for the future by means of a schedule that takes account of natural constraints, but it also allows for the determination of the type of damage to existing property and activities.

Each of the European countries could provide examples of this, but France constitutes a particularly interesting case, in view of its well-advanced legislation in this field. To understand how French legislation is organized with regard to land-use, a certain number of explanations of a general nature are required.

The two main documents in land-use legislation

In France, two main documents concern land-use planning: SDAU (Schémas Directeurs d'Aménagement et d'Urbanisme) or Master Development Plans and Pos (Plans d'Occupation des Sols) or Land-Use Plans. The *SDAU (Master Plans)* define major operations for future regional planning. Land-use plans, as well as all major city planning and public amenity projects for local communities or the State must be compatible with the SDAU. The first SDAUs, which were first drawn up in the 1960s, projected a structured urban planning and development. They have not been updated for 15 years, and it is now evident that there are large differences between these plans and the actual evolution of the areas concerned.

In the light of experience, concerted thinking has led to further renewal of planning ways and means, and to several innovations bearing on the nature of the planning documents and their elaboration. These have now become relatively open legislative texts, that leave some latitude on the issues dealt with. The governing ideas behind such planning, today, are:

(1) 'prospective' planning rather than detailed forecasting;

(2) a project and strategy for urban areas that are based on a few major objectives, capable of mobilizing one and all;

(3) greater flexibility and a new language for graphic expression;

(4) post-Master-Plan management, as a flexible concept of planning.

The *POS (Land-Use Plan)* in France is a town-planning document that authorizes or prohibits construction. POSs are approved by the local Commune after consultation of the competent public authorities, and following a public enquiry among the population affected by the POS.

A POS is more accurate than an SDAU (scale 1:5000 as opposed to 1:50 000) and its approval (after revision) is of much more recent date, anywhere in France. For this reason they reflect the current facts on community urban policies much more accurately. Paradoxically, the SDAUs should now be adjusted to the POSs.

The typical content of a POS is:

(1) definition of settlement procedure;

(2) definition of city planning rule (for example for parking areas or green areas);

(3) preservation of quality of the urban landscape;

(4) adaptation of the shapes of building and of their height;

(5) reservation for public facilities;

(6) definition of the 'COS' (Coefficient d'Occupation des Sols) or land-use coefficient, which defines the ratio between land surface and the authorized area of floor space. These coefficients can vary from 0.1 in a residential suburb (100 m² of floor-space for each 1000 m² of land property) to 5 downtown.

The map that forms the core part of the POS allows the distinction between some main areas, such as:

U   (urban) zones where facilities (roadways, water, sewage) exist. All land may be used for building.

NA   zones (not developed but reserved for future urbanization) where building is prohibited in view of the lack of installations. Urbanization can, however, be authorized in some NA zones, subject to the drawing up of a comprehensive building scheme called a ZAC (Comprehensive Development Areas) or housing estate, where the financing of installations is guaranteed by developers.

NB   zones (natural land with scarce buildings). Very limited.

NC   zones (farming), which can not be used for building.

ND   zones (protected sites), which can not be used for building, except for recreational facilities in some cases.

Taking natural hazards into account for town planning

The State and the municipalities each have a certain responsibility in the prevention of natural hazards.

The State must:

(1) seek and collect data using all available technical means;

(2) point out and make known all potential dangers, in particular to municipalities in the process of drawing up or revising a town-planning document;

(3) check that these town-planning documents effectively take the risk into account once notification has taken place.

The municipalities must:

(1) take the data supplied by the State into account;

(2) in return feed back to the State their own knowledge of hazards.

The POS, drawn up at the Commune's initiative, and under its responsibility, must take the existence of natural risks into consideration. It may be invalidated by the State if this condition is not respected.

The existence of hazard does not necessarily mean that no construction can take place on the land, but it does require the drawing-up of town-planning rules aimed at handling the risk-generating phenomenon or the vulnerability of the property involved. In the case of the ZACs for example (see above for definition), the financing of consolidation or of rehabilitation for the site concerned may be included in the overall development budget.

Various specific prevention documents have led to progressive improvement of POS. The principal documents define the *risk perimeter* and the *risk-exposure plan*.

1. The risk perimeter results from article RIII.3 of the town-planning code. This article enables the Prefet (civil administrator) to set out a zone, regardless of communal boundaries, upon which 'building on land exposed to risks such as floods, erosion, subsidence, landslides, and avalanches may, if authorized, be subject to special conditions'.

2. The risk exposure plan (PER) was instituted by the law of 13 July 1982 concerning the compensation for victims of natural disasters. This document defines the exposed zones and the prevention techniques to be implemented, both by the owners and by the community or other public authorities. The PER thus forms a link between prevention and compensation. Appended to the POS, it can impose corrective or back-up measures for cases concerning property and activities which occur prior to its publication (retro-active effect), and non-respecting the PER is liable to affect the level of compensation.

The PER is drawn up under the guidance of the Prefet, who prescribes it, renders it public, and approves it following public enquiry. The Commune is consulted at each stage of the procedure. A standard plan will comprise:

(1) an introductory report;

(2) graphical documents including in particular:

   (i) a hazard map;

   (ii) a vulnerability map;

   (iii) a risk map with 3 types of zones depending upon the importance of the risk and the vulnerability of existing or future property:

   - *The red zone*, or high-exposure zone: the probability for a hazard to occur, and the intensity of its foreseeable effects, are such that no prevention measure exists which is economically appropriate, other than a ban on construction.

   - *The blue zone* of average exposure: the probability for a hazard to occur and the intensity of its foreseeable effects are average, which means that certain land-occupation or land-use can be authorized, provided that certain guidelines are respected. It is thus defined in such a way that the risk and its consequences are acceptable, provided that these guidelines are respected.

   - *The white zone*, or zone reputed not to be exposed: the risk-occurrence probability and the intensity of its foreseeable effects are negligible here.

The regulations define the prevention measures applicable within each zone or sub-zone marked on the maps. The regulations will, in particular, indicate those measures intended to prevent risks, to reduce their consequences or to render them acceptable, and to define the compatible land-occupation and land-use, given the existence of the risk, taking into account the respect of these preventative measures.

The RIII.3 and PER documents may well be contrary to the (uninformed) wishes of the population.

Other, purely informative, documents dealing with land movements exist. These are, for example, the maps of zones exposed to risks associated with land movements and underground movements.

Once the PER has been drawn up, each landowner knows that he is in either a red, blue, or white zone.

1. If he is in a red zone, he will not get a building permit. However, should he already own a building on the land he will, in the case of damage recognized by the Prefecture as

resulting from a natural disaster, benefit from insurance aid.

2. If he is in a white zone, the above applies both to new and old buildings.

3. The situation in a blue zone is different. The owner is informed of the existence of a risk and economically acceptable measures enabling the risk to be limited are proposed to him. He is free to implement such measures or not, but at his own cost.

   (i) Should he do so, he acts as a responsible citizen and he will benefit from the related compensations.

   (ii) Should he not do so, he has taken the personal decision to ignore the risk and he will not benefit from compensation.

Construction practice—regulations and preventative measures

The PER must include regulations. The Ministry of the Environment through the Major Risks Delegation (DRM), together with professional technicians, has produced a guide for drawing up regulation clauses. This guide proposes models for reports and for recommendations intended to limit risk, particularly in the blue zone, and for both existing and future property and activities.

These rules are proposed for the main categories of land movement: landslides, subsidence, rock falls, collapse, erosion, mud flows, avalanches, lava flows, and underground cavities.

Each refers to a descriptive technical catalogue of recommended work that may be necessary.

Balance sheet and perspectives

The interesting facet of the French experience with regard to taking into account land movements for land-use purposes, is that it provides a clarification in the sharing of responsibilities, and that it provides a link-up between prevention and compensation should damage occur.

Thanks to the considerable work that has been done over the last ten years under the control of the Major Risks Delegation, appreciable progress has been made regarding prevention. The French Geological Survey, BRGM, has played an important role in this work, in particular in defining the strategy for the management of natural hazards, in the evaluation of geological hazards in France, and in the drawing-up of PERs.

Much work still remains to be done, particularly in the training of the various persons involved in the construction process: promoters, developers, architects, design offices, inspection offices, etc. The main reason for this is that, apart from the responsibility of detecting and notifying the risk, and then drawing-up recommendations, there is the additional responsibility of drawing-up special technical specifications for all construction operations on unstable land. These should include procedures, deadlines, calculation methods, and architectural concepts, all of which must undergo specific adaptation.

4.4. CONCLUSIONS

Human society must constantly adapt to an unstable environment and it must recognize that it contributes in no small way to increasing the elements of instability involved. Within this society, the Geological Surveys form part of the multidisciplinary organization that must face and alleviate crises caused by the unstable environment. Such work may be the identification and control of the effects of natural hazards, or the contribution to the foundation of rational urban or regional planning. The selection of development methods that will preserve the environment and, as in the case of climatic change, the evaluation of evolutionary trends, are of vital importance. A geologist is often the best-placed person to appraise such global problems in relation to the time-scale involved.

Within the scientific community, specialists in environmental and earth sciences in the Geological Surveys have been at the forefront of our struggle against natural hazards during the past years. Frank Press, now president of the American Academy of Sciences, speaking to the Eighth World Conference on Earthquake Engineering in 1984, reminded us that the history of life on earth is neatly summarized by the philosophical words: 'Man lives by geological consent, subject to change without notice'.

At Press's initiative, the 42nd Assembly of the United Nations Organization (UNO) decreed in December 1987 that the 1990s would be the *International Decade for Natural Disaster Reduction*. The objectives set for this Decade are mainly:

1. The evaluation of all knowledge in the field of hazard prevention and the identification of any scientific and technological gaps that need filling.

2. The accelerated implementation of the available methods for hazard prevention, particularly as pilot programmes.

3. The development of all scientific and technological know-how presenting a strong potential for the efficient implementation of preventative methods.

The scientific community has been invited to work alongside engineers, sociologists, developers, and planners to achieve these objectives. The UNO has identified this as being the social responsibility of all, underlining that it is more important to predict and control than to react only after the disaster has struck.

The Geological Surveys are ready to contribute and to play a coordinating role in facing the crises that threaten our planet. It is to be hoped that the public authorities and the major international funding organizations, after having fully understood what is at stake, will provide the necessary ways and means for this vital task.

5

Impact of development on the geological environment

H. Aust and G. Sustrac

5.1. GENERAL STATEMENT OF PROBLEMS INVOLVED [65] [74]

5.1.1. Introduction

The requirements of economic development to meet the needs of increasing numbers of people lead to ever-greater pressures on resources and the use of land. Conflicts arising from such pressures are dealt with more specifically in Chapter 6, but here we are concerned with the impacts arising directly from the exploitation of resources and development in general. No such work can be carried out without there being some impact on the environment. This means that every potential development has to be studied in detail and carefully evaluated as to its impact. In most countries of Europe, this type of study is compulsory by law and the identification of actions to minimize the effects of exploitation are part of such studies. This concept applies to even the smallest development, although it is obvious that large-scale developments deserve greater attention. The objective is to achieve the maximum economic benefit at the minimum cost to the environment.

5.1.2. Resource exploitation

Natural mineral resources vary greatly in nature, size, accessibility, and uses (see Chapter 3) and any one or all of these characteristics will place constraints on their exploitation.

The abstraction of fossil fuels, industrial minerals, and metallic ores by opencast methods takes up large areas of land with sites up to several square kilometres in area and reaching several hundred metres in depth. Proper reclamation practices are essential, as is the control of tailings, dust, and noise. Due to the level of economic development in the European countries, and the pressure on space and the environment, impact studies covering all aspects of potential danger to the environment are now the rule, and include consideration of the effects of beneficiation units, haulage systems, utilities, and temporary living quarters. Underground mining has less impact at the surface, but the secondary effects of subsidence may be serious and extensive. In the case of brine pumping, for instance, subsidence is common and the surface areas concerned have to be protected during the lifetime of the mine and long after its closure. Voids are commonly left after the mining of coal, limestone, gypsum, and other industrial minerals. The location and extent of many abandoned workings are inadequately known, and much effort has to be devoted to identifying them so that appropriate remedial measures may be applied.

Problems raised by the abstraction of groundwater are also common. New developments may cause temporary or long-term modifications to the water regime, or changes in the amount of water pumped and in the location of pumping wells, and have an impact on other facilities and land-use. Over-pumping, with a resultant drawdown of the water table, may lead to a reduction of the quantity of water available for use, but a rise in water level can lead to the damage of buildings, degradation of concrete, swelling and shrinkage in clays, and leaching from strata and landfill sites.

5.1.3. Development work

Any development work consumes land, whether it be forest, agricultural, or urban. In each case, an adequate balance must be reached between environmental protection, the cost, and the objectives. The optimum solution is seldom available and compromise is inevitable. The impact is mainly on land-use, stability, and the quantity and quality of water-resources. It is the task of the decision makers to ensure that adequate long-term measures are taken to protect the environment as much as possible.

Transport routes, roads, and railways are essential, but they are so extensive that they intrude on natural landscapes in places, interfere with ecosystems, and affect water regimes. Furthermore, they generate pollution from the vehicles using them, and from such actions as salting to prevent icing and the application of weedkillers. Likewise, the construction of pipelines for oil and gas causes major disruption to the environment, with special problems relating to the effects on wildlife and fishing grounds at the coastline and offshore, and to the potential for explosions resulting from gas leakages.

Sewage networks underlie most urban areas. Many are old and require major repairs to prevent chronic leakage. These may involve extensive and long-term disruption of the environment by civil engineering operations.

Harbour facilities are extensive on both land and water. Many substances are transported and stored in these areas, and factories are commonly adjacent. Solid and liquid wastes may spread from these areas, and careful control of emission sources is necessary together with remedial actions regarding soil, water, and harbour sediments.

The storage of hydrocarbons and various industrial products presents major sources of potential pollution. For many decades they remained uncontrolled, with the result that nowadays more and more pollution is found in and around such sites. This applies to individual petrol stations as well as to large transport facilities (airports, railway shunting yards, military installations, etc.) and industrial areas. Much of this pollution was caused in the past and surveys now show thousands of contaminated sites, some of them very dangerous.

Careful management of all pollution sources, and planning of systematic remedial actions where pollution affects the physical and biological environment, and our everyday life, are now essential.

5.1.4. Soil contamination, erosion, and reclamation

Since the Second World War, agricultural development has been directed mainly towards increasing production yields. This has been achieved by improving seed quality, cultivation techniques (highly mechanized land preparation and harvesting, and efficient irrigation or drainage systems) and agricultural inputs (fertilizers, herbicides, and insecticides). The results have been positive, but the cost to the environment has been in terms of the changing land-use, the degradation of soil qualities, soil erosion, and water pollution. Finding the adequate balance between production requirements and environmental protection requires an adequate expertise, in which the geological sciences must play a significant role in helping to define adequate policies and remedial actions.

Mountain environments are particularly sensitive to changes in land-use and require the development of specific policies for natural hazards prevention, forest management, agricultural preservation, and tourist developments.

Long-term industrial pollution, including atmospheric transport of pollutants, has contributed to soil and water degradation, and to forest decline, although many other contributory factors are involved.

5.1.5. Disposal of waste in the environment

Normal human activity creates much waste. This term applies to a wide variety of organic and mineral substances, which have a highly variable impact on the environment. It is now policy to avoid the leaching into the environment of harmful substances from waste treatment at incineration plants and at storage sites by restricting the disposal of organic material (domestic waste) by landfill, or by introducing the multibarrier (geological and artificial) concept for the control of potential pollution.

Hazardous industrial wastes and radioactive wastes require special attention, due to their long-term effects. For the former, very strict regulations have been set up by the various European countries in order to control their management. Research is ongoing for the storage of the latter, with emphasis on deep storage in the geological environment for high-level waste.

5.1.6. Groundwater pollution

Groundwater is in many cases the final collecting system for pollution, as much of the surface run-off from precipitation ultimately reaches underground aquifers. As groundwater is a key resource for domestic, industrial, and agricultural use, the adequate preservation of its quality is essential for our survival. This requires work on the whole system of pollution transfer, from the source onwards, including surface water (which represents the final collecting system of all run-off), the hydraulics of which are connected with groundwater to a large extent, and sea water.

Water must be protected against all kinds of pollution, whether from point or diffuse sources, including accidental pollution. The application of geological expertise is essential here to provide information on the geological environment and the hydraulic systems, and to assist with adequate modelling of chemical and hydraulic interactions in the process of pollution transfer, and with the general management of groundwater resources.

Pollution control requires a multidisciplinary approach, as well as expertise provided by complementary organizations. Good control will ensure the protection of our environment and a successful future for our economic development.

5.2. IMPACT OF MINERAL EXTRACTION OPERATIONS AND DEVELOPMENT WORKS [104] [74] [65] [99] [102] [113] [115]

In all countries of Western Europe the demands of the population put enormous pressure on the available land, especially in densely populated and highly industrialized areas. It is the task of regional planning to regulate these needs according to priorities. The land requirements for housing, industrial, and transportation facilities seem to be obvious, and provision for adequate water supplies is essential. A proper long-term evaluation of these resources is a priority, together with the assessment of appropriate sites for waste disposal, dams, power plants, and road and rail routes. To maintain an acceptable balance between these developments and the preservation of nature in general, including areas required for recreational purposes, careful management is necessary at all scales, related to the short-, medium-, and long-term.

5.2.1. Impact of mineral resources exploitation

As described in Chapter 3, mineral resources can be exploited by surface mining methods (sand and gravel, industrial minerals, metals, solid fuels), underground mining (metals, solid fuels, etc.), and pumping at shallow (mostly water and salt) or deep levels (hydrocarbons, geothermal water). The impact of exploitation on the environment varies greatly according to the methods employed in exploitation, the type and extent of mineral resources, the size of operation, adequate pollution control and waste management, and proper reclamation. It is appropriate here to consider these matters in more detail and to give examples to illustrate the types of problems involved.

Surface mining

Surface mining involves large tonnages of economic minerals and waste materials. Alongside policies for the management of the mineral and waste produced, operating companies have to be concerned with the acquisition and reclamation of land.

The problems raised are very obvious in the case of sand and gravel mining which has extensively modified the land-use patterns of many valleys, induced erosion effects on the river banks, and changed the water level and often the water quality.

The impact of dust and noise from opencast mining is also important, particularly in all operations dealing with crushed aggregates used in concrete making, road construction, cement, or lime manufacture. Regular watering and the control of emissions from treatment plants are employed to control dust, and in most countries strict regulations control the permissible levels of noise.

In most European countries, an impact study is compulsory for any mining operation, whether it is new or merely an extension of an existing facility. As an example of a small-size operation, the BRGM study for the tailings dam of the Rouez gold mine is described below.

Rouez gold mine [51] [54] [64] [73]

The operation is run by Société Nationale Elf Aquitaine and is located about 30 km NW of Le Mans. The area is gently undulating and the land is used mainly for small cattle farms. The rocks are altered basement with a varied cover of soil.

The objective of the study was to identify the adequate location of a disposal site for more than 200 000 t of waste from a washing plant. During a first phase, an assessment was made of the area's geology, hydrogeology, landscape, climate, ecology, and socio-economic context (habitat, land-use regulation). For the second phase, the geotechnical stability of the tailings dam was studied and two types of dams were considered (Fig. 5.1):

(1) a low elevation dam across the nearby valley;

(2) a basin enclosed by a bund made of mine waste.

In the first case, more than 14 hectares of land would be used; in the second case only about 5 hectares.

Other studies included:

(1) material available for dam construction;

(2) surface and groundwater run-off;

(3) drainage of waste material;

(4) technical, environmental, and financial evaluation of both types of dam;

(5) site reclamation once the mining operation is over.

The whole study was based on the initial proposal that beneficiation would be processed by leaching in the reactor. A decision to employ heap-leaching instead meant that a second impact study had to be carried out for this smaller operation.

Clearly, the Rouez operation is limited in size, located in an area of low population density and relatively secure from the hydrogeological point of view.

Large-scale operations, which in some cases deal with millions of tonnes of material per year and reach several hundred metres in depth, have a far greater impact and require very sophisticated land management. The type example is the lignite mining in the Rhine area which is amongst the largest operations in the world.

Lignite mining [101]

The largest areas of land used for the mining of superficial deposits in Germany are located in the Aachen–Cologne–Mönchengladbach area, where lignite is mined. The thickness of the seams varies between 10 and 100 m; about 35 billion t could be recovered profitably by open-pit mining. At present, the open-pit mines in operation cover an area of about 88 km^2 with a maximum depth of more than 300 m.

Up to now mining activities have used a total area of 236 km^2. Reclaimed areas total 148 km^2 and are used primarily for agriculture and horticulture purposes (65 km^2). Settlements and industry cover 19 km^2. This means that almost two-thirds of the original mining area have been successfully reclaimed (Fig. 5.2).

A matter of particular concern for the lignite mining industry in the Lower Rhine area is the requirement to work few, but very extensive, opencast mines rather than numerous small ones. This means that one is faced with the socio-economic problems of having to resettle thousands of the local inhabitants. It also means that one must accept further large-scale depletion of groundwater reserves and, ultimately, the task of reclaiming land consisting of vast pits. Therefore, geology, hydrology, geomechanics, and soil science will have to contribute substantially towards finding solutions to the problems before final planning decisions can be made.

Impact of water drainage

Opencast mining of lignite requires the water to be drained out of the sediments above and between the lignite seams. The groundwater level has to be lowered sufficiently below the deepest lignite bed to enable the lignite to be mined.

Altogether, 1300 million m^3 of groundwater have been drained per year from 1955 to the beginning of the 1980s. With decreasing groundwater reserves the amount drained annually has fallen back to 800 million m^3.

The impact of the drainage on the natural equilibrium is a matter of continuous concern. The Geological Survey had to investigate whether the lowering of the water table could induce an inflow of high-salinity waters from deeper rocks or cause other harmful changes in water chemistry and temperature.

Protection of natural biotopes

If the planned Garzweiler II opencast mine south of Mönchengladbach (area 66 km^2, depth up to 260 m, 9500 million m^3 overburden, duration 2004–2045) is approved, a highly important nature reserve on the Dutch–German border will be endangered. Humid biotopes with a rich vegetational and faunal life will not survive if the groundwater level has to be lowered for the planned lignite mine. At present, experts from the mining industry, the Geological Survey, and other institutions are investigating which predicted effects of the mining activity have to be considered as unavoidable and which techniques might help to avoid damage to the natural environment. A continuous irrigation scheme using wells or ditches may possibly be the right remedy to preserve important biotopes.

Hydrogeological and geotechnical considerations

The disused mines are refilled with a rather inhomogenous mixture of gravel, sand, clay, subgrade lignite, and ashes from lignite-burning power plants. This will totally change the previous regime of the aquifers in parts of the area. When—several centuries hence—the groundwater level has risen to its former position near the surface, the water will possibly be of quite a different quality and composition due to complex solution processes.

As a rule, withdrawal of groundwater induces geomechanical effects on the surface strata. Sandy, and in particular clayey, rocks suffer additional compaction. This can often be ignored except where abrupt changes in rock composition occur, e.g. along faults where two different rock types are juxtaposed. Here, lowering of the groundwater level may cause damage to buildings, traffic routes, pipelines, or other installations. As faults are rather numerous in the Lower Rhine area, it is one of the essential tasks of the mining industry, the Geological Survey, and other institutions to determine the position of faults and to predict their possible behaviour when the groundwater level is lowered.

So far, the majority of disused opencast mines have been refilled. Some have been used for waste disposal. Future mining may lead to a few deep pits remaining open permanently. In order to stabilize the pit slopes, a flow of water passing from the cavity into the slopes will have to be maintained for a relatively long period of time. Leading water from the Rhine through a tunnel of approximately 30 km length to abandoned pits has already been envisaged.

In the Lower Rhine area, surface mining of lignite clearly interferes with the protection of the environment as we would like to see it. The existence of settlements, as well as famous old forests, humid biotopes, and other objects is at stake. The last two decades have demonstrated a remarkable decline in willingness to tolerate environmental hazards in densely populated areas like the Lower Rhine. Yet, the Lower Rhine lignite deposit represents a low-cost energy source for decades to come. The current policy of working a few vast opencast mines is probably the only one that is economic. In this connection, the Geological Survey's main objective is to provide information and advice which may be useful in limiting the impact on the environment, and if possible avoiding certain harmful effects altogether.

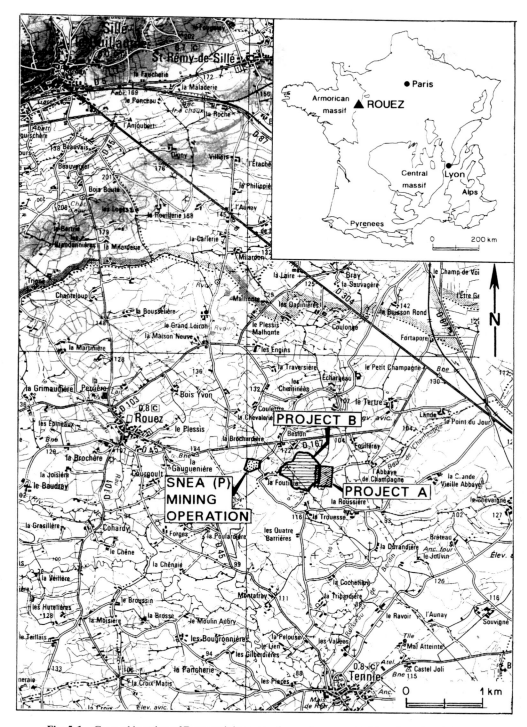

Fig. 5.1   General location of Rouez mining operation and proposed tailings dam sites (France).

The stability of open-pit slopes in the Lower Rhine lignite mining area is currently investigated by the mining company and the results are submitted to the mining authority for approval. The gradients of the slopes derived from stability calculations are of special interest for mine safety. For this reason, the mining authority consults the Geological Survey of Northrhine-Westphalia to obtain expert information on the geology, the structures, and hydrology, as well as on the soil mechanics parameters. The stability of the slopes is determined by means of comparative calculations which are made for selected cross-sections. On the basis of the results, suggestions made by the mining companies—especially with respect to the gradients of the slopes—are either approved or changes are suggested. An opinion on the stability of overburden dumps, which may reach heights of up to 200 m, is also given.

Land reclamation [108]

Lignite has been mined on the Lower Rhine since the eighteenth century. It began in the southern mining district in the Ville area south-west of Cologne. In the meantime, this area has been almost

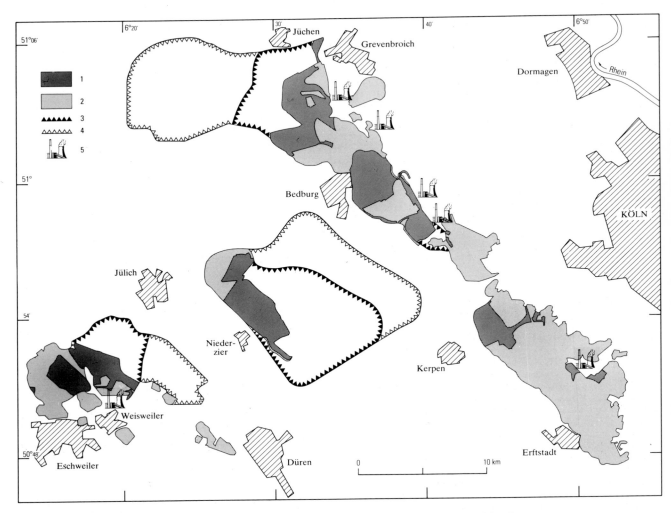

Fig. 5.2   Rhenish lignite mining area (Germany). Location of mined and reclaimed areas.
1. Active mining area
2. Former mining area reclaimed or presently being reclaimed
3. Authorized final boundary of mining
4. Planned boundary of mining
5. Lignite power plant

completely recultivated. In this case there was not much over-burden to be removed. It consisted mostly of sandy–gravelly sediments only about 10 m thick which were almost completely weathered to loam; on the flat parts these were covered by loess loam about 0.5 m thick. As the soils were mostly very compacted and capable of high retention of water, they were used for forestry primarily.

The overburden of the many small open-pit mines was not removed and dumped separately. This led to the formation of new soils which have a higher permeability than the original soils, and provide a better nutrient supply for root growth. These soils were thus quite favourable for afforestation. Commonly, locust trees (*Robinia*) and later white alders and poplars, which grow rapidly and provide protection for the numerous embankments, were the first trees to be planted. Later, these sites were planted with more valuable trees common to the area (e.g. oak, beech, and maple). Thus, the southern mining district, with its large number of lakes resulting from depressions created by the open-pit mining, has developed into a very much frequented recreation area and valuable nature reserve. Agricultural use has returned only on a very small scale.

After 1950, lignite mining in large open pits extended to the mainly agricultural area of the northern mining district (north of the Cologne–Aachen railway line) and the western mining district (near Inden). In the latter district, the overburden consisted of Tertiary marine sands, Pleistocene sand and gravel of the middle and main terraces of the River Rhine and of loess. The loess was mostly 5–10 m thick, and in places more than 20 m.

The overburden in the area of the Hambach open-pit mine east of Jülich differs from that in the northern and western mining districts. The overburden here also consists at the top of 10 m (and more) of loamy Rhine gravels, but the loess overburden is generally only 1 m thick. Loess thicknesses of more than 2 m are unusual. Due to weathering to brown clay during the warm periods of the Ice Age, the soils are mostly very much compacted and sometimes water-logged. Therefore, most of the Hambach open-pit mining area is less suited for agricultural use and is forested.

In general, only Pleistocene sediments are used for reclamation purposes. The quantity and suitability of these sediments depend on the textural composition, lime content, thickness, and the soil type. Because the overburden of loess with its intercalated loamy zones, as well as the loamy sands and gravels of the main terrace, are so thick, it is possible to rehabilitate the areas of the former open-pit mines on a large scale with fertile new soils for afforestation and agricultural use.

Areas for which forestry is planned (in the northern mining district these areas are mostly slopes), are covered with a layer at least 4 m thick consisting of a mixture of loess and loess loam with sand and gravel of the middle and main terraces of the River Rhine. The proportions of the various components of this mixture depend on the angle of the slopes and the quality of the loess (percentage of lime); however, the stability and the water permeability of the slopes must be guaranteed. This mixture develops into new fertile soil which is afforested with a mixture of trees that is often similar to the original indigenous forest.

Due to its grain-size and properties in a mixture with loess loam, loess is particularly well suited for reclamation of an area for agricultural use. It is available in sufficient quantities in thick layers in the northern and western mining districts. It is therefore possible to plan and prepare new fertile cropland on a large scale in these districts. The large amounts of loess in the northern mining district make it possible to rehabilitate large former open-pit mines in the southern mining district for agricultural use, where only little fertile loess soil existed before, and thus reduce the considerable loss of arable land in the mining district as a whole. According to the so-called Loess Agreement of 1961, a cover of at least 2 m of loess and loess loam must be prepared in the northern and western mining districts, whereas in the southern mining district only at least 1 m of loess and loess loam are required because of the long transport distance. The material underlying the loess must be permeable to water and should consist of Pleistocene sand and gravel.

In the northern mining district, loess and loess loam are spread using conveyor belts (dry method). In the southern mining district, the flushing method proved to be very successful. Loess and loess loam are mixed in a 1:1 ratio and transported in a slurry through pipes into areas of about 3 ha surrounded by loess embankments. A number of pipes and short transport distances are necessary to avoid separation of the loess into different grain fractions. Most of the water is drained off, the remainder seeps into the ground or evaporates.

The quality of the soil can be considerably improved by thorough and careful tillage during the first few years. Deep-rooting plants (e.g. alfalfa) are grown during this time, the soil is loosened and enriched with humus, and soil organisms are introduced so that a favourable porous soil structure may develop gradually. Moreover, to compensate for the initial nutrient deficiency, proper crop rotation and proper application of fertilizers are recommended. As these measures require a good deal of experience and specialized equipment, this work is being done by the mining company for the first seven years.

The reclamation of the open-pit lignite mines has provided new soils that, in terms of agricultural yield, are hardly inferior to the original soils.

At the European scale the impact of surface exploitation is very significant and, although the case of lignite mining described above is very exceptional, there are numerous cases of bulk material mining which necessitate careful planning and management. The whole environment is concerned (water, air, natural biotopes, land use, etc.) and the need for appropriate studies and control at the right time can never be emphasized too much.

Underground mining

The impact of underground mining is generally less than that of surface mining. It is usually selective and a significant part of the waste is dumped back underground, depending on the mining method used. The greatest potential risk is from the beneficiation operations which aim at upgrading the ore into a concentrate. This usually requires the use of added chemical

Fig. 5.3 Location of the Chessy mine area (France).

products which largely remain with the waste itself and are often potentially harmful (iron sulphides, arsenic, etc.).

The example of an impact study being developed presently is given by an operation at Chessy, 25 km NW of Lyon (Fig. 5.3).

The Chessy Mine [67]

To satisfy French law, the environmental impact study of the Chessy operation covered:

(1) the mining operation itself and the tailings stockpile, including a tailings dam requiring approval by The Permanent Technical Committee for Large Dams;

(2) the industrial plant located in the mine area.

Annual ore production planned is 300 000 t, i.e. 1000 t d^{-1}, of which 250 000 t is to be transported to the beneficiation plant. Annual production of concentrates is copper (26 000 t), zinc (43 000 t), pyrite (126 000 t) and baryte (51 000 t). On average, the grain-size of these concentrates ranges between 60 and 75 μm (see also case history in Section 3.5, p. 000).

Water requirements, mainly for the beneficiation plant, are 115 m^3 h^{-1}, including partial water recycling (about 25 m^3 h^{-1}). Waste water amounts to approximately 127 m^3 h^{-1}, including 14 m^3 from mine de-watering and 0.3 m^3 of very acidic water from the old mine.

The potential impact of waste water is reduced by several steps:

(1) storage in the tailings dam leading to decantation of suspended material and biodegradation of beneficiation reagents (flotation);

(2) cleaning in 3 basins located downstream from the dam. According to needs, waste material which does not reach required standards can be kept stored. pH will be adjusted as required;

(3) in case of lower river yields, temporary storage in the tailings dam is considered.

Measurements for waste quality control will be by continuous pH control downstream from the dam, and analysis of the major cations present in the ore (cadmium, copper, zinc, iron) will be made once a month.

Rainwater will be collected separately and sent off to the stream network. Workshop and laboratory waste water will also be sent to the tailings dam. Domestic water will join the Chessy municipal sewage system.

Solid waste material from the mine and material ripped from the dam foundations will be used throughout in building the dam. Pyrite from the old mine will be stockpiled together with former tailings and dumped back into mine galleries or re-used for stope filling.

Solid waste material from the beneficiation plant amounts to 80 000 t per year or 38 000 m³. This material will be stockpiled over 20 hectares which will be re-vegetated after mining has ceased. Atmospheric emissions from the ore stockpile, transport, loading and unloading of ore, and crushing will be reduced to a minimum by proper site planning. Noise will also be minimized through equipment siting and operating control and work timing.

The Chessy example shows that the various aspects of environmental impact control have been evaluated and planned in order to achieve maximum safety from harmful substances, minimum disturbance from noise and dust, and careful planning of an industrial installation with waste disposal sites.

Subsidence due to mining

Subsidence caused by mining operations presents a major problem in many industrialized areas of countries like Great Britain, Belgium, France, and Germany, where seams of coal have been exploited for centuries and salt has been leached underground. Shallow mine workings lead to surface subsidence, creating changes not only in the topography, but also in the hydrological cycle. Flooding may occur as a result of aquifers being breached. Deep-mining water from aquifers may flood workings at lower levels. Mining subsidence can have serious structural effects on buildings, transport systems, canals, and even facilities occupying large areas, such as airports.

The collection of geological and mining data, and geological mapping of potential subsidence areas are necessary as a basis for interpreting these man-made hazards. Where insufficient historical records are available to identify areas of potential subsidence, other methods, like the use of remote sensing data and interpretations derived from engineering geological and geophysical surveys, will increase the value of the investigations cited above.

Coal mining in Europe has been so extensive that it is a major cause of subsidence in a number of areas. Impact is a maximum in heavily built-up or industrialized areas. Normally surface subsidence due to mining is unwelcome and costly. There is, however, an interesting example in Germany, where a controlled rate of subsidence was achieved by appropriate mining methods.

Duisburg-Ruhrort port subsidence [104]

The Duisburg-Ruhrort port (Fig. 5.4) at the convergence of the rivers Ruhr and Rhine is the world's largest river port. Progressive river bed erosion caused the level of the Rhine to drop more than 2.7 m during the last 60 years. For many years the level has fallen by 4 cm annualy, and the depth of water in the Duisburg-Ruhrort port has varied at a similar rate. At times of low water a 150-year-old lock ran completely dry. As the utilization of the port was increasingly hampered at times of low water and as dredging was possible only on a minor scale, the harbour authorities reached an agreement with the coal mining industry in 1951: mining of seams underneath the port area was carefully planned to achieve a lowering of the bed of the harbour. Within 15 years the port area was lowered by up to 2 m without inflicting damage on jetty walls or harbour installations in order to ensure the unimpeded future operation of this important river port.

The city of Essen in Germany provides a good example of how a city is involved in investigating former coal mining areas.

City of Essen reconnaissance work [77]

The city of Essen (Fig. 5.4), in the Ruhr area has its own Geological Department which is actively involved in the careful mapping of former coal mining operations. As a result of coal mining over several centuries, the foundation rocks of half of the city, 10 × 15 km in area, are like a sponge cake. The superficial deposits are a few metres thick and of Quarternary and Cretaceous age.

Although most of the former mining companies have merged into a single company, many of the old mine plans are no longer available and reconnaissance work is necessary to locate abandoned workings. Several methods are used:

(1) shallow auger drilling (5–6 m) to locate coal seams below the overburden;

(2) drilling (10–20 m) to locate the extension of coal seams and mined areas at depth;

(3) zoning in 4 classes of potential hazard on 1:2500 cadastral maps.

Owing to the great extent of the hazardous area the reconnaissance work is not carried out systematically over the whole area of potential problem, but rather on a case-by-case basis according to needs. The mapping is then supplemented by recommendations, e.g. for concrete filling of mined sections, special foundation types, and so on.

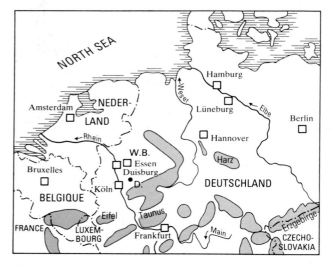

Fig. 5.4 Location of Duisburg, Essen, and Lüneburg areas (Germany). W.B.: Westphalian Bight; D.: Dhünn reservoir.

Subsidence also creates problems in the iron mining area of north-eastern France. Although the mining is carried out 150–250 m below the surface by room and pillar with controlled collapse, subsidence has been noted in a certain number of areas. Land-use plans at the commune level provide specific regulations for these potentially dangerous areas.

The mining of salt by solution is also a tradition in the same area. Two mining methods are used: hydraulic fracturing and the track system. In the former system a central well is used for injection, and brine collection is done in a series of peripheral wells. In the track system the injection and collecting wells are disposed in a profile. Both systems induce ground subsidence and the mining areas are very strictly controlled and protected.

The Lüneburg area, located to the north of the Harz mountains in Northern Germany (Fig. 5.4), provides a type example.

Salt working at Lüneburg [92]

The part of Lüneburg built during the Middle Ages is over a salt dome. The top of the salt is at a depth of 40–70 m. The dome is covered by a cap rock of gypsum. The diapiric Mesozoic strata have been pushed into a nearly vertical position (Fig. 5.5). Brine has been 'mined' from the top of the diapir since the Middle Ages. In this century, brine has been extracted from wells.

This has caused irregular subsidence. In some areas, such as at the margins of the salt dome, there has been horizontal compression and tension at ground level, causing damage to buildings. The problem is as old as the city of Lüneburg, but it has been aggravated by increased extraction of the salt. Subsidence has reached rates of as much as 7 cm yr^{-1}. In addition, sink holes are forming as a result of karstification of the cap rock.

Whereas the ground level is sinking, the water table has remained nearly constant. This has resulted in an apparent rise in the water table and the flooding of basements. More than 170 buildings had to be abandoned between 1640 and 1970. Detailed geological, geophysical, and structural engineering studies by the Lower Saxony Geological Survey (NLfB) in the early 1970s showed that most of the old part of the city can be restored if certain construction measures are adopted. The historic, old part of the city has been extensively restored on the basis of the results of these studies. Future hazards were reduced by the termination of mining in late 1980. Since then, the subsidence rate has been reduced to about one tenth of the previous rate.

The Lüneburg case illustrates the conflict between economic interests and the effects on the surrounding area. The geoscientist can only show the situation as it is. The solution to the problem lies in the political arena. Because areas of potential subsidence are common, geologists, geomorphologists, and engineers should work closely together on both basic and applied research on the mechanics of subsidence so that predictions can be made for structural engineering purposes.

Subsidence due to the former underground quarrying of construction material is also very common, although usually at a smaller scale than described in the examples given above. These quarries are concentrated in certain areas: Ile de France, North Lower Normandy, Aquitaine, and Loire, etc., in the case of France. Quite often their location is unknown or poorly defined. Most of the quarries date back to the Middle Ages, but some of them were still operating in the nineteenth century.

As these quarries are mostly located within urban areas or close by, they are responsible for the deterioration of many buildings. It is the task of Geological Surveys, amongst other bodies, to carry out underground quarry inventories and propose safety controls.

Abstraction of hydrocarbons [218]

Depletion of reservoir pressure by the withdrawal of fluids (hydrocarbon production) may lead to compaction of the reservoir followed by subsidence of the overburden. This is particularly characteristic of shallow, relatively unconsolidated reservoirs, e.g. Wilmington oilfield (California), Orinoco Oil Belt (Venezuela), Lake Maracaibo area (Venezuela), and the Groningen gasfield (The Netherlands). Unexpected sea bed subsidence has also occurred in the Ekofisk oilfield in the Norwegian area of the North Sea.

Reduction in pore fluid pressure leads to an increase in the pressure on the rock matrix as the pore fluid pressure no longer helps to sustain the weight of the overburden. According to calculations, a drop in reservoir pressure of 1000 psi (6.9 MPa) would result in a decrease in the volume of the reservoir of 0.7 per cent. Such compaction is an important mechanism of primary recovery (compaction drive). Contraction of the reservoir is accommodated in the vertical direction simply by subsidence of the overburden. The horizontal stresses generated, however, may result in faulting and seismic shocks. Small shallow earthquakes have been recorded near the Goose Creek oilfield (Texas), Wilmington oilfield (California), Strachan oilfield (Alberta, Canada), and the Pau Basin gasfield (Western Pyrenees, France).

Surface subsidence of greater than 6 metres has been detected over the crests of the Wilmington (California) and Lagunillas (Venezuela) oilfields; in both of these fields the reservoir comprises relatively unconsolidated Tertiary sediments. The subsidence bowl developed above the Wilmington field has the same shape and location as the reservoir closure, and the centre of the bowl lies directly above the crest of the structure. Subsidence was successfully stopped through a major water injection programme to restore and maintain fluid pressure in the reservoir. In addition, the water injection has been a great economic success, increasing production in the old area of the field due to water-flood stimulation.

Subsidence in the Ekofisk oilfield, North Sea

Sea-bed subsidence in the Ekofisk area (North Sea) was first identified in 1984 and by mid-1985 a maximum of 2.6 metres of subsidence was measured. Rates of subsidence have been estimated to be between 0.3 and 0.7 m yr^{-1}. Reservoir compaction as a result of depletion of reservoir pressure by the withdrawal of hydrocarbons is believed to be the cause of this sea-bed subsidence.

The 3 km thick overburden exerts a pressure of 9000 psi (62.1 MPa) on the top of the chalk reservoir in the Ekofisk field. The original reservoir pore pressure was about 7000 psi (48.3 MPa) but this declined to 3500–4000 psi (24.15–27.6 MPa) as a result of oil production. The net pressure on the chalk therefore increased from 2000 psi to 5000–5500 psi (13.8 to 34.5–37.9 MPa). The high-porosity chalk reservoir responded to this increased stress by elastic deformation, resulting in pore collapse. It has been estimated that 65–85 per cent of the Ekofisk reservoir compaction has shown up as sea-bed subsidence, yet effective reservoir permeabilities remain basically unchanged after 14 years of production. The sea-bed subsidence recorded had not been expected in the Ekofisk area since it was thought that the overburden would have had a far greater natural bridging effect. Pressure maintenance is the only way to control compaction in the chalk and thereby limit subsidence. Large-scale gas injection began in 1985 in order to maintain pressure in the Ekofisk Formation and ameliorate sea-bed subsidence; a water-injection programme was scheduled for 1987.

Ground deformation by exploitation of geothermal fields

Subsidence usually occurs in zones directly linked to the areas of extraction or at a certain distance from them. Subsidence rates vary from a few millimetres to a few metres. Consequently, it is important to monitor horizontal and vertical surface deformations by means of topographical techniques on networks of permanent benchmarks covering the entire area under exploitation and the surrounding terrain.

Subsidence from geothermal exploitation, Larderello and Travale (Italy) [128]

In Italy, most of the geothermal activity on an industrial scale is managed by the Italian National Electricity Board (ENEL) and is located in the regions of Tuscany and Latium. Data on ground

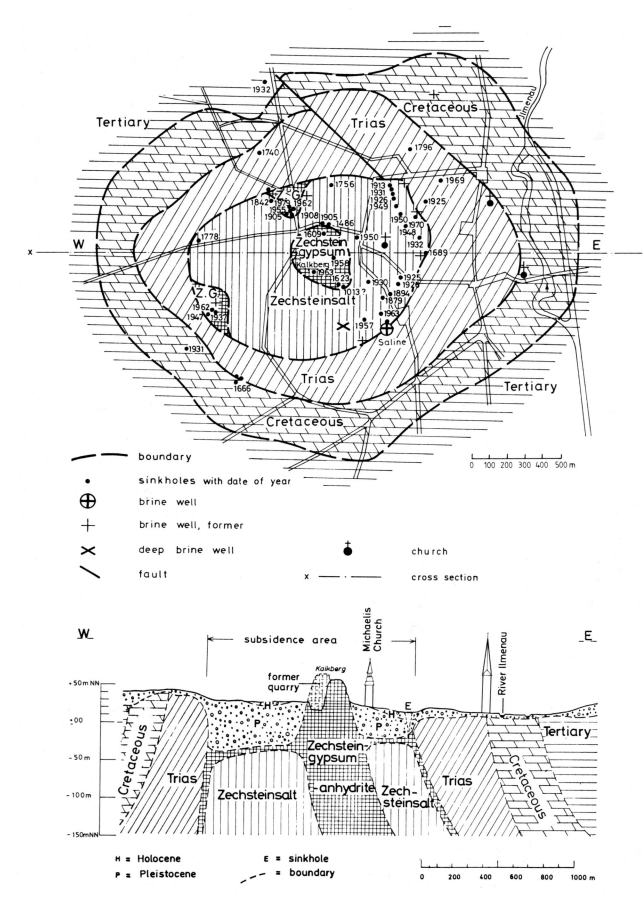

Fig. 5.5   Surface geological map and section of the Lüneburg diapir (Germany).

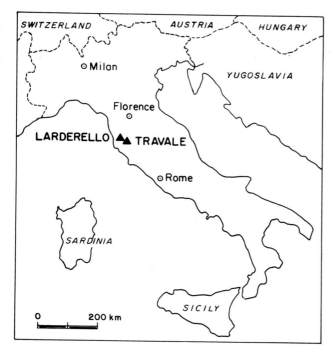

Fig. 5.6 Location of Larderello and Travale geothermal fields (Italy).

deformation have been gathered since 1921 in the Larderello geothermal field and since 1973 in the Travale geothermal field. Since 1985 ENEL has carried out a specific monitoring programme of repeated topographical and gravity precision surveys in all the Italian geothermal fields it manages. So far, meaningful data have been obtained in the Larderello and Travala geothermal fields only (Fig. 5.6).

The networks currently under observation in Italy consist of permanent vertical benchmarks measured at least once a year. The datum point is set up outside the geothermal fields (normally several kilometres from the outermost productive wells) and on geologically more stable terrains. The highest density of benchmarks is in the most intensely exploited areas, decreasing towards the periphery. On the average, there is one benchmark installed for every 300–500 m of levelling line. The lines are mainly run along roads and constitute closed circuits.

Drilling began in the Larderello geothermal field in 1838, but penetrated only a few metres of impermeable cover rock. It was only around 1920, when the wells reached the carbonate rocks reservoir, that fluid extraction from the subsoil posed problems in terms of subsidence. In 1921–23, a series of high-precision topographical surveys was carried out in the Larderello field to monitor any ground deformations caused by intense exploitation, then in its earlier stages. The first control was made in 1985 during the installation by ENEL of a new and more rational network of high-precision levelling in the most densely exploited area of the field, where re-injection of large quantities of water was programmed to recharge the geothermal field. The maximum ground subsidence (about 170 cm), measured in 1985 and 1986, occurred near the village of Larderello in the most intensively exploited area of the field. This corresponds to an average rate of about 2.5 cm yr^{-1} for the period under consideration (1923–86). The phenomenon can be attributed to a drop in reservoir pressure of the order of 2.5 MPa.

The new surveying network described above is gradually being extended to the whole of the Larderello geothermal field. The data gathered from over 140 permanent benchmarks show that, fundamentally, subsidence is continuing, but at a rate of less than 1 cm yr^{-1}, and only in the marginal areas where new wells have begun production. The area that has been exploited for longer now seems to be stable.

Intense exploitation of the Travale geothermal field effectively began in 1973. Prior to that date only a marginal area of the field had been exploited, with relatively small fluid production rates. In 1973 a network of about 180 permanent benchmarks was set up for precise topographical monitoring.

The subsidence trend during the first 15 years of observations shows clearly that there is a coherent and immediate relationship between subsidence and the principal field operations, such as entry into production of new wells and/or the closure of old wells. Fig. 5.7 represents the subsidence trend along a caracteristic SW–NW section crossing the geothermal field.

Exploitation of geothermal fields inevitably leads to ground subsidence of some extent. The phenemenon cannot be eliminated, but it can be reduced by injecting water into the reservoirs and/or regulating fluid production in time and in space.

Abstraction of water and rising groundwater table [56]

The abstraction of groundwater (or surface water) obviously results in the lowering of the water table concerned. The amount of drawdown and its extent depend upon the volumes abstracted and hydrogeological characteristics of the aquifer system:

(1) natural recharge conditions (rainfall, rivers);

(2) conditions of storage and circulation of surface and groundwaters (in particular, voids between soil particles, fractures, or karstic systems in rocks).

The variations of the quantities abstracted induce 'artificial' variations of the piezometric level, which are superimposed upon seasonal or longer-term 'natural' variations (essentially due to variations in rainfall and/or in the water levels of the rivers).

Thus, increasing abstractions lead automatically to the lowering of the average piezometric level (taking account of natural fluctuations), whereas continual reduction of withdrawals eventually causes a rise in that level.

The effects of the variations in groundwater abstraction are very diverse.

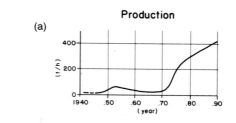

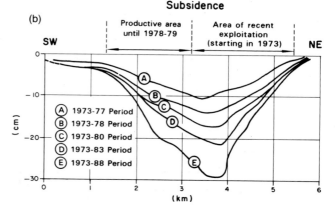

Fig. 5.7 Subsidence in the Travale geothermal field (Italy).

London

The case of London is a good illustration. The basin on which London is built is composed of compact clays (with a thickness of up to 60 m), with overlying beds of sands and chalk, which constitute a deep aquifer, recharged by the Thames and from outcrops to the north and south. In the middle of the nineteenth century, the water level in wells sunk down to this aquifer was close to the surface.

Continued increase in industrial, public, and private pumping led to a lowering of groundwater level, locally by as much as 70 m. This drawdown itself brought about a settlement of the compact London clay of about 10 cm, or even 25 cm locally.

Since 1970, owing to a decrease in abstraction, the level in the deep aquifer beneath the centre of London has continuously risen at a rate of up to 1.50 m per year. This rise causes a tendency for the clays to swell.

A major research programme by the Construction Industry Research and Information Association (CIRIA), completed in 1989, studied the effects of this rise and proposed:

(1) hydrogeological modelling of critical sectors;

(2) reinforcement of foundations which are likely to be affected by clay swelling (estimated cost of £10 m);

(3) repair and reinforcement work over 36 km of tunnels (£2 m–£35 m);

(4) preventative means for new buildings and tunnelling (£4 m);

(5) continuous monitoring of the groundwater level and extending the existing observation network.

Paris

The case of Paris and its suburbs is little different from that of London (except for the hydrogeological context). The combination of several factors had led to a groundwater rise of 10 m or more since about 1970, including:

(1) reduction in industrial, public, and private abstraction;

(2) sealing of existing wells by grouting;

(3) termination of pumping for major construction works in the 1960s (underground car parks, subways, 'Les Halles' market);

(4) improving the watertightness of old subway tunnels.

Many basements built during the 1960s, at the time when the groundwater was at its lowest level, or was still falling, are now suffering damage as a result of cracking and flooding. Some are left flooded, others are de-watered (the pumped water is discharged into the sewerage system, which is thus overloaded). In other cases, costly sealing (or even reinforcement of the whole structure) is necessary. All this occurs mainly in the area of the alluvial plain of the River Seine.

In certain areas, there is a risk of renewed dissolution of soluble beds (gypsum).

Another example can be taken from the district of Hamburg (Germany).

Hamburg [90]

The Hamburg area can be subdivided into two hydrogeological units: the Elbe flood plain and the Geest area. The groundwater in the area of the Elbe estuary is confined under Holocene tidal mud deposits and peat up to 10 m thick. At high tide the hydraulic pressure increases, lifting the flood plain near the Elbe dyke by as much as 3 mm. This complicates considerably the calculation of the hydrogeological balance.

In the Geest areas, the groundwater table is mostly unconfined. Locally there are numerous horizons containing perched ground-water (extending from about 50 m to more than 2 km) affecting groundwater regeneration. Within Hamburg, these perched aquifers are usually residual aquifers remaining after the lowering of the table which was originally at ground level. In some places the water table is still at ground level; these marshy areas are of great ecological value and are often nature reserve areas.

As a result of the long-term abstraction of large amounts of the groundwater, cones of depression 10 m deep and more have developed around wells in the unconfined aquifers of the Hamburg waterworks and various industries. Such local anomalies, as well as lowering of the water table over large areas of urban development, have drastically reduced the area from which groundwater enters the receiving streams.

The water table in Hamburg is monitored in 2272 observation wells. Monitoring began in the 1950s but most of the wells have been drilled since 1965. Recorders are installed at only a few wells now, and measurements are made every two weeks. The hydrographic curves obtained in this way do not present a uniform picture, but help in identifying various relationships.

As a whole, the groundwater regime in an urban region is complicated. Considerable variations in the water level occur in the wells at the waterworks. General management of these variations is nearly impossible for various technical reasons: temporary closure of pumping systems for short periods of time, shifting of production from one well to another, usually with screens at different depths, and uncertainty about the actual amount of water extracted by industry and from private wells. Moreover, the monitoring system itself and the frequency of measurement are still not quite adequate. This has led to new methods of monitoring the hydraulic pressure and improved modelling of the unconfined aquifer at a grid spacing of 25 m. The model is being continuously updated on the basis of new water table data so that up-to-date data is available for large parts of Hamburg.

Another problem of importance for Hamburg is the depth of sewage systems in relation to groundwater. Due to leaks in the pipes, groundwater may be drained into the sewage system or pollutants may spread into the groundwater. The rise of the water table in waste disposal areas is also a problem as it may result in leaching out of pollutants. Significant efforts are made, however, to maintain suitable drainage conditions.

Rural areas

A specific example of impact on agriculture is in the province of Hesse, Germany. The fractured Tertiary basalts of the Vogelsberg constitute an important aquifer. In the 1960s, two main waterworks were developed near the town of Nidda for the potable water supply of greater Frankfurt.

Due to excessive abstraction, there was an appreciable lowering of the groundwater level. As a large number of buildings in or near Nidda experienced water-induced settlements, the abstraction was reduced to the point where the damage was stabilized. During the period of excessive withdrawal, certain areas initially saturated became relatively dry and were used for agriculture, but this came to an end as the water level continued to rise from 1974–80.

Mining areas

The rise in the water table in the coal basin of northern France (and also the Saint-Etienne basin) is considerable as a result of:

(1) mining subsidence: up to 15 m ground displacement;

(2) cessation of mine de-watering, industrial crisis (reduction in water abstraction).

In addition to the flooding of basements in many towns and villages, streams reappear in certain dry valleys, and ground which was marshy before coal mining started but was built upon subsequently, becomes saturated or flooded.

Ponds and lakes created when major subsidence occurs are often developed as recreation areas. Buried structures (foundations, pipes, tunnels) may be corroded by rising groundwater enriched in sulphates.

The cost of pumping to prevent damage to houses amounts to 130 000 FF per year for the commune of Wingles (8500 ha) in Pas-de-Calais, for example, and to 450 000 FF per year in the commune of Lourches (5000 ha) in Département of Nord.

Coastal areas

Large communities situated on river deltas (like Venice, Bangkok, Shanghai) suffer severely from the consequences of groundwater abstraction. Abstraction is increasing rapidly in order to satisfy both industrial and potable water requirements. Furthermore, the subsoil in river deltas generally includes large thicknesses of weak and compressible beds (silts, clays, or peat).

Settlement of up to several decimetres occurs in these weak beds, inducing subsidence with the following consequences:

1. The risk and the extent of flooding by the river or/and the sea increases, especially during spring tides.
2. As the buildings are generally founded on 'floating' piles (with bearing capacity due to the friction between the pile and the ground), the risk of deformation or even rupture of the piles increases because of the 'negative skin friction' effect, i.e. the weak soils which are settling by subsidence cause a drag on the piles, considerably increasing the load.

Various solutions, including the control of groundwater abstraction and the utilization of deep aquifers, have been put into effect in most of the cities concerned, but safeguards must also be arranged against the further risk of aquifers being polluted by saltwater.

In areas of high water table, subsidence may result from the abstraction of surface water to increase the thickness of the unsaturated zone ahead of surface developments. If the strata are weak layers of peat and clay, compaction may take place. Since these impacts on the geological environment are well known, models have been developed to simulate the settlement effect of lowering the surface water levels in rural and urban areas. These are available for conversion into potential costs giving planners and decision makers appropriate tools ahead of development

5.2.2. Impact of development works

Any construction work has an impact on the geological environment, on the ecological system, and on land-use patterns. The impact increases in relation to the scale of development and the vulnerability of the environment. Careful study of impact and planning is necessary, even for small-scale operations.

Large-scale operations range from lengthy corridors such as roads and canals to large surface areas like dam reservoirs or polders. In former times, the tendency was to deal with the geotechnical aspects only of these large-scale operations, but for many years now it has been the common practice to consider all environment aspects (physical and socio-economic) of a new development. The study includes not only the area of the project itself, but also the management of the material extracted and re-used in the construction work. Groundwater

management is also essential for the safety of the construction, for its future operation, and for the protection of the groundwater resource.

A multidisciplinary approach is essential in the management of the impact of large developments. Although it may not be possible to arrange for no impact, it must be minimized even by introducing constraints into the developments.

Railways and roads

Detailed knowledge of the geology and hydrogeology of the foundation conditions, and of the availability of building materials is essential in planning the construction of transport routes, especially roads and railways. Two examples follow.

New railway line—Hannover to Würzburg (Germany) [93]

The new railway line between Hannover and Würzburg extends to 133 km within Lower Saxony (Fig. 5.8). The route includes 33 km in 15 tunnels, 8 km over 96 bridges, 35 km in cuttings, and 45 km on embankments. Extensive site investigations were necessary and were much facilitated by the Geological Survey being involved at an early stage of planning. It was possible to select the best route for the line with respect to groundwater and landscape protection. Extensive geological investigations including stratigraphic correlations and geological profiles contributed to safety during construction. Difficulties arose from hidden faults and ground leaching or subrosion (karst cavities with residual material and collapse breccia), but it was possible to overcome these on site.

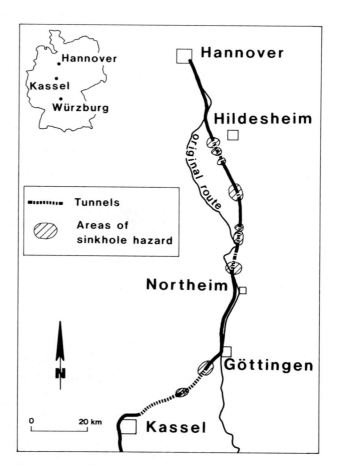

Fig. 5.8 Location of Hannover–Würzburg railway line (Germany).

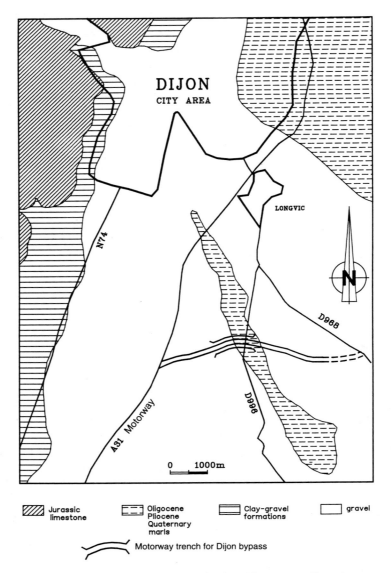

Fig. 5.9   Location of Dijon bypass for the A31 motorway (France).

To give an example of one of the very many environmental studies carried out for road impact evaluation, the case of the A31 motorway in Dijon South (France) has been chosen.

Dijon bypass of A31 motorway [37]

The main topic to be investigated was the vulnerability of the alluvial aquifer of the former Ouche Valley (Fig. 5.9). This aquifer has two layers. The upper one called Perrigny is unconfined, the lower one called Sansford is confined. They are separated by an interval of sandy clay.

The Perrigny aquifer has a general north–south flow and is largely exploited for drinking water. It is very vulnerable to surface pollution. However, the water catchments presently used are located upstream from the freeway and were not to be harmed by the construction. The Sansford aquifer is well protected by the sandy clay layer.

In 1988, digging of the motorway trench reached the top of the Perrigny aquifer creating overflow of the water towards another aquifer (Longvic) situated east of a marly high, and which corresponds to the present day Ouche Valley. Overflow was in the range of 20–30 l s^{-1} and induced lowering of the water table to the west and 30 per cent yield reduction in Sansford spring water.

Early in 1989, construction of a temporary clay dam prevented further seepage and provided the necessary time to supply adequate filling material for the motorway construction. Another potential problem came up at that time, however. After the construction of the dam, the aquifer was at the surface and became more vulnerable to pollution. However, adequate water sampling demonstrated that no pollution was present.

Phase 2 of the programme consisted of finding an adequate replacement for the temporary dam. Two possibilities existed: an impervious screen or a dam using existing material on site. The choice of the former resulted in the following studies:

(1) geological reconnaissance to optimize the location of the screen;

(2) definition of the top altitude for the screen;

(3) installation of a piezometer network to monitor the groundwater level in the vicinity of the screen.

Canals [85] [74]

Canals are of great economic importance as they enable the long-distance transport of voluminous goods at fairly low cost. But they have a high level of impact on the landscape and the hydrological cycle. Forests, farmland, transport corridors, and settlement areas may be interrupted or may have to be abandoned. Prior to construction, the geology, hydrology, and the hydrogeology of the canal routes must be explored accurately. Locks are necessary to overcome topographical barriers. The founding of such installations demands skilled engineering. Furthermore, there is the danger of seepage of water, when canals intersect karstified areas and seals on the canal's bottom and flanks are inadequate. Building canal walls also presents safety problems. Stability must be guaranteed to prevent mass movements into the canal, and also on to the adjacent area. The latter may lead to flooding.

Rhine–Main–Danube Canal

One of the most recent constructions is the Rhine–Main–Danube Canal which passes (Fig. 5.10) through southern Germany. This canal will connect the rivers Main and Danube and also the Black Sea with the North Sea. Countries like Austria and Hungary will get direct access to the sea, thus relieving the overburdened land transport routes to south-east Europe and partly across the Alps.

Detailed exploration by mapping, geophysics, and especially drilling was necessary to determine the most favourable route for the canal. Triassic and Jurassic strata with cover rocks of Quaternary age were the main rock types encountered during the exploration and construction phases. The Bavarian Geological Survey has been involved as an adviser since the 1960s. One of the major geology-related problems of canal construction was the existence of ancient movements of the Opalinuston (a clayey member at the base of the Middle Jurassic) in the area south of Nürnberg. It was necessary to study these former landslides, especially to investigate their behaviour when overlain by younger sediments and therefore subjected to an additional load. Groundwater monitoring and measurement of sliding masses were carried out in relation to excavation and measures for stabilization were proposed. The type, volume, and the duration of stabilizing measures had to be checked in order to stop existing movements and prevent new ones.

The canal is part of a supra-regional water management system, as water from the river Danube and its tributary the Altmühl is delivered to the catchment area of the rivers Main and Regnitz in the north, a region which suffers from a lack of water. The canal's advantages therefore are not only as a means of transport, but also as contributing to the stability of the water regime of the rivers Main and Regnitz, the drinking water supply, the hydrological preconditions for heat producing power plants, the withdrawal of water for irrigation and other production purposes, the water supply for two storage dams, and the tourist facilities of the area.

The impact of the canal on the environment will be great, but various measures have been taken to minimize this as far as possible. These include topographical modifications, the introduction of varied vegetation, the maintenance of wetlands, and the construction of footpaths, bicycle routes, and tourist landing stages.

Dams [114]

Sound investigations are essential for the selection of a dam and reservoir site. Ecological aspects and the economics of water supply and distribution also play an essential role. In every individual case, the basis for the planning, construction, and monitoring of the structures is an engineering geology assessment related to the type of construction and the intended use. The construction and operation of new structures may be impeded by the presence of karstic phenomena,

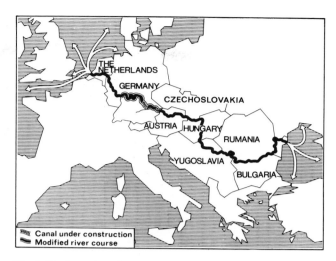

Fig. 5.10 Location of the Rhine–Main–Danube Canal (Germany).

soluble rocks (salt, gypsum, anhydrite), and extensive landslip areas, as well as by abandoned or planned mines and possible earthquakes.

Dhünn Reservoir Dam

The large Dhünn Reservoir Dam (Fig. 5.11) is situated about 25 km north-east of Cologne (Fig. 5.4). With a volume of 81 million m³ it is the largest drinking water reservoir in Germany. The 60 m high rock-fill dam with bituminous core sealing has a crest 400 m long. The rock-fill volume is about 1.2 million m³. The entire project including the main dam, 16 foredams, a 2900 m long bypass tunnel and the drinking water facility were built between 1975 and 1985.

The foredam with a core sealing was completed as early as 1964 to create a reservoir for drinking water. This required 6 core drillings totalling 270 m and water pressure tests. A continuous observation trench was dug out following the course of the cutoff wall. After completion of the main dam this former dam lost its original function.

The reason for the construction of the new Large Dhünn Reservoir Dam was the growing demand for drinking water. The engineering and geological site investigations were carried out during 1971 and 1972. At the major dam site nine core borings with water pressure tests were drilled to a depth of 65 m. Eleven deep trenches were dug out along the future cutoff wall. A further five core borings served for the investigation of appropriate dam fill material. The tunnel was investigated by seven core drills totalling 578 m. The bedrock of the major dam site and of the reservoir area is of Middle Devonian age comprising alternating clay, silt, and sandstone beds with local calcareous intercalations.

The results of the geological investigations carried out by the Geological Survey of Northrhine-Westphalia formed the basis for the layout of the concrete buildings, the dam structure, the location of the quarry and the construction of the bypass tunnel.

During the construction phase the foundation pits were investigated and their geotechnical data documented at a scale of 1:100. Grouting was carried out from the inspection gallery of the cutoff wall in two steps. The first step included a grouting wall 6–15 m deep at pressures of 2–5 bar with a bore spacing of 0.75 m. The high degree of tightness achieved in this bedrock section subsequently permitted higher injection pressures for the deeper rock section without risking displacements of the concrete construction. Thus the grout wall was deepened to 40 m with bore spacing of 1.5 m at a maximum pressure of 10 bar. To monitor any displacements during the grouting work five-point extensiometers were installed in the inspection gallery with a registration range to a depth of 50 m.

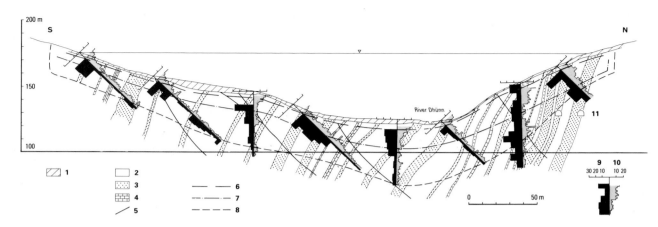

Fig. 5.11   Section in the Dhünn reservoir dam area (Germany).

1. Slope talus and loam, valley fill, terrace alluvium (Quaternary)
2. Siltstone, shale (Middle Devonian)
3. Sandstone, fine-grained (*idem*)
4. Calcareous rocks (*idem*)
5. Fault
6. Foundation depth of cut-off wall

7. Depth of grout curtain (bore spacing approx. 0.75 m)
8. *idem* (bore spacing approx. 1.50 m)
9. Results of hydraulic pressure tests
10. Number of separation planes per metre
11 Bottom outlet

The assessment of stability and function of the main dam was supported by a large monitoring programme. The testing stage of 1984 and 1985 was also monitored and finally assessed from a geotechnical point of view.

A speciality of the Large Dhünn Reservoir are the so-called eco-basins which have a purely ecological function. As their constant water level cannot be disturbed by varying levels of the main reservoir they serve as refuge biotopes for water birds and lacustrine plants.

In addition to the involvement in the construction of new reservoirs, a large amount of time of the Geological Survey is dedicated to stability assessments of old structures and recommendations for strengthening or partial reconstruction work.

Out of 70 dams in operation in Northrhine-Westphalia, 22 were constructed before 1917. Of these, 20 are concrete gravity and gravity-arch dams. As a result of the law concerning the water resources in Northrhine-Westphalia dated July 4th 1979, dams have to be built, maintained, and operated according to the general standards of engineering. When constructions do not fulfil these regulations, necessary modifications have to be made.

To study the foundations, knowledge of the subsurface is necessary. For old dams the influence of the load of the dam and reservoir must be considered together with possible changes in the ground stability and hydraulics. For most of the 20 old dams, old and fragmentary documents concerning the geology and the foundation system were the only ones in existence.

Since 1980 fifteen sites have been investigated by the Geological Survey of Northrhine-Westphalia. From two dam sites boring samples with a diameter of 400 mm were drilled in order to determine the deformation potential of the foundation rock.

To determine the rates of subsoil deformation and the location of rock discontinuities a large triaxial shear-test device was developed by the Geological Survey. It facilitates the examination of cylindrical test bodies with a diameter of 400 mm and a length of 1000 mm. This new test allows adequate study of the behaviour of the bedrock underneath

the dam foot and general evaluation of the present state of the dam.

Polders

Polder reclamation raises complex problems connected with the stability of the ground and changes in the hydraulics: hydrodynamics evolution, salt water–freshwater interface. A typical case history is the impact of the construction of the new polder 'Markerwaard' in Lake Ijssel, in Central Netherlands (Fig. 5.12).

Markerwaard polder [141]

The Markerwaard polder extends over an area of more than 100 km². The reclamation work carried out had specific effects:

(1) drop in groundwater pressure and change in groundwater flow-direction;

(2) ground subsidence associated with compaction of soft clay and peat beds due to vertical drainage;

(3) impact on the foundations of houses from the small cities bordering Lake Ijssel; the damage was evaluated at US$200 m;

(4) decrease of moisture content of the topsoil inducing lower crop yields; the damage was evaluated at US$125 000 annually;

(5) damage to ecology: changes in species distribution.

Fortunately, modern hydrogeological engineering techniques enable mitigation of these negative effects of polder construction. The most promising measures all have in common the compensation of the drawdown of the piezometric levels by artificial recharge of the topmost aquifers. Three potential countermeasures were studied in detail:

(1) injection wells/recirculation systems;

(2) infiltration grooves;

(3) infiltration wells.

The effects of these countermeasures on the hydrogeological system and the costs of damage to buildings were calculated using a combination of models. From these calculations it was proved that injection wells and recirculation systems are the best options.

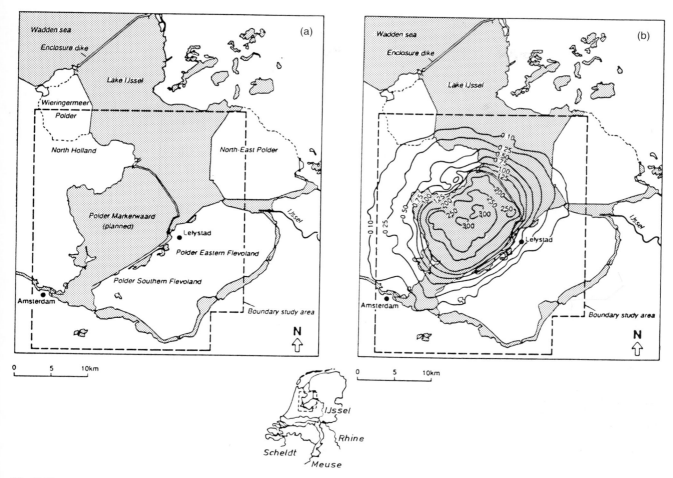

Fig. 5.12 (a) Location of the planned Markerwaard polder, which is part of Lake Ijssel, in relation to the earlier reclaimed polders in The Netherlands (Wieringermeer Polder, North-East Polder, Eastern Flevoland Polder and Southern Flevoland Polder). (b) Contour map of the predicted drawdown of the piezometric level in the first aquifer after reclamation of the Markerwaard Polder.

The detailed geological model provided by the Geological Survey of The Netherlands has been very valuable for the calculations of damage costs and the effects of countermeasures. However, the Dutch government has decided recently to delay the construction of this polder.

Megacities [74]

In 1800 only 3 per cent of the world's population lived in cities. By the turn of the present century the urban population will be at least four times greater than that for 1950. Indeed, by the year 2000, the urban population will outnumber those of rural areas in Western Europe. These demographic changes have led to considerable environmental impacts especially with the development of megacities.

In former times, residential areas were predominant in the centres of cities. Living in the central parts of megacities became progressively less attractive with the increase in trade and rising industrial development, and as a result their suburbs have become of increasing importance for housing and recreation. Thus, the function of the city centre has changed drastically to become a locus for administration, production, and consumption of goods and in providing a multitude of services, including a broad range of cultural activities.

To ensure easy access, communication networks have had

to be planned and constructed, frequently over and through complex subsurface geological conditions. The construction of such as subways and tunnels has considerable impact on groundwater regimes, as has the prevention of natural seepage of surface waters as a result of the sealing effect of buildings and roads. The temperature of groundwater rises and the quality is adversely affected. Such problems lead to the costly construction of drainage systems.

Adjacent to the sea, coastal protection and flood prevention must be an integral part of town planning. The water supply must be protected from the intrusion of salt water. Similarly, megacities situated close to or above areas of underground mining are prone to subsidence hazards, and careful planning and the designation of zones of protection are necessary.

Another hazard with severe environmental impact is the seepage into the subsurface of a great variety of effluents including the leakage of fuels and sewage, and percolation from inadequately controlled production sites. The latter involves many liquids, often toxic, containing dangerous hydrocarbon compounds and a variety of heavy metals. The generation of large volumes of solid waste and sewage sludge makes the establishment of geologically-safe disposal sites absolutely essential (see Section 5.4.1).

In the Mediterranean area, megacities are subject to earthquakes and volcanism, hazards which must be taken account of in development planning.

The planner must evaluate all these elements in balancing the functional and spatial requirements of a megacity, and in minimizing its impact on the environment. The overlap of development involving the multiple use of resources such as groundwater and aggregates must be avoided, as should unnecessary demands on farmland, forest, and nature reserves.

Environmental geology and regional planning for Greater Hannover

Greater Hannover (Lower Saxony) is situated on the border zone between the German midlands and the north-German lowlands (for location see Fig. 5.8). It is at the stage of developing into a typical megacity. It has an area of 2287 km^2, a population of 1.1 million, and is responsible for 20 per cent of the gross income of Lower Saxony. A special institution, the Zweckverband Grossraum Hannover (Administration Union for Greater Hannover) was established in 1980 to deal with the diverse pressures of industrial, commercial, and residential developments.

The targets to be considered are the growth expectations for population and employment as well as the changing requirements of population and economy. Additionally, all planning considerations must be integrated into a network of decisions which will be linked to environmental protection. The Geological Survey of Lower Saxony was asked to advise from its extensive databases on the use of natural resources.

As great parts of the area (up to 60 per cent) are agricultural land, the way of using the soil, the intensity of fertilization and the possible change from agricultural use to such as forestry and housing, were the main points of decision-making. Areas which are sensitive to erosion and flooding had to be checked both with regard to countermeasures and for their potential as nature reserves. Such problems had to be integrated into planning activities, supported from soil surveys under the heading of 'consolidation of farmland'.

Two-thirds of the area of Greater Hannover is covered by Quarternary sediments, the remainder by Mesozoic strata. The Pleistocene, with its considerable reserves of sand and gravel, is the source of raw materials and a major reservoir for drinking water. A system of evaluating raw materials was created. This resulted in establishing areas of first and second priority for the extraction of mineral resources. These reserves are of great importance for the future, in supporting the growth of development in the urban area.

The same can be said for the protection of water resources. The surface run-off has to be controlled for water quality, which can be severely affected when sewage treatment plants are working inefficiently. Rising water consumption is a constraining factor, together with the discharge into the sewage network of a variety of pollutant chemicals from households and commerce and public institutions. Groundwater resources are important for the supply of drinking water, because they are protected against pollution by cover strata with low permeability. Within Greater Hannover, about 80 per cent of groundwater is supplied from local waterworks. The rest is imported from surface reservoirs in the Harz Mountains. The task of planners, together with geologists, was to establish where the abstraction of groundwater had priority over the extraction of mineral resources. As a result, special protection areas for groundwater were established.

The efficiency of soil and groundwater protection depends to a large extent on how waste disposal is managed. Some 356 abandoned waste disposal sites contain materials ranging from garbage to special waste. For the latter, special measures are required to stabilize and neutralize harmful components. Today there are three central installations to which waste has to be delivered. This is mainly garbage from households and commerce, sewage sludge, and building rubble. Recycling will be favoured to reduce the volume of waste, because surface disposal is limited for environmental reasons and by the shortage of hydrogeological barriers underground.

In summary, the environmental and economic development of Greater Hannover has been possible as a result of significant input from the Geological Survey of Lower Saxony. There is enough space to guarantee the controlled expansion of housing as well as the careful use of mineral resources and groundwater. The need for the construction of new highways, and a new railway to the south, and the selection of a favourable site for Expo 2000 south of Hannover, is covered by the recently established 'Regional Planning Programme 1990'.

RECOMMENDED FURTHER READING

Arendt, F., Hinenweld, M., and Van den Brink, W. J. (eds) (1990). *Contaminated soil*. (Kluwer, Dordrecht).

Bosse, H. R., Brinkmann, K., Lorenz, W., Roth, W., Barth, W., Pauly, E., *et al.* (1982). *Karte der Bundesrepublik Deutschland 1:1 000 000, Gebiete mit oberflächennahen mineralischen Rohstoffen*. (Bundesanstalt für Geowissenschaften und Rohstoffe, Hannover).

Browne, M. A. E., Forsyth, I. H., and McMillan, A. A. (1986). Glasgow, a case study in urban geology. *Journal of the Geological Society*. **143**, 509–20.

Canter, L. W. and Hill, L. C. (1981). *Handbook of variables for environmental impact assessment*. (Ann Arbor Science, Michigan).

Chadwick, M. J. and Nindman, N. (1982). *Environmental implications of expanded coal utilization*. (Pergamon, Oxford).

Collective Authors (1983). Utilisation de l'espace souterrain. *Annales des Mines*, Nos. 5–6.

Deutsche Gesellschaft für Erd- und Grundbau (ed.) (1973). IAEG Symposium on sink-holes and subsidence. Engineering-geological problems related to soluble rocks. (International Association of Engineering Geologists, Essen).

Eggert, P., Hübner, J., Priem, J., Stein, V., Vossen, K., and Wettig, E. (1986). Steine und Erden in der Bundesrepublik Deutschland— Lagerstätten, Produktion und Verbrauch. *Geologisches Jahrbuch, D82*. Hannover.

Legget, R. F. and Karrow, P. F. (1983). *Handbook of geology in civil engineering*. (McGraw-Hill, New York).

Mitsch, W. J., Bosserman, R. W., and Klopalek, J. M. (eds) (1981). *Energy and ecological modelling*. (Elsevier, Amsterdam).

Rau, J. G. and Wooten, D. C. (1979). *Environmental impact analysis handbook*. (McGraw-Hill, New York).

Solomans, W. and Forstner, U. (1986). *Environmental impact and management of mine tailings and dredged materials*. (Springer, Berlin).

UNEP (1983). Environmental guidelines for the restoration and rehabilitation of land and soil after mining activities. *UNEP Environmental Management Guidelines No. 8*. (UNEP, Nairobi).

Union Nationale des Producteurs de Granulats (1986). L'eau continentale et les carrières. *Compte rendu de la journée nationale du 4-12-1986; Coll. technique No. 5.*

Usbeck, H., Herrmann, H., and Leistner, F. (1990). Structure and dynamics of land-use in metropolitan regions: the example of Leipzig. *GeoJournal* **22**, 2, 143–51, (Dordrecht, Boston, London).

Whiteside, P. G. D. (ed.) (1987). The role of geology in urban development. *Bulletin Geological Society of Hong Kong*. No. 3.

5.3. SOIL CONTAMINATION, EROSION, AND RECLAMATION [65] [71] [70] [74]

5.3.1. Overview

Since the human race passed from the stage of hunting and collecting to that of agricultural production, the impact of agricultural practices on the environment has progressively increased. This situation is due not only to changes in demography which required increasing quantities of food, but also more and more to changes in agricultural practices and to increased production yields. These trends have grown significantly since the Second World War.

The increase in yields has gone in tandem with the increasing use of fertilizers, herbicides, pesticides, mechanization, increasing surface area of cultivated plots, etc. It is only during the past few decades that we have become aware of the negative effects of this policy.

The increasing difficulties for the storage and marketing of agricultural products and the impact on the environment of agricultural practices is particularly relevant at the EEC level. Modern methods lead to excessive requests for land, or even degradation of its internal qualities with regard to agricultural production and to the quality of groundwater resources.

Under the term 'land', which includes the notion of soil within a specific landscape and climate, we must consider the whole of the physical and biological surface environment, which is characterized by complex interactions between the mineral base, gaseous and liquid fluids, and the biological component. At the biological level, the interactions between these different media lead to specific ecosystems which differ mainly as a function of climate, water regime, availability of nutrients, and existing living species.

In current use, the term 'soil' is used for the medium giving crop support. It is more restrictive than the term 'land'. It is characterized at the physical level (profile morphology and texture), as regards associated water regime, and from the chemical point of view (pH, redox potential, capacity of exchangeable bases, etc.).

Land is also affected by industrial pollution, localized, diffuse, or precipitated in a dry or humid state after being transported by wind.

In addition, toxic substances are found in the hydrological cycle (surface and groundwater), partly in the climatic cycle and in the food chain. Adequate management of our agricultural land is thus of particular importance and management concerns in particular:

(1) a well thought-out policy for the spreading of fertilizers, herbicides, and pesticides;

(2) an adequate policy for agricultural practices together with consideration of the best layout for agricultural areas, to prevent soil erosion: the soil, it must be emphasized, is a non-renewable resource;

(3) the control of industrial emissions.

Throughout this chapter, we shall discuss how our knowledge is still inadequate in relation to the nitrogen and phosphorus cycles (and associated metals such as cadmium), the degradation and persistence of pesticides and herbicides, and the management of land in compatibility with the needs of economic development and the requirements of environment protection.

The management of agricultural land must be integrated into the general management of the global environment, particularly to avoid desertification due to the uncontrolled development of agricultural land by the destruction of stabilizing forests, and to prevent sterilization or perturbation of agricultural land through urban or infrastructure development.

The trend toward integration is having an influence on the selection of parameters used to describe the quality of agricultural soils. Soil scientists have traditionally used soil classifications orientated towards genesis. These are now being replaced by reference systems based on soil morphology and its relation to geology. Thus the notions of gley, rendzina, podzol soils are tending to disappear to be replaced by more general criteria of thickness, general composition (clay, sand, etc.), associated water regime, acidity, and so on.

This evolution is related to the strong wish to integrate a specific agricultural land area within its broader physical environment. Nowadays the requirements of production increase at any cost are being replaced progressively by optimized quality criteria, with fewer inputs, a better economic balance, together with greater concern for environmental protection.

5.3.2. Management of agricultural and afforested land

For this integrated topic, comments related to agricultural practice, surface management, and the supply of substances needed by agriculture are grouped in the following section. In practice, various procedures are carried out simultaneously with the objective of optimizing agricultural production. The management of forestry raises specific problems which will also be considered.

Problems related to the use of agricultural input substances: mineral fertilizers, manure, pesticides and herbicides

The primary objective in the use of these substances is to provide an increase in the crop volume and quality. The fertilizers supply nitrogen, phosphorus, and potash, three components essential for plant growth. Other elements can be added in minor amounts for specific or high yield crops. Pesticides and herbicides are used to eliminate harmful species (fungi, bacteria), destructive pests (insects mainly), and weeds. Some products are designed to have a general effect, but others are specific to species. The choice of the quantities to be applied is, from the technical point of view, a function of the physical and chemical state of the soil and of the nature of expected crops. In advising farmers, the manufacturers, professional bodies, and official laboratories rely on a long tradition of agricultural practices and experimental trials.

Organizations involved in regional management and planning are not concerned with the evaluation of the quantities used per type of product and administrative unit (commune, county, department, etc.). Studies aimed at the determination of a global input–output balance for a specific substance or a specific element are even less common.

Nitrogen balance in the agricultural areas of Northrhine-Westphalia [111]

Such a study was carried out by the Geological Survey of Northrhine-Westphalia (Germany). It became apparent over the past 30 years that the quantity of fertilizers used in this area had increased three-fold whereas yields had improved by only 50 per cent.

During 1987 alone, the 1.6 million hectares of agricultural area received 207 kg nitrogen per hectare, two-thirds being mineral fertilizers and one-third manure. The needs of most cultivated crops* are in the range of 130 kg per hectare, which leaves a surplus of at least 70 kg per hectare. This situation led the Geological Survey to evaluate the risk of nitrogen loss in the soils.

---

* 27 crops covering 97 per cent of agricultural land, of which 47 per cent for cereals, 30 per cent for grass, and 20 per cent for silage.

The quantity of mineral fertilizers has been evaluated from the Agricultural Chamber's statistics and questionnaires. The manure supply was calculated from statistics of land occupation per type of cattle. The supply of atmospheric nitrogen has been estimated at an average of 50 kg N per hectare per year. The asymbiotic fixation of nitrogen by algae and bacteria was estimated at 30 kg per hectare per year. It has been calculated from the surfaces allocated to each plant that the largest quantity of nitrogen relates to plant uptake. Study of denitrification shows a nitrogen loss of 30 kg per hectare per year in gaseous form.

The overall calculation is based on the assumption that the organic nitrogen balance of the soils is in equilibrium.

Despite the uncertainty regarding various parameters, the evaluation conducted in Northrhine-Westphalia has the great merit of proposing a tentative balance evaluation confirming a surplus of nitrogen supply. Caution is necessary, however, concerning the conclusions drawn from such a study. Optimum supply of nitrogen does not necessarily coincide with precise plant uptake evaluation. Furthermore it is quite obvious that such a calculation must be integrated into a global economic balance, which takes into account not only the requirements of agricultural production, but also the needs for environment protection.

In considering a variety of largely conflicting interests this concept of economic balance is of particular importance. In the first place, the balance takes account of the prices for agricultural products, which provide a basis from which the farmer tries to optimize his input expenses and to anticipate a range of yields.

It is obvious, however, that the scales to be considered for agricultural production or environment protection are not the same. At the level of the individual farmer the evaluation is made on a seasonal basis, even if for certain inputs like phosphorus and potassium fertilizers the accumulation factor in the soil is fundamental. For the sake of environmental protection, it is necessary to estimate the short- medium- and long-term effects of adding substances.

The problem is clear in the case of nitrates for which the cycle is essentially seasonal or annual, with reference to a specific supply, but which persists in groundwater for several decades. This problem is dealt with more extensively in Section 5.5.

The pollution problems originating from animal manure, mainly from pigs, are of particular importance in certain areas (The Netherlands, NW Germany, Brittany). Manure is both an animal waste and a nitrogen fertilizer. Although its storage is more difficult than in the case of mineral fertilizers, the present-day management policy is mainly based on seasonal storage and spreading, including manure banks. Several trials of manure treatment have been made, though they have remained mostly at the pilot stage. The presence of copper and zinc in manure has also led to specific regulations, in the US for example, as to the content of those elements in animal food.

Domestic waste water treatment results in water which is usually sent off to the natural drainage system and sludge which has to be disposed of. Quantities involved are quite large and much of it is spread over agricultural areas. The CEC has set up regulations as to sludge disposal. These mainly concern the balance between the heavy mineral content of the sludge (cadmium, mercury, lead, zinc, chromium, nickel, copper) and that of the soil on which the sludge is spread. Part of the sewage sludge is landfilled.

Phosphate fertilizers also create problems for the environment. They contribute to water eutrophication together with the phosphates derived from detergents through waste water. Phosphates are also rich in cadmium, a pollutant which to some extent may be taken off by plants. Hence the tendency in certain countries to try to decrease significantly the level of cadmium tolerated in phosphate ores. This creates market difficulties as cadmium removal processes (calcining or humid extraction) are not extensively used.

In the field of agricultural inputs, it is therefore very important that the different partners involved, among them the Geological Surveys, coordinate their work so as to optimize the use of these substances.

This is particularly the case for pesticides and herbicides, whereas the problem raised by excessive use of nitrogen fertilizers has been clearly identified, and has even led to an EEC norm for drinking water (50 mg l^{-1}). The knowledge of the behaviour of pesticides in the environment is still primitive. This knowledge only partly overlaps with the results obtained from standardized product authorization procedures and corresponds mainly with laboratory tests and short-term effects.

Furthermore, despite the efforts of EPA (USA), OECD, UBA (Germany), and the European Council, we are still far from a standardization of procedures for evaluating the impacts of the use of these products. This is particularly the case in regard to the soil ecosystem which controls the first steps of their dispersion in the environment.

Today it is apparent that among the several thousands of pesticide and herbicide molecules existing in the market, only several hundreds are referred to in monitoring reference lists (of which several tens are forbidden), and several tens have been selected for detailed studies (still partial). The three main criteria for evaluating the environmental impact of pesticides and herbicides are toxicity, persistence, and mobility. Therefore, only some of the several thousands of molecules create problems.

Our knowledge of the degradation and *in situ* transfer processes is still largely insufficient, and there is still a great lack of analytical work to confirm the presence of pesticides in the environment. The question to be addressed is the establishment of an overall balance of selected pesticides and herbicides in relation to sorption–desorption processes in soil (clay, organic matter) and water particles, degradation, volatilization, leaching, etc. Equally poorly understood are the effects on terrestrial or aquatic faunas and microfloras. The promotion of clearly biodegradable products is now advocated, and this would bring remedies to some of the problems presently raised.

Soil erosion and agricultural practices

The topic of soil erosion was not dealt with in Chapter 4. It is actually not a risk in the same way as the landslide risk, for example. However, the impact of human activity is fundamental with regard to soil erosion, even if this phenomenon has a natural component.

Following continuous rainfall, sudden storms, or snow melt, torrents of surface water cause widespread soil erosion in mountainous areas. The dynamics of the associated fluid system have been described in Section 4.2.5, which deals with flood hazards.

In the present chapter emphasis will be on erosion related to agricultural practices, together with the precipitation regimes. Traditionally, studies have been carried out in mountainous and hilly areas, where slopes are fairly steep, with particular reference to the problems of vineyards.

During the last 10–20 years, much work has been done in less steep areas, which has also led to the identification of significant erosion. It is therefore appropriate to distinguish areas with marked relief in central and southern Europe, and specific erosion problems occurring in the extensive agricultural plains of central and northern Europe.

Erosion areas of marked relief of central and southern Europe

These areas are characterized by a relative spatial coincidence between run-off and erosion surfaces. This means that the surfaces undergoing erosion are also the ones on which run-off occurs.

This is the hypothesis formulated in the Universal Soil Loss Equation (USLE) proposed by Wischmeier and Smith. In this equation, the soil loss is a function of soil erodibility (K factor), climatic aggressiveness, topography (intensity, slope length, and shape), vegetation cover, and agricultural or anti-erosion practices.

Following these guidelines, Schwertmann and his team carried out research into relevant methodology in Bavaria. This project was carried out on 32 reference soils on which the sensitivity to erosion factor has been studied since 1974 over four test zones. The result of this project was the adaptation of USLE to the case of Bavaria and the estimation of a tolerance factor in tons per hectare and per year on the basis of a steady production potential over a period of 300–500 years. A further study on 900 soil profiles, scattered over the whole of Bavaria, showed the close relation between ploughing depth and erosion.

On the scale of the European Community (EEC), the experimental programme CORINE (Coordination of the Information Concerning the Environment), carried out between 1985 and 1990, included a project on 'soil erosion and important soil resources'. With reference to the applicability field of USLE, this project concerned only southern Europe: the Iberian Peninsula, southern France, Italy, and Greece. The southern France operation is taken as an example of this programme.

CORINE experimental programme [58]

The French Group (INRA, BRGM) worked on two test areas, Lodève in the Montpellier area and Nice, and in the whole of southern France south of Bordeaux. The general programme coordination was by AQUATER (Italy). Each of the test areas cover approximately 250 km². On each of them, four thematic maps were established: slopes, vegetation, soil erodibility, and climatic aggressiveness index. In the original CORINE methodology, the combination of the various factors was made by a cumulative process. After a preliminary phase, it was apparent that parameter aggregation was more useful and this was carried out in two phases:

(1) establishment of a potential risk map by multiplying erodibility classes, climate and slope;

(2) establishment of an actual erosion risk map by using the vegetation parameter.

This method was adapted from USLE standards and the results were presented at the scale of 1:1 000 000. The soil resources evaluation was more specifically performed by INRA at the scale of 1:1 250 000 over the whole of southern France, with subsequent scale reduction to 1:1 000 000.

The methodological choices and the work scale were based on the availability of basic data in the various countries involved. The results show that this approach is useful for a global zoning, the identification of the most sensitive areas and the awareness of the economic factors. Further work should be carried out at more detailed scales (1:25 000 to 1:100 000) in sensitive areas and using adequate information at this scale: geology, pedology (interpreted in terms of soil erodibility), land use, and so on. Such maps rarely exist over large areas.

Soil erosion in the large plains of central and northern Europe

The areas concerned support intensive agriculture and have a smooth relief and extensive loessic cover. In these areas the formation of surficial sealing and crusting decreases infiltration and induces significant run-off leading to downstream soil erosion and degradation of water quality (turbidity and chemical elements supplied). This impact is very different in nature from the erosion processes described for areas of marked relief, and is based on the spatial separation between run-off and erosion areas. Despite weak relief contrasts the effects are significant in relation to the soil and water resources.

The areas to which these processes apply are not located in northern Europe only. There is some overlap with areas in which the USLE predominantly applies. From a global point of view the difference between these two types of erosion is intimately connected with slope differences in relation to the precipitation regime. The USLE is not well adapted to soil erosion problems of large agricultural plains for the following reasons:

(1) emphasis on factors weight rather than interrelationships: for example the progressive degradation of the soil by rainfall;

(2) standardized measurements on 22×4 m plots do not allow risk evaluation due to concentrated run-off;

(3) length of time necessary to adapt the USLE parameters to a context different from that of the US (case of Bavaria already mentioned);

(4) degradation not limited solely to the soil loss.

Soil erosion in large plain areas leads to a range of degradations:

(1) at the level of crop plots: plant extraction, seed losses, seed coverage by transport of material, decrease of the soil water reserve, and soil losses;

(2) in the field of infrastructure: accumulation on roads, filling of sewage networks, and undermining of roads;

(3) as to water quality: turbidity resulting on filling and blocking of streams, eutrophication by phosphorus, and nitrogen pollution.

The question of responsibility for degradation commonly arises. The problems raised are not necessarily of the same nature; neither are they of simultaneous occurrence. There is room for debate, for example, on the allocation of responsibility between upstream agricultural developments and downstream urban areas.

In order to prevent erosion, pilot projects have been conducted in the northern part of France in particular with three main objectives:

(1) an agronomic objective, through research and promotion of the most adapted agricultural practices: allotment, machinery traffic, soil-working methodology, intercropping methods, and liming;

(2) a hydraulic part, concerned with finding remedies for problems raised by concentrated run-off and gullying;

(3) a landscape component, based on soil erosion analysis in connection with landscape evolution. Specific management systems may be proposed such as the digging of water ponds.

Remedial approach

The solution of this general problem of erosion has to be considered at various levels. At the scale of individual plots, appropriate plot, distribution, and crop allocation may lead to erosion reduction. It is therefore appropriate to limit the length of surface run-off, avoid surface channels, alternate infiltration and run-off plots, minimize soil consolidation and vehicle tracks, etc.

The adequate management of periods between crops is also essential and three main types of system are at present on use:

(1) surface conservation of crop residues (straw);
(2) ploughing after cropping so as to regenerate soil infiltration capacity;
(3) sowing of 'green fertilizer'.

The last method has the advantage of minimizing the leaching of nitrogen fertilizers not used by the previous crop.

Furthermore, the structural stability of soils, which is mainly related to soil texture and the content of organic matter, may be increased by the addition of lime, which reduces soil acidity.

In order to reduce erosion problems it is thus necessary to deal with problems at all scales, from the individual cultivated plot to a whole catchment area. The disadvantages of re-alottment operations are clearly underlined, as in the case of France. Here the main objective considered was the consolidation of agricultural land surface, not the protection of the physical environment, in particular the preservation of regulating ecosystems (protected biotopes, ecological zoning) and the main water circulation channels.

Marginal land set aside, without re-use for other purposes, leads to the creation of fallow land, and of forest in the long term. In the meantime this may facilitate the introduction of undesirable flora and fauna in the neighbouring cultivated areas. Proper management of such marginal lands is therefore a necessity, particularly in mountainous areas. An equilibrium must be found between cultivated and non-cultivated areas, including marginal land and agricultural land used for other purposes. 'Set aside' does not mean mere withdrawal as in the case of fixed fallow; it may involve a rotation system in which fallow is only one part.

These general considerations are valid for all types of agricultural land. Although the following case history is specific to the peatlands of northern Germany, it provides a good example of the problems involved in relation to infrastructure policy (re-allotment, drainage, etc.).

Impact of cultivation techniques in northern Germany [96]

Cultivation techniques have a long tradition in Germany. Although they were initially used for reclaiming arable land, they were also applied for soil improvement when land became increasingly scarce. The objective of soil amelioration up to and during the early 1970s had the general effect of intensifying agricultural use (yield increase and improvement of the load capacity of the soil for machines). There is now increasing use of cultivation techniques for the rehabilitation of soils.

Cultivation of former coastal marshlands

Owing to the steadily increasing sea levels, dikes have been built along the North Sea coast since the tenth and eleventh centuries. In the meantime they have been raised to about 7.5 m above sea level. In Germany, dikes protect about 800 000 ha of marshland and bogs at elevations between 2 m above and 1 m below sea level.

Cultivation techniques are applied even on the seaward side of the dikes: ditches are dug at intervals of 10–20 m with the excavated material piled up between them. This has been done to drain and desalinify the sediments which, in time, and with continual deepening of the ditches, will finally mature into soil. This cultivation technique has been applied for hundreds of years. The long, narrow, raised strips of land between shallow ditches is still characteristic of marsh landscape, which is used primarily as grassland (pastures).

After the Second World War, cultivation measures were promoted within the Regional Development Programme of the federal Government. A pre-requisite for planning the successful large-scale implementation of these measures was a pedological map (scale 1:25 000) covering all marsh areas, and thematic maps with recommendations for land-use and amelioration. These maps were compiled by the Geological Surveys.

Electric pumping stations were installed, improving drainage to such an extent that pipe drains could be laid. This made it possible to fill the ditches and level the strips of land between them. Although intensive agricultural utilization has profited considerably from these measures, they have had damaging consequences for the natural environment.

Since the beginning of the 1980s, priority has been given to protection of the environment. In the middle of the 1980s, all projects for the reclamation of arable land were stopped, and since the end of the decade the government-promoted measures for drainage of marshy areas have also ceased.

Reclamation of peatland

While nutrient-rich fens (lowmoor) have been cultivated since the Middle Ages, systematic settling of nutrient-poor raised bog (highmoor) areas began only during the eighteenth century. At present, 98 per cent of about 700 000 ha of the lowmoor areas in Germany and about 65 per cent of the 450 000 ha of highmoor are cultivated.

Fens can be cultivated immediately after the first open-ditch drainage. In contrast to nutrient-rich fens, nutrient-poor acidic highmoors can be cultivated only after application of lime and fertilizers for amelioration. Lime and nutrients are mixed with the peat to a depth of only 20 cm. Using the 'German raised-bog cultivation method' developed by Bremen Peatlands Experimental Station (Moorversuchsstation Bremen), about 300 000 ha of highmoor were reclaimed during the last hundred years, mostly before the Second World War.

Although this peatland cultivation technique was applied primarily to improve the soil for agricultural use, soil protection was also a consideration in the development of the German highmoor (raised bog) cultivation method. The objective of this method was to eliminate the destruction of peatland caused by burning off peat before cultivation, a common practice at that time.

Peatlands are sensitive systems which inevitably degrade with increasing duration of cultivation, especially with the intensifying of drainage and agriculture. The degradation processes (Table 5.1) can be slowed, though not completely prevented, by low-intensity agricultural use (grassland, grazing animals less than the equivalent of two cattle per ha) and leaving the groundwater table as high as possible.

Another method for stabilizing organic soils is to mix the peat cover with mineral soil. Sand-covering methods require special machines which dig up sand from the substrate (down to 3.5 m) and distribute it as a 10–20 cm thick layer at the surface.

In the case of highmoor soil, the sand layer is mixed with peat at a ratio of 1:2. Peat is not mixed with the sand cover for lowmoor sand-cover cultivation, avoiding rapid oxidation of the nutrient-rich peat.

Table 5.1 Impact of cultivation of lowmoor and highmoor soils (within the table, increasing drainage intensity and intensity of use are from top to bottom)

Lowmoor	Highmoor
Loss of thickness due to settlement	
up to 30 per cent	up to 50 per cent
Decrease in permeability due to settlement	
Loss of peat due to biological oxidation	
Up to 1 cm yr^{-1} if used as grassland	up to 0.5 cm yr^{-1} for grassland, low-intensity use
up to 2 cm yr^{-1} if used as cropland	up to 1 cm yr^{-1} for grassland, high-intensity use
	up to 1.0 cm yr^{-1} for cropland
High nitrogen mineralization (up to 1600 kg N ha^{-1}) and denitrification; Risk of groundwater pollution by nitrate; Pollution of the atmosphere with N_2O	Risk of elevated rate of leaching of phosphorus
Formation of horizons with a crumb structure due to shrinkage and increase in resistance to wetting lead to deterioration of the water balance in the soil (e.g. due to interruption of the supply of groundwater via capillary action)	Considerable decomposition of the topsoil leads to the formation of pellicular water. Decrease in load capacity of the soil.

The above-mentioned cultivation techniques result in cultivation areas of varying stability in the following order: fen cultivation < German raised-bog cultivation < lowmoor sand-cover cultivation < application of sand by machines < deep-ploughing sand-cover cultivation < German deep-ploughing cultivation. A general summary of impact of cultivation of lowmoor and highmoor soils is given in Table 5.1.

Drainage and irrigation

Numerous examples show that drainage is one of the main operations in preparing for agricultural use. In the case of West Germany, it has been estimated that 51 per cent of the potential agricultural land should be drained. In France, about 100 reference drainage sectors, scattered over the whole of the country have been studied between 1980 and 1986. This programme has contributed to the definition of drainage operations on the basis of a well-documented methodological approach.

The objective of drainage is not only to produce an increase in crop yields, but also to achieve a better crop diversity and above all, a greater flexibility in crop system management.

The combined effects of an optimum level of agricultural production, profitability of drainage equipment, and environment protection (safe biotopes) may significantly alter the potential for drainage operations.

Irrigation is used in a climatic context essentially where there is a deficiency in natural precipitation. Management of such systems also induces environmental impacts. Although very limited in terms of erosion, the effects can be far greater in relation to groundwater quality. This is particularly true in the Mediterranean environment and for crops requiring high levels of irrigation and fertilization (e.g. vegetable crops, fruit trees).

Management of afforested land

The problems raised by the management of afforested land are very different from those related to agricultural activity. Forest cultivation has three principal objectives:

(1) wood production;

(2) land conservation and the prevention of erosion;

(3) equilibrium with other uses (agriculture, urban development, leisure areas, etc.) in land-use management.

Generally speaking, the forest is considered to be a natural environment, for which, contrary to agriculture, the input of fertilizers, pesticides, and herbicides is non-existent or limited. Although forest fertilization is now tending to increase, forest management is principally by planning, pruning, and wood cutting. The time-span separating planting and cutting is in the range of 20 years for quickly growing trees (e.g. conifers), and a century or more for slowly growing trees (e.g. oaks).

Forest management is beyond the scope of this book, but some comment on its relevance to soil erosion is appropriate. Deficiency and excessive degradation in plantations, and extensive fires lead to land erosion, particularly in mountainous regions, and in southern Europe where the vegetation cover is often reduced due to climatic conditions. Another impact is the interaction between the type of forest cover and the soil chemistry. Conifers and eucalyptus plantations for example, often induce soil acidification, with decrease of exchangeable cations, and increase of toxic aluminium.

The principal remedies are the selection of appropriate species, fertilization, and general husbandry. Forest degradation is described in the next section, which concerns the specific case of mountainous areas.

5.3.3. Mountain cultivation

Introduction [65] [72] [69]

Mountainous areas are much more sensitive than flat-lying areas to a wide range of factors. Ecological vulnerability is due to high relief, heavier precipitation, and short periods of growth. It is related to more severe climatic conditions, seasonal activity, and distance from consumption markets.

In mountainous regions, there is a tradition of forest being managed with many objectives:

(1) protection of the physical environment; reduction of hazards from floods, avalanches, erosion, and landslides;

(2) protection of the biological environment;

(3) development of the wood industry;

(4) development of tourism.

Maximum benefit is achieved by applying flexible methods of silviculture.

Considering the Alps region as a key European example, the historical record shows clearly the present situation: heavy woodcutting in the thirteenth and fourteenth centuries, overgrassing in the sixteenth and seventeenth centuries, and uncontrolled wood exploitation of the nineteenth century. The resulting problems in the 20th century are:

(1) marginal areas abandoned;

(2) intensive exploitation in the most suitable areas;

(3) soil erosion;

(4) air and water pollution;

(5) large areas developed for tourism;

(6) depopulation of mountain areas, together with an influx of urban and foreign manpower.

Any exploitation system must ensure that ecological stability is preserved, productivity is kept assured, a variety of natural species is maintained, and the natural state of the landscape is kept.

Two elements are constantly in conflict:

(1) measures to ensure the survival of natural ecosystems;

(2) socio-economic developments to consolidate mountain societies.

Only careful balance between these two strategies can meet the needs of modern management systems. As part of the ecosystem strategy of France, an inventory started in 1982 resulted in subdividing the country into:

(1) limited areas with protected or threatened species, or outstanding ecosystems;

(2) large areas which underwent weak modification only and with significant potential.

The former represent 7.5 per cent of the territory, the latter 19.4 per cent. The two categories total 50 per cent of mountainous areas. Various types of protected areas have been established ranging from national parks (including 350 000 hectares of high and medium altitude) to natural reserves (over 100 in France, of which more than 30 are in mountainous areas), protected sites and regional parks (11 out of 26 in mountainous areas).

In France, of a total of 31 million hectares of useful agricultural land, 4 million are in mountainous areas. Another 2 million must be added to the 4 million to include high altitude grass, recreation, and access routes. The main threat to the environment is, therefore, the reduction of agricultural areas. Hence various measures are necessary to increase productivity, quality standards, and maintenance of the farming population. Pilot experiments alone are not enough, rather it is necessary to promote generalized management systems. Although not satisfactory, the end result was a significant reduction in the total loss of agricultural development in mountainous areas compared to the national mean, and also an increase in the number of young farmers compared to the situation in other areas. These results concern a period between 1979 and 1988.

In the meantime the agricultural policy of the EEC has been largely concerned with specific production (milk, butter, corn, wheat) and not so much with an integrated policy for specific areas. This situation is changing and this evolution should be strongly supported.

One of the key results of degradation in mountainous areas is erosion and this topic, for which geological expertise is essential, deserves specific comment.

Erosion [86] [87] [65] [152]

In mountainous areas, erosion is one of the major factors threatening man's environment. It occurs in the form of natural process such as:

- mass movements;
- boulder streams;
- landslides;
- mud-flows;
- avalanches (in the alpine areas of central and northern Europe);
- collapses and subsidence due to karstification.

These phenomena may be triggered by:

- heavy rainfall;
- rapid melting of snow;
- earthquakes;
- human activities.

In mountainous areas, understanding the dynamics of erosion involves studying the dynamics of water yields and solid yields. Beyond a certain limit, which corresponds to catastrophic phenomena, this distinction is no longer valid and both dynamics intermingle. The study of torrential erosion, which must include hydrology, may be by natural observation, or by experimental or empirical methods. Information is still too scarce, and pilot schemes are too few and too restricted to specific geological environments.

However, as suggested in the introduction, the study of torrent erosion is only one aspect of the integrated approach which is so important in mountainous areas owing to their great vulnerability.

Remedial actions against erosion should integrate:

(1) agricultural methods minimizing erosion, i.e. ploughing directions;

(2) surface-water control by maintaining adequate drainage;

(3) flood control.

As an example, the case of the Bavarian Alps is presented below. In the Alps, the continuous destruction of forests and the steady spread of built-up areas, as well as the extension of agricultural land up to elevations of 2500 m, have changed the landscape considerably. Deforestation was practised even by the Celts and Romans and was intensified during the Middle Ages when demand for agricultural land compelled man to cultivate less suitable areas. Forest areas were used as pasture land. Felling of trees and overgrazing encouraged erosion. With the exploitation of mineral resources and the beginning of mining activities, there was additional demand for wood for mine timbering and charcoal for smelting the ore, which caused further destruction of the forests.

Further, in the Mediterranean countries, human activities, both past and present, are directly or indirectly responsible for exacerbating erosion in mountainous areas; this situation has been aided by the fact that the soils and climatic conditions over much of this region are conducive to erosion thus making these countries particularly vulnerable. Even in very early times, man also used his resources in a rational way and took measures to combat soil erosion and other erosional hazards such as landslides and floods. Evidence of terracing and stone walls that were constructed in an attempt to control soil erosion on arable land can still be seen in mountainous regions; similarly, there are areas of pasture in which the causes of accelerated, widespread, and irreversible erosion are activities such as agricultural practices, e.g. farming in unsuitable areas which are later abandoned, excessive felling of trees, overgrazing, the periodic burning of scrub and undergrowth, and forest fires.

Growing risks of this kind have increased the rate of migration of the rural population to the towns. It is the responsibility of the geoscientist to recognize and assess such risks, particularly in the long-term. Remote sensing data in combination with detailed ground control can be used to produce maps showing areas endangered by erosion and natural hazards. Land-use which is not compatible with the environment should be avoided or modified. Environmental protection measures should include reforestation programmes and the protection of ski slopes against erosion. In these regions of highly diverse morphology, where access and communication are difficult, thorough consideration during regional planning must be given to potential hazards to ensure that their short-term and long-term impact on man and the environment is kept as small as possible.

Land-use and slope erosion risks in the Bavarian Alps seen from the geological and pedological points of view [86] [87]

Risks caused by slope erosion

Compared to other European regions, the Alpine area is, from many points of view, a special case. Natural factors such as the geology, geomorphology, and climate determine the typical features of soil formation and the water budget, as well as the flora and fauna. Special prerequisites must be fulfilled for the cultivation of this area, especially with respect to agriculture; some of them are intended to protect the soil especially.

Many slopes in the Alps covered with soil and loose rock debris, with rock outcrops on the valley sides too, are in a more or less unstable equilibrium. In general, erosion takes place gradually and is in no way spectacular; seen as a whole, however, it causes major

ecological changes, particularly since it damages or even destroys the protective cover of vegetation

The impact of 'catastrophic' natural events is generally of local extent. Mass movements, such as landslides or creep, may cause damage to settlements located further down the mountain side, cut communications or ruin agricultural and afforested areas. Mass movement also takes place as soil- and mud-flows, which may cover valuable land with mud and detritus, and by avalanches, which convey wood and unconsolidated superficial materials down the valley.

For the Alps to be a natural environment as well as a recreation and economic area, intensive conservation of the landscape is essential, especially since the mountainous area loses part of its water-storing capacity with increasing soil erosion. This increases the danger of floods and causes the yields of springs to fall more rapidly during dry periods. Because they possess a high water-storage capacity, soil and unconsolidated material are of great importance in meeting the demand for drinking and industrial water. Increasing slope erosion in the Alpine region also increases certain hazards that may threaten the forelands, e.g. the danger of flooding.

The assessment of mass movements and the planning of protective measures should take account of the fact that the Alps are comparatively young mountains with complicated structural geology and an extremely high relief energy, and erosive forces are therefore dominant. From the geoscientific point of view, slope erosion must be regarded as a natural process subject to the laws of nature. It must be borne in mind, however, that the natural habitat of the Alps has not escaped the impact of man's activities.

Through agriculture and forestry, man has steadily increased his influence on the natural environment over the last several thousand years; often, this has caused irreparable damage to the soil and the landscape. However, during the pre-industrial period, man did succeed in minimizing the risks to the Alpine region; where it was possible, he avoided settling in danger zones, and great efforts were made to combat soil erosion which occurred especially on Alpine pastures after the mountain forests had been cleared. The social changes that have taken place in agriculture since then preclude such soil conservation measures nowadays.

The effects of excessive agricultural use and subsequent abandonment of the land in the Alps have been the subject of numerous studies, mainly because the necessary protective measures have to be planned with great care and expert knowledge, and general measures are not adequate for dealing with every aspect of the extremely complex process of slope erosion.

Numerous studies have demonstrated the outstanding role played by healthy stands of trees as a stabilizing factor in Alpine regions. Although they offer no absolute protection to the unstable soil of the slopes, they nevertheless represent the most effective protection that is possible. The intensive utilization of trees (salt extraction, iron smelting, glass making, and as a domestic fuel) over many centuries led finally to the planting of ecologically unstable conifer monocultures on many mountain slopes, instead of the indigenous mixed forest. Nowadays, the forestry and water resource authorities take great pains to impress the natural stability of the mountain forests (general amelioration programmes for protective forests). Owing to damage caused by the actual grazing and the hooves of the animals in woodland pastures (woodland grazing is still practised locally), as well as by deer and other game animals nibbling the bark off young trees, it is uncertain whether these endeavours will be successful. Overpopulation of hoofed game, which is tolerated for hunting, is held responsible for much of the damage. Other aggravating factors are: (1) a new type of forest damage known as 'Waldsterben', the actual consequences of which cannot yet be assessed; (2) extreme weather conditions, such as the storms in the late winter of 1989/90.

Tourism brings with it further serious detrimental effects; however, it has become an absolutely essential economic factor for the majority of the Alpine population. New roads and mountain railways, which make areas at high altitudes accessible, levelled and prepared ski-slopes for mass skiing, as well as the popularity of climbing during the summer months all have a considerable impact on the Alpine ecology, modify the water budget and run-off behaviour and are thus causing soil erosion on a great number of mountain slopes.

The fact that there are only limited areas available and that there is enormous demand for building land, has led to the housing development areas being extended into endangered zones and thus to increased risks for mountain dwellers and holidaymakers alike. Building permits for residential buildings and recreation facilities were often granted without considering potential hazards caused by mass transport from rock faces, steep slopes, avalanche-prone gullies, and mountain torrents, and often by slope failure even of relatively shallow slopes.

A variety of geoscientific maps are absolutely necessary as a tool for research on soil protection and protection against the mass transport phenomena mentioned above, as well as for planning suitable measures.

Soil inventory

In the course of work on the soil cadastre of Bavaria by the Geological Survey of Bavaria, soil catenas and individual soil habitats are being investigated with the objective of recording the most important soil types, determining the constituents and characterizing the habitats. In the Bavarian Alps, apart from problems of environmental pollution, soil erosion must be regarded as the most serious problem. The mechanism of soil erosion can be derived almost completely from the geology, the topography, the soil profile, and the distribution of certain soil series along toposequences.

5.3.4. Industrial pollution

Industrial activity creates emissions to the environment and more specifically to air, water, and soil. Although significant efforts have been made to decrease emission sources, there is still much work to be done. We must also consider historical emissions that have accumulated over the years in soils and in lake or river sediments. In addition, large-scale atmospheric pollution affects entire regions by humid or dry fall-out on to the soil.

Two distinct types of pollution are distinguished: local in the immediate surroundings of polluting plants (a few km² to 10 km²) and regional transported pollution, which contributes to forest degradation and soil acidification.

Localized pollution [11]

The European metallurgical plants, which are all relatively old, pollute the surrounding soils. This phenomena is well known regarding plants treating base metals (lead, zinc, copper), associated metals like cadmium of well-known toxicity, and aluminium (pollution by fluor). Pollution is not restricted to metals but also includes organic substances from industrial processes (chemical industry in particular) and energy production, which cause more than 50 per cent of soil pollution.

During recent years there have been an increasing number of studies of the impact, to find remedies (soil decontamination) and, often, to discriminate between primary 'geological' pollution and the effect of human activity. This latter target raises the question of how to define pollution or to establish pollution norms on a sound scientific basis.

Inventory for heavy metals in soils in Germany [94]

An example of research to discriminate between natural pollution and that resulting from human activity is a systematic inventory of heavy metals in soils (Schwermetalle in Böden) being carried out in Germany since 1984. This inventory follows the request of various German ministries for more reliable data concerning the heavy metals background in soils and tolerance limits. The work is therefore mainly oriented towards quantification of heavy elements per type of soil and/or crops.

As a first stage, a systematic sampling by auger drilling of all soil types in Germany was performed by the various Geological Surveys, leading to approximately 2120 sampling points. At each location four or five samples were collected representing the whole profile. The total number of samples was about 9700. Analyses for lead, copper, zinc, cadmium, and nickel were done by BGR and NLfB in order to avoid discrepancies between laboratories or analytical methods. As a result, 526 points were selected for further sampling. In each location, a pit was dug and soil samples were taken from all horizons.

The profiles were chosen to cover as wide a range of rock and soil types as possible. This material was analysed for the following elements: lead, copper, zinc, cadmium, nickel, cobalt, mercury, iron, manganese, lithium, and arsenic.

Those samples which were considered to be unaffected by anthropogenic contamination were selected for statistical analysis. The mean (defined here as the background) and the threshold value for the above elements were calculated for the following rock types: limestone, marlstone, sandstone, claystone, mica schist, basalt, granite, picrite, and peat. The threshold in this case is that value above which an element can be considered to be abnormally enriched, i.e. values above this suggest anthropogenic contamination might be present, although the anomaly might be due to concealed mineralization.

It is possible to distinguish between anthropogenic and lithogenic concentrations of an element in a soil using the following methodology. For soils derived from the weathering of consolidated rock, each sample is separated into a fine soil fraction (<2 mm) and a coarse fraction (rock fragments, >2 mm). The degree of contamination by a particular element can be obtained from the ratio of that element in the fine fraction to that in the coarse fraction. For soils formed from unconsolidated material, it can be determined whether an element concentration is anthropogenic or lithogenic/pedogenic by studying the element distribution in the soil profile.

In order to have a better appraisal of the pollution origin, 1100 samples underwent different chemical attacks before being analysed. Finally, five locations were more heavily sampled in order to study the influence at the plants level of the application of increasing quantities of cadmium, calcium, nickel, lead, and zinc. These experiments were carried out by the Soil Technology Institute of Bremen.

To summarize, the programme had two main objectives:

(1) the distinction between primary pollution and that derived from human activity (526 sites); this operation is done mainly by the separate study of the upper and lower parts of the soil profile and comparison between the fine (<2 mm) and the coarse fraction (>2 mm) of the soils;

(2) the assimilation capacity of crops for various contamination levels (five sites).

The first phase of the programme was terminated in 1988. Phase 2 has two main objectives:

(1) the study of the mobility of heavy metals in soils;

(2) the preparation of general instructions for assessing effects of heavy metals in soils.

This inventory provides a reference network of heavy metal distribution in soils, with particular emphasis on the industrial impact.

The criteria to be used for the establishment of soil pollution norms are being studied, revised, or discussed in various European countries and the problem of the evaluation of regional variability is far from being solved, in particular in northern France, Belgium, and Germany. The case history of Stolberg in the Aix-la-Chapelle region provides an example.

Contamination of soil by lead and cadmium in the area around smelters and former mines at Stolberg near Aachen [110]

The lead–zinc ore in the Devonian and Carboniferous limestones near Stolberg on the northern Eifel region was mined as early as the Roman period. Mining made this area a centre of the European brass industry for centuries. Large amounts of mining wastes, tailings, and slag were dumped during this period, usually unregulated. Owing to the low level of development of beneficiation and smelting processes, the tailings and slag had a rather high metal content. Wind and water erosion of the bare waste heaps distributed the heavy metal bearing material throughout the area. Material from the waste heaps was used for paving and other construction purposes. Smoke from the smelters also contained heavy metals. Thus, the heavy metal enrichment in the soil has many sources. When mining ceased in 1919, smelting in Stolberg continued, since imported ore had been smelted here for centuries.

Considerable environmental pollution was evident in earlier centuries. It was accepted as unavoidable, partly due to ignorance of the causes. Not until increasing lead production led to serious environmental damage, and the population and politicians began to understand the need for environmental protection, were measures taken to remedy the existing conditions and to prevent future pollution.

The death of grazing cattle in the Stolberg area resulting from ingestion of plants containing heavy metals had been known for a long time. Following a sharp increase in cattle deaths in 1972 due to lead poisoning, causing fears also that people could be poisoned, studies were ordered by the state government. Analyses were made of heavy metals in emissions, soils, drinking water, sewage water, fodder, and foodstuffs, and medical surveys were conducted.

About 2000 soil samples were taken at various depths at 480 sites in the Stolberg area and 175 samples from mining waste and slag dumps. X-ray analyses for lead, zinc, cadmium, copper, and nickel were made by the Geological Survey of Northrhine-Westphalia.

Lead contamination of the soil is shown on a map (1:50 000) of the 130 km^2 study area. Lead concentrations in the slag dumps are particularly high (more than 1 per cent = 10 000 ppm). In the three main mining areas, lead concentrations range from 5000 to 10 000 ppm. These areas are not, for the most part, used for agricultural purposes. Surrounding areas show concentrations of 2000–5000 ppm lead in the soil. Concentrations of similar level also occur in the valleys of several small streams draining these former mining areas.

The vertical distribution of the lead in the soil depends on the geological origin of the soil and changes resulting from agricultural use. The humus-rich topsoil of the largely undisturbed soil in woods and forests has a high natural lead enrichment, owing to the affinity of humus for lead. With increasing depth the lead concentration decreases, reaching a nearly constant value at a depth of about 50 cm. Soils consisting of re-deposited surface material, e.g. colluvium and recent stream deposits, have high lead concentrations to considerable depth.

Zinc and cadmium concentrations in mine dumps and former mining areas show elevated concentrations similar to lead. A map showing areas recommended for growing fruits and vegetables in the Stolberg area (Fig. 5.13) has been made by the Geological Survey of Northrhine-Westphalia on the basis of a map of cadmium distribution in the soil. Studies on the cadmium content of fruits and vegetables were conducted by the Office for Protection from Emissions (Landesanstalt für Immissionsschutz Nordrhein-Westphalia). It is recommended that private gardeners do not plant certain vegetables or fruit in areas with cadmium concentrations of more than 10 ppm.

The natural lead concentrations in the soil in the Stolberg area are derived from galena and cerussite. Chemical and mineralogical studies have shown that lead oxide, mainly from smelter emissions, is also present. The percentage of lead from emissions is too low to determine. However, compounds of heavy metals present in

smelter emissions are more soluble than those in ores and tailings and are thus more easily assimilated by plants and animals.

A series of studies has confirmed that lead is more soluble in acid soil and thus easily assimilated by plants.

Several mine dumps in Stolberg represent a hazard in that heavy metal bearing dust blown from them settles on fodder plants, as well as being breathed by humans and animals. This hazard can be reduced by covering the dumps with unpolluted soil and then sowing grass on them. This has been done already for many mine dumps. The dumps are not used for agricultural purposes even after being covered.

Efficient filters now prevent the emission of heavy metal bearing smoke from the smelters. Thus, further accumulation of heavy metals has practically ceased. Agricultural products are continually tested and thus no longer represent a hazard.

Various techniques have been developed for soil decontamination, both on site and after extraction. Incineration, leaching and biological degradation have been used on extracted soils. Biological degradation on site is increasing, but substances like chlorinated hydrocarbons, which are heavier than water, still present many problems.

Soils can be polluted by the dumping of muds dredged from rivers, channels, and harbours. These muds are frequently strongly contaminated by metals and they raise problems of storage within the requirements of general environmental protection. These problems are dealt with singly, but spreading along banks or stockpiling is the general rule. Various stabilization trials have been conducted on stockpiles of harbour muds.

Diffuse transported pollution

Concern with transported pollution has increased considerably over the last few decades. Better access to the countries of Eastern Europe has revealed the extent of the catastrophes resulting from an industrial development totally lacking in environmental concern. But the problem also exists in Western countries and, for a long time, the Nordic countries, due to their sensitive soil systems, have promoted studies in that field and installed monitoring networks and experimental sites. Monitoring networks also exist in other European countries, in each case with specific operational characteristics. The UN Economic Commission for Europe network of reference stations for long-range transport of pollutants also exists. This network covers the whole of Europe.

Current methodology is to establish a regional network of experimental sites at which the whole air–water–soil cycle is studied. As an example the Gardsjön site in southern Sweden is an international reference centre with 20 years of experience. Presently studied at this site are the effect of various quantities of nitrate spreading (EEC NITREX programme) and the reversibility of acidification. Other less extensive sites exist in Germany (Black Forest) and in France (Vosges).

Polluting agents involved in long-range atmospheric pollution are nitrogen compounds (NO_x), sulphur dioxide (SO_2) compounds, and ozone (O_3). The first originates from thermal power stations, industrial plants and domestic heating. Emissions from coal power plants and domestic heating are the key sources of SO_2. Emissions from vehicles mainly contribute NO_x. Ozone is formed at low altitudes through ultraviolet action on NO_2 and O_2 and then trapped above inversion temperature layers. Quite often mountainous areas reach above these layers.

Climatic change can be an important factor in forest degradation, and drought periods particularly affect tree growth and soil acidity. In fact, forest degradation may be the

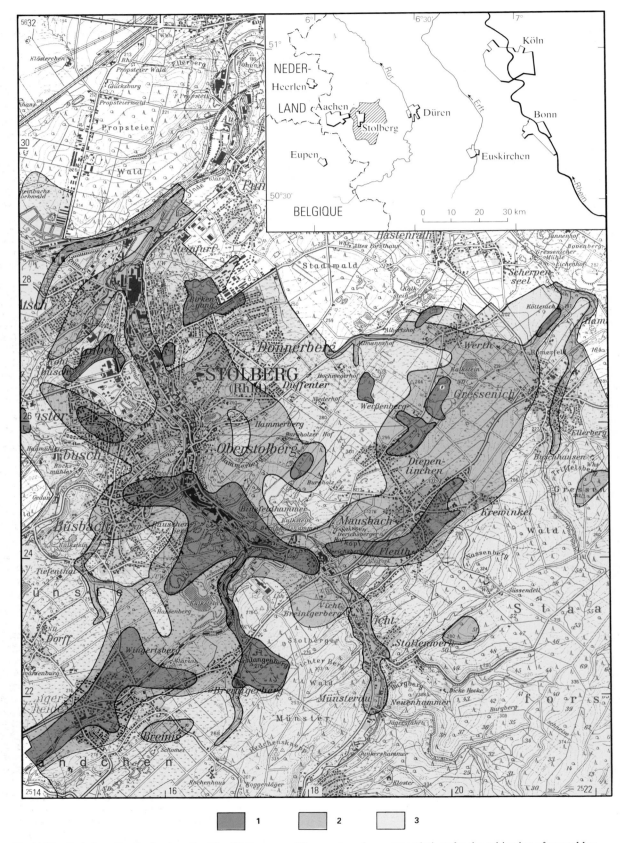

Fig. 5.13 Cadmium distribution in soils of the Stolberg area (Germany), and recommendations for the cultivation of vegetables.
1. > 30 ppm cadmium in the soil; celery, parsley, lettuce, spinach, kale, leek, red beets, carrots, and potatoes should not be cultivated.
2. 10–30 ppm cadmium in the soil; celery, parsley, lettuce, spinach, and kale should not be cultivated.
3. < 10 ppm cadmium in the soil; no restrictions.

result of a complex system of interactions involving climate, soil behaviour, and the presence of transported harmful substances. The situation is often more critical in acidic soils which have low buffering capacity. In those areas acidification leads to the formation of toxic aluminium which destroys life in rivers and lakes, and to eutrophication from NO_3. In some cases forests may be poisoned by SO_2 fallout.

Transported radioactive pollution also deserves a specific comment. The Chernobyl catastrophe emphasized the problem. The dynamics of radio-elements in the soil–plants–water system is still not understood adequately to cope with a critical situation properly. Increased R&D in those fields is certainly recommended. The similarity of the behaviour of caesium and potassium has already been underlined, but other radio-elements, such as strontium, deserve more specific attention.

Let us now examine more closely the impact of diffuse pollution in forests and underlying soils. As shown by long-term studies, acidic pollutants contribute considerably to the deterioration of our forests. These substances cause:

(1) leaching of nutrients from needles and leaves;

(2) leaching of nutrients from the topsoil as a consequence of acidification of the soil;

(3) disruption of nutrient assimilation as a consequence of root damage.

Mineralization of organic substances is impeded by acidification of the soil. This results in the accumulation of organic nitrogen in the humus layer of forest soils with a low base content, i.e. a low buffering capacity. Application of fertilizer or lime can lead to increased nitrification which results in increased leaching of easily soluble nitrate into the groundwater.

Nitrogen from the atmosphere (NO_x) can cause acidification of the soil. Nitrogen that is not absorbed by the plants may be transported by percolating water into the groundwater. This process causes leaching of cations such as potassium, magnesium, calcium, and manganese. Some nutrient-deficiency symptoms appear to be due to this process.

Heavy metals are accumulating in increasing quantities in forest soils as a consequence of air pollution. They are bound to organic substances in the upper humus layer. Liming and fertilization of forest soils to increase the pH value leads to a relatively rapid degradation of the humus layer. As a consequence, heavy metals are mobilized and, at least on a sandy subtrate, percolate into the groundwater.

Liming, particularly using easily soluble quicklime, can lead to the displacement of exchangeable cations. Potassium, an essential plant nutrient, specially tends to be easily leached out of soils with a low clay content. Liming is only suitable for neutralizing highly acidified, already podsolized soils in impaired forests. Lime from dolomite is the most suitable due to its low solubility. Whether application of fertilizer is necessary can be determined by the analysis of needles, leaves, and soil. Application of fertilizer is useful to stabilize and strengthen the resistance of slightly affected stands of trees.

If fertilizer containing sulphate is used to improve forest soils, there is a danger that in deeply weathered areas the surface may be leached into the groundwater together with the exchangeable aluminium, and then transported into surface waters, where it may have a toxic effect on fish.

The appraisal of regional forest degradation requires an understanding of the processes leading to degradation, and also the execution of systematic inventories. An example of such an inventory is given below.

Inventory on the present state of forest soils— a representative investigation of random samples taken from forest soils in Germany [106] [109]

In the western part of Germany, the physical condition of forest trees has been checked annually since 1984 taking an 8 × 8 km grid as a basis. In 1986, similar studies were also introduced at the level of the European Community, where a 16 × 16 km grid was used. The federal state of Northrhine-Westphalia is compiling an inventory of forest damage. The studies include about 500 forests and are based on a 4 × 4 km grid. These representative investigations of random samples are used as a basis for the annual reports on forest damage. In Northrhine-Westphalia they are supplemented by analyses of leaves and needles.

As yet, no study has been made of the present state of the forest soil, although it is known that forest soils have become exceedingly acid since the beginning of industrialization and are depleted in calcium, magnesium, and potassium, while at the same time substances such as aluminium and heavy metals, which are toxic to the roots, have accumulated in the soil. Contamination and modification of the soil vary according to its composition and the amount of toxic substances that have permeated it.

To obtain representative data on the present state of the soil, as well as statistically reliable data on the relations between the physical conditions of the trees and the soil properties, an inventory of the present state of the forest soil is being compiled in Germany. Data on soil properties are collected in the immediate area in which forest dieback is being investigated. In this way it is guaranteed that the chemical condition of the soil and the physical condition of the trees can be correlated. The inventory on the present state of the soil is intended to provide comparable data on all forest areas in Germany, particularly with respect to:

(1) the present state of the forest soils and their gradual modifications, with regard to the present condition of the tree tops;

(2) improving the applicability of forest dieback data to large forest areas;

(3) identification of the causes of changes in the condition of the soil, as well as the effects of pollutants;

(4) assessment of the threat posed by the present state and future forest generations;

(5) assessment of risk to the quality of groundwater and spring water;

(6) planning and implementation of measures to conserve and improve the condition of the soil, as well as to increase the supply of plant available nutrients in the soil and the uptake of nutrients by the tree roots.

Soil acidification is not distinctly marked in morphological soil characteristics but has to be assessed using soil analysis. Losses of calcium and magnesium, and increases in aluminium are characteristics of soil acidification.

In Germany the inventory of soil parameters is carried out by the State Soil Survey of forest habitats.

In Northrhine-Westphalia, this comprehensive task is carried out by two government authorities. Forestry-related data are compiled and finally evaluated by the Landesanstalt für Ökologie, Landschaftsentwicklung und Forstplanung (Institute of Ecology, Development of the Natural Environment and Forestry Planning). All pedological studies, both in the field and in the laboratory, and their evaluation and correlation are carried out by the Geological Survey. Such studies are of great importance for environmental policy.

In Northrhine-Westphalia, about 500 sites were selected to compile a soil condition inventory based on the 4 × 4 km grid already mentioned. At the site where a group of trees is examined for damage, a soil map over an area of about 20 × 20 m is compiled at the scale of 1 : 200. The central point is permanently marked and its position determined accurately. There, the soil is excavated and the macroscopic characteristics are described in detail, i.e. soil horizons, texture, colour, structure, whether the soil is influenced by groundwater or perched water, root penetration, etc. To obtain data that reflect the range of variation of chemical parameters over this area, composite samples are taken from the excavation as well as at 8 sampling sites radially at a distance of 10 m from the centre.

This inventory has the great advantage of providing a sound basis for long-term monitoring of forest changes. It also enables better definition of the Research and Development work to be carried out. International cooperation to improve and standardize methodology and develop monitoring areas for long-term follow-up is certainly warranted.

5.3.5. Concluding remarks

This section clearly underlines the importance of the adequate study and monitoring of a large spectrum of pollution problems.

The problem is not simply the quantification of various pollutants. From the research point of view, the discrimination between primary (geological) and anthropogenic pollution must be done on a sound basis, together with the detailed study of the transfer and degradation mechanisms involved. From the economic point of view an adequate balance must be found so as to preserve activities of vital importance for man and protect the environment.

The various institutions concerned with geosciences must be key contributors in meeting these challenges.

RECOMMENDED FURTHER READING

Anderson, F. and Olsson, B. (1985). Lake Gärdsjön. An acid forest lake and its catchment. *Ecological Bull.* **37**, Swedish Research Council.

Auerswald, F. and Schmidt, F. (1986). Atlas der Erosionsgefährdung in Bayern, Karten zum flächenhaften Bodenabtrag durch Regen. Bayerisches Geologisches Landesamt, Munich.

Auzet, A. V. (1981). L'érosion des sols par l'eau dans les régions de grande culture: aspects agronomiques. *Minist. Env. Minist. Agric. CEREG*, Organisation Env. Oct. 87.

Boiffin, J. and Marin-Lafleche, A. (ed.) (1990). La structure du sol et son évolution: conséquence agronomiques, maîtrise par l'agriculteur. *Colloques INRA* **53**, 9 janvier 1990. Centenaire Station agronomique de l'Aisne, Laon (France).

Bonneau, M. (1989). Que sait-on maintenant des causes du 'dépérissement' des forêts. *Revue forestière (France)*, XLI, 5, 367–84.

Calvet, R. (1990). Nitrates—Agriculture—Eau. *Intern. Symp. Paris La Défense*, INRA.

Chisci, G. and Morgan, R. P. C. (1986). Soil erosion in the European Community. Impact of changing agriculture. *Proc. Seminar on land degradation due to hydrological phenomena in hilly areas: impact of change of land-use and management*, ed. A. A. Balkema, Cesena 9–11 October 1985.

Collective Authors (1982). La transformation des terres. Bases méthodologiques, exemples français. *Recherche Environnement*, n° spécial. Ministère de l'Environnement. Mission des Etudes et de la Recherche, centre national Français du Comité Scientifique sur les problèmes de l'environnement.

Collective Authors. *Contaminated soil* KfK-TNO conferences.
1—1985—Delft. Martinus Nijhoff, Dordrecht, 1986.
2—1988—Hamburg. Kluwer Acad. Pub., Dordrecht, 1988.
3—1990—Karlsruhe. Kluwer Acad. Pub., Dordrecht, 1990.

Ferin-Westerholm (ed.) (1990). *Convention on long-range transboundary air pollution. Pilot programme in integrated monitoring.* 1 Annual synoptic report. 1990. Environment Data Centre, National Board of Waters and the Environment, Helsinki.

Gregor, H.-D. (1988). Critical levels for the effects of air pollutants on plants, plant communities and ecosystems. Conference Discussion Document, 'Air Pollution in Europe'; *Environmental*

effects, control strategies and policy options, I. Norrtälje and Stockholm.

Hindel, R. and Fleige, U. (1990). Geogene Schwermetallgehalte in Böden der Bundesrepublik Deutchland. *VDI-Berichte*, **837**, 'Wirkungen von Luftverunreinigungen auf Böden', pp. 53–74, Düsseldorf.

Jamet, P. (ed.) (1989). Aspects méthodologiques de l'étude du comportement des pesticides dans le sol. *INRA Versailles*, juin 16–17 1988. INRA.

Kuntze, H. (1986). Soil reclamation, improvement, recultivation and conservation in Germany. *Z. Pflanzenernähr. Bodenk.*, **149**, 500–12.

McLeod, A. R., Skeffington, R. A., and Brown, K. A. (1986). Effect of acid mist, ozone and environmental factors on conifers: chamber and open-air studies. *Air pollution research report*, **4**, (ed. O. Lökeberg). CEC Directorate for Science Research and Development. Environmental Research Programme.

Schneider, F. K. (1982). Untersuchungen über den Gehalt an Blei und anderen Schwermetallen in den Böden und Halden des Raumes Stolberg (Rheinland). *Geol. Jb.*, **D53**, 3–31.

Schwertmann, V. *et al.* (1987). *Bodenerosion durch Wasser. Vorhersage der Abtrags und Bewertung von Gegenwasserahmen.* Ulmer.

Ulrich, B., Benecke, P., Khanna, P. K., and Mayer, R. (1980). Soil processes. In *Dynamic properties of forest ecosystems*, (ed. D. E. Reichte) Int. Biol. Progr. vol. 23. Cambridge University Press.

5.4. DISPOSAL OF WASTE IN THE ENVIRONMENT

5.4.1. Disposal of domestic and non-radioactive industrial wastes [92]

Introduction [65] [28]

The term 'human activity waste' covers a large spectrum of materials and substances. Among these, domestic waste refers to that produced by direct human consumption (food, cleaning, etc.). In the statistics, large-volume domestic waste is often separated from such as market or garden wastes. All these categories are grouped under domestic waste here.

Disposal of domestic waste is still largely by landfill (20–100 per cent according to country), followed by incineration (0–80 per cent), recycling (0–18 per cent), composting (0–8.7 per cent). In Europe, Denmark, Sweden, and Switzerland use incineration mainly. Landfill is dominant in Finland, Germany, and the UK, as it is in the USA and Canada. An overview of disposal practices per country is given in Table 5.2. Re-use and pre-treatment before dumping should be given priority.

Apart from domestic waste, there is a large range of other waste matters which may be disposed of as follows:

(1) trade and industry waste (except hazardous waste): landfilled, incinerated, or recycled; food industry waste is classified as industrial waste;

(2) construction and demolition waste: in general landfilled or reused in construction work;

(3) hospital waste: generally incinerated;

(4) tyres: landfilled, incinerated, partly recycled;

(5) used cars: in general recycled in steel-making;

(6) phosphogypsum: stockpiled, dumped at sea and, to a small extent, recycled in the plaster industry;

(7) gypsum from desulphurization plants: stockpiled or reused in building work;

(8) fly ash from power stations: stockpiled or reused in construction work (roads for instance);

Table 5.2 Proportion (in percentage) of municipal waste managed by different methods (after Carra and Cossu 1990)

Country	Landfilled	Incinerated	Recycled	Composted	No service
Austria[1]	64	20	—	16	—
Canada	95	4	1	—	—
Denmark	31	50	18	1	—
Finland	95	2	3	—	—
France	47.9	41.9	0.6	8.7	—
Germany (West)	74	24	—	2	—
Italy	83.2	13.9	0.6	2.3	
Japan[1]	29.6	67.6	—	2.8	—
Netherlands	51	34	15	—	—
Poland	99.9[2]	—	—	0.1	—
South Africa[3]	69.2	20.8	3.1	3.8	3.9
Sweden	35	60	5[4]	—	—
Switzerland	20	80	—	—	—
UK	88	11	1.05[5]	—	—
USA	83	6	11	—	—

[1] Figures do not take recycling into account
[2] Includes waste disposed of in controlled and uncontrolled dumps; less than 1 per cent of the dumps are true sanitary landfills
[3] 225 of 564 landfills are uncontrolled
[4] Separation/composting plants
[5] Mostly waste-derived fuel

(9) agriculture waste: in general recycled in agriculture itself;

(10) contaminated soils: left on site or extracted, then treated and redisposed on site or landfilled;

(11) dredged muds: stockpiled on site or in landfills, or dumped at sea;

(12) sewage sludge: re-used as fertilizer or landfilled.

For EEC countries, waste production is estimated at 2200 Mt of which 1100 Mt is agriculture waste, 400 Mt extraction industries and power stations, 230 Mt sewage sludge, 160 Mt industrial waste (of which 20 Mt is toxic waste), 90 Mt domestic waste, 1.9 Mt used oil and 160 Mt construction and demolition waste.

Apart from different approaches to landfilling and incineration in the various countries, there are also differences in landfill methodology. Some countries are in favour of total encapsulation of waste to prevent leaching, others support open-air storage for waste stabilization purposes. In Germany the multibarrier concept is prevailing (artificial and geological barrier). In the UK, management of mixed industrial and domestic waste (co-disposal) has been adopted commonly, but is now much under debate. Usually, there are different landfill classes (three in France) according to waste type, or separated landfill of different waste within the same site.

In the field of incineration, the present tendency is to develop emission systems with high levels of control so as to avoid contamination by harmful substances (e.g. dioxine). Organic waste composing residues are limited except in certain countries like Spain. Recycling is increasing despite difficulties associated with the collecting systems.

A scenario for the future, based on the Dutch strategy involving strong environmental dynamism, would be:

(1) minimum waste production leading to control at the pollution source;

(2) incineration (and/or composting) of domestic waste;

(3) landfill restricted to waste of ultimate domestic or industrial origin, with the exception of hazardous waste which requires specific disposal techniques.

It is obvious that the present-day situation, although evolving constantly, is far from being satisfactory with regard to existing storage sites.

Background policy

Wastes are produced not only during the manufacture of consumer products, but also when the products have reached the end of their useful lives. Wastes normally have no immediate value as a useful material, but can often be utilized after intermediate processing.

It is not possible to eliminate waste completely, even when a large amount of material is recycled. In fact it is quite likely that as environmental regulations are tightened up, the quantity of waste will increase. In West Germany, for example, the introduction of new regulations governing large combustion plants has led to a reduction in the amount of sulphur entering the atmosphere from 1.3 Mt to 0.22 Mt per annum. As a direct consequence 3.4 Mt of gypsum are produced annually, and as the market cannot consume this amount, a considerable proportion of it has to be disposed of as waste.

Waste disposal sites should not pose a threat to the environment, although objective criteria are not yet available which can be used to determine when a threat actually exists. Waste, once disposed of, is extremely difficult or even impossible to recover and remove to a new disposal site (e.g. liquid wastes which have been pumped into aquifers) and thus persists as a very long-term potential hazard. For this reason, waste disposal sites must have the highest safety standards. Since it is not possible to modify the subsurface to act as a barrier to pollutants, the wastes themselves should be in a chemical form suitable for disposal in an environmentally compatible manner. This is only possible if the waste is inert or is rendered non-toxic by reaction with the host rock, or if the surrounding rock is impermeable (i.e. permits no migration of liquid), e.g. rock salt.

Wastes can usually be rendered more or less inert by incinerating them. Organic substances, which normally cause problems at waste disposal sites, are removed by incineration. The inorganic components of the waste, which are often present as complex mixtures, are transformed into inert glass slag and water-soluble salts. If these substances cannot be used, then they must be disposed of as waste; and this can be done in a way that has minimum impact on the environment. It is, therefore, not so much a matter of 'incineration or disposal' but 'incineration, then disposal'.

The volume reduction achieved by incineration depends on the proportion of combustible components in the waste. Normally, unsorted domestic waste, for example, undergoes a volume reduction of more than 60 per cent on incineration. Incineration is expensive. Therefore, it is important to avoid uncontrolled mixing of different kinds of wastes, whether this takes place during its production, collection, or transport. This permits the problem wastes to be isolated so that only these need to be subjected to thermal treatment. The remaining wastes can then be dumped directly at the disposal site, after undergoing any necessary physical or chemical treatment.

If the organic components in domestic waste are composted and not incinerated, then the inorganic components mostly retain their original composition, although their chemical behaviour is sometimes modified, e.g. reduction of solubility of metal compounds due to the formation of metal sulphides.

In spite of all the efforts made in that direction, incineration and composting of selected wastes does not lead to the required zero impact on the surroundings of the disposal site. For this reason it is all the more important to pay special attention to the geological setting of waste disposal sites.

The treatment and disposal of wastes are the final steps in a long chain involving production and consumption. In most cases the processes involved in this step have relatively little influence on the volume of waste produced. Only if there is a systematic movement towards reducing the amount of waste products, in some cases avoiding them altogether, or making use of them, will there be less waste, which in turn will require less space for its disposal. An alternative solution often discussed, by creating an artificial shortage of waste disposal land, is irresponsible.

It is possible, however, to place certain limits on the chemical composition, solubility, and stability of the waste; these, of course, would be compatible with the conditions at the disposal site, i.e. the geology and type of lining. In this way it is possible to allow the requirements of the disposal site to influence the properties of the waste, and this itself could lead to a reduction in the amount of waste in some cases.

Ground contamination problems can be serious, but ground which has been contaminated with crude oil derivatives can be cleaned up, using microbial methods, to such an extent that it can be put to general use. The contaminated soil should not, therefore, be removed to a general waste disposal site since it will take up the already limited available space, and in addition it might come into contact with contaminants (e.g. heavy metals), which might inhibit the effectiveness of the microbial processes or even arrest them altogether (Fig. 5.14).

Storage problems

The chief objective is to protect the biosphere. For this reason, disposal sites should release no pollutants, or at most controllably small amounts, depending on the toxicity of the waste.

Dispersal of pollutants may take place via water or the atmosphere, predominantly the former. The best protection is to keep water away from the waste. Depending on the toxicity of the waste material, water must be prevented from entering the facility by a sufficiently impermeable host rock or a suitable liner system (to prevent infiltration of rainwater) or a water collecting and drainage system (for surface water and groundwater). These liner systems are part of the so-called multibarrier concept, whereby several independent barriers (liner, waste, geological barrier) prevent pollutant from entering the surroundings, or at least reduce spreading to a minimum.

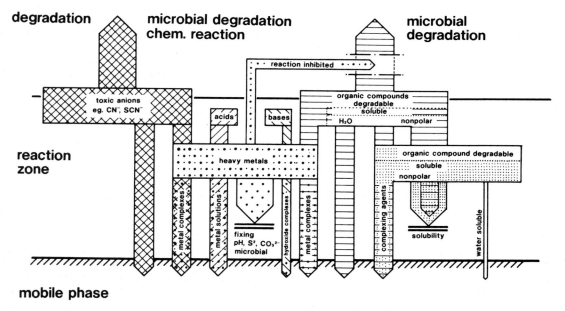

Fig. 5.14   General diagram showing pollutant degradation in a waste disposal plant (after Göttner, J. J. (1985). Mögliche Reaktionen in einer Sonderabfalldeponie. Folgerungen für das Deponiekonzept. *Müll und Abfall*. **7**, 29–32).

Since the early 1980s the Danish Geological Survey has experimented with the construction of a surface cover as an integral part of a barrier system for a waste disposal facility. The DGU introduces the idea of a capillary barrier that makes use of hydromechanical subsurface flow, and shows how a reduction in the infiltration rate can be achieved.

Research on the Capillary Barrier system [11]

The Capillary Barrier reacts as an umbrella above a waste dump, and is intended to prevent or limit generation of leachate to groundwater and surface water, and thus to act as a protection against pollution. The Capillary Barrier consists of two layers of sand, a coarse-grained layer below and a fine-grained layer above. The interface has a slope against the periphery of the dump, and the recharged water flows laterally within the fine-grained layer under unsaturated conditions (Fig. 5.15(a) and (b)).

The method was tested at Botterup in Northzealand, where five testing sections were installed. Results from measurements (two sections) made between April 1986 and August 1989 show that the capillary barrier has been able to drain off 44 per cent of the precipitation. However, 45 per cent of the precipitation has penetrated the barrier. Further measurements (four measurements in each section) performed after covering experimental sections with 0.5 m of top soil and grass, showed that leachate passing through the barrier was less than 5 per cent.

It seems reasonable to expect that a capillary barrier covered with top soil and sown with grass is able to limit the amount of recharging water to rather low values. A more exact value can be given only after longer periods of observation.

A similar experimental study of a capillary barrier system was carried out by BRGM in the 1980s on the Chevilly Municipal Waste disposal site located 10 km north of Orléans (France).

The multibarrier principle is a largely accepted state-of-the-art process. If all other barriers fail, then the host rock or geological barrier must be able to prevent the pollution from spreading into the biosphere. This means that, at a given disposal site, only certain types of waste can be accepted for which the surrounding rock is either sufficiently impervious or can adequately retard and retain the pollutants.

Reduction of the rate of spreading assumes that the transport mechanisms underground are known, taking into consideration buffering, precipitation, sorption, desorption, and degradation. These migration processes are governed by the mineralogical and petrographical composition of the rock and the hydrochemistry of the pore fluids, in addition to charge and energy gradients. These properties also affect the permeability of the rock on a large and small scale, as well as its buffering, precipitation, and sorption capacities.

The migration of substances can be simulated using mathematical flow models based on diffusive and convective displacement mechanisms. Such numerical models take the binding, remobilization, and breakdown of the migrating substances into account. Owing to the interaction of the different substances one with another and with the rock, it is currently only possible to establish migration models for individual substances. To be able to make a reasonably reliable estimate of the possible impact on the area around a waste disposal site, the groundwater system in the surrounding area is included in the model.

Special importance is given to protection of the groundwater. How much importance depends on the present and possible future use to which the groundwater may be put. A waste disposal site should never be located in a wellhead protection area or even in a proposed groundwater protection area. The groundwater recharge area is especially sensitive to detrimental influences because it has a direct connection to the biosphere. Models simulating the spread of pollutant must therefore deal with this problem in particular.

With respect to the particularly sensitive groundwater system in contact with the biosphere, it was thought that highly toxic waste should not be allowed to be dumped on the surface, but underground, where it cannot come into contact with the biosphere. However, this is possible only if the waste cannot be disposed of in any other way (e.g. incineration), or if the waste is compatible with the environment.

If the objective is for the disposal site to have little or no impact on the surroundings, and if, on grounds of environmental protection, there is no alternative, then it is possible that the idea of 'concealing' the problematic waste underground must be abandoned and the whole problem of waste disposal re-thought. A guiding principle of any solution to the problem must be: the more hazardous the material is, and the less possible it is to control spread of the pollutant, the more important it is to ensure that the waste has a neutral effect on the surrounding rock. Waste that is incompatible with the surrounding rock should be 'parked' on or near the surface, where it is easily accessible, until suitable processing methods are available to deal with it.

Figure 5.16 shows various possible methods for dealing with industrial wastes: surface disposal, underground repository, or disposal in a deep aquifer, for example. The German Federal Ministry for the Environment has produced Technical Guidelines for Special Wastes (TA-Abfall), in which the most suitable processing and/or disposal procedures are given as a function of the chemical composition of the special wastes. Thus, sufficient criteria are available for the planning of waste facilities of these kinds and for the authorization procedures connected with them (Table 5.3).

The waste at the disposal site is subject to the same alteration processes as the surrounding rock. The nearer the waste is to the surface, the more rapidly would one expect pollutants to escape into the environment. Non-degradable toxic substances must be prevented from spreading into the biosphere over long periods of time.

In view of the expected reactions between the waste and the host rock, the types of ground or rock that can function as host (for surface or underground disposal sites) should be determined for given types of waste. It should be borne in mind that different types of waste can be disposed of together only if this does not lead to an enhanced degree of mobilization of certain groups of substances. Figure 5.14 shows which reactions can take place in a special waste disposal site, and thus provides a guide to which groups of substances should be kept separate, to avoid increasing the degree of hazard by non-recommended mixing of certain types of waste.

Waste disposal sites are essential but unpopular components of our industrial society. A waste-free society is an illusion and will remain so in the future. The best kind of protection against environmental pollution directly from waste disposal sites is to ensure that the waste itself is environmentally neutral. Wastes should therefore undergo pre-disposal sorting, de-toxification, incineration, consolidation, etc. to make them as nearly neutral as possible with respect to the host rock. This does not mean, however, that no leachate will ever spread from any waste disposal site (an economically unattainable situation). The neutral behaviour of wastes becomes evident only when the waste or leachate interacts with the geological barrier. A geological barrier is, therefore, an essential feature of every waste disposal site.

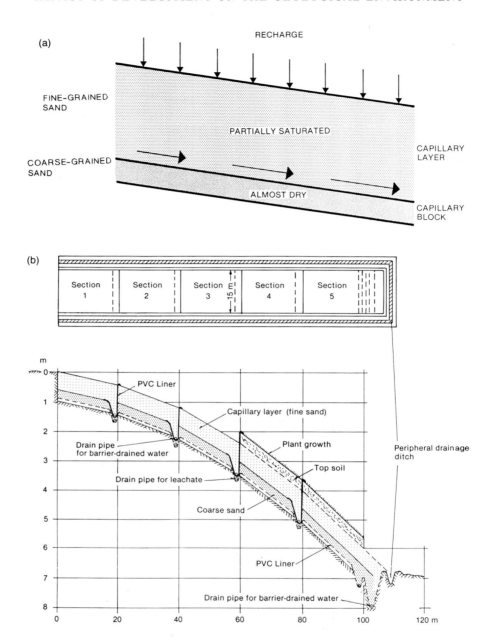

Fig. 5.15   (a) Sketch of the principle of the Capillary barrier (after Andersen, L. and Madsen, B. (1985). Die Kapillarsperre, eine alternative Methode zur Deponiesicherung. (*Nationale Tagung Ing.—Geol., Kiel*); (b) capillary barrier experimental system of the Danish Geological Survey.

As with mineral resources, rocks that are suitable as geological barriers are of local occurrence and cannot be reproduced elsewhere. For the sake of the health of future generations, it is, therefore, one of the priority tasks of the geoscientist to find suitable sites for waste disposal, and of political bodies to make decisions ensuring that this kind of site is reserved for future waste facilities.

Specific problems related to the surface storage of solid wastes

The quality and suitability of a geological barrier is determined chiefly on the basis of the type of waste to be disposed of. If the waste is very heterogeneous, then more stringent

requirements on the properties of the ground are necessary. Waste containing organic compounds are always problematical, since they can act as complexing agents, leading to solution of inorganic substances, such as heavy metals. This means that, strictly speaking, waste containing organic compounds should not be disposed of. This also applies to domestic waste. In principle, these wastes, depending on their composition, should either be incinerated or composted, at least enough to reduce the organic substances to a tolerable residual concentration.

Before disposal, materials which could increase the degree of mobilization of pollutants should be removed. Wastes which, because of the conditions at the waste site, are subject to rapid degradation, increased solubility and subsequent

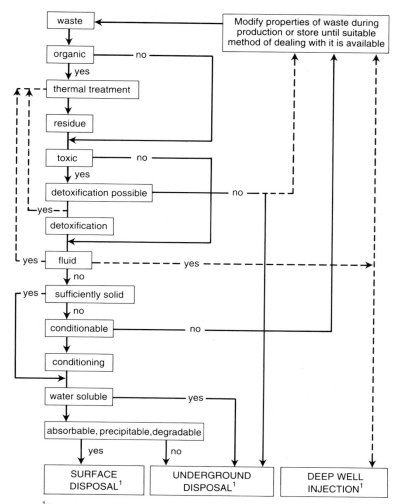

Fig. 5.16 General flow diagram for industrial wastes.

precipitation (e.g. lime-rich rubble in domestic waste), should be disposed of separately. If this is not done, the drainage system at the base of the waste site may rapidly become clogged.

The spreading of pollutants from a waste disposal site, as has already been mentioned, is mainly in the form of a leachate. It is therefore essential to have an efficient seepage collection system from which the leachate is led off to a waste water treatment plant. The best way to ensure that all leachate is collected is to install a bottom lining. This consists of a lining (in Germany normally a composite system consisting of a mineral lining overlain by a plastic liner) and a drainage system. Should the last-mentioned fail then there will be a build-up of leachate in the waste; the leachate may seep uncontrolled into the geological barrier. For this reason it is necessary to guard against failure of the drainage system firstly by processing the waste before disposal, and secondly by careful management of the actual filling of the site. Mistakes made during planning of the disposal site or as a result of unauthorized waste disposal and/or treatment techniques lead to overloading of the geological barrier. These mistakes cannot be corrected by the geology.

In principle, the different types of waste do not require different types of geological barriers at a disposal site, but there are slight differences in the minimum requirements to be met. As yet, no mandatory specifications exist as to the mineralogical and/or petrographical composition and minimum thickness of a geological barrier. It is unlikely, in fact, that it will be possible to set up specifications for geological barriers in the near future for disposal facilities containing waste of highly variable composition. However, it is a well-known and a well-tested fact that clays are efficient inhibitors of the movement of liquids. They are considered to constitute good barriers owing to their low permeability and the high absorptive potential of the clay minerals.

The type of clay minerals in these soils (e.g. kaolinite, mixed-layer clay, and smectite) can provide additional advantages; however, it is rarely the case that one is in a position to choose areas with a particular type of clay for a waste disposal facility. In general, kaolinite clays possess a lower sorption capacity but a higher sorption energy than smectite. On the other hand, smectites, owing to their property of swelling, are often able to seal open fissures; thus these clays possess a lower permeability than kaolinitic clays.

Apart from different types of disposal facilities, the various State Geological Surveys of Germany, together with the Federal Institute for Geosciences and Natural Resources, have compiled a list of properties a geological barrier must

Table 5.3 Catalogue of special wastes (TA—Abfall, Technical Instruction for waste)

Code no.	Description	Origin	Bulk waste	Method of disposal or processing to be used if utilization assessment (No. 4.3) is negative						
				CPB	HMV	SAV	HMD	SAD	UTD	Other
593 04	Substances contaminated with chemicals	Chemical industry, trade research organizations, plant laboratories, schools		1	1	1		2	1	
594	Waste detergents and washing powders									
594 01	Residues from washing powder manufacture	Chemical industry, producers of washing powder, cleaning and polishing agents				1		2		
594 02	Tensides	Chemical industry, producers of washing powder, cleaning and polishing agents, textile industry				1				
594 04	Sulfosoaps, sulfoacids	Oil refineries, producers of washing, cleaning and polishing agents								
595	Catalysts									
595 07	Catalysts and activators	Chemical industry, oil refineries		1			2	1	1	

CPB = chemical/physical/biological treatment
SAV = incineration of special waste
SAD = special waste disposal site (surface)
HMV = incineration of domestic waste
HMD = domestic waste disposal site
UTD = underground waste disposal
1, 2, etc.: order of treatment

Table 5.4 Recommendations of the working group on waste disposal sites set up by the Geological Surveys of the German Länder and the Federal Institute of Geosciences and Natural Resources. Requirements of geological barriers for waste disposal facilities (the need for liner systems is not considered here)

Type of waste facility	Surrounding rock		Thickness d(m)$\geqslant$	Permeability of rock Kf(m s^{-1})$\leqslant$	Specific discharge[1] (m^3 m^{-2} s^{-1})	Type of groundwater utilization
Rubble	Unconsolidated rock	Cohesive soil	Suited to type of waste and site	1×10^{-6}	1×10^{-8}	Not in TGG or HQSG
	Consolidated rock	No karst	Suited to type of waste and site	1×10^{-6}	1×10^{-8}	
Domestic waste and mono-/oligo-waste facility type 1[2]	Unconsolidated rock	Cohesive soil	5	1×10^{-7}	1×10^{-9}	Not in TGG, HQSG, or TVG
	Consolidated rock	Claystone (siltstone), marlstone, plutonic, and metamorphic rocks	20	1×10^{-7}	1×10^{-9}	
Domestic waste and mono-/oligo-waste facility type 2[3]	Unconsolidated rock	Cohesive soil	10	1×10^{-8}	1×10^{-10}	Not in TGG, HQSG, or TVG
	Consolidated rock	Claystone (siltstone), marlstone, plutonic and metamorphic rocks	30	1×10^{-8}	1×10^{-10}	
Special waste for underground disposal	Consolidated rock	Rock salt, plutonic rocks, claystone not in earthquake zones 3 and 4 (DIN 4149)	Suited to type of waste and site	1×10^{-9}	—	Efficient geological barrier to aquifers

[1] See DIN 4049 Pt. 1 No. 4.55.
[2] Requirement less strict than the Federal Ministry for the Environment's technical guidelines on special wastes
[3] Requirement stricter than the Federal Ministry for the Environment's technical guidelines on special wastes

TGG = Drinking water extraction zone
HQSG = Medicinal-spring protection area
TVG = Area proposed for groundwater extraction

have. The list is neither official nor compulsory; it represents a compromise between the scientifically desirable conditions and those dictated by the geological conditions (Table 5.4).

The German Federal Ministry for the Environment is currently compiling technical guidelines on special wastes; these will be the regulations for those authorities responsible for enforcing legislature on wastes. The States' joint working group on wastes (LAGA) is revising guidelines for establishing disposal sites for domestic waste. Both of these guidelines assume a geological barrier of low to very low permeability soil with a minimum thickness of 3 m and a coefficient permeability $Kf \leqslant 1 \times 10^{-7}\ ms^{-1}$. The highest predicted groundwater level should not be higher than 1 m beneath the bottom liner. If these conditions are not met, then both guidelines permit a course of action to be taken involving the replacement of a non-existent or inadequate geological barrier by an engineered barrier. An example is shown in Fig. 5.17.

Close interdisciplinary cooperation, including comprehensive geological, hydrogeological, and geotechnical investigations, possibly supplemented by geochemical studies, is necessary for assessment of proposed waste disposal sites and thus of the geological barrier.

The scope of the investigations depends basically on the geology of the proposed area. For this reason it is not possible to quote a mimimum number of sections that have to be examined or boreholes drilled. In any case the scope of the investigations should be sufficient to obtain a detailed picture of the stratigraphic succession, the geological structure and distribution of the various rocks, the geometry of any aquifer present, depth to the water table, flow directions of the groundwater, and permeabilities of the rocks. Field tests should be carried out to determine permeability. They should be done for the different beds or formations separately and must provide a comprehensive picture of the groundwater movement, if any.

It is only possible to assess a waste-disposal site as suitable when all relevant geological factors have been considered, e.g.:

(1) low permeability (rock and bulk permeability);

(2) sufficient thickness and homogeneity of barrier rock;

(3) high potential for retention of harmful substances (sorption, precipitation, buffering, degradation);

(4) long-term stability of the barrier rock with respect to chemical changes;

(5) sufficiently stable subsurfaces;

(6) minimum effect on the groundwater;

Naturally, the special requirements of the waste that will be disposed of there must also be taken into account. It is not sufficient to base a positive assessment merely on random field and laboratory tests of permeability.

If large areas are being investigated with the aim of selecting a suitable site, and this is always essential for an adequate basis for selection, then a systematic step-by-step approach is recommended. This will be illustrated with an example of site investigation for a solid waste disposal site in Lower Saxony.

Waste disposal site investigation in Lower Saxony [92]

At the end of the 1970s, the Lower Saxony Ministry of Agriculture and the Lower Saxony Geological Survey (NLfB) agreed to produce maps showing the distribution of suitable clay areas representing suitable sites for disposal of special wastes. These maps (Phase I) were to form the basis of Lower Saxony's special waste disposal plan.

At that time no information was available as to what was meant by 'suitable' ground on which to establish a special waste facility nor on how to determine the most suitable material. In 1981, a broadly-based investigation of the suitability of clays was initiated at the suggestion of the NLfB. The main part of these investigations (Phase II), which were carried out by an interdisciplinary group of geoscientists from the NLfB (geologists, hydrogeologists, soil physicists, geochemists, engineering geologists, and geophysicists), were concluded in 1986. External research institutes and universities were also involved in the project, particularly for certain specialized investigations.

For various reasons, which are not considered here, the comprehensive field, laboratory and model studies were concentrated on a 2000 m thick, predominantly clay succession of Lower Cretaceous rocks in the area of the Lower Saxony Cretaceous basin.

Apart from boreholes sunk to study the relevant parts of the Lower Cretaceous succession, hydraulic tests were carried out in the field, as well as detailed laboratory work (mineralogy, geochemistry, hydrochemistry, and soil mechanics), and specialized investigations such as SEM element distribution studies and field tests on the dispersion of dissolved substances. An important aspect of these studies was to discover what effect dissolved organic and inorganic substances have on the stability of clay structures. Special attention was paid to possible changes in the strength and permeability of the clays, and also to precipitation and sorption effects. Numerical models were used to predict migration behaviour as a function of precipitation, sorption, desorption, and degradation.

The results of laboratory work are summarized in an assessment scheme for Lower Cretaceous clay sequences, as well as on maps of the different areas studied on which the suitability is represented taking account of the hydrological situation.

The individual stages of the investigation are listed in Table 5.5.

Phase I   Production of a map from archive data showing the distribution of rocks apparently suitable as geological barriers.

Phase II   Investigation of the rock successions selected in Phase I; identification of potentially suitable parts of the succession and the corresponding sub-areas.

Phase III   Studies at definite locations within the sub-areas in order to identify possible sites for waste disposal facilities.

Phase IV   Detailed investigation of the selected sites to provide a sound basis with which to support a possible application for authorization at a later date.

It will be noted that the investigated areas become successively smaller from Phase I to Phase IV due to the selection process used.

The area studied during Phase I included almost all the clay areas in Lower Saxony (Fig. 5.18). This area was reduced to 2000 km² for Phase II; the relevant parts of the Lower Cretaceous succession were drilled and mapped. Detailed investigations (stratigraphy, soil chemistry, mineralogy, and petrography) were also carried out. This led to a general order of suitability for the four last sub-areas. The most suitable was found to be sub-area B.

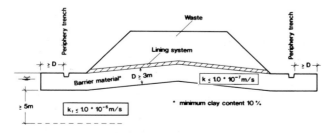

Fig. 5.17 Idealized waste disposal site showing improvement of an inadequate natural barrier system by the addition of an engineered barrier beneath the waste. D = thickness, k = coefficient of permeability for the artificial clay barrier, kf = coefficient of permeability for substrate.

Table 5.5 Investigation and assessment of special waste disposal sites in Lower Saxony

Process	Phase			
	I	II	III	IV
Selection of suitable rocks (clays)	x			
Preparation of maps showing the distribution of these rocks (clays)	x			
Selection of areas for detailed study taking into consideration nature reserves, groundwater protection, built-up areas etc.			x	
Evaluation of geological maps and existing borehole data	x	x		
Drilling of further boreholes		x	x	x
Stratigraphical studies		x	x	x
Determination of: • the thickness of the clay beds		x	x	
• the lithological sequence		x	x	
• the bedding and joint geometry		x	x	x
• stratigraphical discontinuities		x	x	x
• fracture zones		x	x	x
• the structure		x	x	
Geophysical well logging		x	x	x
Mapping of aquifers around the prospective site			x	x
Equipping boreholes with monitoring equipment		x	x	x
Monitoring the groundwater table			x	x
Location of receiving stream(s)			x	x
Setting up test plots to determine permeability		x	x	x
Experiments to study the dispersal of dissolved substances		x	x	x
Chemical and physical analysis of the groundwater		x	x	x
Temperature and conductivity cross sections		x	x	x
Age determination of the groundwater		x	x	x
Description of clays and siltstone		x	x	x
Determination of the strength of the clay		x	x	x
Mineral analysis of the clay		x	x	x
Swelling, buffering, and adsorption capacities		x	x	x
pH, carbonate, and buffering capacities		x	x	x
Permeability		x	x	x
Development of simulation models for estimating the spread of pollutants		x	x	x

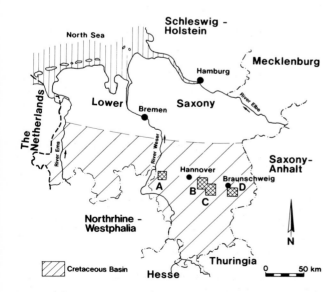

Fig. 5.18 Location of investigations into waste disposal in Lower Saxony. A, B, C, and D are areas of investigation.

Following the above assessment, a start has been made on the investigations of Phase III at five drilling sites in sub-area B. This work aims at achieving a sufficiently detailed coverage to select individual sub-areas, each about 80 ha, as definitely suitable for waste disposal. These would then be subject to detailed site investigation during Phase IV. The results of Phase IV studies will constitute part of the final appraisal of the documentation supporting the application for a licence to operate one or more facilities.

Another approach to waste disposal siting was carried out for the city of Málaga in southern Spain.

Waste disposal siting for the city of Málaga [156] [163]

The city of Málaga has a population of 610 750 people in winter (245 days) and 830 750 people in summer (120 days). Solid waste production varies from 0.836 kg per inhabitant per day in winter to 0.674 kg per inhabitant per day in summer. Málaga is the capital of the Costa del Sol, which is an important tourist area on the Mediterranean coast. The estimated growth rate of the population is 3 per cent annually, which results in land shortage problems within the municipal district.

The Málaga City Council has contracted the ITGE to carry out a site study for the new controlled solid urban waste dump. The anticipated life-span of the dump is 20 years and it is to be located within an area of about 30 km² inside the municipal district. The purpose of the study is a comparative evaluation of three possible dump zones.

The ITGE's technical team has proposed a dump method using grinding, intensive compacting, and periodic filling (high-density), based on the requirements of the dump's life-span, maximum space saving, and minimization of environmental impacts.

Two groups of factors were used for siting the dump:

1. Technical and construction factors: useful volume to absorb the expected volume of waste generated during the projected life-span, access to the zone, proper morphology, land movement, material for filling and sealing, geotechnical characteristics of the substratum, and other infrastructure (water, electricity).

2. Environmental factors: distance to inhabited areas, geology and geomorphology, surface and underground water pollution, climatology (temperature, rainfall, wind evaporation and relative humidity), soil and vegetation, fauna, land-use and urban planning, degree of human involvement, natural areas or areas of cultural and/or scientific interest, geological risks (floods, slope movement, erosion, seismicity), quality of the landscape, and visual impact.

All these factors were used to define a feasibility index based on the respective weight of these factors using a coefficient range of 0.5 to 2. Less favourable conditions clearly prevailed in one zone. The selection between the other two was to be done using economic factors such as infrastructure, construction, or environmental protection costs.

Fig. 5.19   Location of Sarpsborg (Norway).

Difficulties in the assessment of the requirements for such disposal sites arise mainly because the relevance of geoscientific knowledge for problems of waste disposal was not sufficiently recognized for a long time, and therefore properly made measurements of the migration of pollutants from waste disposal sites have not been carried out. Reliable data, however, can be obtained only in long-term field studies under natural conditions. Laboratory tests can provide only provisional information.

Sometimes these field studies, which had been omitted earlier, are carried out in connection with current assessments of the hazards emanating from abandoned waste disposal sites. It is not possible, however, to verify the initial data in such cases or retrace the migration of pollutants in the underground system and thus obtain accurate data for future planning.

Such investigations for the assessment of hazards emanating from abandoned waste disposal sites require a great amount of time and money, as well as close interdisciplinary cooperation of geoscientists, chemists, hydrologists, and disposal site specialists. Since each case is different, no general procedure can be recommended. In any case, the investigations should be planned in a way that, in addition to information on possible hazards, data are obtained that are useable for planning new waste disposal sites. A typical example of such studies is given by the contamination by mercury in the area around the chlorine factory of Sarpsborg (Norway).

Mercury pollution at Sarpsborg (Norway) [147]

The Borregard chlorine factory of Sarpsborg was founded in 1949 (Fig. 5.19). The production was based upon the amalgam-method in which a considerable quantity of mercury is used. The total amount used at any one time in the process was about 100 tons. In the period 1949–87 a quantity of 130 tons of mercury was lost from the factory. Some of the loss was directly to the atmosphere by vapour through ventilation systems and chimneys. The mercury

also found its way into the ground through cracks in the floor or via leaking sewage pipes. Some unwanted amalgam formed during the process also had to be removed. At the graphite electrode, mercury-containing dust is produced, and this also had to be removed. The products contained Cl_2, and NaOH and the water used for washing the chlorine gas also contained some mercury.

Mercury loss to water has been evaluated at 70 tons and to air 20 tons. Twenty-five tons remained with the final products and 15 went off to the waste landfill, close to the Glomma River.

In 1987, the Geological Survey of Norway was asked by the Norwegian State Pollution Control Authority to investigate the mercury pollution. This was done by bedrock topography, soil mapping, and hydrogeological and geochemical investigations. Eighteen wells were dug for groundwater study and 22 for waste study.

Soil pollution totalling 10–11 tons of mercury was found in more than two hectares around the factory. Compared with normal soils from the Sarpsborg area the mercury content should have been 25 kg. Normal mercury contents of the top-soil were found at a distance of 1000 m from the factory. At the waste site 6–7 tons of mercury were found. The normal amount in an equivalent mass of soil would have been 30 kg. Under the factory the mercury had penetrated through 10 metres of clay and into the underlying bedrock. Dispersion of mercury from the factory area and waste site has taken place through the groundwater. The dominating migration has occurred in the upper zone. Dating of the water in the lower zone gave an age of 1000 year BP.

Each year, 1 kg of mercury is transported with groundwater from the factory area to the River Glomma and 0.1 kg at the waste site. Careful sampling and analysis of surface water from the River Glomma shows an annual transport of 400 kg of mercury before the river flows through the industrialized area. The source of these 400 kg of mercury is considered to be mainly natural.

In conclusion, significant soil pollution was discovered in the factory area and at the waste site. However, due to very favourable geological conditions, the risk of pollution spreading to the River Glomma is very small.

An important factor for the safety of surface disposal sites is the mechanical stability of the waste itself. Mechanical instability is not only a hazard to the immediate vicinity (landslides, flow of disposed material from the landfill, etc.).

The liner may also be damaged, releasing pollutants, and contaminating the surrounding area.

Landfill must, therefore, be mechanically stable with respect to both the site itself and the surrounding area. Like any other construction, sites must be properly planned and designed, and supervised—each individual component (e.g. liners and embankments), the disposal site as a whole, as well as the filling of the landfill.

Calculations must be carried out to establish deformation and stability to be expected during construction and operations (disposal of the wastes), as well as after the closing of the site. The consolidation and deformation behaviour of the wastes, which is different from that of materials normally used for the construction of earthworks, must be taken into consideration.

Underground storage of solid industrial wastes

Underground disposal may have the following objectives:

(1) complete isolation from the biosphere;

(2) large distance from groundwater in contact with the biosphere;

(3) utilization of caverns if the aforementioned prerequisites for environmental protection can be fulfilled.

The disposal method depends on the type and degree of danger resulting from the waste. No matter which method of disposal is selected, the waste:

(1) may not contain substances that react to produce more hazardous substances;

(2) must not have any undesirable effects on the host rock;

(3) must not be inflammable;

(4) must remain sufficiently stable after disposal.

Liquid and organic wastes must be excluded.

Complete isolation from the biosphere

Rock salt deposits and unjointed solid rock, such as granite, are most suited for this purpose. The galleries, as well as access shafts, tunnels, or boreholes, must remain dry despite any deformation, not only during the time of operation, but also during the post-operation phase. Because a hermetic seal must be guaranteed in the geological time-scale, it is possible to store in such galleries solid toxic wastes that cannot be stored at other disposal sites. It is essential to guarantee long-term safety with respect to this type of waste.

Caverns in rock salt created by solution mining or in normal salt mines can be used for disposal. Caverns in rock salt created by solution mining are particularly suited for the irreversible disposal of problematic solid waste or waste which will solidify after disposal. Due to the restricted access via the borehole, and because the waste in the cavern should have as few voids as possible (so that little deformation will occur later), only fine- to coarse-grained waste can be disposed of in this way. The caverns must be filled without supporting liquids, since serious problems with waste water will occur otherwise. The absence of supporting fluids, however, leads to a high rate of convergence. Thus, the rate of disposal must be faster than the expected convergence rate; such a cavern should be filled within five years, depending on the character of the salt deposit. Thus, the maximum size of such a cavern is about 200 000 m³.

The anticipated solidification of the waste after disposal depends on the geostatic conditions. It must be guaranteed that significant deformation of the surrounding rock after

disposal is excluded. Any fluid used for hydraulic transport of the waste must be completely chemically bound to the waste after setting.

The following case history describes geotechnical studies for a project of solution mining caverns in Lower Saxony. Residues from waste incineration plants, salts, flue gas desulphurization residues, and non-ferrous metalliferous sludges are to be stored in these caverns; this is part of the waste disposal concept of the federal State of Lower Saxony.

Salt cavities for underground disposal of hazardous waste in Lower Saxony (Jengum Project) [75]

The geological conditions in Lower Saxony provide particularly favourable conditions for the permanent underground disposal of special wastes which, even in the long-term, have no potential for utilization and should be kept away from the biosphere. The Jengum Project is to test the large-scale utilization of caverns formed by solution mining in salt for the disposal of certain special wastes. In Germany, such caverns have been successfully used for the storage of oil and gas for quite a long time. For the storage of solid and solidified waste, however, the mechanical interaction between the wastes and the salt rock need to be studied, as well as the long-term safety.

In 1986, the Geological Survey of Lower Saxony prepared a geological assessment for the initial selection of suitable salt structures for underground disposal sites. In 1988, it was supplemented with expert opinion on the suitability of the Lower Saxony salt structures for groups of caverns for the disposal of non-radioactive toxic wastes. A total of 36 salt structures near the coast and more than a hundred other salt structures were considered.

Evidence had to be found for the existence of sufficiently thick, quasi-homogeneous, and statistically isotropic rock salt of moderate depth, the main geological prerequisite for the construction of large caverns or groups of caverns. Further requirements were: that a sufficiently thick roof of rock salt must be present to guarantee the stability and seal of the disposal caverns; that the distance of the caverns from each other and from the surrounding non salt rock should be several hundred metres; and that thick intercalations of potassium salt and/or insoluble rocks should be absent to ensure mechanical stability.

Saline rocks, sometimes with a high percentage of rock salt, occur in various geological formations in Lower Saxony:

(1) Permian (Upper Rotliegende, Zechstein),

(2) Triassic (Upper Bunter, Muschelkalk, Keuper),

(3) Upper Jurassic (Münder Marl).

The salt rocks of the Upper Bunter and Muschelkalk, however, do not have the thickness necessary for creating cavities by solution mining. Although sufficiently thick rock salt deposits have been demonstrated locally in the Keuper and Münder Marl, they must be regarded as little suited for creating caverns by solution mining, because they sometimes contain a high proportion of insoluble layers or finely distributed impurities. Therefore, the main interest is being given to the saliniferous formations of the Permian, where more than 120 cavities have been constructed for the storage of hydrocarbons.

The Federal Institute for Geosciences and Natural Resources (BGR) was commissioned by the BMFT (Federal Ministry of Research and Technology) to assess, from the geoscientific and geotechnical points of view the above-mentioned research project on the safe disposal of solid toxic waste in cavities created by solution mining in rock salt and to provide expert support to this research project.

One objective of the project is to develop a concept for the disposal of solid waste (e.g. flue ash and filter dust from incineration plants for domestic wastes, sewage sludge, toxic wastes, and solid reaction products from flue gas desulphurization systems) in such cavities.

A second objective is to demonstrate the technical feasibility and mechanical stability of such a cavern. The project is being conducted by a group of firms under the technical and scientific direction of the Lower Saxony Association for the Ultimate Disposal of Special Wastes (Niedersächsische Gesellschaft für die Endlagerung von Sonderabfällen—NGS). The project is divided into three parts: (1) processing and solidification of the wastes, (2) filling of the cavity, and (3) geotechnical aspects of solution mining. The work will be done in three steps: in the first and second steps, tests will be done at laboratory and pilot scale, respectively. In the third phase, the results will be checked in full-scale trial operations on site.

Salt mines offer the advantage that they are more easily accessible and thus they can be used for both disposal and storage of wastes. The size of the individual waste containers or other objects to be disposed of or stored, is restricted by the existing hoisting facilities. Disposal of fine- and coarse-grained waste via a pipeline is possible. Depending on the composition of the waste, the transport facilities, and the size and shape of the cavities to be used as repositories, the waste can be disposed of in containers or in bulk. Because deformation must be expected otherwise, the cavities should be completely filled. The waste must be sufficiently solid or it must solidify after disposal. Hydraulic methods used for filling the cavities must not lead to any excess transport fluid after setting. The degree of solidification of the waste that is required must be sufficient to minimize deformation.

If parts of operating mines are used for waste disposal, a strict partition with a sufficiently tight seal is necessary between the section used for disposal and the part that is still being mined. Ventilation and operating facilities must be adapted to the specific requirements of the two very different modes of operation. The Herfa-Neurode mine in Germany is typical of this kind of mixed operation. It is located in the northern part of the Werra-Fulda rock salt basin in the Zechstein 1 (Werra Series). The bedding of the Werra Series is almost horizontal with a total thickness of about 250 m covered by sandstone (Bunter) more than 500 m thick. The Werra Series contains two potash seams some 2–3 m thick, which are being worked. The upper potash seam (Hesse seam) is overlain by the Upper Werra rock salt, about 100 m thick. The two potash seams are separated by about 50 m at the Middle Werra rock salt, whereas the lower potash seam (Thüringen seam) is underlain by the Lower Werra rock salt, about 100 m thick. A great variety of special waste is disposed of here. It must be at least semi-solid; liquid waste is not allowed.

Recoverability of delivery and storage containers is one of the requirements for obtaining a permit for waste disposal. Also stored here are residues and objects, such as transformers, that contain hazardous substances for which decontamination methods have not yet been developed or for which insufficient capacity is available.

The waste is transported to the mine, an identification check is made, and then it is taken below ground via a shaft (separate shaft for waste transport). About 120 000 t of waste were disposed of in the Herfa-Neurode mine in 1989.

Almost complete isolation from the biosphere
Isolation from the biosphere must be ensured almost completely if there is a possibility that water will enter the cavities after the termination of waste disposal, even in the long-term. This applies also to cavities in argillaceous rock. Such cavities must not be filled with non-degradable, non-sorbable, highly toxic waste if they are soluble in water and thus capable of diffusive or convective migration. The Heilbronn salt mine in Germany, in which waste incineration residues are disposed of in worked-out underground chambers, is an example of such an underground disposal site.

The lower 20 m of an almost horizontal rock salt deposit of the Middle Muschelkalk, about 40 m thick are mined here. The total thickness of the overlying rock (Muschelkalk, Keuper, Quaternary) is 170–210 m. The rock underlying the worked-out chambers consists of the basal dolomite of the Middle Muschelkalk, which is in turn underlain by limestone and marls of the Lower Muschelkalk. The waste is transported and stored in containers. The same shaft is used (in different shifts) for taking the wastes into the mine as for removing the mined salt. About 9000 t of ash from the incineration of wastes were stored in the mine in 1988.

Increase in the distance to the groundwater system in contact with the biosphere
For only slightly hazardous, partly soluble wastes it may be sufficient to dispose of them at depths where the groundwater circulation system has little contact with the biosphere. Caverns left by underground mining may be used for this purpose. In Germany it was planned to use two abandoned coal mines in the Ruhr area for waste disposal: Zollverein of Bergbau AG Lippe and Minister Stein/Fürst Hardenberg of Bergbau AG Westfalen. Both are located in the Upper Carboniferous of the Variscan marginal depression and are mainly in argillaceous shale, greywacke, and sandstone.

The galleries of the Zollverein mine (about 150 000 m^3) would have been filled within about two years. However, while operating this site, it was hoped to gain experience for further similar projects. Only a few types of waste (ash, furnace slag, waste from lime and gypsum production) were approved for disposal in this mine; the wastes were transported into the mine in slurry form in pipes up to 3 km long. Disposal operations, started in 1988, have been given up for economic reasons.

The application for a licence for disposal in the planned underground disposal site Minister Stein/Fürst Hardenberg was withdrawn by the applicant for the same reasons. The usable volume was about 550 000 m^3 at five levels between depths of 450 and 1150 m. On the basis of the experience with the Zollverein mine, the list of acceptable waste was extended to rubble and spoil, waste from foundries and stone and concrete production.

When the wastes are transported in a slurry, measures must be taken to guarantee that the excess water does not cause problems. Chemical binding of the excess water is not required.

Geological and geotechnical investigation and assessment
The hydrogeological conditions of the surrounding area must be included in the geological and geophysical assessment of any waste disposal site as well as for the use of existing caverns for this purpose. This is especially the case for cavities that are not part of a closed hydraulic system.

Groundwater migration and the accompanying pollutant transport must be simulated with mathematical models in order to forecast effects on the surrounding area. Detailed analyses of the geomechanical stability of the cavities must also be made because instability may directly influence the functioning of the host rock as a geological barrier.

All results of the investigation must be included in the assessment of possible effects on the biosphere. This assessment must cover the creation of the cavities, the filling operations, and the post-operation phase. For this purpose, the barriers of the underground disposal site (host rock, the waste itself, walls, backfill, and seals of shafts and boreholes), the surrounding rock, the cover rock, and scenarios involving the total system must be simulated using suitable models, and must be assessed on the basis of site data or sufficiently conservative assumptions. The geochemistry and hydrogeology (e.g. groundwater movement), and their influence in barrier effectiveness must be taken into consideration. Due to their limited lifetime, containers and cavity liners must be disregarded when calculations on long-term safety are made. The creation of underground cavities for waste disposal requires careful weighing of the ecological impact resulting from the creation of these cavities against the expected decrease in pollution from the wastes disposed of.

Problems specific to the disposal of mine waste

Waste from mines presents a special problem since it is generally produced in large quantities and therefore requires large amounts of space for disposal. When disposal sites are assessed, the impact of previous mining must be taken into consideration, e.g. on the homogeneity of the ground, its stability, or the groundwater system. The last must be investigated with particular care, and related to the time when mining and thus the draining of the mine has ended. As mentioned in Section 3.5, waste from mining and waste from beneficiation raise different problems. In the latter case, the problem may arise from contamination by chemical substances added to the process.

In general, it is necessary to determine the extent to which the original geogenic conditions can be restored by the selected disposal method. This is largely done by backfilling as much of the mining waste as possible into the mined-out parts. When a company is planning to enlarge an existing mine or start a new one, it should become obligatory for it to provide evidence that it can dispose of the mine spoil with a minimum of impact on the environment.

If pollutants can be released from the wastes as a result of weathering or processing, the disposal sites must meet the same requirements as those for other wastes, i.e. they must be lined and, depending on the degree of solubility of the wastes, must not be below the groundwater table. Contact with surface water and groundwater must be avoided. The disposal site should have a sufficient geological barrier.

The type of seal to be used depends on the pollutants in the waste, the effects of weathering with respect to the formation of soluble substances, the selected disposal method, the expected rainfall, as well as the general hydrogeology of the site.

Pollution from mine tailings in Sweden [166] [169]

The Swedish Geological Survey carried out a number of studies in the field of pollution control from mine tailings and had the opportunity of comparing three different methods for restricting leachate production:

(1) draining the tailings with ditches;

(2) sealing/covering the top surface of the tailings to reduce leachate production;

(3) covering the tailings with water to reduce weathering and leaching.

The choice between these methods is a case-by-case operation which involves a proper geological investigation and adequate water balance estimation.

Pollution from mining sites in Greenland [119]

The Geological Survey of Greenland carried out monitoring studies of lead dispersion in sea water at two mining sites, where tailings and mine waste were dumped in the fjord.

The lead content in dated cores of the bottom sediments has likewise been determined in order to trace the level of lead dispersion through time. In addition, the content of copper, zinc, and cadmium was established by analysis.

Dumping in the sea was not considered to pose an environmental risk. In one case, where mining activities started in 1858 and terminated in 1987, limited tonnages of mine waste were disposed of along the coast of the fjord and subjected to tidal erosion which led to dispersion of both dissolved and particulate lead in the environment.

In the second case, permission to dump tailings from the concentration plant of a lead–zinc ore was given by the local authorities in 1972. Soon, dissolved lead heavily polluted the bottom water layer in the fjord, at a depth of 40 m, presumably because of the relatively high oxygen content of the sub-arctic sea water. This case is particularly well-documented and monitoring of the lead pollution continues even after closure of the mine in 1990.

It is clear from these studies that any dumping of both mine waste and tailings in fjords in the sub-arctic region should be avoided, as even relatively minor sources of heavy metals are subjected both to oxidation and dissolution, with dispersion of particulate lead in the tidal zone.

In the case of 'dry' disposal of mining waste with only a low to medium solubility, a surface cover or lining is generally sufficient to reduce the input of rainfall to such a degree that the migration of pollutants remains within tolerable limits. If, however, waste is hydraulically filled into the disposal site and if it consolidates there, a bottom liner is required for the disposal site to meet an increased input of seepage water. In the case of highly water-soluble waste, such as rock salt obtained in potash mining, an adequate bottom liner at the disposal site is imperative. Salt is neither immobilized by substances in the ground nor decomposed, and therefore nothing can hinder its migration in the groundwater.

Also when solid mine-waste is disposed of, care must be taken that the solubility is not increased by the mixing of the substances e.g. by contact with toxic heavy metals whose migration mobility may thus be considerably increased.

A special problem is the stability of such waste disposal sites. The effect of water on the solidity of the waste and on the distribution of stress within the waste itself is of particular interest as a function of the disposal rate. In order to cope with hazards such as changes in solidity, failure of drainage systems, earthquakes, etc., stability calculations must be made for each case. These calculations must include the load-bearing capacity of the substratum, as well as provide information on the deformation of the wastes.

Disposal of liquid waste

The disposal of liquid waste is not allowed at or near the surface, but underground disposal is possible (and allowed in certain countries) if certain conditions are fulfilled. At present, the only possibility is in cavities that are not subject to reduction in volume due to convergence. These are represented by pore, joint, and karst systems which can be open, closed, or partly closed. The long-term sealing of entries (boreholes, tunnels, shafts) to caverns and mines is still too unreliable for such cavities to be used for disposal of liquid waste.

In principle, any fluid waste can be disposed of, provided it does not react unfavourably with the host rock, does not readily form a gas, and does not produce sediment or precipitate, or lose its fluidity. Such circumstances would reduce the capacity of the cavity or make it impossible to fill it.

This means that the fluid waste must be inert with respect to the host rock and the groundwater. If this is not taken into consideration, the safety of the facility may be affected directly or indirectly in a decisive way. Organic liquids or liquids containing organic components should be excluded from underground disposal, except when they are being put back into the formation they were taken from, e.g. formation water obtained during oil or gas production.

In general, a site under consideration for underground disposal should have no hydraulic connection with the surrounding rock, or the spread of pollutants into the biosphere must be delayed by thick cover rock. This requires a very large aquifer system to guarantee the necessary mean residence time.

Site reconnaissance studies should provide information on:

(1) the geological and tectonic structure of the system to be used for waste disposal;

(2) the petrography and mineralogy of the host rock and the rock surrounding the waste disposal site;

(3) the hydraulic system (closed, partly closed, open);

(4) the possible waste disposal capacity and pressure conditions;

(5) seismicity of the site.

These data, together with information on the type of liquid to be disposed of, are used to assess:

(1) inertness of the geological environment with respect to the waste;

(2) quality of the seal between the disposal site and the surrounding rock;

(3) impact on the groundwater circulation system in contact with the biosphere;

(4) impact of the waste on the mechanical stability of the surrounding rock;

as well as

(5) safety of the site during and after operations;

(6) pressure required during operation;

(7) acceptable waste supply rates;

(8) factors that may influence operations (formation of gas, clogging of cavities due to precipitation and settling of sediments);

(9) impairment of the mechanical stability (changes in the shape of the cavities, movement along faults (earthquakes), subsidence or uplift),

(10) sealing of the entries to withstand efficiently pressures that may build up within the repository.

In addition to the evaluation of various data on the structural geology of the area, the identification of aquifers for waste disposal requires comprehensive geological, geophysical, and hydrogeological studies in the field and in the laboratory. Test drilling, including hydraulic experiments for the investigation of the actual behaviour of the aquifers (conductivity, porosity, pressure, etc.), are absolutely essential.

The input of liquid wastes should take place only when systems have been installed for monitoring the hydraulic environment. Changes in the chemical composition of the groundwater must be recognized in time to exclude hazards to the groundwater in contact with the biosphere. Therefore, a network of observation wells should be installed at different depths, taking the geology and hydrogeology of the area into consideration.

RECOMMENDED FURTHER READING

Aust, H. and Kreysing, K. (1985). Hydrogeological principles for the deep-well disposal of liquid waste and wastewaters. *IHP/OHP-Berichte*, Sonderheft 1. Deutsches IHP/OHP Nationalkomitee, Koblenz.

Baccini, P. (ed.) (1989). The landfill. Reactor and final storage. In *Swiss workshop on land disposal of solid wastes, Gerzensee, March 14–17, 1988*. (Springer, Berlin).

Bonomo, L. and Higginson, A. E. (eds) (1988). *International overview on solid waste management*. (Academic, New York).

Collective Authors (1989). Landfill concepts, environmental aspects, lining technology, leachate management, industrial waste and combustion residues disposal. *2nd International Landfill Symposium. Porto Corte (Alghero, Sardinia), October 1989*.

Carra, J. S. and Cossu, R. (eds) (1990). *International perspectives on municipal solid wastes and sanitary landfilling*. (Academic, New York).

Forester, W. S. and Skinner, J. H. (1987). *International perspectives on hazardous waste management*. (Academic, New York).

Franzius, V., Stegmann, R., and Wolf, K. (eds) (1988). *Handbuch der Altlastensanierung*. (R.v. Deckers, G. Schenk/Economica, Heidelberg).

Woods, R. D. (ed.) (1987). *Geotechnical practice for waste disposal: Proceedings of Specialty Conference*, University of Michigan, Ann Arbor, June 15–17, 1987. (American Society of Civil Engineers (Geotechnical Special Publication 13), New York).

5.4.2. Disposal of radioactive waste [75]

Overview [75] [41] [60] [52] [42]

One of the major causes contributing to the degradation of the geoenvironment is the disposal of wastes from industry, agriculture, municipal services, and mining. Among these, radioactive waste is a highly sensitive one, particularly due to the radioactive contamination risk in the disposal of various types of waste. Radioactive wastes are usually subdivided into short-lived waste (about 10 years) and long-lived waste (about 1000 years). The German regulation, for instance, differentiates three categories:

(1) low activity (level) waste (LLW). Surface activity reaches 200 mrem/hour;

(2) medium activity waste (MLW) = intermediate level radioactive waste (ILW), which requires distant handling and protection. It is stored in unused chambers or drill-holes;

(3) high activity waste (HLW) for which final storage temperatures in salt can reach approximately 200 °C.

The French classification of A, B, and C categories is comparable.

A characteristic of radioactive waste is that the quantity which has to be stored is relatively small. In the case of France, which is the European country most dependent on electricity of nuclear origin (Table 5.6), storage requirements per year are approximately 300 000 m³ for A category waste, 300 m³ for B and 200 m³ for C. The storage conditions differ in that LLW can be stored in deep or shallow facilities, whereas deep disposal (600–1000 m) is internationally agreed upon for

Table 5.6 Comparative production of electricity of nuclear origin (French Atomic Energy Commission (CEA) 1990)

Country	Production gross generation (TWh)		Cumulative total to end 1989 (TWh)	Nuclear share in electricity production 1989 (per cent)
	1988	1989		
Argentina	5.8	5.0	57.1	11.4
Belgium	43.1	41.2	346.1	60.7
Brazil	0.6	1.8	9.3	0.7
Bulgaria	16.0	14.6	145.8	32.9
Canada	85.6	83.2	799.7	15.6
Czechoslovakia	23.3	26.4*	133.8*	27.6*
Finland	19.3	18.8	181.5	35.4
France	275.2	303.9	2131.5	74.6
Germany, East	11.7	12.3	152.7	10.9
Germany, West	145.3	149.5	1207.2	39.5
Hungary	13.4	13.8	58.4	49.8
India	6.1	4.0	64.1	1.6
Italy	0.0	0.0	93.1	0.0
Japan	175.0	186.7	1630.6	27.8
Mexico	0.0	0.4	0.4	0.1
Pakistan	0.2	0.1	5.2	0.2
Sweden	69.4	65.6	624.8	45.1
Switzerland	22.7	22.8	258.1	41.6
South Africa	11.1	11.7	48.7	7.4
South Korea	40.1	47.4	208.3	50.2
Taiwan	30.7	28.3	232.3	35.2
USSR	215.6	212.6	1686.1	12.3
Yugoslavia	4.1	4.7	32.6	5.9

* Estimated value

HLW, and this is being investigated in various European countries at present.

In nuclear medicine and nuclear research, in nuclear power plants, in the nuclear fuel cycle industry, and in other industries, there occur various types of radioactive wastes which can be divided into different categories according to their specific activity and physical-chemical state. Their potential for radiological danger, to which a range of protective measures are applied, does not depend only on the above-named properties, but also on the activity inventory, the type of radiation, and the half-life of the radionuclides contained in the waste. For long-term exclusion of unacceptable radionuclide concentrations in the biosphere, the radioactive wastes must be transformed into a sufficiently erosion- and leach-resistant form and disposal must take place in a suitable geological medium.

Important steps for establishing a radioactive waste repository are site selection and site investigation. These do not only produce the necessary data for safety analysis, but also for the planning and operation of the repository.

From a geological point of view, any disposal site for radioactive wastes must be chosen so that the transport of dangerous quantities of radionuclides into the biosphere via circulating groundwater can be avoided. In principle, all final disposal arrangements should be safeguarded by a system of parallel or interlocking natural and technical barriers (multibarrier principle), although the effectiveness of such technical and natural barriers may receive different weighting in different disposal concepts.

The waste package and any artificial sealing materials provide technical barriers, which reduce the leaching rate to a minimum, and hinder chemical reactions between the waste materials and the soil or rock which could lead to the disintegration of either of them. Natural barriers consist of the soil or rock formations surrounding the waste. Above all, the host rock should ensure that the inflow of water, capable of leaching, and/or the outflow of contaminated water from the final repository facility into the biosphere is either fundamentally impossible, or at least can be kept within acceptable limits, as a result of the natural hydraulic condition.

A number of geological formations have been studied in recent years to assess their potential as host rocks for deep underground repositories. The formations studied include evaporites (generally rock salt), hard crystalline rocks (e.g. granite, gneiss, basalt, schist), and argillaceous formations (clay, shale). Extensive research has been carried out into the possible use of some of these formations, including repository design, waste-emplacement techniques, and long-term safety assessments.

Since 1975, several CEC research programmes have led to the preliminary appraisal of a significant range of potential geological repositories. At the scale of individual countries the strategy varies. Germany has chosen to store highly radioactive waste in rock-salt. In other countries, such as Sweden, Finland, UK, and Spain, a granite environment has been preferred. In France granite, metamorphic schist, clay, and rock-salt are being considered, but the final choice has not yet been made.

The suitability of a geological medium for final disposal can only be demonstrated if a comprehensive safety analysis has shown that the interaction between the waste product, the disposal facility, and the geological environment can maintain the predetermined level of protection. This level is in fact discussed in various international organizations and there are national-specific interpretations of choices. The products to be disposed of and the engineering concept for the disposal facility (on the one hand) and the geological conditions (on the other hand) place requirements on each other.

The proof of the safety of a disposal facility is the ultimate consideration. This leads directly to questions on the principles and criteria for such a demonstration. Since, in the case of final disposal in geological formations, the load-bearing capacity of the soil or rock, the protective effects of the surrounding rock formations over extended periods, and the geological stability of the area around the facility are all important, such a demonstration of safety cannot be undertaken solely from the viewpoint of construction engineering, but must also take geological factors into account. Safety factors as applied to civil and structural engineering are not sufficient in this case.

A safety concept which takes into consideration the above-mentioned requirements and is based on the principle of multiple barriers should include the following main features:

1. Separate analysis of the effectiveness of individual barriers as technical systems (waste package, sealing material), as rock mechanical systems (boreholes, mines, shafts), and as geological systems (hydrogeology, tectonics).

2. Analysis of the physical and geochemical processes which may result from the mutual interactions of the barriers of the various systems, and evaluation of their significance for the unwanted transport of harmful materials, both in the immediate vicinity of the final storage facility and over greater distances.

3. Comprehensive analysis of the safety of the final facility, performed by identifying and evaluating the mutual interactions of all barriers for the case of particular theoretically possible events (accidents, failure) which could lead to the release of radionuclides, e.g. judgement of the possible paths of release and the ensuing damage (scenario analysis, failure analysis).

The following case histories give some examples of how the Geological Surveys contribute to solving environmental problems connected with radioactive waste disposal.

Surface disposal of short-lived waste

The Soulaines-Dhuys (France) example [41] [60]

Low and medium activity waste contains mainly short-lived radionuclides, the half-life of which is less than 30 years. The maximum accepted content in long-lived radionuclides is fixed at a low value laid down by the Safety Authorities. After 300 years the potential risk is considered to be negligible.

The decision to build a second disposal facility for LLW in France, to replace the Hague (Manche) site which operated for nearly 20 years, was made in 1984 following approval by the French government of the National Radioactive Waste Management Programme prepared by l'Agence Nationale pour la gestion des déchets radioactifs (ANDRA). Following a phase of preliminary selection and evaluation, the Soulaines-Dhuys site was chosen and detailed investigations carried out, mainly by BRGM. Site location is shown on Fig. 5.20.

Fig. 5.20 Location of La Hague and Soulaines storage sites for radioactive waste.

Preselection phase

Area selection was carried out in 1984 and 1985 on the basis of the fundamental safety criteria which characterize this type of disposal:

(1) well-defined hydraulic boundaries;

(2) homogeneity of lithology;

(3) stable climatic and hydrological conditions;

(4) weak to medium regional seismicity;

(5) good protection from ground movements;

(6) good topographic protection from water flow and flooding;

(7) proximity of water drainage outlet with an adequate dilution capacity.

Screening of available information by ANDRA led to the choice of two basic types of site:

(1) sand or clay-sand formation overlying a clay formation. Respective minimum thicknesses are 10 m and 20 m;

(2) massive and impermeable Permian pelite formation.

Crystalline igneous or metamorphic basement, mountain areas, and strongly seismic zones were excluded.

The preselection phase led to an initial selection of more than 70 potential sites located in about 20 zones which were finally reduced to three: Montmorillon (Vienne), Lignac (Indre), and Soulaines-Dhuys (Aube), all belonging to ANDRA's site type 1.

Geological and hydrogeological surveying was done on the three sites (1:25 000 scale geological map, subsurface reconnaissance by boreholes). Later, work was concentrated on the 100 km² Soulaines-Dhuys area.

Geological mapping at the 1:25 000 scale, drilling for geological reconnaissance and piezometer setting enabled the selection of the 100 ha Pli area, which in 1987 was officially confirmed as the final site selected and renamed 'Aube Storage Centre'.

Detailed investigations on the Aube Storage Centre

Investigations started in 1987 and some studies are still continuing, while construction of the Centre is progressing. Studies can be

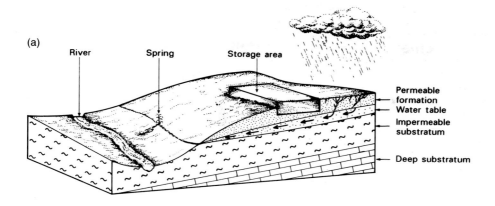

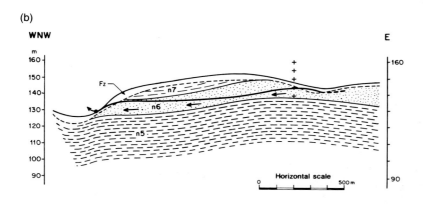

LEGEND

LITHOLOGY		HYDROGEOLOGICAL CHARACTERISTICS	STRATIGRAPHY
Fz	Silt and sand	Semi-permeable to permeable	Surficial Quaternary formations
n7	Clay ± sandy	Very slightly permeable	Lower Albian
n6	Sand	Permeable	Upper Aptian
n5	Clay	Very slightly permeable	Lower Aptian

Spring	Flow direction	Ground-water level (Sept. 1985)	Ground water divide

Fig. 5.21 (a) Ideal section for ANDRA; (b) geological section across the Soulaines site.

grouped under three headings, which more or less correspond to the chronology of investigation:

(1) site characterization,

(2) geotechnical control,

(3) water flow modelling.

Site characterization included complementary regional geological mapping (1:25 000) with the help of geo-electrical sounding, detailed geological mapping (1:5000) and systematic boring for geological purposes and piezometer network setting. Geophysical logging (mainly γ-ray) was used to help in locating the lithological boundaries precisely.

The typical geological sequence includes, from top to bottom (Fig. 5.21):

(1) a few metres of clays;

(2) 10 m sands;

(3) 20–25 m clays;

(4) 60 m mixed sands and clays;

(5) Jurassic limestone, considered to be the geotechnical basement.

Site characterization drilling was mainly concentrated on the first sand layer. The geotechnical control phase included core-drilling down to the Jurassic basement followed by laboratory geotechnical tests on cores, dynamic geotechnical tests from the surface (penetrometer, pressiometer), and cross-hole geophysics for the determination of dynamic modules to be used in seismicity risk calculations. A piezometer network was set up for groundwater monitoring during the site construction and exploitation phases. Piezometer location was optimized by geostatistical methods.

Water flow modelling was aimed at understanding in detail the water movement in the geological strata underlying the site and building a model to simulate a large variety of scenarios. Tracer tests were used for measuring water flow velocities and dispersive properties of the aquifers. The final objective is a fully operating coupled model, including water flow dynamics and substance dispersion and transport.

At the time of writing, this phase was still in progress.

Concluding remarks

The background strategy for selection of the Aube Storage Centre venture is original in that the objective was to choose a specific site which could be characterized with great accuracy, with respect to the detailed geology, site stability control, and water flow monitoring.

Although the studies concerning this last point are still in progess, a wide range of scenarios have been tested for site safety purposes. According to the Declaration of Public Utility for the site, disposal capacity is 1 000 000 m³ and the annual supply of waste 30 000 m³. Operating period of the site is thus of the order of 30 years.

Deep disposal of long-lived waste [75]

There are a number of reasons why disposal in deep-mined repositories or boreholes can meet concerns about high-level radioactive waste. Firstly, it is clear that burial in deep geological formations provides adequate shielding against radiation from the waste. Suitably chosen formations can provide the required isolation and can also absorb and disperse the heat produced by high-level waste. A deep underground location also decreases the likelihood of inadvertent intrusion by man, and essentially rules out the possibility of malicious intrusion. Geological disposal is also an entirely passive system of disposal which does not depend on human involvement for continuing security.

In deep disposal of long-lived and highly-radioactive wastes the geological barrier plays the essential role regarding the multibarrier system. Therefore, the evaluation of the efficiency of the natural barriers is an important part on the safety assessment of an underground waste repository.

Over the lengthy periods for which some of the nuclides remain radioactive, the possibility of their movement away from the repository site and their return to the biosphere must be considered. Consequently, safety assessments of geological disposal systems are based on the premise that natural processes combined with slow geological evolution may eventually lead to the release from the original site of some portion of the waste. Such assessments include modelling of these processes to determine in what concentrations and over what time periods particular radionuclides might be released, and if they reach the biosphere whether they will then constitute any biological hazard.

Site selection and characterization

The overall criteria for choosing a site are generally agreed as being:

(1) safety criteria: geodynamics: stability
                               hydrogeology: very low
                               permeability

(2) potential for development: geomechanics
                               (hydrogeology): economic
                               and stability assessment.

The process of site selection has been carried out in each of the European countries concerned with the deep storage of radioactive waste. In each case, the Geological Survey played an essential role in connection with the Atomic Energy

Authorities. The site selection procedure for the deep disposal of intermediate level radioactive waste in Great Britain is given as an example.

Site selection in Great Britain [195]

Introduction

It is estimated that by the year 2030 civil sources in the United Kingdom will have produced about 250 000 m³ of intermediate level radioactive waste (ILW) with an additional 20 per cent of defence-sourced waste.

In 1987, plans to develop a shallow disposal facility for low level radioactive wastes (LLW) were abandoned in favour of a combined disposal facility with ILW. Thus, since this time the United Kingdom has been cncentrating on defining a suitable location for the construction of a deep repository to accommodate an estimated combined volume of 1.3×10^6 m³ of ILW and LLW wastes, and remain operational for upwards of fifty years.

Since 1982, the United Kingdom Nuclear Industries Radioactive Waste Executive (UK Nirex) has held responsibility for the implementation of government strategy for the management of radioactive waste. UK Nirex has followed the three-stage process suggested by the International Atomic Energy Agency (IAEA): (1) national survey and evaluation to define potentially suitable geological sites; (2) site identification within areas defined by (1) to produce a small number of suitable sites for further study; and (3) site confirmation by detailed geological and geophysical study to test site suitability in detail. The British Geological Survey (BGS) in its capacity as geological advisor to UK Nirex, has been instrumental in developing the geological rationale for the repository site selection.

At present UK Nirex is at state (3) above, investigating the potential of two sites, one at Sellafield in Cumbria, England, and the other at Dounreay in northern Caithness, Scotland.

This overview outlines the geological aspects of the site selection process (1) and (2) above, which has led to the present position and describes the involvement of the BGS in this process.

National survey and evaluation: the definition of suitable disposal environments

The UK, in common with many other countries, has been through the exercise of defining suitable host rocks for high-level radioactive waste (HLW) disposal, using factors such as thermal stability and low permeability of the host rock. Much of the approach to defining suitable hydrogeological environments was based on well-understood guidelines applied in earlier exercises. However, these were applied to defining aspects of the larger scale geological environment, namely, the presence of (a) predictable groundwater flow paths, preferably long and resulting in progressive mixing with older, deeper waters or leading to discharge at sea, and (b) very slow local and regional groundwater movements in areas with low regional hydraulic gradients and/or low hydraulic conductivity.

In addition to these factors any proposed environment would have to demonstrate conformity with the many accepted restrictions regarding the regional and local significance of mineral deposits, geothermal gradients, seismicity, formation depth, structural complexity, etc. From the geotechnical viewpoint, the concept would have to meet construction criteria to allow for economic repository design.

By applying these concepts to the growing body of deep geological and regional hydrogeological studies for the United Kingdom, five hydrogeological environments were defined as the most suitable for the disposal of long-life wastes in Great Britain.

Inland basinal environment

Deep sedimentary basins containing mixed sediments with a high proportion of poorly permeable formations (mudstones, evaporites, volcanics, etc.).

Seaward dipping and offshore sediments

Similar in concept to inland basinal environments, with groundwater movement expected to be very slow at depth, and towards and under the coast.

Low-permeability basement under sedimentary cover (Fig. 5.22)

Basement rock of low intrinsic permeability (principally hard shales, mudstones, slates, quartzites or volcaniclastics, with some hard crystalline rocks) occurring under more recent sedimentary cover.

Hard rock in low relief terrain

Hard rocks have the advantage of making the construction of the underground facility easier.

Small islands

Small islands were considered worthy of investigation, almost regardless of rock type, since a deep repository could be sited below the sea water-freshwater interface, where groundwater conditions are essentially stagnant.

The conceptual environments above were ranked in order of preference. Three, with simple and consequently more predictable hydrogeological structure were initially considered to show the best potential in the United Kingdom and initial efforts were directed towards these, namely; (a) offshore dipping sediments in areas of simple structure, (b) low relief hard rock, and (c) small islands. It was thought likely that these three environments would yield a sufficient number of sites suitable for further investigation.

Location of suitable environments in Great Britain (Fig. 5.22)

A desk study was initiated reviewing the available geological database to define the location of suitable environments within Great Britain.

Inland basins, seaward dipping and offshore sediments

Five stratigraphic intervals were considered under this heading as potentially suitable for further investigation in Great Britain, their lithologies being dominantly argillaceous or evaporitic. These were, (a) the Permian sequence, including the basal clastic sediments; (b) the Triassic, Mercia Mudstone Group; and three Jurassic mudrock sequences, namely (c) the Liassic, (d) the Oxford Clay, and (e) the Kimmeridge Clay (including the upper Corallian and Ampthill Clays).

Areas of interest were defined for each of the above by applying depth and thickness guidelines to the BGS deep geological database. These guidelines, although somewhat arbitrary, proved to be a useful means of narrowing down the areas of interest, namely, (a) where any part of the formation is more than 200 m below the surface, (b) where the base of the formation is 1000 m below the surface, and (c) where the formation thins to less than 50 or 100 m thick.

In addition to the geometric considerations applied above, tectonic features, such as the presence of major fault zones or zones of structural complexity, were used to reduce further the areas of interest.

Of particular interest was the area known as the Eastern England Shelf, where all the stratigraphic intervals considered are presented in an area of minimal structural complexity, which extends to the Southern North Sea Basin.

Most of the areas concerned are licensed for oil and gas exploration and production or have a potential for geothermal energy.

Hard rocks in low relief terrain

Hard rocks in low relief areas, comprising crystalline, igneous or metamorphic rocks, well indurated argillaceous rocks and some clastic sediments are dominantly found in the north and west of Great Britain. Potentially suitable environments were found in Wales and Scotland.

Small islands

Over one hundred potentially suitable small islands were identified, the majority of which lie off the west coast of Scotland, or in the Orkneys and Shetland, although a few are located around the coast of England and Wales. Islands of less than about 0.5 km², those with extreme topography, and those not sufficiently far from the mainland to necessarily possess independent hydrogeological regimes were excluded.

Low-permeability basement beneath sedimentary cover

A large area of potentially suitable Precambrian and Palaeozoic basement, is present at relatively shallow depths forming part of the London Platform.

The region was delineated by the 800 m structure contour on the pre-Cambrian surface, the presence of Coal Measures, which are thought to be a local source of gas, and the presence of major aquifers in the overlying sedimentary sequence, namely, the Triassic age Sherwood Sandstone Group.

Site identification

After the initial definition of potentially suitable areas on geological and hydrogeological grounds, the site identification followed a process of progressive evaluation and sieving in which many non-geological factors were taken into consideration.

The areas were reduced by the application of two basic planning and environmental filters which excluded (a) areas with high population densities, and (b) areas designated as nationally important for their landscape and/or nature conservation value.

Perhaps the most significant constraint, was the reduction of the area of search to specific sites available to Nirex in the absence of compulsory land purchase powers, largely those in public and government ownership. Some sites were rejected as being too small for repository construction.

Site-specific geological and hydrogeological assessments were performed for all remaining sites (around one hundred and twenty), utilizing data available in the BGS national geological data archives. Further sites were eliminated on geological and hydrogeological grounds by applying expert judgement on factors e.g. suitability of local stratigraphy, structural complexity, local hydrogeological conditions, the presence of public supply aquifers, and potential hydrocarbon and mineral resources.

A number of stages of progressive evaluation reduced the number of sites to around forty. During this process increasingly detailed assessments were made of the planning, transport, construction, radiological protection aspects and availability of the specific sites. About twenty sites were identified as 'front-runners' on the basis of expert review. These sites could be categorized into twelve options, which comprised coastal, inland and island sites underlain by hard rock; coastal and inland sites underlain by basement rock under sedimentary cover; coastal sites underlain by seaward dipping sediments; and generic offshore hard rock and sedimentary disposal sites.

To compare the merits of the different options a decision analysis methodology of weighted multi-attribute analysis was developed following methods similar to those applied previously in the USA. A mathematical model was produced to describe the system of identified attributes, which were grouped under four main headings:

(1) *Safety*: the pre-closure or operational safety, and post-closure safety including possible future human intrusion;

(2) *Socio-economic and environmental impact*: the impact of a repository on the local community, environment, and economy;

(3) *Robustness*: the extent to which the site performance could be sustained and verified in light of current uncertainties and identified constraints;

(4) *Costs*: these were estimated from conceptual repository design and transport studies of the sites.

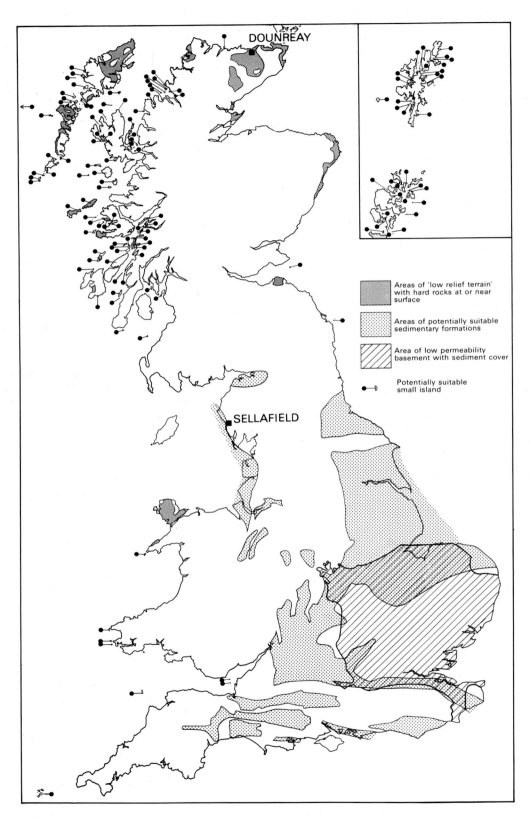

Fig. 5.22   Location of potentially favourable environments for deep storage of radioactive waste in Great Britain.

The geology and hydrogeology of the site has a large effect on the radiological safety, specifically, in defining groundwater fluxes and potential radionuclide migration pathways to the biosphere, and potential future inadvertent human intrusion. Geology also influenced repository design and construction, and hence costs. In addition, two gross geological attributes were considered under robustness:

(1) *Predictability*: the geological and hydrogeological predictability of the environment in which the site is situated;

(2) *Investigability*: the likelihood of a site investigation being able to provide the relevant data with a sufficiently high degree of confidence.

A number of weightings were applied to reflect differing opinions on the relative importance of the different attributes and a series of sensitivity analyses were performed. Overall, land-based locations in basement under sedimentary cover and in hard rock with low relief were judged to show the best potential. In light of this a shortlist of sites thought worthy of more detailed appraisal was identified.

Taking into account the views expressed during a public consultation exercise, UK Nirex decided to limit initial investigations to sites from the shortlist with a measure of public support, resulting in the choice of sites at the existing nuclear establishments of Sellafield and Dounreay.

The Sellafield works of British Nuclear Fuels Limited is located on the Cumbria coastal plain in north-west England. Basement host rock of the Borrowdale Volcanic Series of the Skiddaw Slate Group is overlain by a mixed marginal sedimentary sequence of Carboniferous Limestone, Permian sandstones, marls and evaporites, and Triassic sandstones. North–south to north–north-west trending faults divide the region into a series of tilted fault blocks.

The Dounreay Nuclear Power Development Establishment is located on the low relief coastal plain of north Caithness, Scotland. About 300 m of Devonian, indurated siltstones and sandstones of the Caithness Flagstone Group overlie a crystalline basement host rock of either diorite, Moine metasediments or granite.

As part of the site confirmation phase detailed geological, geophysical, hydrogeological, geochemical, and geotechnical investigations are now under way at the two sites with the eventual aim of demonstrating the suitability or otherwise of the sites for disposal of ILW. If these sites prove to be unsuitable then further possible locations will be defined.

Planning and construction [75]

From a general definition of stability and the particular protection aims of radioactive waste disposal, it follows that a disposal mine can be considered stable when the following separate conditions can be satisfied:

1. For the construction and operation phase (open or partly backfilled mine):

   (a) The load-bearing capacity of the rock will not be impaired as the result of stress changes (including thermal stresses) to such an extent that the load-bearing capacity of the system is exceeded, either suddenly (rock burst) or slowly (creep failure);

   (b) During or after the mining operations, no inadmissible deformation and/or creep occurs either in the underground openings (convergence, creep failure) or on the surface (subsidence) which may critically influence the utilization of the mine and/or the safety of above-ground contructions;

   (c) The extent of decomposition of minerals (e.g. by dehydration) due to thermal loads must not lead to the detriment of the integrity of the host rock;

   (d) Uncontrolled leaks of water, brine, and/or gas can be avoided.

2. For the post-operative phase (completely backfilled mine):

   (a) The integrity of the surrounding rock will not be impaired, even over extended periods, as the result of stress changes (possible crack formation), decomposition of minerals, and/or corrosion,

   (b) Thermally-induced brine migration and groundwater movement must not initiate any unacceptable release of radionuclides.

From these criteria it can be seen that the repository design and the evaluation of the long-term effectiveness of the host rock barrier must include a consideration of the following additional factors:

3. Natural factors:

   (a) Geological conditions (e.g. sequences, petrographic composition of the rock, tectonic features, joint texture);

   (b) Hydrogeological conditions (e.g. pore water, joint water, permeability, groundwater movement);

   (c) *In situ* stress conditions;

   (d) Effects of earthquakes;

   (e) Occurrences of gas and brine;

   (f) Temperature and time-dependent deformation characteristics of the rock and/or rock mass (e.g. elasticity, creep behaviour, plasticity);

   (g) Temperature and time-dependent failure of the rock and/or rock mass (e.g. compression, shear and tensile strength, creep failure parameters, maximum load bearing capacity).

4. Technical factors:

   (a) Properties of the radioactive waste (e.g. heat generation, radioactivity, half-lives of radionuclides, long-term physical and/or chemical effects);

   (b) Mining techniques (e.g. blasting, excavation by cutting machines);

   (c) Geometry of underground opening (e.g. shape and size of galleries and pillars, construction of dams within the mine);

   (d) Operational data (e.g. arrangement of boreholes for final storage, maximum admissable temperature in accessible galleries, safety measures).

5. Factors related to the system:

   (a) Failure modes of the rock and/or support;

   (b) Limiting values for temperature increase in the groundwater (overburden);

   (c) Limiting values for temperature increase in minerals;

   (d) Change of rock properties due to radioactive irradiation;

   (e) Adsorption properties of rock (e.g. adsorption of radionuclides in clay minerals);

   (f) Long-term geochemical processes;

   (g) Long-term geological processes (e.g. tectonic processes).

The exhaustive list of factors described above does not apply to all geological contexts. Some of the factors may be more relevant to specific geology.

The practical demonstration of stability of a final disposal facility can only be done using a combination of various

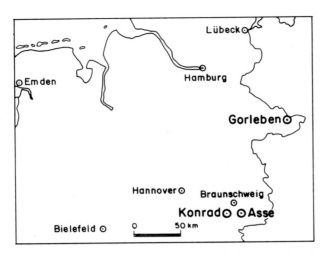

Fig. 5.23 Location of Asse, Konrad, and Gorleben sites in Germany.

investigations and calculations. Engineering geological and geotechnical investigations, rock-mechanics measurements, static calculations, the monitoring of technical parameters, and mining experience must all receive consideration.

As an example, the stability analysis done by the Federal Institute for Geosciences and Natural Resources of Germany (BGR Hannover) for the Konrad Mine project is described (Fig. 5.23).

Konrad Mine [75]

In the former iron ore mine in the vicinity of Salzgitter-Lebenstedt it is planned to dispose of non-heat-generating radioactive waste. The licensing procedures for this project have been initiated. The assessment of the stability of the storage cavities and the effectiveness of the barrier of the surrounding rock strata are considered to have favourable values. The overlying strata consist for the most part of clay/marl stones of low permeability. Mining data on the area have been collected now for more than 20 years. Measurements have shown that the mine, including the overlying formations, has been stable throughout its existence, and that mining has not produced subsidence at the surface. For planning procedures it had to be proved that this stability would also apply for the storage facility.

The following geological factors were of decisive importance in selecting the Konrad location as a repository for non-heat-generating radioactive waste (Fig. 5.24).

1. The former iron ore mine, at depths of 800–1300 m, is well covered by a Lower Cretaceous claystone/marlstone sequence of low permeability and several hundred metres thick.

2. The ore horizon does not crop out anywhere in the vicinity of the Konrad pit; for this reason there is no direct contact between the deep groundwater and the unconfined groundwater near surface level of the biosphere.

3. The barrier represented by the Lower Cretaceous is not cut by any fault zones. Large fault systems usually terminate below the Cretaceous transgression.

4. More intense tectonic stresses on the ore horizon are limited to some few sections in the underground structure of the mine and are controllable using straightforward constructional measures.

The storage cavities are arranged along the strike of the ore bed. The finite element model used describes a section through an area of 170×135 m. Three chamber cross-sections of approximately

40 m² were specified with a separation of 28 m which represents a pillar/chamber ratio of $4:1$, using a finite element grid. The results of the calculations can be shown as contours of equal safety. The coefficient indicates the ratio between maximum permitted shear stress to calculated shear in the rock mass.

Long-term safety

As already mentioned, in relation to the multibarrier principle, the geological environment must be able to contribute significantly to the isolation of the waste over long periods. The assessment of the integrity of the geological barrier can only be performed by making calculations with validated geomechanical and hydrogeological models. As an example, some work of the Federal Institute For Geosciences and Natural Resources of Germany (BGR) for the Gorleben site is given in the following case history.

Gorleben [75]

In February 1977, the government in the State of Lower Saxony nominated the Gorleben salt dome (Fig. 5.23) as a candidate site for a waste repository. PTB (Physikalisch-Technische Bundesanstalt), which at that time was the acknowledged federal agency for the construction and operation of repositories in the Federal Republic, in cooperation with BGR, organized an extensive site investigation programme to be carried out from the surface. This programme was carried out between 1979 and 1983 and had three main goals:

(1) hydrogeological investigations;

(2) geological investigations;

(3) shaft pilot holes in preparation for underground exploration by mining methods.

The site investigation from the surface will be completed by a final phase of an underground exploration programme that has just started. This programme consists of:

(1) salt petrographic investigations;

(2) geological structural investigations;

(3) geochemical investigations.

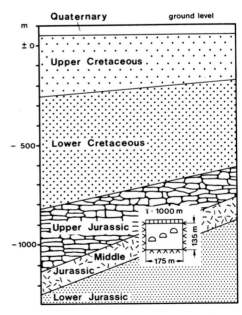

Fig. 5.24 Type section of the strata around the Konrad mine.

Geological investigations of the salt dome

From January 1980 to March 1981, four deep geological investigation boreholes, Gorleben 1002, 1003, 1004, and 1005, were drilled and cored on the flanks of the dome to a depth of about 2000 m. They encountered all of the Zechstein strata with the exception of Zechstein 1. The evaluation of the results from these boreholes has been documented in several geological profiles published by BGR. All boreholes were plugged with clay and various special cements.

The Goreleben salt dome is situated in north-west Germany in the eastern part of Lower Saxony near the Elbe river, which here forms the border with the former German Democratic Republic (DDR). The dome is about 15 km long and 4–5 km wide. During Late Permian (Zechstein) time, evaporites with a total thickness of about 1300 m were deposited in this part of Germany. This sequence of evaporites consists of six cycles (z1–z6), the second and third of which contain potash layers; the first and last two cycles (z5 and z6), are very incomplete. In the Gorleben area, a diapir, formed by Keuper time, subsequently developed into a salt dome. During Late Jurassic and Early Cretaceous times, the dome broke through the enclosing sediments. From Late Cretaceous to Quaternary times, a cap rock formed, above which Upper Cretaceous, Tertiary, and Quaternary sediments were deposited.

During the Quaternary an erosion channel formed beneath the ice sheet covering the Gorleben salt dome. The channel belongs to a system of glacial channels that can be traced throughout north-west Germany. The channel above the dome originated during the Elsterian period of glaciation.

The thickness of the ice sheet covering the salt dome reached a maximum of 1000 m during Elsterian time. The combined action of erosion by ice and meltwater, and subrosion (subsurface dissolution of the salt) by meltwater, led to the formation of a deep channel above the salt dome. This channel cut its way downward a maximum of 300 m through the cover and cap rock of the dome and into the salt itself.

Within the channel, the cap rock was broken up by the erosive strength of the meltwater, and a breccia formed at the bottom of the channel above the salt. The breccia consists of various types of rock debris from the cover and cap rock. Thus, fragments of gypsum and anhydrite several centimetres in size are enclosed in a matrix consisting of clastic material from Upper Cretaceous, Tertiary and Quaternary sediments.

The structure of the salt dome is shown in Fig. 5.25. An examination of the drill cores shows that the fold axes suggest a gentle plunge parallel to the longitudinal axis of the dome. Thus, there is no evidence in the study area for steeply plunging curtain folds.

Hydrogeological programme

From April 1979 until the end of 1984, 145 hydrogeology exploration holes, 326 monitoring wells, and two boreholes were drilled over an area of about 300 km² as part of a programme of hydrogeological investigations. In addition, 37 boreholes were drilled as much as 80 m into the salt itself to explore the surface of the dome. Pumping tests were also carried out.

This programme has produced a detailed knowledge of the strata, which consist of a rather complicated system of Tertiary and Quaternary sands, silts, and clays. A protective clay layer, which may prevent subrosion of the salt, is weakened by the channel mentioned above and is missing in places. The occurrence, salinity, and flow of groundwater in the sediments overlying the salt dome are now sufficiently well known, and the results have been used in modelling investigations for a preliminary safety assessment evaluation.

A groundwater flow rate of several m yr^{-1} has been computed for a depth of 170 m below mean sea level. The flow field mirrors the hydrogeological setting of the salt dome. Outside the structure of the dome, the groundwater flows in the surrounding Miocene sands. Above the dome, high velocities have been computed for the Quaternary subglacial channel system where the main channel crosses the dome from south to north-east. At greater depths, the model indicates that flows of more than 1 m yr^{-1} occur only in the main channel and in the Tertiary sands north-west of the dome.

Planned underground exploration

The starting point for the underground exploration programme is to identify the problems to be investigated. At the same time, one must determine how the data will be used to solve these problems, and how important the problems are in order to set priorities and decide which methods should be used. It is also necessary to determine which answers will provide new information. Uncertainties in the data precision can be justified only when this is done. The detail to which the underground exploration is to be carried out is governed in the end by the degree of certainty required of the data. This can vary depending on the problem to be solved.

The identification of problems and their priorities requires a systematic examination of the measures necessary for planning,

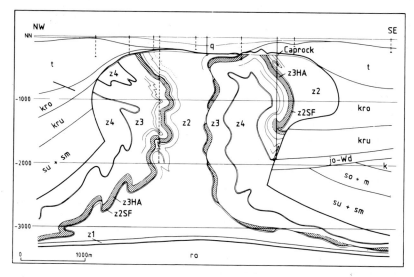

Fig. 5.25 Geological section across the Gorleben salt dome (after Bornemann, O. (1987). Die geologische Erkundung des Salzstocks Gorleben. *Kerntechnik 50, 21*). q = Quaternary; t = Tertiary; kro = Upper Cretaceous; kru = Lower Cretaceous; Wd = Wealden; jo = Upper Jurassic; k = Keuper (Upper Triassic); m = Muschelkalk (Middle Triassic); su, sm, so = Buntsandstein (Lower Triassic); Z1–4 = Zechstein (Upper Peruvian); Z2SF = Staßfurt Potash Seam, Z3HA = Zechstein 3 Anhydrite, ro = Rotliegendes (Lower Permian).

construction, and operation of the repository. This examination shows that the underground investigations have three essential tasks:

(1) assessment of the thermomechanical load capacity of the rock salt so that emplacement strategies can be determined for the site;

(2) determination of the safe dimensions of underground openings (e.g. cavern stability and mine safety);

(3) evaluation of geological barriers and long-term safety analysis to satisfy authorization procedures.

Geotechnical and geological data are needed to answer the various questions associated with these tasks. The following are examples of some of these questions:

1. Are the homogeneous parts of the rock salt large enough for the repository?

2. Which failure scenarios cannot be excluded on the basis of the structure of the salt dome?

3. Can the required width for the safety zones between the repository and unfavourable layers be observed?

4. What is the load capacity of the rock under the expected thermomechanical stress changes (e.g. rock burst and creep rupture)?

5. Will the caverns become unusable as a result of convergence, dilatancy, or subsidence?

6. Can the thermal load induce decomposition of the minerals (e.g. loss of water of hydration) or a hazardous migration of brines?

7. Can an uncontrolled generation of gases or brines be avoided?

8. Will the integrity of the salt dome be jeopardized by long-term changes in stress (e.g. fracture formation)?

An extensive and detailed characterization of the salt dome that is based partly on underground exploration must be performed to answer these and other similar questions. This characterization, combined with the requirements given in the repository concept, the design of underground drifts, and the failure analysis, will be used as a basis for deciding the suitability of the salt dome as a host formation for a permanent repository for radioactive waste. This process will be documented for the authorization procedures.

Calculation of integrity of the salt dome barrier

The question of how well the salt formation can sustain thermal loading was raised early for the Gorleben site and has been the subject of many investigations in Germany. Thermally induced deformations and the related stress fields, as well as their influence on barrier integrity, have been studied from the geomechanical point of view.

Heat generation resulting from radioactive decay decreases with time. Temperatures are increased initially in the vicinity of the repository causing thermal expansion. This, however, results in thermally induced stresses because the rock builds up a transient resistance to deformation. As temperatures decrease later, the rock mass tends to contract and stresses are reduced in the repository area.

An assessment of the integrity of the geological barrier can be based on calculations only. Preliminary design calculations have been oriented towards problem identification and an indication of possible trends. At the present stages of exploration at the Gorleben salt dome, the final input data for the computational model are not yet sufficiently available.

Temperature distributions with time and the related far-fields stresses have been computed for specific disposal conditions. A maximum temperature of 455 K (i.e. about 170 °C) after 150 years was obtained, and the maximum uplift at the surface after 2000 years amounts to 3.3 m. Under certain conditions tensile stresses may develop in the cap rock. All numerical calculations are performed using the ANSALT code.

For the analysis of the future development of the geological barrier, special methods of predictive geology have been developed. The National Geological Survey of France (BRGM) procedure is described as an example.

Predictive geology applied to the deep storage of long-lived radioactive waste [41] [60]

General

The predictive geology approach consists of simulating the future evolution of geological factors on the basis of knowledge of the past evolution. The objective is to quantify the phenomena in the terms of relative importance as to their impact on a deep storage site area and the time scale of impact. For long-lived waste the usual period considered is 100 000 years.

Although the equivalent previous geological period is the best known, the quantification of the variety of parameters which influence the evolution of a specific site is difficult, mainly due to the complexity of quantifying the dynamics of the individual factors and the interaction between them, and to performing adequate time generalization.

The predictive geology studies carried out by BRGM started about 10 years ago and have been continuously supported by the Commission of the European Communities (DG XII) for the part concerned with methodology.

Parameter selection

Before attempting any modelling approach, the first task was to identify the parameters which have a potential impact on the future evolution of the site. These parameters are grouped into two categories, internal (IG) and external geodynamics (EG).

IG parameters refer to mechanisms related to subsurface dynamics of the Earth: fault movement, seismicity, volcanism, etc. The EG group is directly connected with the impact of climate evolution: erosion, glacier movement, sea level changes, etc.

A tentative listing of these various parameters is given below:

Internal geodynamics	External geodynamics
Tectonic constraints	Morphology
Fracturing	Alteration
Vertical movements	Erosion
Permeability variations	Sedimentation
Salt domes	Glaciation
Volcanism	Sea level changes
Seismicity	

Tentative global modelling

The initial stage of definition of parameters was followed by a tentative approach to global modelling: the CASTOR programme (Construction Automatique de scénarios d'évolution d'un site STOskage de déchets Radioactifs).

Forty-eight variables concerning 13 processes can be simulated in CASTOR.

Simulated processes
Sedimentation
Erosion
Fluvial erosion
Isostatic adjustments
Epeirogenic vertical movements
Site morphology
Constraint variations
Permeability variations

Processes described by standard values
Glaciation
Ice melt
Diapir upward change
Vertical movement change
Fracturing change
Water transfer

Global simulation

The parameter selection phase was based on available data. The setting up of the CASTOR model was the logical following step, as an attempt to correlate a fairly large number of variables. During the following phase two areas were selected for model control purposes. These are not sites chosen for actual disposal but only for methodological purposes.

As a preliminary test on the Fougères (Brittany) area, three different types of scenario were simulated. If no major geological or climatic event occurs during 100 000 years, the main process to deal with is erosion, which would not lead to any major changes in water flow at depth. The impact of a new glaciation period would be mainly reduction of permeability due to the ice burden, and fault reactivation to some extent.

A further trial was performed on the site of a borehole drilled in the Pays-de-Caux, east of Rouen, in the north-western portion of the Paris basin. The end result was a simulation of the hydraulic system at depth using the River Seine as a base downstream level and several scenarios including erosion, sea level variations, and vertical movements. Modelling was compared with the real site evolution described by the geological evidence. Another conclusion reached was that the site safety was preserved in general terms within the limits of uncertainty due to lack of information, in particular on actual flow systems.

Complementary factor modelling

To improve simulation by CASTOR, new models were developed on specific parameters: coupling of erosion and alteration (HERODE model), glacial cap advance, and reconstitution of former climate using the confirmed relationship between plant associations and aridity, using average yearly temperature indexes.

By comparison with the initial version of CASTOR, the introduction of HERODE, enabled erosion alteration combination within a systematic grid system.

Available information suggests that major glaciation is very likely to occur within the coming 100 000 years. Modelling of this feature is of major importance. A first attempt should be followed by more detailed studies concerning the various impacts of ice movement on soil compaction, lateral and footwall erosion, the freezing of significant thicknesses of ground (permafrost), and the effects of glaciation and deglaciation.

Although significant results have been achieved over the last ten years in terms of quantification of geological and climatic phenomena, a lot of work remains to be done to reach a level of scientific precision to allow adequate predictive scenarios. Studies presently carried out in BRGM and other institutes continue.

Underground laboratories [75]

In most countries a facility for *in situ* tests (Underground Laboratory) is considered to be indispensable for preparing a repository concept, for site characterization and for safety demonstration. To date, Belgium, Germany, Switzerland, Canada, USA, and Sweden have developed such facilities underground.

In Belgium [4] [5], the decision to build an underground laboratory (High Activity Disposal Experimental Site, HADES) is largely connected with the first CEC R&D programme (1975–79). The laboratory is a joint CEC–ONDRAF (Belgium Atomic Energy Commission) operation and the site selected is at Mol (Fig. 5.26) where the Belgian Nuclear Energy Research Centre (CEN/SCK) is located. France and Japan also participate in the experimental programme. Before the construction started and during con-

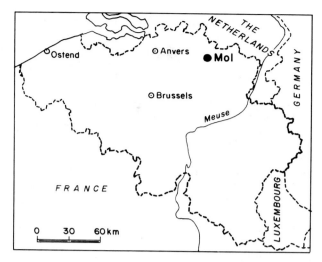

Fig. 5.26 Location of the underground laboratory at Mol (Belgium).

struction, the Belgian Geological Survey assisted in providing accurate geological and hydrogeological information.

The R&D activities are concentrated mainly in the laboratory tunnel and the experimental drift constructed between 1978 and 1984. The main topics investigated are:

(1) interactions between waste and the Tertiary Boom Clay;

(2) migration studies within the clay;

(3) hydrogeology of the Mol site;

(4) geomechanical behaviour of the clay;

(5) conceptual studies;

(6) backfilling and sealing;

(7) safety and performance assessment.

After the demonstration drift was constructed (1987), further experiments were carried out including:

(1) geotechnical modelling of tunnel construction and behaviour in plastic clay;

(2) evaluation of the combined effect of radiation and heat on the clay;

(3) testing of the interactions with the clay of relevant radioactive waste materials;

(4) simulation of the behaviour of a concrete-lined gallery structure and surrounding clay in a temperature field.

Now the Boom Clay is well documented: hydrological and geotechnical parameters are well known at the Mol site. By estimating some values, semi-quantitative scenarios were also attempted for some types of site, but a thorough quantification still has to be done.

The Swiss Grimsel Test Site (GTS) has operated since 1983–84 (Fig. 5.27) [76]. It has been developed in granitic rocks and consists of a 1000 m long laboratory tunnel and a central building which houses the whole infrastructure. The diameter of the tunnel is 3.50 m. Following an agreement signed in 1982 between Switzerland and Germany, four institutions collaborate on the site: the National Cooperative for the Storage of Radioactive Waste (NAGRA) which operates the site, the Federal Institute for Geosciences and Natural Resources (BGR, Germany), the Research Centre for Environmental Sciences (GSF, Germany), the Department of Environment (DOE, USA).

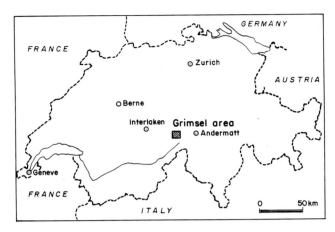

Fig. 5.27 Location of the underground laboratory at Grimsel (Switzerland).

The BGR was mostly involved in fracture system flow tests and rock stress measurement. GSF performed mainly tilt measurements and heating tests. NAGRA covered migration tests, prediction ahead of tunnel face, fracture-zone investigation, and near-field hydraulics.

During phase 1 (1980–87), the validity of *in situ* measurements was demonstrated together with their adequate use in modelling systems. During phase 2 (1988–1990) emphasis has been put on the hydrogeology around underground constructions, the relationship between geophysics and hydraulics, nuclide transport, and excavation effects. The US DOE participated in the geophysical borehole testing (seismic tomography).

Apart from Konrad, Gorleben (Germany) and Grimsel (Switzerland), Germany has been working in the former Asse salt mine, some 20 km southeast of Braunschweig (Fig. 5.23). Most of the work is carried out by GSF (see above). The more important investigations carried out in Asse were:

(1) experimental emplacement of HLW in boreholes and study of its interaction with the salt;

(2) retrievable emplacement of ILW in boreholes. The aim was to develop and demonstrate solutions to waste handling, transport and disposal problems;

(3) experiments on direct disposal of spent fuel elements.

The studies also included a large spectrum of hydrogeological, geophysical, and rock mechanics characterization of the site and problems of borehole scaling.

The Stripa Project (Fig. 5.28) [75]

The first experiments at Stripa started in 1976 when the reserves of iron ore in the mine were exhausted. The international cooperation at Stripa began with the joint Swedish–American Cooperative (SAC) programme in 1977, followed in 1980 by the ongoing autonomous OECD/Nuclear Energy Agency International Stripa Project. The OECD/NEA has a broad international programme focusing on the safety of nuclear installations and radioactive waste management in its 23 member countries. Seven countries are currently participating in the Stripa Project: Canada, Finland, Japan, Sweden, Switzerland, United Kingdom, and the United States. The project is managed by the Swedish Nuclear Fuel and Waste Management Co, SKB, under the direction of a Joint Technical Committee with representatives from each of the participating countries.

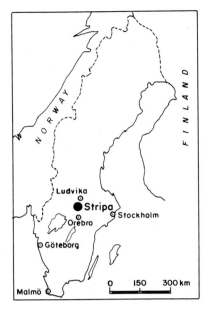

Fig. 5.28 Location of the Stripa (Sweden) mine and underground laboratory.

Phase 1 of the project was carried out between 1980 and 1985 and phase 2 between 1983 and 1988. Phase 3 started in 1986 and is scheduled to be completed in 1991. The Joint Technical Committee coordinates the scientific review and decides on programmes in the following fields: geology, geophysics, hydrogeology, numerical modelling of fracture flow, hydro-geochemistry, rock mechanics, chemical transfer, and engineered barriers. The Geological Survey of Sweden played a central role in the geoscientific studies.

During phases 1 and 2 specific tools and methods have been developed to study the distribution and extent of the fracture zones, investigate the thermo-mechanical properties of the granite, assess the migration in single and multiple fractures of sorbing and non-sorbing tracers, and to understand the groundwater hydraulics of the Stripa area, and test the behaviour of sealing materials.

On the basis of these results, phase 3 has developed in two main research directions:

(1) fracture flow and nuclide transport;

(2) groundwater flow-path sealing.

Table 5.7 indicates the main research topics to be investigated and the various stages of the programme.

Conclusions

The problem of radioactive waste disposal has now become a major environmental issue. Short-lived waste is mostly stored on production sites. For long-lived (or high level) waste, the general tendency is presently towards deep siting, although no final disposal site has been chosen. The main studies being developed mainly concern geological site selection, safety analysis, and long-term perspective. Several ground laboratories have been developed in various geological contexts and these serve to test various methods.

The Commission of the European Communities accepts that every state has to dispose of its own radioactive waste, but in fact some countries are treating the spent fuel of others. The recent past (Chernobyl) has proved that nuclear calamities cross frontiers and that damage may result outside the country of origin.

There is legitimate concern, therefore, about the capacity for the disposal of nuclear waste in all countries. Some are

Table 5.7 Research programmes and stages for phase 3 of the Stripa project

---

Fracture flow and nuclide transport

Site characterization and validation

The size of the selected volume of rock is $125 \times 125 \times 50$ m.

Stage 1: Preliminary site characterization.

Stage 2: Preliminary predictions: establishment of a conceptual model based on the geometry and physical properties of the major fracture sets.

Stage 3: Detailed characterizations and preliminary validation: additional boreholes and tests.

Stage 4: Detailed predictions: establishment of deterministic/statistical mathematical models based on detailed prediction of fracture network geometry.

Stage 5: Detailed evaluation: validation by tracer experiments of detailed predictions of stage 4.

Improvements of site assessment methods and concepts

Further experiments concerning:

(1) identification of disturbances (fractures etc.) in the host rock (high-resolution radar and borehole seismics);

(2) study of water flow through specific channels within the fracture planes;

(3) feasibility study of the modelling of the fracture network.

Groundwater flow-path sealing

Stage 1: State-of-the-art survey of sealing materials: emphasis on durability.

Stage 2: Determination of the long-term stability and function.

Stage 3: Field pilot-tests using selected sealing materials.

Stage 4: Large-scale sealing tests.

---

probably less able than others to arrange for the storage of HLW. It is likely, therefore, that a compromise will have to be found between the EEC policy and the potential for storing all the waste in a few well-chosen sites which may offer greater levels of safety. Both possibilities are still very much under debate.

The national Geological Surveys have been involved in most of the different research phases and in a wide spectrum of study techniques. In those countries in which sites for radioactive waste disposal are already selected, the relevant Geological Surveys are engaged in site characterization and safety assessment. Their expertise is a major contribution to the understanding of the problems to be solved, the solutions to be found, and the development of appropriate policies for Western Europe.

RECOMMENDED FURTHER READING

CEC (1988). PAGIS: Performance assessment of geological isolation systems for radioactive waste. *CEC report No. EUR 11775.* (Luxembourg, Euroffice).

CEC and OECD-NEA (1984). Geological disposal of radioactive waste: an overview of the current status of understanding and development. (OECD Press, Paris).

CEC, IAEA, and OECD-NEA (1990). Safety assessment of radioactive waste repositories. *Proceedings of the Paris Symposium, October 1989.* (OECD Press, Paris).

Cecille, L. (ed.) (1990). Radioactive waste management and disposal. *Proceedings of the third European Community Conference, Luxembourg, September 1990.* (Elsevier Applied Science, London).

Chapman, N. A. and McKinley, I. G. (1987). *The geological disposal of nuclear waste.* (Wiley, New York).

Chapman, N. A., Black, J. H., Bath, A. H., Hooker, P. J., and McEwen, T. J. (1987). Site selection and characterization for deep radioactive waste repositories in Britain: Issues and research trends into the 1990s. *Radioactive Waste Management and the Nuclear Fuel Cycle.* **9**, 183–213.

Collective Authors (1988). Hydrogeology and safety of radioactive and industrial hazardous waste disposal. *Doc. BRGM No. 160.*

Collective Authors (1989). The nuclear fuel cycle: review on R&D policies in the Member States of the European Community. *CEC report No. EUR 12380.* (Luxembourg, Euroffice).

Combe, B., Johnston, P., and Müller, A. (eds) (1985). Design and instrumentation of *in situ* experiments in underground laboratories for radioactive waste disposal. *Proc. Workshop CEC-OECD, Brussels, 15–17 May 1984.* (A. A. Balkema, Rotterdam).

Kaelin, J. L. (1990). Région de Soulaines (Aube, France). Géologie, hydrogéologie et géotechnique. (ANDRA, Paris).

Langer, M. (1989). Waste disposal in the Federal Republic of Germany—concepts, criteria, scientific investigations. *Bull. IAEG.* **39**, 53–8.

Langer, M. (1989). The impact of waste disposal on the geoenvironment. In *Proc. Int. Symp. The impact of mining on the environment*,
vol. 1. (CIP-GUNT, Tallin, Moscow).

Lieb, R. W. and Müller, W. H. (1988). *NAGRA Bull.* Special issue on
Grimsel test site and other underground laboratories. (NAGRA).

Morfeldt, C. O. and Langer, M. (1989). Problems of underground
disposal of waste. Report of the International Association of
Engineering Geology. *Commission No. 14. Bull. of the IAEG
No. 39, April*. pp. 1–58.

OECD-NEA (1990). *In situ* experiments associated with the disposal of radioactive waste. *Proceedings of the 3rd NEA/SKS
symposium on the International Stripa Project, Stockholm,
October 1989*. (OECD Press, Paris).

Orlowski, S. and Schaller, K. H. (1989). Euradwaste Series No. 1:
objectives, standards and criteria for radioactive waste disposal in
the European Community. *CEC report No. EUR 12570*. (Luxembourg, Euroffice).

Simon, R. (ed.) (1986). Radioactive waste management and disposal.
*Proceedings of the second European Community Conference,
Luxembourg, April 1985*. (Cambridge University Press, Cam-
bridge).

Simon, R. and Orlowski, S. (eds) (1980). Radioactive waste manage-
ment and disposal. *Proceedings of the first European Community
Conference, Luxembourg, May 1980*. (Harwood Academic Press,
Cambridge).

United Kingdom NIREX Limited (1989). Deep Repository Project:
preliminary environmental and radiological assessment and preliminary safety report. (UK NIREX).

5.5. GROUNDWATER POLLUTION [39]

Human development affects all phases of the water cycle, the
greatest apparent impact being on readily observed surface
waters in rivers and lakes. The greater concern and involvement of the geoscientist is with groundwater, and this section
concentrates mainly on the behaviour of pollutants in the soil
and subsurface and the implications for groundwater regimes.

5.5.1. Local pollution

When the French playwright, Jean Giraudoux wrote, on the
subject of groundwater, '. . . thou, so pure in the bosom of the
Earth . . .', he was only repeating, more poetically, certain
official texts that recommended giving priority to the use of
groundwater, when it was available. The popular concept of
groundwater, from pagan mythology to the blessing of springs
by Christian clergy, has been that it is naturally pure. Unfortunately, the facts are not so simple. The subsurface, which is
very varied and complex, does not provide an absolute
guarantee for the protection of groundwater from ordinary
domestic contamination. The most vulnerable areas are the
catchment points themselves.

*Quality and natural purity of groundwater before
industrial pollution*

Groundwater suffered from pollution long before the industrial era. This contamination was bacterial, resulting from
faulty sanitary installations, manure, and all types of human
and animal faecal infections, generally carried by rain and run-
off water directly towards the wells. This kind of problem,
which may cause typhoid epidemics, recurs at times with
dramatic consequences, reminding us that, even at the dawn of
the twenty-first century, bacterial pollution is still to be feared.

These are vital considerations in the formation of the various
policies adopted to protect groundwater tapped for human
consumption.

With the arrival of the ideas of Pasteur at the end of the
nineteenth century, and the discovery of microbes, it was
recognized that limpid, crystal-clear water could be seriously
contaminated, and that the tiniest particles were not naturally
filtered out during the passage of water through coarse-
grained, fractured, or karstic media.

Later, it was also discovered that the subsurface has the
capacity to reduce very considerably the pathogenic bacteria
and the 'ordinary' faecal or coliform bacteria which generally
accompany them, and which are tracers of anthropic contamination. The bacteria, 'ill-at-ease' in the cold groundwater and
poorly nourished, do not reproduce and usually disappear
within a few weeks. The foundation and experimental development of a protection policy was based, therefore, on the
knowledge that groundwater circulates slowly in many geological environments. Although somewhat passive measures,
the necessary extent of any protection area has to be calculated with reference to the rate of water movements in relation
to the bacterial life-span.

Groundwater pollution, prior to the industrial and chemical
era, was thus mainly bacterial and local and, in particular,
linked to media with rapid circulation such as plateau and
mountain karsts. In these media, where the chasms and
swallow-holes are often considered to be providential dumping grounds, the water moves very quickly—several kilometres per day—and can carry pollution over long distances,
in particular towards the largest resurgent 'springs' which have
been tapped since Roman times.

With the development of the tourist industry in mountain
areas, bacterial pollution of faecal origin still occurs from time
to time. The tapping of springs, even at high altitudes, does
not, *ipso facto*, give 'pure mountain water', as the permanent
inhabitants of the valleys might expect. Tourists, not im-
munized by the constant use of this water, discover this to their
cost! In Austria and indeed throughout the Alps, and in the
Pyrenees, pollution originates from chalets and cattle on the
high pastures, and from mountain huts and hostels for
climbers and hikers. The rapid downward movement of water
towards catchment works and on to villages can therefore
bring bacterial pollution to the taps and standpipes in the
valleys. Very few gravitational mountain water supplies have
disinfection systems. It is essential that care is taken to avoid
the pollution of these resources, as they are difficult to protect
by the methods described below. It is mainly by adopting
simple sanitary techniques, such as compost latrines, that this
type of pollution can be overcome.

Thus, general considerations on filtering, transit times, and
the protection of aquifers by 'impermeable' surface layers
resulted in the appearance of an activity specific to hydrogeologists—assessing the feasibility of establishing protection
areas for tapped water. A concept of concentric 'circles of
protection' around the wells has been adopted by all countries, with a few variations as shown in Table 5.8.

In an outer protection zone there are only mildly restrictive
directions, recommendations, and incentives, whereas near
the water point all activity is forbidden and the space is fenced
off: this immediate area surrounding a well is owned outright
by the water authority. In the intermediate or 'close at hand'
zone, which may be hundreds of metres wide, land ownership
is not acquired by the authority and, under these conditions,
the various prohibitions, such as manure spreading or the
storage of hazardous materials, lead to the partial deprivation
of the rights of the owner. Administrative measures have to be

Table 5.8 Catchment protection areas in various European countries (after Lallemand-Barrès, A. and Roux, J. C. (1989), *Hydrogéologie 2*, pp. 119–24)

Germany	Austria	Finland	The Netherlands	France	Switzerland	Belgium	Norway	Sweden	United Kingdom	Rep. of Ireland
Zone I (10–100 m)	Immediate protection zone	Catchment zone 20–25 m	Vicinity of the well	Immediate protection	Zone I (5–20 m)	Zone I 24 hours (100 m)	Well zone (10–30 m)	Well zone	Immediate protection zone	Zone I A (10 m)
Zone II	Protection zone	Proximity protection	Catchment area	Proximity protection	Zone II	Zone II		Proximity protection	Protection adjusted to context	Zone I B (10–300 m)
50 days	50 days	60 days	(>30 m 50 to 60 d)	?	10 days >100 m	60 days (100–300 m)	60 days	60 days >100 m		Zone I C (300–1000 m)
Zone III A	Partial protection zone	External protection zone	Protection zone 10 days (>800 m)	Far protection zone	Zone III >200 m	Zone III	Infiltration zone	External protection zone		
2 km			Protection zone 25 days (1200 m)		Zone A	2000 m				Zones 2–3–4
Zone III B			Far basin recharge zone		Zone B					
Slope limit										Catchment basin protection

Note: Transit-times are for matrix permeability rocks only and not for karstic or fissured media.

taken with respect to the owners, in some instances in the form of compensation.

The form of protection is not available in the karst regions, where circulation can be very rapid, and where the water can be of distant origin. Protection 'zones' are irrelevant and it is necessary to study the vulnerability of the whole system of aquifers over very large areas (whole hydrogeological basins) in some cases. The technique of mapping the vulnerability has led to the establishment of vulnerability maps in many countries in Western Europe, some on a systematic basis.

A very pragmatic approach to such protection measures, adopted in Ireland, is presented below.

Protecting groundwater in Ireland: a pragmatic approach [124] [123]

Introduction

The protection of groundwater in Ireland is influenced by several factors:

1. The geology and hydrogeology of the country are complex. The bedrock aquifers have a secondary or fissure permeability only and are very unpredictable. The unconsolidated Quaternary deposits—sands, gravels, tills, and clays—are very variable in thickness, extent and lithology.

2. Diffuse pollution sources such as fertilizers are not causing serious pollution problems.

3. The main threat to groundwater is from point sources—farmyard wastes (silage effluent and soiled water mainly), septic tank effluent, and some leachate from waste disposal facilities, leakages, and spillages.

4. Detailed geological/hydrogeological knowledge is lacking for many areas.

5. The number of people with specialized training in groundwater protection rests mainly with engineers and planners.

6. Pollution of wells and springs is occurring widely.

7. There are unlikely to be extra financial resources in the near future for research on groundwater protection.

8. Groundwater pollution risk depends on the interaction between:
   (a) the natural vulnerability of the aquifer;
   (b) the pollution loading that is, or will be applied to the sub-surface environment. Vulnerability is an intrinsic characteristic of an aquifer, whereas pollution loading can be controlled or modified.

The Geological Survey of Ireland recommends a pragmatic approach to groundwater protection using available information and expertise.

The approach involves several stages:

(1) preparation of an initial groundwater protection scheme;
(2) vulnerability assessments using water quality data and available geological and hydrogeological data;
(3) assessments of groundwater pollution loading;
(4) site investigations;
(5) preparation of a final groundwater protection scheme.

Stage 1: Preparation of an initial groundwater protection scheme

The scheme consists of a groundwater protection map and a code of practice.

Groundwater protection map
The map is based on the division of an area into four groundwater protection zones according to the degree of protection required,

Zone 1 requiring the highest degree of protection and Zone 4 the least.

Zone 1: Source protection zone around each designated groundwater supply source (public and group scheme supplies and important industrial supplies). It is subdivided into three subzones; 1A, 1B and 1C:

Zone 1A is the area within 10 m from source;
Zone 1B is the area between radii 10–300 m;
Zone 1C is the area between radii 300–1000 m.

The distance can be varied to suit the hydrogeological conditions in any area and can be modified and improved as more up-to-date information becomes available.

Zone 2: Major aquifers.
Zone 3: Minor aquifers.
Zone 4: Poor and non-aquifers.

These zones are shown on a map at 1:10 000 scale or 1:50 000 scale.

Code of practice
Acceptable and unacceptable activities for each zone are listed and the controls recommended for potentially polluting developments.

Zone 1A: It should be fenced and potentially polluting activities forbidden.

Zone 1B: The following developments would normally be prohibited:

(1) septic tank treatment systems;
(2) the spreading of slurry, manure, and sewage sludge;
(3) the establishment of burial grounds;
(4) waste disposal sites;
(5) industrial developments using, producing, or storing toxic or potentially polluting material;
(6) intensive agricultural activities, e.g. intensive rearing or housing of livestock and the construction of silage pits;
(7) construction of main foul sewers;
(8) construction of sewage and trade effluent treatment works;
(9) excavation of minerals extending below the water table.

Zone 1C: All the developments listed for Zone 1B, with the exception of (1), (2), and (3), should not be allowed.

Zone 2: Activities which should normally be prohibited include:

(1) waste disposal sites intended to receive hazardous wastes, including domestic wastes and sewage sludge;
(2) major industrial and agricultural developments which use, store, or handle toxic materials, unless adequate protective measures are agreed with the local authority.

Zone 3: A balancing of interests between the need to protect groundwater resources and the need to locate potentially polluting developments may often be necessary. Waste disposal sites and major industrial and agricultural developments are discouraged and only allowed if adequate precautions are taken and risks reduced to an acceptable level.

Zone 4: Protection of existing groundwater sources is the only restriction.

At present sufficient hydrogeological information exists to enable this first stage. The GSI is recommending that the local government bodies should prepare these basic schemes and several have done so. However, it is emphasized that these schemes have no statutory authority but are a guide for engineers and planners.

Stage 2: Vulnerability assessments

Two main approaches to vulnerability assessments are adopted:

(1) vulnerability mapping using geological and hydrogeological data;

(2) using water quality data.

Vulnerability mapping

The vulnerability of groundwater to pollution varies greatly depending on a number of interrelated geological and hydrogeological factors:

(1) type of subsoil;

(2) permeability of subsoil;

(3) thickness of subsoil over bedrock;

(4) type of permeability—intergranular or fissures;

(5) thickness of unsaturated zone;

(6) attenuating capacity of subsoil and bedrock;

(7) type of recharge—whether diffuse or by sinking streams.

Vulnerability mapping assesses all available geological and hydrogeological data and ranks the vulnerability of groundwater on a map in an understandable and useful manner. Hydrogeological expertise is essential in preparing these maps.

Vulnerability assessment using water quality data

The vulnerability of wells and springs can be placed within decreasing pollution risk categories based on available groundwater quality data, e.g. hardness, conductivity, temperature, and bacteriological data. The variations in hardness and/or conductivity due to recharge between groundwater sources are indicators of vulnerability to pollution. The more rapid the response to recharge and the greater the variations the more vulnerable the source is to pollution. Regular measurements—fortnightly is usually considered adequate in Ireland—are needed for this technique. Wells or springs can be ranked on whether bacterial pollution is regular, occasional, or absent and priority given to protecting the more vulnerable.

Vulnerability rating and hydrogeological setting

Vulnerability ratings for typical Irish hydrogeological settings are shown in Table 5.9. These ratings are subjective and qualitative as data are normally inadequate for quantitative and impartial ratings to be determined.

Stage 3: Assessment of groundwater pollution loading

Vulnerability assessments by themselves do not provide a system of groundwater protection. They have to be linked to investigations of existing potentially polluting activities, and a code of practice for controlling developments.

One method is to overlay the vulnerability map with a map showing the location of existing potentially polluting sources. This allows an assessment of pollution risk and the initiation of appropriate monitoring.

Stage 4: Site investigations

The information provided by stages 2 and 3 enable an assessment of the existing data and of the pollution risks. Depending on this assessment and on available financial resources, site investigations may be carried out.

Stage 5: Preparation of final groundwater protection scheme

The final stage is to refine the code of practice and the zonal boundaries prepared in stage 1 based on the information collected in stages 2, 3, and 4. The final product is a map showing four zones—source protection zones (Zone 1), major aquifers (Zone 2), minor aquifer (Zone 3) and poor or non-aquifers (Zone 4). The map scale should preferably be 1:50 000. Large-scale vulnerability maps—1:25 000 or 1:10 000—are recommended for the source protection zones. The maps include warnings that they have no statutory authority and should not be used alone to determine site-specific susceptibility of groundwater to pollution from specific sources.

Use of groundwater protection schemes and vulnerability assessments

A well-prepared groundwater protection scheme has the following benefits:

1. It brings the groundwater interest to the attention of decision makers and makes information widely available.

2. It assists planners and engineers in locating and regulating potentially polluting activities.

3. It assists local bodies in making water treatment investment decisions.

4. It enables more detailed and expensive investigations to be directed where the threat is greatest.

5. It enables industrial users to assess the pollution risk either before or after developing a groundwater source.

It is important that the limitations of vulnerability maps are clearly understood by those using them. A prominently displayed warning should be printed on the maps stating they are not to be used for specific siting purposes and that a hydrogeologist be consulted in their interpretation. Vulnerability maps generalize very variable geological conditions and should be regarded as a guide only, leading to specific site investigations. Their use needs a realistic and flexible approach.

Conclusion

The Irish approach enables regulatory authorities to take account of groundwater when locating potentially polluting developments in situations where expensive site investigations and environmental impact assessments are not practicable or possible. It may be applicable in other countries where groundwater is not yet highly developed but where there are pollution risks.

With development, urban and industrial pollution appears and becomes prominent (Fig. 5.29). For a century now, the incidence of various types of chemical pollution has overtaken bacteriological pollution in industrial areas. The reason for this is that a systematic disinfection of supply networks in large towns has led to the bacterial pollution remaining exclusively as a problem 'at the tap' in rural areas, where, locally, the most primitive chlorinating plants are still lacking.

With the progressive build-up of health standards for potable water, periodically improved and reinforced by directions concerning trace chemical substances, the alarming degree of pollution of surface and, subsequently, underground water was recognized.

National, EEC, and World Health Organization (WHO) standards have progressively multiplied and become stricter*

---

* It is neither the duty of hydrogeologists nor is it within their competence to argue the validity or otherwise of standards set by public health experts; their sometimes constraining severity promises an uncertain future, but their interpretation must not lead to a conclusion of catastrophe. Nevertheless this is why the new notion of guide-level introduced by the EEC is more satisfactory. It should be remembered, just the same, that a dish of radishes contains more nitrate than a litre of supposedly polluted water.

Table 5.9　Vulnerability of hydrogeological setting

Rating	
Extreme	1. Outcropping bedrock aquifers (particularly karst) or where overlain by shallow (less than 3 m) subsoil.
	2. Sand and gravel aquifers with a shallow (less than 3 m) unsaturated zone.
	3. Areas near karst features such as sink-holes.
High	1. Bedrock—major, minor, and poor aquifers and non-aquifers—overlain by 3 m + sand and gravel or 3–10 m sandy till or 3–5 m low-permeability clayey tills or clays.
	2. Unconfined sand and gravel aquifers with 3 m + unsaturated zone.
Moderate	1. Bedrock—major, minor, and poor aquifers and non-aquifers—overlain by 10 m + sandy till or 5–10 m clayey till, clay, or peat.
	2. Sand and gravel aquifers overlain by 10 m + sandy till or 5–10 m clayey till, clay, or peat.
Low	1. Confined bedrock aquifers overlain by 10 m + clayey till or low-permeability bedrock such as shales.
	2. Non-aquifers and poor aquifers overlain by 10 m + clayey till.
	3. Confined gravel aquifers overlain by 10 m + clayey till.

with the advance of medical science in the fields of epidemiology and human eco-toxicology, and of the ability to analyse very small amounts of chemicals (microgram for heavy metals, Pb, Zn, Hg, etc.; nanogram for the innumerable organic micro-pollutants, especially pesticides and solvents)—see Table 5.10.

Faced with these types of pollution, protection methods for sanitary zones of the 'passive defence' type no longer suffice. Filtration and life have no longer the same meaning for most of the chemical substances, especially if their emission is maintained from a permanent centre. Nevertheless, the notion of protection zones must not be dropped. They still have an important role to play with respect to accidental pollution.

Accidental pollution

The spilling, infiltration and underground migration of undesirable or harmful substances in the unsaturated zone and in the actual groundwater will have very different consequences according to whether they are incidental or continuous in the long-term.

Accidental pollution—for example, a traffic accident involving the transport of a toxic chemical—even when large, has a fixed volume and is of fixed duration; partial recovery of the spilt product is therefore possible. Surficial deposits, even when not 'impermeable', will only release the substances by vertical percolation towards the underlying groundwater with attenuation and delay. This can be taken advantage of to remove the impregnated ground and decontaminate it. This presupposes, naturally, that the accident is known of in time, before significant movement in the groundwater gets under way.

This is why the protection zones have their role to play, at least in giving time to intervene or, at the worst, time to connect the water distribution system to another source of supply. Awareness of potential hazard can lead to particular measures being taken along roads near catchment points—for example impervious ditches and an invitation to motorists to slow down: 'Reduce speed—wells!'.

Vulnerability mapping is useful for this type of mapping, especially when the resulting map is used for planning and decision-making. Such maps have greatest impact at the stage of preliminary study. They are of little use later when constraints and opposition are being analysed. Geological Surveys can do much to educate those concerned with the application of such maps in planning the prevention of pollution.

If, in spite of prevention measures, an accident takes place, all is still not lost if the contaminated source cannot be completely eliminated in time. In such instances protection zones prove their value. Commonly they achieve the reduction of toxic substances carried by the distributed water to levels within the limits of potability.

A variety of phenomena, physical, chemical, and water/rock interactions, delay the dispersion of pollutants. The modelling of such phenomena, and the forecasting of delays and concentrations, are now tools at the service of water management (Fig. 5.30), as a result of research and development carried out in many Geological Surveys. When the pollution is not an accident, but rather results from a source regularly emitting harmful and non-degradable substances, we are in the dark realm of endemic local pollution.

Endemic pollution

If we caught someone dropping a few pinches of cadmium salt or a few drops of nitrobenzene into customers' glasses on the terraces of cafés, the villain would undoubtedly be brought before the criminal court. What, then, about all those who commit the same crime in relation to groundwater by dumping large quantities of pollutant indiscriminately? How many of these offenders have been brought to justice?

The real impact on groundwater pollution of the dumping of household waste has long been underestimated. The end-product of waste degradation appears to be trivial if only the main constituents are considered. The effects on dump material by leaching by rainwater are poorly known, but

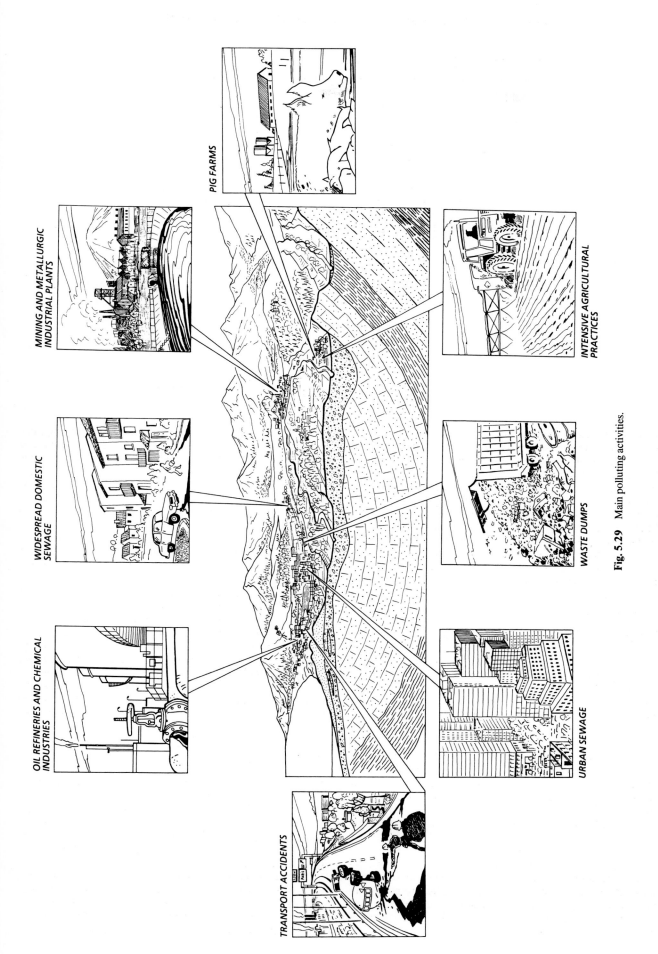

OIL REFINERIES AND CHEMICAL INDUSTRIES

WIDESPREAD DOMESTIC SEWAGE

MINING AND METALLURGIC INDUSTRIAL PLANTS

PIG FARMS

TRANSPORT ACCIDENTS

URBAN SEWAGE

WASTE DUMPS

INTENSIVE AGRICULTURAL PRACTICES

Fig. 5.29 Main polluting activities.

Table 5.10 European Community drinking water standards (*European Community Gazette*)

Parameters	Results form	Community values			Observations
		Guide level (NG)	Maximum acceptable concentration (CMA)	Minimum required concentration (CmR)	* Exceptional Maximum Acceptable Concentration (EMAC standard) ** MAC: Maximum Acceptable Concentration
pH	pH units	6.5–8.5	9.5	6.00	$pH_{i.s.} = 0$ (saturation index)
Conductivity	$\mu S\ cm^{-1}$	400	1250		possible appeal to EMAC* resistivity correspondant values: Ω cm: 2500–800
Total mineralization	dry residue: $mg\ l^{-1}$		1500		possible appeal to EMAC*
Total hardness	hydrotimetric degree D°F	35		10	
Calcium	Ca: $mg\ l^{-1}$	100		10	
Magnesium	Mg: $mg\ l^{-1}$	30	50	5	
Sodium	Na: $mg\ l^{-1}$	<20	100		possible appeal to EMAC*
Potassium	K: $mg\ l^{-1}$	≤10	12		possible appeal to EMAC*
Aluminium	Al: $mg\ l^{-1}$		0.05		possible appeal to EMAC*
Alkalinity	HCO_3^-: $mg\ l^{-1}$	30			
Sulphates	SO_4^{--}: $mg\ l^{-1}$	5	250		possible appeal to EMAC*
Chlorides	Cl^-: $mg\ l^{-1}$	5	200		possible appeal to EMAC*
Nitrates	NO_3^-: $mg\ l^{-1}$		50		<15 mg l^{-1} for water, bottled or not, for babies bottles possible appeal to EMAC*
Nitrites	NO_2: $mg\ l^{-1}$		0.1		
Ammonium hydroxide	NH_4^+: $mg\ l^{-1}$	0.05	0.5		
Kjeldahl nitrogen	N^+: $mg\ l^{-1}$ (N of NO and NO_3 excluded)	0.05	0.5		
Cadmium	Cd: $\mu g\ l^{-1}$		5		
Cyanide	CN: $\mu g\ l^{-1}$		50		
Total chromium	Cr: $\mu g\ l^{-1}$		50		
Copper	Cu: $\mu g\ l^{-1}$		50 1500		possible appeal to EMAC* 1500 $\mu g\ l^{-1}$ after 16 hours' contact with user's tap
Fluorine	F: $\mu g\ l^{-1}$		700–1500		MAC** variable according to the average temperature of the geographical area considered
Iron	Fe: $\mu g\ l^{-1}$	100	300		possible appeal to EMAC*
Mercury	Hg: $\mu g\ l^{-1}$		1		
Manganese	Mn: $\mu g\ l^{-1}$	20	50		possible appeal to EMAC*
Nickel	Ni: $\mu g\ l^{-1}$	5	50		
Phosphorus	P: $\mu g\ l^{-1}$	300	2000 after sequestration treatment		
Lead	Pb: $\mu g\ l^{-1}$		50		
Hydrogen sulphide	S: $\mu g\ l^{-1}$		absence		
Antimony	Sb: $\mu g\ l^{-1}$		10		
Selenium	Se: $\mu g\ l^{-1}$		10		
Zinc	Zn: $\mu g\ l^{-1}$		100 2000		2000 after 16 hours' contact with user's tap
Mineral oils	residue: $\mu g\ l^{-1}$		10		
Hydrocarbons polycyclic aromatic	residue: $\mu g\ l^{-1}$		0.2		
Phenol index	C_3H_6OH: $\mu g\ l^{-1}$		0.5		
Anionic detergents	Lauryl sulphate: $\mu g\ l^{-1}$		100		
Pesticides and allied products:	$\mu g\ l^{-1}$				By pesticides and allied products are meant: persistent organochlorides organophosphorus carbamates weedkillers
total:			0.5		
by individual substances			0.1		

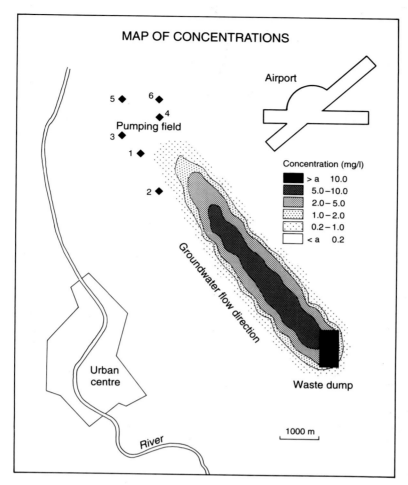

Fig. 5.30 Groundwater modelling (example of computer program Viking developed by BRGM) enables hydrogeologists to give precise contours for the migration of pollutants.

plumes of water spreading out in the direction of groundwater flow are made non-potable by high contents of chloride, sulphate, and nitrate.

An area of overlap exists between household and handcraft-type pollution; that of solvents, dilutants, and acids of all kinds. The organochlorine products used by do-it-yourself enthusiasts and unscrupulous companies, thrown away on household waste dumps (or in sewers) are toxic and somewhat carcinogenic. Their poor bio-degradation in groundwater, high mobility (they move through clay more quickly than water), and poor viscosity associated with a density well over 1 g cm^{-3}, make them particularly formidable. These peculiarities of the heavy organochlorine solvents make them responsible for the surprising pollution of groundwater around a hundred metres deep.

The most serious dangers result not from domestic refuse but from all the products of our consumer society, along with those of industrial and trade activities which themselves pollute the household waste. In poorly ventilated dumps, anaerobic and reducing conditions are maintained by organic matter (methane, not carbon dioxide, is produced here), conditions favourable to the solution of most of the toxic metals and to the conservation of organic molecules. Lead, zinc, chromium, cadmium, mercury, chlorinated solvents, etc. are the real dangers of the dumps.

In locating appropriate sites for dumping waste, the hydrogeological aspect is always the main preoccupation, since water is the vehicle of almost all forms of pollution (apart from smells). Hydrogeology is an essential input in the construction of maps showing the suitability of land for dumping, and in all studies of the potential impact of the development of landfill sites.

A typical example of what is involved is provided by the case history of the Fretin landfill site in France.

Impact of the Fretin landfill site on the chalk aquifer [63]

The problem

The Fretin landfill site, near Lille in the North Department in France, was opened in 1967 by the household waste disposal service of Lille in an old chalk quarry excavated down to the water table. Waste fill continued until 1982 at an average rate of 225 t per day, bringing the total to 3.5 Mt occupying an area of 7.5 ha to a depth of 12 m. The waste tipped on the site comprised household waste: 25 per cent, sludges: 2.2 per cent, normal industrial waste: 57.7 per cent, and inert materials and bulky objects: 12.3 per cent.

The removal of the protective surface layer of loam, the high rate of percolation of rainwater through the dump (160 mm yr^{-1}) and the submergence of the base of the dump, to a depth of as much as 3 m at the high point of the seasonal water table, led to severe groundwater pollution by the landfill. As the chalk aquifer is the sole source of potable water for the entire area, it became clear that protective measures and redevelopment were essential.

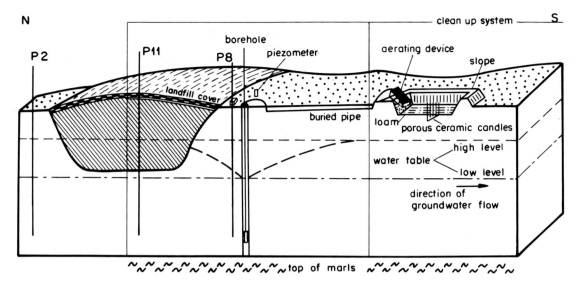

Fig. 5.31   Diagrammatic section of the Fretin landfill site showing the system for purification of the chalk aquifer.

The landfill operator and the regional authorities approached BRGM's Nord-Pas de Calais Regional Agency, which was then entrusted with the tasks of proposing a redevelopment plan, studying the effects of the pollution caused by the landfill on the chemistry of the groundwater in the chalk, and designing and implementing a means of decontaminating the groundwater.

The redevelopment plan (Fig. 5.31)

The plan was to cover the waste to a depth of 1–1.5 m with a dome-like layer of clayey soil, to limit percolation through the landfill, and so decrease the amount of pollutants introduced into the aquifer.

Study of contamination in the chalk aquifer

Observations were made and groundwater chemistry monitored for 13 years, from 1975 to 1988, by analysis of water samples from some ten observation wells, and by measurements of temperature and humidity profiles in the landfill site:

1. Very high total mineralization (dry residue 2850 mg l^{-1} below the dump and 1700 mg l^{-1} on its downstream edge compared with 300–400 mg l^{-1} on its upstream side), resistivity 320 Ωcm beneath the dump as against 250 Ωcm on its upstream side, acid pH, high bicarbonate content, and temperatures between 21.5 and 40 °C instead of the normal 10–11 °C groundwater temperature (see Table 5.11 and Fig. 5.32) were observed.

2. A strongly reducing environment was shown to be associated with the dump by a decrease in nitrite, nitrate, and sulphate contents (0–60 mg.l^{-1} instead of 150 mg l^{-1} upstream), and an increase in ammonia (324 mg l^{-1} instead of 50 mg l^{-1} upstream), in oxidizability, in free carbon dioxide, and in bicarbonate.

3. An increase of chlorides was observed from 40–60 mg l^{-1} above the site to 400–600 mg l^{-1} below it. On the other hand, no undesirable toxic compounds were detected, apart from traces of lead (0.182 mg l^{-1}).

4. Rises in the level of the water table cause increases in Cl, NH_4 and K contents.

The pollution front is moving southwards, the direction of flow in the aquifer, at a maximum rate of 50 m per year. This rate could be explained in part by the domed form of the protective cover limiting percolation and thus the leaching of the landfill, and in part by the downstream increase in thickness of the aquifer from 5 to 10–15 m. Nevertheless, the initial equilibrium is re-established about 500 m downstream from the dump.

Purifying the groundwater

The long-term observations allowed a means of purification to be designed. On the one hand the water-table was lowered by pumping to prevent leaching of the landfill, and on the other a ponding basin was built to enable the treatment and percolation of the polluted water.

Pumping programme

Step-by-step and long-term pumping tests were undertaken to define the characteristics of a borehole that was drilled to a depth of 28 m immediately downstream from the dump; the hole was cased from 19 m to the bottom. The results were: transmissivity 3.9×10^{-3} m^2 s^{-1}; maximum flowrate 41 m per day, calculated for convergent radial flow using a sodium iodide tracer.

The information from these tests was used to establish a pumping programme to limit the downstream spread of the pollution (Fig. 5.31).

Aeration and ponding

The pumped polluted water is piped underground to a ponding basin where it is aerated before being returned to the ground.

The basin was partly filled with a filter layer of clayey loam. Porous porcelain candles at different depths in the landfill were used to collect water for analysis. Chemical profiles beneath the ponding basin showed decreased levels of chlorides, ammonia, and bicarbonates between the surface and the chalk, indicating the efficacy of the aeration and filtration; the improved quality of the groundwater in the aquifer in 1988 indicated that the system is effective.

Gas drainage

The anaerobic conditions resulted in the bacterial production of gases (65 per cent CH_4, 35 per cent CO_2) from the organic material in the landfill. In 1983 the site was fitted with a system of gas drainage and collection with a view to carrying out production and utilization tests on the gas produced by the dump.

In 1983, BRGM supervised the siting, drilling and equipment of seven boreholes at 100 m intervals and 33 observation wells for the Agence Nationale pour la Récupération et l'Elimination des Déchets (ANRED: National Agency for Waste Recovery and Disposal), and performed gas production tests (yield, pressure, composition, and temperature).

Table 5.11 Variations in groundwater chemistry in the chalk aquifer upstream and downstream from the Fretin landfill site (April, 1980: mg l^{-1})

	100 m upstream	Landfill	150 m downstream
Dry residue	789	2339	593
pH	7.25	7.20	7.20
Oxidizability	3.20	65	1.40
Cl$^-$	47	420	33
NO$_2^-$	0.14	2.30	<0.05
NO$_3^-$	90.50	83.10	52.10
SO$_4^{--}$	170	185	115
HCO$_3^-$	500	1296	349
Ca^{++}	212	182	158
Mg^{++}	28	55	3.20
NH$_4^+$	<0.10	220.50	<0.10
Na$^+$	29	220.80	24.80
K$^+$	3.10	160.40	6.80

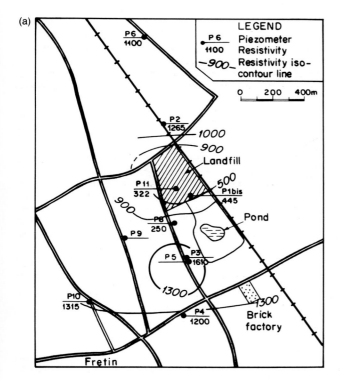
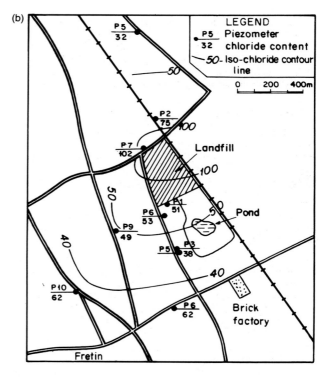

Fig. 5.32 Measurements at the Fretin landfill site.
(a) Resistivity map of the groundwater in the chalk aquifer, April 1980;
(b) map of the chloride content in the chalk aquifer, April 1977.

Industrial pollution is the most dangerous, however, being the greatest, the oldest, the most harmful to the environment, and the most expensive for society to cope with. Originally it resulted from ignorance and carelessness, but now it arises from lax supervision of the regulations, from a reluctance to meet the costs of appropriate processing, and from clandestine, illegal, subsurface dumping.

An inexhaustive list of the main types of industrial pollution is as follows:

(1) waste heaps and tailings in the mining industry;

(2) clinker and slag heaps in the metallurgical industry;

(3) sludge and residues from decanting of various kinds;

(4) leakages and neglect in the petrochemical industry;

(5) clandestine injection of chemical and pharmaceutical products into abandoned wells;

(6) burying or dumping barrels and cans containing highly toxic products amongst household waste.

Such pollution in the long term has already caused the spread of plumes of contamination over wide distances locally—several tens of kilometres in the Alsace plain in France. In places whole aquifers have been permanently sterilized, particularly in the case of industrial wastelands where clinker and other wastes have replaced the normal land surface for more than a century where pollution is long-established.

Steady-state regimes have been set up in which the mass of waste present liberates enough material to contaminate the water for centuries to come.

Decanting ponds, which commonly have permeable bases and permeable walls which breach the protective fine-grained surface deposits, allow massive inputs of chemical pollutants: this is the case for the heavy metals from ore-processing industries in particular, but also for the sludge from water treatment plants in which the food chain excreta and household waste are concentrated.

Decontamination, which is always costly, takes various forms: the removal and treatment of the contaminated ground, dumping millions of tonnes of polluted earth in a safe place is the extreme solution. The hydrogeologist can offer less drastic remedies; some consist of confinement by the injection of waterproof barriers, others use hydraulic trapping which diverts the polluted water towards extraction by pumping. *In situ* treatment—microbiological degradation or the more recently developed air or steam stripping—now makes it possible to envisage the purification of groundwater contaminated by organic micropollutants.

As a result of a century and a half of neglect, new disciplines have developed in the attempt to repair the damage. The greatest contributions are made by modern studies in hydrogeology in collaboration with chemistry and microbiology. Nevertheless, the preventative role of hydrogeology is more important for the future; but this will be successful only if the senior decision makers become more aware of industrial pollution in groundwater. In this area the essential task of hydrogeologists in national Geological Surveys is to provide information, explanation, and education.

5.5.2. Diffuse or areal pollution

Areal pollution, by definition less concentrated than the various types of local and direct pollution, was recognized later. Groundwater is affected by it in three possible ways:

(1) by water from streams which are themselves polluted, when they recharge the groundwater;

(2) by rainwater polluted by substances in the atmosphere or during the flow over the surface;

(3) by water from natural percolation, irrigation, or manure spreading.

The general and repetitive character of these types of pollution, especially the last, which affect the renewable resource from its origin onwards, prohibits the practical use of protection areas. The level of pollution is so serious that its economic consequences will be felt well into the twenty-first century.

Linear pollution spreading

The considerable amounts of groundwater tapped from alluvium (see Section 3.2), and the ease with which this kind of abstraction can be made (in practice, a recharge is brought about by the depression of the water table caused by pumping), are responsible nowadays for two hazards, accidental and endemic. Nevertheless, the hydrodynamics and hydrochemistry of the banks still enable adequate protection of the water tapped from alluvium, even when only a few tens of metres away from heavily polluted water.

Accidental spillage of harmful substances into streams (resulting, for instance, from a fire in an industrial plant during which the water used for extinguishing the fire has leached storage areas, or from navigation accidents on rivers), leads to the worst cases of pollution. In addition to the obvious potential for ecological disaster, these accidents may have serious effects on alluvial aquifers, the pumping of which may have to be stopped, and for several days a whole town may be deprived of its water supply.

The endemic discharge of domestic waste water, dirty urban rainwater, and badly purified industrial water leads to the serious contamination of many streams, especially during periods of low water when the flow is insufficient to dilute the pollutant load. These causes of organic and mineral pollution make water-purification more difficult and costly. However, groundwater tapped some distance from river banks does not contain all the pollutants present in the surface water. The filter effect of the partly clogged banks is capable of intercepting various substances. Turbidity and micro-organisms are removed by mechanical filtering; various organic substances, particularly hydrocarbons, undergo biodegradation in this bacteria-rich zone; and heavy metals become fixed on clay particles.

This providential filter effect of clogging, a phenomenon long considered harmful to quantitative recovery, is precarious, however. There is no guarantee that it will not fail at some time with a change in hydraulic or chemical conditions leading to the remobilization of heavy metals, for example. Considering what is at stake, high priority should be given to hydrogeological research in the area of filter effects.

Leaking sewage pipes are another form of areal pollution, which may lead to the pollution of wide areas between a town and its sewage treatment plant. Useful aquifers beyond the already sterilized urban groundwater may be fouled.

Excessive pumping of groundwater may cause pollution by sea water over vast areas adjacent to coasts where the socio-economic impact is great because of limited national resources.

The socio-economic impact of marine intrusion as a result of overpumping aquifers on the Spanish coast [160] [154] [161]

Background

The mildness of the climate on the coasts of Spain, especially on the Mediterranean coast, has favoured a considerable development in socio-economic activity.

A lack of suitable planning has caused a deterioration in the quality of groundwater in coastal aquifers in these areas. This is a major problem because it has led to a reduction in available resources and to a restriction in their use for drinking water.

Inventory of coastal aquifers: characteristics

In order to draw up an inventory of coastal aquifers, information existing in different public and private organizations was collected.

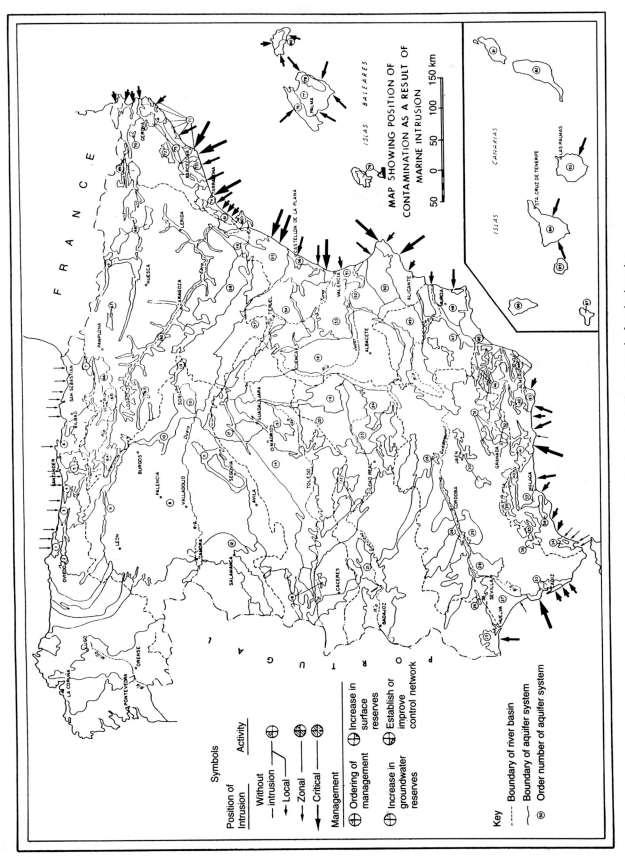

Fig. 5.33 Map showing position of contamination as a result of marine intrusion.

Table 5.12 Distribution of the degree of marine intrusion

River basin	Total surface area (km²)	Surface area with general intrusion (per cent)	Surface area with zonal intrusion (per cent)	Surface area with local intrusion (per cent)	Surface area without intrusion (per cent)	Surface area with no intrusion data (per cent)
Eastern Pyrenees	2870	13.5	49.6	36.9		
Ebro	440					100
Jucar	2679	13.6	33.4	46.8		6.2
Segura	2050	77.1		22.9		
South	1768	0.2	35.3	18.3		46.5
Guadalquivir	2895	3.1		96.9		
Guadiana	600		100			
North	3589				100	

The results of analysis of the information provided by all the documentation consulted, is as follows (Fig. 5.33):

Length of peninsular and insular coastline	6 121 km
Peninsular aquifers examined	59
Surface covered	16 891 km²
Water Resources	3 194 Hm³ yr⁻¹*
Volume pumped	1 540 Hm³ yr⁻¹

Two distinct zones have been identified:

1. Northern Zone: The intrusion of salt water does not cause major contamination problems. This is largely explained by the low degree of groundwater used, this being a result of the high precipitation yield, in many cases greater than 1000 mm per year on average. In recent years, however, the irregularity of precipitation and the lack of storage reservoirs have caused supply problems, and this means that the use of groundwater is on the increase.

2. Mediterranean and Atlantic Zones: In this case, the intrusion of salt water constitutes a major contamination problem. The irregularity of precipitation, mainly in the Southern Mediterranean zone (Almeria below 250 mm yr⁻¹), and the impossibility of attaining surface control because of adverse geological conditions, mean that the use of groundwater is considerable. Approximately 1540 Hm³ yr⁻¹ are pumped and 258 173 hectares are under irrigation.

The 16 892 km² of surface area that are occupied by all the coastal aquifers of the peninsula are distributed in the following way: 5913 km² (35 per cent) have local intrusion problems; 3538 km² (21 per cent) have zonal intrusion problems; 2423 km² (14 per cent) have general intrusion problems; 3589 km² (21 per cent) show no evidence of intrusion; and for the rest, 1428 km² (9 per cent), data are not available.

The distribution, by river basins, of the degree of marine intrusion as a percentage of the surface area of the aquifer is presented in Table 5.12.

Use of groundwater in the coastal zones

Figure 5.34 shows, in terms of river basins, the use of groundwater from the aquifers of the Spanish peninsular coast, the population supplied, the areas under irrigation, and the reserves pumped.

It is estimated that the total reserves of these aquifers are about 3194 Hm³ yr⁻¹, of which 1539 are pumped out and 1655 flow into rivers, springs, and the sea.

By comparing the values obtained with data on a national scale, it becomes clear that of 3 000 000 hectares of irrigated agricultural land, about 30 per cent (900 000 ha) are supplied with groundwater and, of these, just under 30 per cent (258 173 ha) lie in coastal zones.

* HM³ = one million m³.

Assessment of the economic repercussions

To assess the economic repercussions of marine intrusion as a result of over-pumping, the cost per unit of one m³ of water has been calculated and the agricultural production per irrigated hectare has been evaluated.

Figure 5.35 indicates, in millions of pesetas per year for each basin, the potential value of the pumped extractions, the potential value of the resources and reserves managed, and the evaluation of possible economic costs for alternatives.

Conclusions

There is a considerable degree of deterioration in the quality of the groundwater in coastal aquifers, and a great dependence on these resources in social and economic activity.

The methodology used offers a quantified approximation of the problem concerned and, although not rigorously, it does provide an estimate of the economic and social scale being dealt with.

The general results obtained are: 59 aquifers studied; the total resources of the coastal aquifers are 3194 Hm³; 151 Hm³ of the reserves are managed; 5 252 571 inhabitants are supplied with groundwater; the surface area under irrigation is 258 173 hectares; the potential value of the resources and reserves managed, reached about 98 949 million pesetas per year; the annual production of agriculture under irrigation is 189 393 million pesetas per year; the alternatives, in the event of total salinization of the aquifers, would mean a cost between 105861 and 129793 pesetas per year.

Pollution by acid rain

The combustion of fossil fuels—oil, coal, and lignite, mainly in chemical plants, oil refineries, and power stations—leads to the release into the atmosphere of acid-forming chemicals, in particular sulphur and nitrogen dioxide. These substances, leached from the atmosphere by rain, contaminate all potentially usable water at its origin—that which runs off as well as that which percolates into the ground.

The direct impact of this kind of pollution (which affects Scandinavia and central and eastern Europe in particular and also large areas around the Mediterranean) on forests, historical buildings, and human health are well known. Industrial environment control services are now concentrating on reducing these emissions, and the modernization of factories allows much hope for improvement. Nevertheless, for decades rain has been far from being distilled water and this situation will surely continue for some years.

When soaked up by limestone, the acids form salts—sulphates, which are not too harmful, and nitrates—which, though they are in small quantities, are nevertheless significant

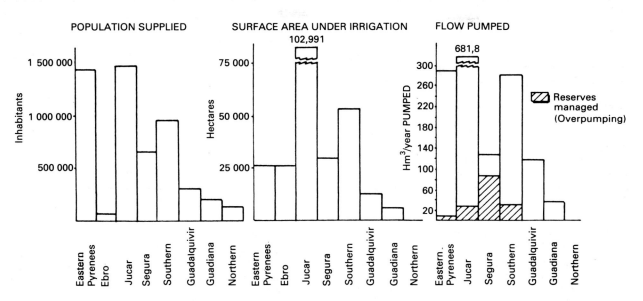

Fig. 5.34 Use of groundwater in coastal zones of Spain.

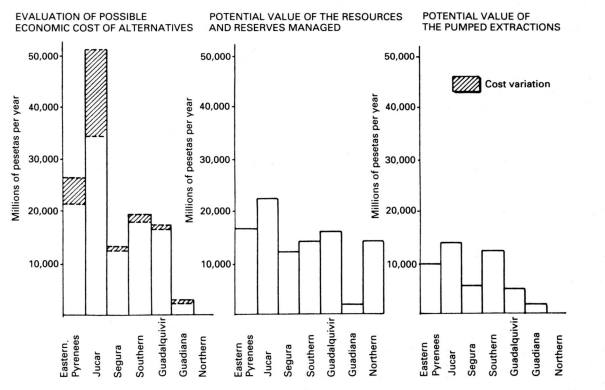

Fig. 5.35 Potential values of water extraction in coastal zones of Spain.

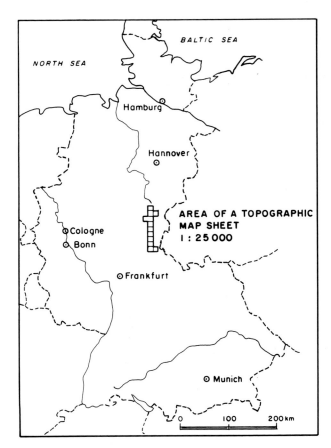

Fig. 5.36 Location of the study area in Hesse, Germany.

since they are additional to other types of pollution. Denmark, with a few tens of mg l^{-1} of nitrate in its rain, at present receives more nitrate from the air than it did by the spreading of fertilizer before the Second World War!

When acid rain falls on siliceous soils rich in trace heavy metals, induced pollution of the groundwater can occur. A good example of this is provided by the following case history from the Geological Survey of Hesse.

Investigation of the acidification of groundwater by immissions ('acid rain') in Hesse, Germany [91]

Introduction

The largely forested mountainous study area of about 1200 km^2 is located in North and East Hesse (Fig. 5.36). Average annual precipitation varies between 600 and 900 mm due to the varying altitudes between 250 and 600 m and prevailing wind direction. The rock sequence is dominated by horizontal or gently dipping layers of the Lower and Middle Bunter (Lower Trassic). As a result of high tectonic fracturing the Bunter formations are an excellent groundwater recharge and resource area. Consequently, central water supply services take their water requirements almost 100 per cent from Bunter groundwater resources. However, Bunter aquifer bedrocks consist chiefly of silica-cemented quartz sandstone extremely poor in basic cations (i.e. sum of calcium, magnesium, potassium, and sodium).

Shallow groundwater especially shows low concentrations of basic cations and low neutralization capacity (measured as hydrogen carbonate), and is very sensitive to man-made acid

deposition, quantitatively measured at two sites in the study area since 1983. The most common strong acids in acid precipitation are sulphuric and nitric. Regarding groundwater acidification the first important question is whether spring water is acidified by anthropogenic deposition (acid rain) or whether it is acid for more natural reasons (see below).

Then the question arises as to how spring waters, acidified by atmospheric deposition, can be identified by means of simple chemical models (see below).

Historical reconstruction of changes in groundwater systems

About 3000 water analyses—chiefly annual control analyses of drinking water between 1960 and 1976—are stored in the water data bank of the Geological Survey of Hesse (HLB). On the basis of the uniform hydrochemical data the influence of geogenic (terrigenous) and atmospheric components on groundwater quality was investigated by geostatistical analyses. An iterative cluster analysis identified a cluster 'acid groundwater' including 83 water samples. Next, a hierarchical cluster analysis was carried out to classify the 83 acid water samples (Fig. 5.37). The dendrogram structure of Fig. 5.37 models the process of acidification from an original natural acid groundwater to an acidified groundwater caused by acid atmospheric deposition in three successive phases:

Phase 1:   (= subcluster B in Fig. 5.37, including 41 water samples) is characterized by an equivalent ratio of geogenic hydrogen carbonate (HCO^{-3}) to airborne sulphate (SO_4^{-4}) significantly greater than 1.0, demonstrating the state of an original natural acid groundwater.

Phase 2:   (= subcluster A2, including 24 water samples) shows an increasing influence of airborne sulphuric acid (H_2SO_4), with an equivalent ratio (HCO^{-3}): (SO_4^{-}) now lower than 1.0.

Phase 3:   (= subcluster A1, including 19 water samples) symbolizes the dominant role of SO_4^{-} in the hydrosphere with absolute concentrations of SO_4^{-} significantly higher than 50 mg l^{-1}.

The water of numerous springs in the Bunter formation reached the Phase 2 already at the end of the 1960s.

Actual scale of shallow groundwater acidification

To quantify the actual dimension of groundwater acidification, water samples were collected monthly in 1988 at 20 representative spring sites of the cluster 'acid groundwater'. The calibrated spring catchments are almost completely forested and without habitation upstream. The scale of spring water acidification (Fig. 5.38) can be proved by the relationship of basic cations (Ca^{++} and Mg^{++}) to strong acid anions (SO_4^{-} and NO_3^{-}). Corresponding to the results of the cluster analysis (Fig. 5.37) there are three successive phases in actual spring water acidification:

Phase 1:   The ratio ($Ca^{++} + Mg^{++}$): ($SO_4^{-} + NO_3^{-}$) greater than 1.5 refers to spring waters belonging to the hydrogen carbonate buffering system with sufficient buffer capacity to neutralize anthropogenic acid impacts. The pH levels of these natural acid spring waters remain above 5.5.

Phase 2:   Lower values of the ionic ratio less than 1.5 cause sudden strong pH depression down to pH 4.5, expressing the beginning of spring water acidification.

Phase 3:   With ionic ratios lower than 1.0, acid deposition of air pollutants can be neutralized by aluminium (Al) only. Increasing the SO_4^{-} concentration decreases the buffering concentration of HCO_3^{-}. As a result the dominant acid input of H_2SO_4 reacts with soil and aquifer minerals to form potentially toxic aluminium species. Specifically the proportion of the (in ecological terms) dangerous free ionic Al^{+++} will increase because of the relatively

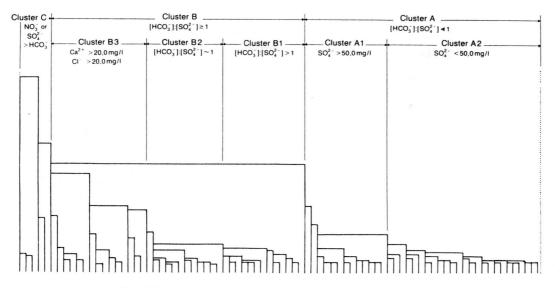

Fig. 5.37 Dendrogram structure of acidification (83 water samples).

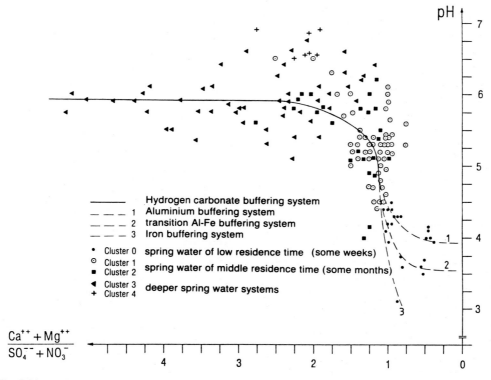

Fig. 5.38 Relationship between pH and the equivalent ratio ($Ca^{++} + Mg^{++}$): ($SO_4^{--} + NO_3^-$) in spring water.

low pH (<4.3). Consequently, there is a significant relation between the concentrations of the potentially toxic heavy metals Zn^{++}, Mn^{++}, Cu^{++}, Cd^{++}, Cr^{++}, Pb^{++}, and the pH of the spring water.

Figure 5.39 demonstrates the advance of the acidification front downwards. Spring waters of type (a) with low residence time (some weeks) are strongly acidified. Especially after snowmelts the concentrations of SO_4^{--}, Al^{+++}, and of the heavy metals (Zn^{++}) increase. The limit values for drinking water are 0.2 mg l^{-1} Al, 0.1 mg l^{-1} Zn, and 0.05 mg l^{-1} Mn. This means that after toxic snowmelts the toxic concentrations of these elements are 10–15 times higher than the limit value of drinking water.

Spring waters of type (b) with middle residence time (some months) show light acidification without mobilization of Al^{+++} and heavy metals. Effects of increasing SO_4^{--} and Ca^{++} and concurrently decreasing HCO_3^- refer, however, to the first stage of airborne acidification.

Deeper spring water systems (type c) actually do not become acidified by 'acid rain'. The longer groundwater flow path (residence time over one year) in deeper spring water systems causes higher mineral weathering and basic cation exchange processes to buffer the incoming 'acid rain' effectively.

Conclusion

As described (historical and actual evidence for changes in groundwater systems), the significant high input of acids by emission led to the contamination of shallow groundwater. Severe problems arise for water supply (aluminium, heavy metals), especially for small rural waterworks without high-technology resources for water treatment. To prevent further harmful effects of acid deposition on shallow and deeper groundwater systems a drastic reduction in emissions of SO_x, NO_x, and heavy metals is recommended as a regulatory measure.

Pollution by 'acid rain' also causes technical problems which, although not affecting the natural environment, do affect potable water. This occurs in two ways:

(1) lead salts form from old lead pipes, fortunately scarce now in modern networks;
(2) pipes in the potable water distribution and domestic networks and installations become corroded and prematurely destroyed. Correcting this fault, by passage over powdered limestone, is an extra, costly operation which increases the cost of the water unnecessarily.

Rain can also introduce substances from high altitudes that are carried by air currents on a continental or planetary scale. Since the 1950s, hydrogeologists have been able to date percolation water, thanks to a number of gigantic blasts of tritium from the atomic weapons tests of the nuclear military powers. Rainwater stamped in this way has been found in groundwater and its percolation dated, bringing great progress in hydrogeology. Since atmospheric tests have stopped, this is a pollution that is fortunately no longer at a dangerous level, due to the short half-life of tritium. However, dangerous radioactive elements can be produced by accidents at nuclear power plants, such as the recent catastrophe at Chernobyl.

Large surface pollution spreading

Nevertheless, with the extension of urban areas, and the multiplication of points of emission, pollution is now widespread. In suburban areas it is from domestic sanitation, sundry leakages, and petrol stations that the groundwater is contaminated.

In very large cities, the thermal impact, rising water tables, and water transfers brought about by the extent of water extraction and unauthorized dumping, lead to different disorders. Not only is the groundwater no longer potable, but it is harmful to underground structures by corroding pile or caisson foundations below the water table, for example.

The main cause of areal pollution, both as regards the level of undesirable chemicals in the water and the extent of the contaminated areas, is without doubt the introduction of intense farming methods.

The search for higher annual crop yields for all crops, the disappearance of cheap labour, and the constraints of the market, have caused farmers to adopt practices which are harmful for the environment. They do this with the support of the consumer, despite the significant resulting contribution to diffuse pollution.

Two pollutants, nitrates and pesticides, are predominant.

The main origin of the nitrates is the excessive levels of fertilizer applied with the intention of guaranteeing high cereal yields (up to and over 10 tonnes per hectare). They also result from the spreading of liquid manure from industrial stock-rearing or from harmful farming practices such as leaving the soil bare. All the great crop-growing regions are affected, the cereal areas of the plains of Western Europe, and the intensive stock-rearing parts of Brittany, The Netherlands and Denmark, and the intensive fruit-growing, market-gardening, and horticultural areas of the Mediterranean coastal plains.

The intense nitrogen pollution of the cereal-growing plains of northern Europe is well known, but the following case history shows the impact of nitrates on other vulnerable environments. Low groundwater recharge by effective rainfall contributes to high nitric nitrogen contents, with accumulations which cause concern. The mechanisms of the leaching of nitric nitrogen by percolating water, and its slow journey through the unsaturated zone, have been the subject of much research. The conclusions are rather pessimistic and tend to show that the increase in nitrate contents will continue in the long term. The nitrate in much of the groundwater shown to have contents approaching the statutory limit of 50 mg l^{-1}, is likely to increase, and in many places it has been shown that the benefits of an immediate reduction in the use of nitrate-based fertilizers will become available only after several decades (Fig. 5.40).

A case-history from Germany describes a specific evaluation system to assess vulnerability of soils to nitrate pollution. Another case history describes the situation in Spain.

Evaluation of vulnerability to nitrate pollution in Northrhine-Westphalia [112]

The Geological Survey of Northrhine-Westphalia was charged by the Ministry of Environment, Regional Planning and Agriculture with producing a map concerning the probability of nitrate leaching out of soils in Northrhine-Westphalia.

The resultant map is based on a high resolution soil map at a scale of 1:50 000, which exists in digital form for more than 80 per cent of the area of Northrhine-Westphalia (32 000 km^2). The soil map also describes the depth of the groundwater table.

To describe the relief, the Digital Terrain Model from the Ordnance Survey of Northrhine-Westphalia is used. By combining these rasterized relief data with the vectorized soil data it is possible to calculate mean heights and slopes for any single area on the soil map.

To describe the climate a third database is used. It contains daily records of about 600 stations of the German Meteorological Survey covering the last 30 years.

The data obtained are part of the soil information system GLABIS, forming the basis for a soil water simulation model. This

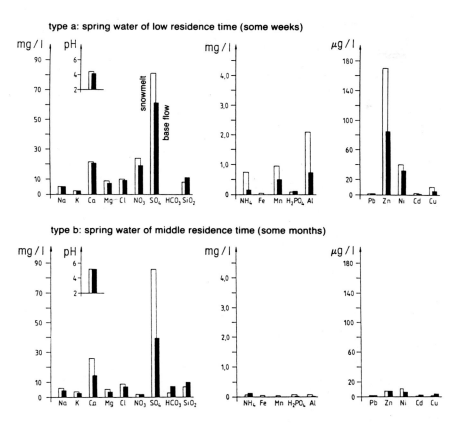

type a: spring water of low residence time (some weeks)

type b: spring water of middle residence time (some months)

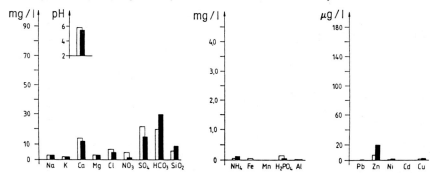

type c: deeper spring water systems with residence time over 1 year

Fig. 5.39 Concentrations of major chemical components and heavy metals after snowmelt and after a dry season (base flow) in shallow groundwater.

Type a: spring water of low residence time (some weeks).
Type b: spring water of middle residence time (some months).
Type c: deeper spring-water systems with residence time over 1 year.

model calculates the water influx as well as the percolation out of the soil, depending on:

(1) precipitation (reduced by the portion of direct run-off);

(2) potential evapotranspiration;

(3) capillary rise from groundwater;

(4) soil moisture;

as well as depending on the site-specific

(5) depth of groundwater table;

(6) water conductivity;

(7) water storage capacity.

This includes a correction of the potential evapotranspiration to the actual evapotranspiration with respect to the soil moisture values.

The results will be completed in mid-1992 as the first set of country-wide uniformly evaluated spatial data with a resolution equal to the soil map at a scale of 1 : 50 000.

The main importance of the soil information system introduced above is given by the fast and clear arrangement of evaluations and estimations, which integrate many expert interpretations and complex simulations. Consequently, the system supports quick decisions in regional planning, e.g. defining areas of water protection or laying down adequate manure application schemes.

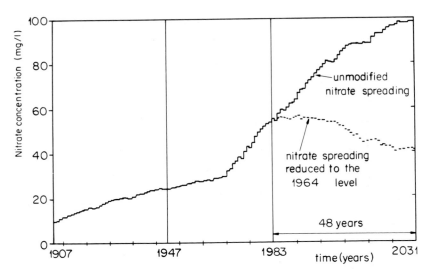

Fig. 5.40 A model predicting NO$_3^-$ content in groundwater. Spring of Provins, Paris Tertiary Basin, France. After slight growth until 1970, the nitrate content increases rapidly. Simulation, using the BICHE model (BRGM) shows that NO$_3^-$ content will reach the 100 mg^{-1} level in about 50 years; even if land-use and farming practices are changed, the decrease will be very slight.

The contamination of groundwater by nitrates in Spain [162] [161] [159]

Source of contamination

An increase in the presence of nitrogenous compounds in groundwater, at times in quantities that constitute a health hazard to the consumer, is an undeniable fact.

In Spain, as in other countries with a long agricultural tradition, nitrate contamination is closely linked to this activity. The presence of nitrogen compounds in the water can also be a consequence of other factors (industrial and urban waste, etc.). These effects may be more limited, but sufficient to be taken into account in assessing the origins of such contamination.

From 1936–87 the consumption of nitrogenous fertilizers in Spain increased eightfold (Fig. 5.41). Nevertheless, within the EEC, Spain was fifth in the total consumption of these products and the last as regards consumption per hectare (Fig. 5.42).

Situation in Spain

The diverse programmes of work developed in the field of nitrate contamination, especially over the last ten years, has made possible an awareness of the situation in Spain.

Since 1974, when the ITGE brought into service the systematic groundwater surveillance network in Spain, a large amount of periodical information has become available regarding water quality. There is also a considerable amount of data provided from specific studies.

Recently, in view of the development and trends recorded, a synthesis and study of the relevant information has been made for the period 1976–87 (12 years and 24 field studies). A total of 14 000 items of analytical data were processed. These were taken from 3042 points, spread throughout 84 aquifer systems in the large river basins of the peninsula and the islands.

The study reveals a whole series of situations and levels of influence, which range from aquifers with optimum quality to those that are severely contaminated. In general terms, those of the Northern Basin (Asturias and Cantabria) are in good condition, while along certain areas of the Mediterranean Coast, stretching from Gerona as far as the north of Alicante, there is very high contamination. Among the inland areas, the Guadiana Basin, in general, and the Province of Ciudad Real (Plains of La Mancha), in particular, have deteriorated considerably, as has the central part of the Douro Basin, and certain stretches of the Provinces of Toledo and Madrid, in the Tagus Basin.

A similar situation occurs on Mallorca, and on the northern parts of Tenerife and Gran Canaria.

The details for the Plains of La Mancha are shown in Figs 5.43 and 5.44. Two zones, one to the east, and the other to the south-west, have contents which exceed 100 mg l^{-1} amongst a background of between 25 and 30 mg l^{-1} of nitrates. Over the years there has been an increase in the percentage of samples with the highest content, and also an increase in the ranges of values in the aquifer. Generally, it has been observed that the regions most affected are those where intensive farming and large-scale extraction of groundwater takes place. Recycling through irrigation takes place and, as a result, the water is progressively enriched in nitrogenous compounds.

Tables 5.13 and 5.14 show the results obtained concerning the groundwater situation in the river basins and on the islands, and the development of the nitrogen content for the period 1976–87.

The development during this period shows a rising trend in nitrate content, both in the aquifers most affected and in those which although still in a satisfactory condition, are beginning to show the first signs of contamination. The future trends are pessimistic. If the present growth rate in the use of nitrogenous fertilizers continues without any modification in farming practices, there will be an increase in the number of areas affected and a rise in the

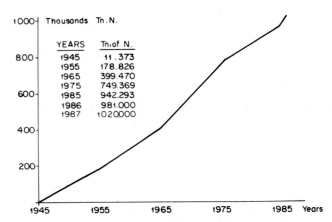

Fig. 5.41 Consumption of nitrogenous fertilizers in Spain.

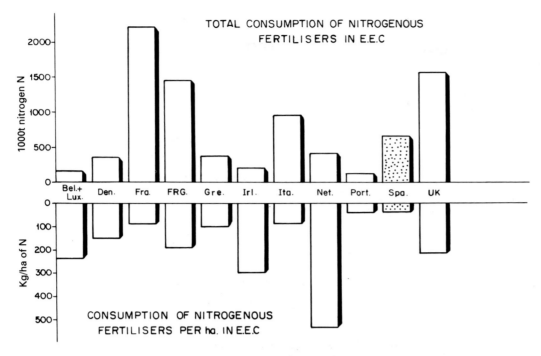

Fig. 5.42 Consumption of nitrogenous fertilizers in the EEC (1982–83).

degree of contamination in the aquifers already affected. This situation, although not as serious as in many other European countries, must be dealt with both by the bodies in charge of the protection of water resources, and by those responsible for agriculture. Joint action is essential to find solutions to this disturbing problem, in order to coordinate the consumers' interests with those of the farmers.

Over the past ten years or so, governments of the EEC countries have been vigorously undertaking research, notably in the United Kingdom, Denmark, and The Netherlands. The research has been directed both towards assessing, understanding, and modelling the extent of the problem, and at finding possible ways of solving it. Although the main hopes are based on agronomy and the processing of liquid manure and plant genetics, hydrogeology, in close liaison with the other disciplines, has its contribution to make.

To quote an example, modelling mineral contents and forecasting their evolution enables protection strategies to be proposed that are no longer based on zones of restriction, but rather on the agricultural management of an entire basin. It is now technically possible to allocate areas of farmland for the production of potable water, and actions tending in this direction, such as leaving land fallow, the drawing up of codes to regulate the use of fertilizers, and various incentives for the employment of extensive farming practices, are already being promoted in several countries.

Another example is that of natural or induced denitrification. This process enables good quality water to be obtained even under conditions such that nitrate contents could be expected to be above statutory levels. In the reducing conditions within confined aquifers, bacterial activity takes place by breaking down the nitrate ion to obtain the oxygen required by the organisms; the nitrate is thus destroyed, leaving a residue of nitrogen gas. This phenomenon, which has been confirmed by the use of isotopic techniques, can be reproduced and magnified by doping. Various industrial techniques are being

developed to apply this under suitable conditions to wellfields, or to effect a final treatment through the natural environment downstream from an industrial plant.

Pesticide pollution, spread by various plant-health products such as herbicides and insecticides (either chlorine or phosphate based), is not caused by agricultural use only: it is also a consequence of their use along railway tracks and roads. It is therefore predominantly of diffuse type. The amounts permitted by legislation are very low and, taking into account the technical and financial difficulties (such as sample-taking, conservation, and analysis) inherent in very high quality studies, our knowledge of this phenomenon remains incomplete. It should be noted also that several hundred different molecules are currently on the market. The behaviour of these substances in the natural environment and what becomes of them when they break down partially in the unsaturated zone is little known. Although they are organic molecules their degree of biodegradation is highly variable: their life (expressed in terms of the time taken for half of the substance to break down—analogous to the half-life of a radionuclide), is very variable. Another worry is that residual fragments of the molecular chain, the long-term toxicity of which is unknown, remain after the initial molecule has 'disappeared'.

Some products have already been banned, and it is highly probable that research will turn towards creating very shortlived molecules.

The main role for hydrogeology in this context is to identify the transit times in the unsaturated zone that are greater than the life of the product. Here is another possible use of vulnerability maps, with the time factor included in this case.

5.5.3. Conclusion

Both the long- and short-term future of water is seriously threatened by a combination of factors that can intervene at any point in the cycle. As far as groundwater is concerned the

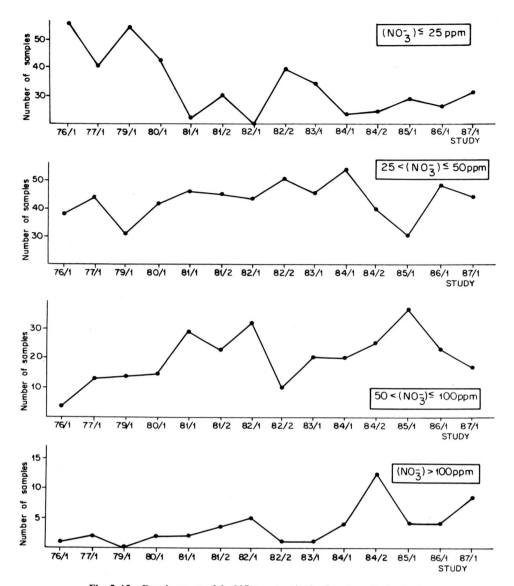

Fig. 5.43   Development of the NO$_3^-$ content in the Guadiana Basin, Spain.

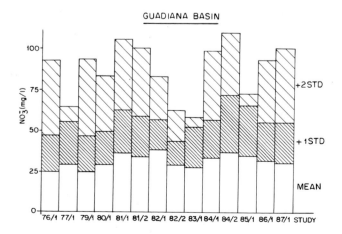

Fig. 5.44   Development of NO$_3^-$ content in Aquifer system No. 23 of the Guadiana Basin, Spain.

factor which constitutes both its strength and its weakness must be emphasized. The advantage of groundwater lies in its low rate of flow and the inertia of the large capacity systems that regularize flow over annual or longer periods, and thus help to overcome drought. This property can also be considered to be a major disadvantage, however, as it means that any pollution will be of long duration. In most countries, except those of the Mediterranean, it is certain that the quantity of useable water available will be restricted not so much by the overall resource available, but rather by a reduction in its quality.

To ensure a water supply to entire populations that meets accepted standards and is adequate for all intended purposes, it is conceivable that, with the help of technological research and enormous financial backing, virtually 'artificial' water could be derived from any polluted resource and then transported over great distances. This solution, however, is neither economically viable nor acceptable to environmental geologists as it would encourage the abandonment of

Table 5.13 Present situation (1987) of the nitrate content in Spanish groundwater (percentage distribution of the concentration in samples for river basins)

NO_3^- (ppm)	North	Duero	Tagus	Guadiana	Guadalquivir	Southern	Segura*	Jucar	Ebro	Eastern** Pyrenees	Baleares	Canarias
≤25	94	58	53	31	39	25	70	38	49	46	38	67
25–50	5	21	31	44	12	15	16	18	33	14	32	–
50–100	–	13	12	17	16	36	14	13	17	9	22	5
>100	1	8	4	8	13	24	–	31	1	31	8	29
No. samples	116	77	134	124	79	84	63	114	179	140	117	21
Maximum concentration (ppm)	108	420	422	175	320	740	81	500	113	825	193	800

* Data 1985
** Data 1986
Source I.T.G.E. 1988

Table 5.14 Evolution of the nitrate content in Spanish groundwater (percentage distribution of the concentration in samples)

NO_3^- (ppm)	1976 (1)	1976 (2)	1977 (1)	1977 (2)	1978 (1)	1978 (2)	1979 (1)	1979 (2)	1980 (1)	1980 (2)	1981 (1)	1981 (2)	1982 (1)	1982 (2)	1983 (1)	1983 (2)	1984 (1)	1984 (2)	1985 (1)	1985 (2)	1986 (1)	1986 (2)	1987 (1)
≤25	67	100	56	54	58	32	55	44	58	64	57	48	42	45	59	63	48	47	52	49	45	39	49
25–50	21	–	25	21	18	19	25	27	26	19	23	27	30	31	22	24	28	28	22	15	25	20	26
50–100	8	–	14	14	17	19	16	19	13	11	14	19	20	20	14	9	17	18	17	18	19	23	15
>100	4	–	5	11	7	30	4	10	3	6	6	6	8	4	5	4	7	7	9	18	11	18	10
No. samples	265	4	451	236	327	292	490	201	574	242	1050	736	1188	758	1154	555	1173	552	1265	782	940	348	1141
Max. value (ppm)	149	4	244	376	260	526	165	142	231	210	346	450	560	332	276	212	657	615	540	621	816	825	740

Source: I.T.G.E. 1988.

measures for the protection of resources and the prevention of pollution in the natural environment. The only lasting solution to the problem is, therefore, to take steps to protect against and prevent pollution, and the hydrogeologist must educate the general public in this respect, and take his place alongside others in the management of the quality of water.

RECOMMENDED FURTHER READING

Carra, J. S. and Cossu, R. (eds) (1990). *International perspectives on municipal solid wastes and sanitary landfilling*. (Academic, London).

Collective Authors (1978). *International symposium on groundwater pollution by oil hydrocarbons, Prague, June 1978*. (Stavebni Geologie, Prague).

Collective Authors (1984). *Guidelines for drinking water quality*. (World Health Organization, Geneva).

Collective Authors (1987). *International conference on the vulnerability of soil and groundwater to pollutants, Noordwijk-aan-Zee*. (Netherlands Organization for Applied Research, TNO, The Hague).

Forester, W. S. and Skinner, J. H. (eds) (1987). *International perspectives on hazardous waste management*. (Academic, London).

Förstner, U. and Wittmann, G. T. W. (1979). *Metal pollution in the aquatic environment*. (Springer, Berlin).

Foster, S. S. D., Bridge, L. R., Geake, A. K., Lawrence, A. R., and Parker, J. M. (1986). The groundwater nitrate problem. A summary of research on the impact of agricultural land-use practices on groundwater quality between 1976 and 1985. *Hydrogeological Report, British Geological Survey No. 86/2*.

Hites, R. A. and Eisenreich, S. J. (eds) (1987). *Sources and fates of aquatic pollutants*. (American Chemical Society, Washington).

Jamet, P. (ed.) (1989). *Aspects méthodologiques de l'étude du comportement des pesticides dans le sol. Workshop INRA, Versailles 1988*. (Institut de la Recherche agronomique, Paris).

Lallemand-Barrès, A. and Roux, J. C. (1989). *Guide méthodologique d'établissement des périmètres de protection des captages d'eau souterraine destinée à la consommation humaine*. (BRGM, Orléans).

Salomons, W., and Förstner, U. (eds) (1988). *Environmental management of solid waste: dredged material and mine tailings*. (Springer, Berlin).

6

The role of geoscience in planning and development

E. Stenestad and G. Sustrac

6.1. INTRODUCTION

In the preceding chapters the role of geoscience in the study and management of solid and fluid mineral resources, and the control and mitigation of natural and man-induced hazards has been heavily emphasized. In the present chapter it is argued that an integrated view of the problems concerned with development and land-use is necessary.

In this context, geoscientific data and expertise can make an important contribution to the identification of possible solutions to the problems concerned. Coupled with other data related to the physical environment and the socio-economic context, they have a leading role to play in the choices to be made. The interaction of all these aspects and facts leads inevitably to the exposure of potential conflicts and offers a basis for necessary compromises and plans concerning changes in land-use, which may include new uses fundamentally different from those practised previously.

6.2. KEY ROLE OF GEOSCIENCE IN RESOURCE MANAGEMENT AND RISK CONTROL

6.2.1. Optimizing resource management

Natural resources of any kind are inevitably sterilized in many ways. It is essential that the potential for man-induced sterilization be recognized and that conscious decisions to effect such sterilization be made only on the basis of the best possible knowledge of the socio-economic and environmental consequences, and of the possible side-effects. Analysis of the information held by Geological Surveys is of considerable importance in making such decisions.

In the first place, it is essential that the concept of resource be taken in its wider environmental sense. The opinion commonly held today is that the wider definition of environmental resources, in addition to the traditional elements of raw materials, water, fossil fuels, and ground for development purposes, also includes arable land, forests, wetlands, lakes and rivers, sea, scenic landscape, underground space, biotopes, air, soil, the buffer capacity of soil, and the durability of systems of geological processes. This understanding of resources means that 'non-use' is not an option, but rather that everything is being used or being held in reserve for use for the benefit of mankind. It is obvious that major geoscientific elements form parts of all types of natural resources.

As a result of much experience, some general principles and standard procedures are routinely deployed in the various phases of work involved in the assessment of resources in the geological environment. The first rough estimate of which resources exist in the area in question, where they are located, and in what quantity, is normally available from existing data.

But even at this stage, limited investigations may be necessary to supplement existing data.

A second phase in the assessment of environmental resources would normally include the detailed mapping and evaluation of the resources identified. This should be done for each resource present, for minerals and ores, groundwater, areas of scenic landscape, unpolluted environment, or for whatever group of elements is appropriate.

The most important criteria to consider are the geometry and size of the resource and its economic value. The latter will be assessed with reference to prevailing market conditions and to the current importance of the resource to society.

Understanding the value of any kind of resource in the geological environment is essential in resource management. Political and other decision-makers would often need some measure of the relative value of natural resources when choices have to be made between apparently similar options.

In general, this evaluation cannot be done in a simple or objective way. The value of an area for the production of some raw material, as a water reservoir, as scenic landscape, or for urbanization, depends on market forces and contemporary needs, and it inevitably varies with time. In some cases the relative value of natural resources can best be illustrated by associated parameters, such as the cost of restoration or replacement, the increased value of real estate in desirable surroundings, and the available savings in transportation costs because of the proximity of essential resources (raw materials, water, areas suitable for waste disposal). On the other hand, the value of environmental resources (such as clean air, unpolluted sea with good fishing, healthy mountain forests) may be measured by the potential reduction of socio-economic or environmental long-term costs, such as of health services, unemployment indemnities, and compensation for loss of life and property.

The limited extent of a resource or its vulnerability are also issues to be kept in mind. The general understanding of the fact that there are limits to nature's resources, and of the reaction of the environment to human interference, is now growing rapidly. Also, the natural resources themselves are much more widely understood now than they have ever been. The idea of vulnerability of certain types of geological conditions or natural processes is, in many cases, merely another way of expressing how long it takes to reach the limits of availability or exploitation [11].

The consequence of such observations is that the extent of sustainable development must be based on the recognition of the total potential or the 'natural environment's potential' (NEP), which includes all natural materials and processes which affect or can affect the human population and its environment [79].

The case histories from different parts of Western Europe strongly support the common understanding that natural resources consist of much more than just minerals and water, and that they include all elements constituting the environment. All natural resources are finite. Neither natural landscapes once destroyed nor extinct wildlife can be replaced. On the other hand, the position with respect to mineral resources, water, and underground space may be better. Used raw materials can be recycled, other materials can be substituted and production methods may be changed. Underground space can be used for secondary purposes even if, in some cases, it may be very expensive to backfill the voids and to re-arrange the existing installations and facilities. Polluted water can be purified, and reservoirs can be restored. But it all takes time and is extremely costly. It makes sense, therefore to conserve natural resources carefully and to use them in an informed way. Such deliberate policies are essentially political, but the national Geological Surveys have a major role to play in providing the best possible infrastructure of geological information and expertise for the planners and decision makers.

6.2.2. Controlling natural and man-induced geological hazards

Planners should seek and take account of geoscientific advice in relation to problems produced by natural geological hazards and human activity. Natural geological hazards are and always have been present in the environment. Landslides, volcanic eruptions, earthquakes, flooding, and permafrost are examples of natural hazards, although some of them, under certain conditions, may be induced by human activities. Pollution, on the other hand, is more commonly a result of human activities, but some cases of natural pollution are known.

The geosciences are a key source of information on natural geological hazards. Risk areas can be shown in thematic maps, natural geological processes producing background 'noise' in monitoring operations can be described and quantified, and unstable natural conditions which are vulnerable to the impact of certain types of human activity, can be identified. The understanding of natural processes is often crucial for evaluating the potential of an area onshore or offshore. The detailed monitoring and registration of the environmental changes over time is the research field of environmental history, which, in concert with other elements, produces retrospective *geological time-series* enabling the planner to understand better the meaning of current processes and trends in such areas as pollution, vegetation, climate or coastal zone development.

An example of a geological time-series is the long-term Danish study of bogs, which has demonstrated the level of lead pollution before and after the industrialization, the increase of diffuse lead contamination as the combustion engine was introduced, and the successive decrease as the lead content of petrol has been reduced.

A further example is the geoscientific analysis of strata showing climate and sea-level changes over millions of years. In the case of climate, for example, analysis made on lake sediments suggests that the climate may be getting colder, but that the rate of cooling seems to be lower than had been anticipated. The impact of possible changes in sea level should also be taken into account, along with local geological conditions and the long-term environmental history of the region. For example, the consequences in coastal areas of a rise in sea level will be markedly different in a subsiding region, such as The Netherlands, from those in such as Finland, where the regional isostatic movement is upwards.

Natural hazards

Natural hazards are present on Earth and will continue to be so until the planet ceases to exist in its present form. They are mainly the results of surface and deep-seated processes in the Earth's crust, and are experienced by man as earthquakes, landslides, volcanic eruptions, transgressions and regressions of the sea, glaciations, regional subsidence, floods, desertification, and other phenomena. Geoscientists know some of the reasons why they occur and can forecast that they will happen again sooner or later. They can also identify areas or regions where certain hazards are most likely to occur but, as yet, there is no means of forecasting accurately either the time of occurrence or the degree of severity.

In areas known to be at high risk for natural hazards, regular monitoring of the geological environment is essential to detect any minor physical changes in advance of catastrophic events. Such monitoring should include high-risk subsidence areas in order to reveal incipient subsidence features which may develop into major collapse structures, akin to natural phenomena.

In the light of the serious economic and social impact of geological hazards, some governments have established forecasting systems with a view to understanding the incidence and impact of such events. In Spain a systematic and quantitative approach to the economic and social impact of geological hazards is being carried out by the Instituto Tecnologico Geo-Minero de Espana (ITGE) for the period 1986–2016. Forecasts have already been made of economic losses, of the geographical distribution of losses, and of the potential loss of life in the population exposed to high levels of hazards. For each type of hazard, maps at the scale of 1:50 000 have been prepared of the hazard itself and of the expected economic losses (Fig. 6.1).

The work has revealed that the most damaging hazard in Spain is flooding. The potential economic losses due to flooding are greatest in the most populated regions along the Mediterranean coast. Whilst about 25 per cent of the population is exposed to a high level of risk from flooding, only 2–3 per cent of the population is exposed to risk from volcanic eruptions or great sea waves produced by submarine earthquakes or volcanic eruptions (tsunamis). These latter risks, however, may account for the highest losses of life (approximately 15 000 and 4000 respectively) if they occur at maximum intensity. At medium intensity the losses will be substantially smaller than potential losses from medium level flooding (~ 750) or landslides (~ 35) (Fig. 6.2). The results of the study have been adopted as the most important basis for the Civil Defence Policy in *Natural Hazards in Spain*, an action that could usefully be copied by other countries [152] [164].

Seismic hazards, volcanic hazards, and landslides are very well understood by the public. Other types of natural hazard are much more difficult to define and to locate. This is the case with the accumulation of explosive concentrations of methane by outgassing from groundwater. The scale of this hazard has been fully appreciated only in recent years. The usual sources of gas in tunnels and mines, such as coal, hydrocarbons, oil-shales, etc. are well known, but methane, at considerable depths and under pressure, can dissolve in groundwater, be transported considerable distances, and then be released into tunnels or other excavations where the pressure is lower. The danger is that, if the geology of the tunnel or excavation does not include an obvious methane source, the potential hazard may not be fully considered at the planning stage or investigated during construction.

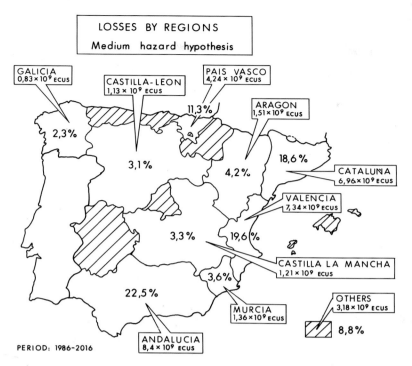

Fig. 6.1 Estimated economic losses in Spain through medium-level natural hazards (mostly flooding; see Fig. 6.2) over the period 1986–2016.

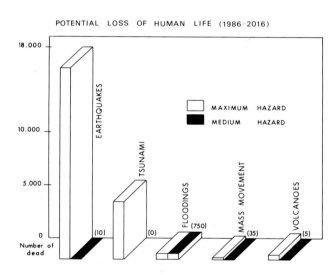

Fig. 6.2 Estimated potential loss of human life through maximum- and medium-level natural hazards in Spain over the period 1986–2016.

A case in point occurred at Abbeystead, Lancashire, in the UK. All 44 people on a visit to the valve house of a water transfer tunnel were either killed or injured by a methane explosion. Subsequent investigations showed that the gas had accumulated by outgassing from groundwater leaking into the partly empty tunnel from a sandstone in the core of an anticline [220].

The cultural heritage of Europe includes remarkable ancient monuments, some of which are threatened by natural hazards in addition to the well-understood ravages of air pollution and disrepair from lack of financial support. In Scotland many of the sites of ancient and historical monuments have been chosen fortuitously for long-term investigations of stability. Elements such as weathering, erosion, changing groundwater levels, urban development, and deep mining have all been found to be responsible for geotechnical and structural problems. A particular problem is that associated with medieval fortifications on prominent rocky outcrops in urban areas, which are sometimes prone to rock falls. They are possibly the most hazardous ancient structures from a safety viewpoint.

At Dunbar Castle in East Lothian, terrestrial photogrammetry was used to survey accurately the unstable rock outcrop and cliff-tops on the coastal slopes next to the castle. The unstable slopes in a basanite intrusion below the castle were examined in detail, in particular their planes of structural weakness. The joint orientations in the rock were plotted on a stereographic projection and a number of possible failure modes were considered. The methods proved to be particularly apposite for the assessment of the potential problems at this ancient monument. On the basis of the survey remedial measures were taken to ensure public safety [202].

One important question being addressed at present is the scale of the impact on the environment and on man's well-being, if the scenarios suggested for climatic change resulting from the so-called 'greenhouse effect' are borne out. For this reason most governments and many international organizations have funded research into climate change, and a number of scenarios and models have been developed. Further studies of possible effects of climate change are being encouraged. Here again, there is a role for geologists who, unless persuaded otherwise, tend to leave the development of climate models to colleagues in other disciplines. But studies of the geological environment yield historical evidence of rates of climatic changes and of the effects and impacts of such changes. It is well known that climate changes very fast indeed in a geological time-scale and may lead to dramatic changes in the distribution of land and sea, in precipitation, in vegetation and in general living conditions. Such evidence will be essential to the development of global scenarios for the future.

Most current climate models assume a global warming due to the greenhouse effect leading to a sea level rise which has been estimated to be in the range of 0.5–3.5 metres over the next 100 years [140]. A rise in sea level of that order will have a serious impact on coastal lowlands and on many big cities and their industrial complexes located near the coast. Higher water levels combined with changes in average and maximum wave height and the possibility of stronger and more frequent storms will increase coastal erosion and flooding. Coastal protection structures will have to be reinforced and harbour facilities may have to be modified. Flooding may lead to the overflow of drainage and sewerage systems and cause widespread pollution of the sea with untreated waste water. In some areas, rising groundwater may wash out phosphorus, pesticides, or other pollutants from the unsaturated zone and in this way contribute further to the pollution of the sea. Salt water may invade groundwater, rivers, bays, and farmland and may destroy water supplies and arable land. Fishery may be disrupted and wildlife resorts and recreational beaches may be lost. Rising sea level may lead to rising levels of groundwater giving rise to associated problems. In some regions a change in the pattern and amount of precipitation may add to such problems.

The preparation of remedial actions for the possible adverse effects of climatic change is the common responsibility of governments, the financial world, and a wide range of technical specialists, including geoscientists. What geologists can especially contribute is the evaluation of the single and combined effects of elements of the models, taking into account regional or local geological conditions. Geologists must be involved at all levels in supplying information, analyses, and expertise to assist in establishing standards, in identifying significant variations from that norm, and in forecasting the possible effects of any scenarios for the future development of the natural environment.

Man-induced hazards

Unfortunately, man's activities often disturb the natural environment. Delicately balanced ecological systems are in many cases upset by human interference and may be difficult or impossible to re-establish. It is now understood, for example, that the spreading of large amounts of nutrients and pesticides in agriculture, and pollution from industry and the home, are responsible for a general degradation of the environment and loss of diversity in wildlife. Every day several species of plants and animals disappear. More specifically, in the area of geoscience, examples of negative effects of human activities are abundant. Forests are cut and not replaced, resulting in increased erosion. Badland topography and unstable soil conditions develop factors, which, in mountain areas, cause landslides and avalanches destroying lives and property. Lowering of the water table in coastal areas may lead to the intrusion of salt water.

A striking example of a man-induced hazard related to the lowering of the groundwater level has been reported from The Netherlands. A major study calculated the effects of the construction of a new polder in the Lake Ijssel. It was shown that the land reclamation would change the dip of the piezometric surface towards the new polder. This would cause a drop in the groundwater pressure in the aquifer below the adjacent land area and a change in groundwater flow direction. The soft clay and peat beds which constitute the sub-

surface of the present Province of North Holland would be drained and land subsidence would occur. This, in turn, would cause foundation problems, especially for the historic houses in the small cities bordering the Lake Ijssel. Geological, hydrogeological and geotechnical advisory work led to the calculation of potential damage to the existing infrastructure. This has shown that damage claims due to the construction of the new polder would amount to about US$200 million. It has been suggested that some mitigation of these negative effects could be achieved by compensatory measures, especially by means of groundwater injection wells [141].

Abandoned mine-workings often cause hazards for many different reasons. When pumping is stopped and the groundwater rises, much damage to the environment can occur. The risk of pollution from mine water increases and subsurface structures and foundations may be affected by flooding and associated problems of instability. Mines also present problems of subsidence as a consequence of progressive collapse of old workings. Commonly, local earthquakes induced by coal mining are seen on seismographs. In Scotland, a series of tremors with magnitudes of up to 2.8 on the Richter scale have been correlated with current mining activities. They are strongly felt locally, even to the extent of causing slight damage to a 15th Century chapel at Rosslyn near Edinburgh [217].

Depending on the local geological setting it is sometimes possible to identify areas where the risk of general subsidence and crownhole formation is high. Knowledge of the exact location of old galleries and mine shafts is important, of course, but this information is not always readily available because of poor documentation. Various geophysical surveying methods have been tested to help locate the cavities, but no single method will provide the answer to all problems. On the other hand, by combining standard geophysical methods, it is often possible to detect a cavity whose depth of burial is less than twice its effective diameter (the cavity and the affected surrounding zone) [213] [208] [198].

In certain cases insufficient caution and investigation leads to damage, the cost of which is far greater than that of proper preventative studies and work.

A good example is provided by the 37 metre high Carsington dam in Derbyshire (UK), which slipped just before it was completed. The failure was very expensive for the owners and highly embarrassing for all those involved. The subsequent government-sponsored investigation revealed that progressive failure arising from brittleness of the soils and geometry of the section was the predominant element in the slide. Further contributory causes were the weak clay foundation and the existence of ancient shear surfaces in the strata. The site is underlain by mudstones of Upper Carboniferous age, altered by near-surface weathering. The report from the British Geological Survey also concluded that dangerous accumulations of carbon dioxide found at the site were due to the reaction of limestone used in the construction with groundwater made acidic by the microbic alteration of pyrite in the mudstones. From a socio-economic point of view it is unfortunate that the failure was not prevented. The general geological conditions at the site were known in detail before construction started from the previously completed geological survey of the district. The subsequent maps and associated data were available to the Water Authority. The benefits which would have been available if a full detailed report on the geology of the site had been made before the construction and not after the failure, are obvious. This is a clear example where the knowledge and skills available within a Geological Survey were not employed soon enough to prevent a costly disaster [187] (Fig. 6.3).

In general, man induces two main types of hazard; those generated by modification of the physical environment and those associated with the addition or removal of material to or

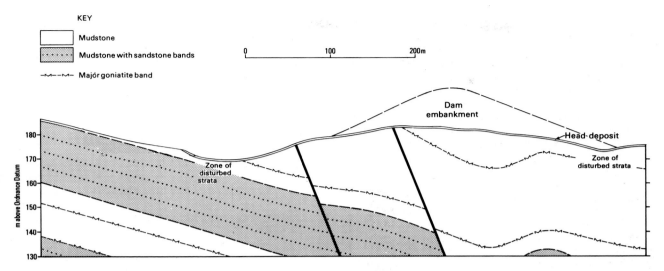

KEY

☐ Mudstone

▨ Mudstone with sandstone bands

–◠–◠– Major goniatite band

Fig. 6.3 Geological cross-section at Carsington Reservoir (UK). The only *in situ* geological stratum that was directly implicated in the embankment failure was the thin deposit of 'yellow clay' (Head) that was left in place beneath an extensive area of the embankment.

from the environment. The first category includes a wide range of construction work and agricultural practice. The second category embraces the agricultural input of fertilizers, pesticides and herbicides, and the disposal of a great variety of domestic and industrial waste.

The impact is evaluated in terms of economic risk, the extent, and the duration of the phenomena. Sometimes, the risk is difficult to identify, as shown in the Loscoe case history, which underlines the importance of caution in planning and development.

> At Loscoe, Derbyshire, UK, a house was completely destroyed in 1986 by a methane gas explosion, badly injuring the three occupants. At a Public Inquiry, the sequence of events leading up to the incident was established and evidence produced to ascertain the origin of the methane. Some years ago, ground heating about 90 metres beyond a nearby landfill had been misinterpreted as being due to the burning of a shallow coal seam. It had, in fact, been caused by the oxidation of methane accumulating from the landfill. The landfill occupies a former quarry and much of the supporting scientific evidence related to the careful recording of the strata in the quarry which had been made during the routine geological survey of the district in 1962 [187] [222] (Fig. 6.4).

Finally we should mention the problem of waste disposal, which every human community has to solve. This ranges from the domestic garbage we produce every day, to highly dangerous waste of industrial origin, including radioactive waste. For each case, proper solutions based on proper geological expertise must be found. The integration of relevant geoscientific data and expertise is essential in the planning of every waste disposal project.

6.3. MANAGING CONFLICTING INTERESTS AND PLANNING THE AFTER-USE

6.3.1. Supply of geoscientific information

Background

The essence of good planning is founded on a sound basis of relevant information, and most enlightened planners will probably agree that geoscientific information should be inte-grated into the planning of development on and under the Earth's surface. Planners probably also know that geoscientific information can be supplied by the national Geological Surveys, but it is recognised by the Surveys that planners and decision makers are in general not fully aware of the resources available, and of all the benefits they can obtain by using data and expertise held by these institutions. Commonly, data related to economic geology are the only type of geoscientific information to be considered.

The main reason why geoscientific information has not been better integrated into planning is the physical separation between Geological Surveys and organizations concerned with planning. To overcome this impediment it is recommended that it becomes a statutory obligation of such organizations to consult Geological Surveys and the data held by them whenever appropriate. This is particularly necessary in the early stages of planning, as many case histories have demonstrated that early contact and information could have avoided later difficult problems and costly remedial actions.

The Geological Surveys know that planners need easily accessible, up-to-date, high-quality information. They are well aware that an increased flexibility of access to the information is needed if the existing data are to be better integrated into planning. Traditionally, the information has been available in analogue form as maps, reports, and files. Today, many of the data are in digital form and organized in databases from which they can be retrieved and integrated, as necessary, with data from other disciplines, by means of database management systems such as Geographic Information Systems (GIS). Such developments are beginning to provide a flexibility of access to data of all kinds, which is important as it allows the choice of the best tools for the solution of any problem. The Geological Surveys recognize their responsibilities in this matter and consistently work towards increasing the flexibility of access to information and achieving an even better quality of data. This implies that it is necessary to have minimum common standards for data and data sets. The establishment of standards in geoscience is not merely for data transmission, but is even more important in creating the codes of practice and standards for the types of information to be provided in relation to the planning of different kinds of land-use.

While the lack of relevant information or the failure to integrate existing information may be a problem, the reverse

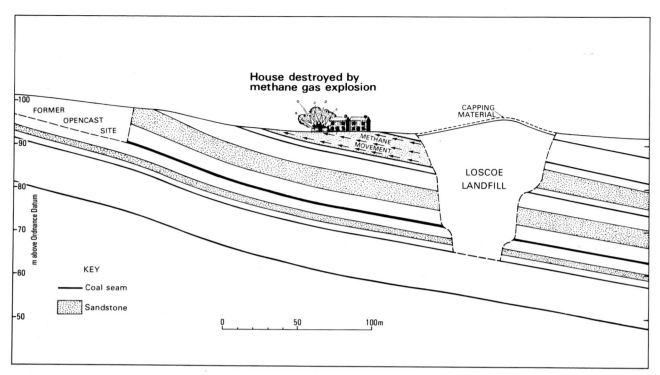

Fig. 6.4   Geological cross-section through Loscoe landfill (UK), showing the migration path from the landfill to the ill-fated house at number 51.

problem also exists. In some cases, data may be used for purposes other than those for which they had been prepared, and false conclusions may be drawn as a result.

Experience in Scotland, for instance, indicated that a series of thematic and environmental geology maps portraying known and inferred abandoned underground workings was of great benefit to planners and civil engineers concerned with land-use and ground stability. In addition they could be used successfully for resource assessment. While it was generally agreed that it is valuable to show on maps the extent of mining known from mine plans, opinions differ on the desirability and usefulness of showing inferred areas of mine workings. This latter information is important in national, regional and urban planning, but there is some concern about the release of such information on open file to the general user, who is not fully informed of the background and nature of mining maps. Areas designated as 'undermined' should not automatically be subjected to planning blight, but rather it should be understood that engineering problems in urban areas occur only if property is not planned, designed, and constructed with reference to the state of undermining [214] [191].

One of the more difficult problems in planning is the political nature of many planning issues. It is relatively easy to make plans, but much more difficult to have them carried out. In many cases success is conditional on political courage and public support, which may be interconnected. For the political and administrative decision-making in land-use planning, it is often a major problem to find methods which can be used as 'tools' for the assessment of the value and importance of different components of the environment. Such Environmental Value Assessments (EVAs) should supplement traditional Environmental Impact Assessment (EIAs). Commonly, problems arise in quantifying the cost of a development due to the lack of appreciation of all its effects, including both the unexpected and the undesirable.

Since many natural systems pose a threat to society, they should be taken into account during planning. The monetary costs may be less tangible, but insecurity of life, disruption to production, and destruction of property all reduce the well-being of society. Whether the causative hazards are the result of either planned or unplanned changes to the environment, or due to natural processes, they can only be successfully predicted or avoided if their mode of interaction with human activities is understood [198] [197].

The development of new tools for evaluation is thus of considerable importance to society. Such development will inevitably be rather complex and it is recommended that it should be made through close cooperation between geoscientists and decision makers, in order to reach an agreed common evaluation of trends and future needs.

Providing adequate information

We feel it appropriate to present here the main analytical tools which the Geological Surveys keep at the disposal of planners, decision makers, and the layman. From one Geological Survey to another, the availability is largely the same, even if differences can be observed in data presentation or the level of computerization. Globally, the data are provided in two main forms: maps and databases.

Maps

The Geological Surveys publish different types of maps at various scales. A common problem in map production is the choice of scale and the density of information presented on the map. The misuse of information presented in maps cannot be completely avoided, but Surveys do try to inform potential users about the proper use of different map types. All users should carefully examine all marginal information printed on the maps to identify possible warnings against misuse. It should not be necessary to warn that a map presenting marine geology data is not meant for navigation and that a map giving a general overview should not be enlarged and used for detailed planning, but such misuse is known to take place.

Most Geological Surveys publish at least one geological map series for the whole country at scales (1:25 000, 1:50 000, and smaller) which give an overview of a major area or a region on each sheet. Such maps are general documents and are most useful as a first source of information for such items as general geology and structural conditions.

Besides these general maps, more specialized 'thematic' maps are produced to shøw specific subjects such as geo-chemistry, groundwater resources, potential resources of raw materials, and so on. Thematic maps are published at various scales meant for general overview (1:50 000 to 1:1 000 000) or for detailed use (1:5000 to 1:25 000). Recently, the Geological Surveys have been producing Environmental Geology Maps (EGMs) which show the distribution of a single geological element (e.g. the distribution of lead in topsoil) or combined elements (e.g. the type, location, and quality of a raw material). To provide the basis for work on a problem related to environmental geology, it will often be necessary to make use of a set of environmental geology maps (an EGM set) which together yield the relevant information.

EGM sets, combined with data and maps produced by other organizations concerned with topics related to land-use, may form the basis for the development of so-called geo-potential maps in which all major aspects relevant to land-use are brought together. The mobility of data between databases and all these types of map can be readily achieved today through the use of Geographical Information Systems (GIS).

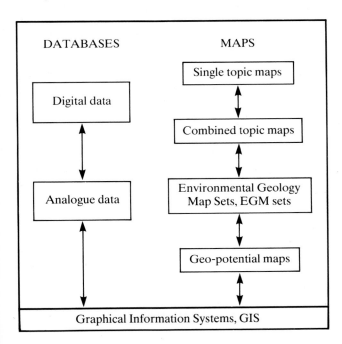

DATABASES   MAPS

The large set of maps used as sources for the Geoscientific Map of the Natural Environment's Potential (GMNEP), developed in the 1970s by the Geological Survey of Lower Saxony, incorporates EGMs, and non-geoscientific data and maps produced and held by other organizations [78].

Using these techniques many Geological Surveys produce maps which present relevant information for land-use planning at a scale and in a simplified form which can be readily understood and used by non-geologists.

Single topic maps Typical single topic maps (maps of elements) are the geochemical atlases of countries or regions, such as the geochemical atlas of Germany [94] and of the Nordkalott part of Norway, Sweden, and Finland [173] [149] [23]. Other single topic maps may show the distribution of features such as brick clay, karst sinkholes, mine shafts, water wells with certain characteristics, and many other features. All such maps represent inventory work, which is a major task for the Geological Surveys.

A good example is the mapping and assessment of aggregate resources in the south Midlands and Welsh Borderland carried out between 1969 and 1983 by the British Geological Survey with funding from the Department of the Environment. Although there are extensive spreads of sand and gravel, most of the area had not previously been geologically surveyed in detail. The first and most important work therefore, was a geological survey of the area at the scale of 1:10 000 including studies of the superficial deposits and solid rocks. In addition to these surveys, boreholes were drilled to supplement existing data and to give data for a resource assessment in one selected area of apparent potential. Also, several geophysical investigation techniques were used. The whole assessment was carried out at the rate of one square kilometre per day. The results were published in some 100 reports containing maps showing reserves at the scale of 1:25 000. Classification criteria included thickness of overburden, areal extent of the deposit, location of boreholes and accessibility for further exploration. A major point of interest of this programme was the systematic approach providing data and maps that can be directly used in planning documents [188] [193] [215].

Combined topic maps A direct combination of thematic elements is demonstrated by the Cyclogram-Basic-Data-Map which was developed at the Geological Survey of Denmark in the early 1970s. The cyclogram map contains geological, hydrogeological, and technical information and achieves the dual purpose of giving an overview together with detailed information about each single well. Maps of this type are now available for 75 per cent of Denmark (Fig. 6.5), and are automatically produced from digital data in the ZEUS database, using an interactive graphic system and a colour raster plotter [11] [13].

New planning regulations in the municipalities in Sweden have resulted in cooperation between the Geological Survey and the authorities of some municipalities on the development of a new 'map of vulnerability to infiltration of pollutants' into groundwater. The map is meant for planning the protection of water wells and groundwater reservoirs, and to assist with the pinpointing of polluting industries and waste deposits. The map can also be used as a basis for the planning of routes for the transport of goods dangerous to the environment [167].

In Belgium a Profile Type Map at the scale of 1:25 000, based on parameters distinguishable in the field, was developed to assist with the perception of the complexity and variability of Quaternary deposits in the Flemish coastal lowlands. Profile Types consist of associations of facies units and these form the real mapping tool. A complex is composed of one or more of such Profile Types. To avoid an unnecessary increase in complexity of the Profile Type Map, the deposits are classified according to their origin (eolian, mass-wasting, and fluvial). The map adds a third dimension to the presentation of information on soil-types and increases the utility and applicability of the information. In view of the economic importance of sand, a special geo-economic map at the scale of 1:38 500 was drawn to describe in more detail the quality/ characteristics of the sands in combination with the depth/ thickness of the deposits [4].

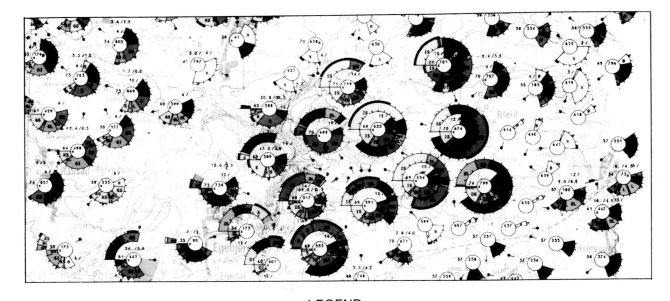

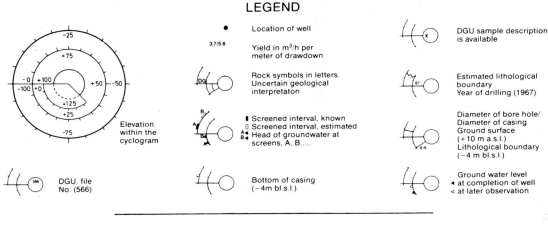

<h2>LEGEND</h2>

- • Location of well
- 3,7/5.8 Yield in m³/h per meter of drawdown
- Rock symbols in letters. Uncertain geological interpretaton
- ▮ Screened interval, known
- ▯ Screened interval, estimated
- Head of groundwater at screens, A, B.....
- Bottom of casing (−4m bl.s.l.)
- Elevation within the cyclogram
- DGU. file No. (566)
- X DGU sample description is available
- Estimated lithological boundary Year of drilling (1967)
- Diameter of bore hole/ Diameter of casing Ground surface (+10 m a.s.l.) Lithological boundary (−4 m bl.s.l.)
- Ground water level ◄ at completion of well ◁ at later observation

---

<h3>ROCK LETTER SYMBOLS</h3>

B	Dug well	GV	Oligocene – Miocene alternating thin beds	LL	Eocene Clay, plastic clay
BK	Danian bryozoan limestone	HI	Postglacial salt-water silt	M	Mull
C	Brown coal	HL	Postglacial salt-water clay	MG	Glacial gravelly till
DG	Glacial melt-water gravel	HP	Postglacial salt-water gyttja	ML	Glacial clayey till
DI	Glacial melt-water silt	HS	Postglacial salt-water sand	O	Fill, waste
DL	Glacial melt-water clay	HV	Postglacial thin salt-water beds	P	Gyttja
DS	Glacial melt-water sand	I	Silt	PL	Paleocene clay
DV	Alternating thin melt-water beds	ID	Interglacial diatomite	PV	Alternating thin Paleocene beds
FS	Post-glacial fresh-water sand	IL	Interglacial fresh-water clay	S	Sand
G	Gravel, sand and gravel	IP	Interglacial fresh-water gyttja	SL	Eocene marl
GC	Miocene brown coal	IS	Interglacial fresh-water sand	U	Clay, sand and gravel
GI	Oligocene – Miocene mica silt	KG	Miocene quartz gravel	V	Alternating thin beds
GL	Oligocene – Miocene mica clay	KS	Miocene quartz sand	X	No information
GS	Oligocene – Miocene mica sand	L	Clay, marl		

<h3>LITHOLOGY (interpretation)</h3>

- Post-glacial fresh-water sand, -gravel
- Post-glacial salt-water sand, -gravel
- Post-glacial salt-water clay, -silt, -gyttja, -peat, -alternating beds
- Late-glacial fresh-water sand, -gravel
- Late-glacial fresh-water clay, -gyttja, -peat, -alternating beds
- Glacial melt-water sand, -gravel
- Glacial melt-water silt
- Glacial melt-water clay, alternating beds
- Glacial Clayey till
- Interglacial fresh-water sand, -gravel
- Interglacial fresh-water clay, -silt, -gyttja, -peat, -diatomite, alternating beds
- Oligocene - Miocene sand, gravel, sandstone
- Oligocene - Miocene clay, silt, brown coal, alternating beds
- Paleocene - Eocene clay, silt, diatomite, volcanic ash
- Danian limestone

Fig. 6.5 Map excerpt from Denmark, showing cyclograms providing information on borehole/well characteristics, hydrology, and geology. As layers at the same depth are shown at the same 'hour' in the diagram the map yields a 3-D impression of the geology [31]. 'Gyttja' is organic ooze.

The landslide vulnerability map of Italy at the scale of 1:500 000, produced by the Italian Geological Survey, also combines environmental geology elements. It describes the known and potential vulnerability of the country to land-mass movements and divides the area into three risk zones. It is based on information on lithology, geomorphology, landslides, earthquakes, and volcanic eruptions. [129] [127]. In Section 4.2.3 the very interesting experiments by the regional geological surveys of Piemonte and Emilia-Romagna are described in some detail. In Norway, thematic maps were produced on the basis of a classification of ground-movement types [145].

Urban geology is an important area of work for environmental geoscientists and most Geological Surveys provide information for development in the field of civil engineering. Since 1980 the Geological Survey of Northrhine-Westphalia (Germany) has been publishing engineering–geology maps at the scale of 1:25 000 for urban areas in the Rhine Valley and the Ruhr. Such maps are excellent for the general overview of an area while maps at the scale of 1:5000 or larger will have more practical use for detailed land-use planning [103].

Environmental geology map sets (EGM sets) In Denmark, the EGM set for groundwater consists of three maps at scales of 1:25 000 and 1:50 000: (1) basic geological data, (2) piezometric surface and transmissivity, (3) hydrochemistry.

In Ireland a suite of Environmental Geology Maps, an EGM set, is prepared as part of the scheme for groundwater protection. The scheme is a pragmatic approach which may be applicable to the economically less-developed areas of Western Europe. The plan of work involves several stages:

(1) preparation of an initial groundwater protection scheme;
(2) vulnerability assessment using water quality data and readily available geological and hydrogeological data;
(3) assessment of groundwater pollution loading;
(4) site investigations;
(5) preparation of a final groundwater protection scheme.

The EGM set includes bedrock geology, Quaternary geology, depth-to-bedrock, groundwater vulnerability, location of potentially polluting sources, and finally a groundwater protection map; The EGM set and the accompanying reports are deliberately set at a level which is understood by a non-geologist user. This ensures that the information is widely disseminated and that groundwater interests are brought to the attention of planners and decision makers [123] [124].

Geo-potential maps The geo-ecological stocktaking for landscape planning in Baden-Württemberg, Germany, is a good example of what is meant by geo-potential maps. Legislation places 'the natural fundamentals of living', 'the capacity of natural balance', and 'the usefulness of natural resources' under federal and communal protection. Landscape planning is an important means of achieving these aims. Investigations started predominantly in urban landscapes, but enquiries were continued as ecological stocktaking in the planning districts of Reutlingen/Tübingen and Freiburg-im-Breisgau. The investigations were conducted by the Geological Survey of Baden-Württemberg in cooperation with the German Weather Service, and geological departments of universities, together with numerous authorities at state and communal level. Basic scientific maps at the scales of 1:25 000 and 1:50 000 were prepared and were supplemented by maps comprising data relevant to planning from different scientific domains.

Essential to such an interdisciplinary map series are geological maps, from which all geo-potential maps are derived. Maps on soils, natural resources, mineral resources, geomorphology, hydrogeology, and engineering geology can only be relevant and useful if they are based on the geology of the area in question. In Baden-Württemberg, the EGM set was accompanied by maps of local climate, maps of nature conservation and landscape cultivation, and maps of open-space functions which should be protected. The latter map defines areas which contain scientific phenomena of such importance that they have to be particularly well protected (Fig. 6.6) [80] [84].

Data banks
The Geological Surveys are the national custodians and co-ordinators of geological information. They interpret and maintain the data, and in this way add to the value of the information for society. A most important result of the expert evaluation and interpretation of the data is the conversion of point information (observations from a single point, e.g. a borehole or an analysis) into two-dimensional or three-dimensional models.

Databases are established at different scales for local, regional, or national use. The topics encapsulated may be part of a systematic approach or may be chosen case by case. The data banks and archives held by the Geological Surveys contain two major groups of data:

(1) official data which are systematically produced, collected, interpreted, and maintained as an official mission;
(2) inventory data produced for special purposes, such as assessments of mineral resources or the monitoring of groundwater quality, made at the request of authorities or private enterprise.

Many databases hold basic geoscientific information which may be used for various purposes, while other databases are limited to more specialist data. Like all surveys, the British Geological Survey has established a computer framework for assembling applied geological data. Borehole data and results of geotechnical and hydrogeological analysis, a digital terrain model (50 m grid), geological cartographic data (lines, points, and boundaries from the 1:10 000 scale map), and a scanned topographic map have been assembled in the computer system. A Geographic Information System (GIS) carried out much of the work of assembling the information for an applied geological map, even though the area covered by such a GIS map is, as yet, limited in extent [211].

A representative example of an integrated GIS is provided by the Geological Survey of Lower Saxony (Germany). The system has been given official support by the Lower Saxony ministries for Economy, Technology and Transport, and for the Environment. It includes two major branches: a number of data banks concerning methods and thematic topics, and a central management system covering methods and data. In Lower Saxony, two-thirds of the data managed by the system are soil data. The others concern geology, raw materials, hydrocarbons, hydrogeology, waste disposal, engineering geology, geophysics, and geochemistry [98].

Another example of a land-use related database is the soil inventory in Bavaria (Germany) [88]. The core of the system is represented by the basic soil inventory (Bodengrundinventur). The number of reference soil profiles is about 5000, and the system is sufficiently flexible to be able to incorporate additional data on a case-by-case basis, according to needs. This inventory has the advantage of covering the whole of

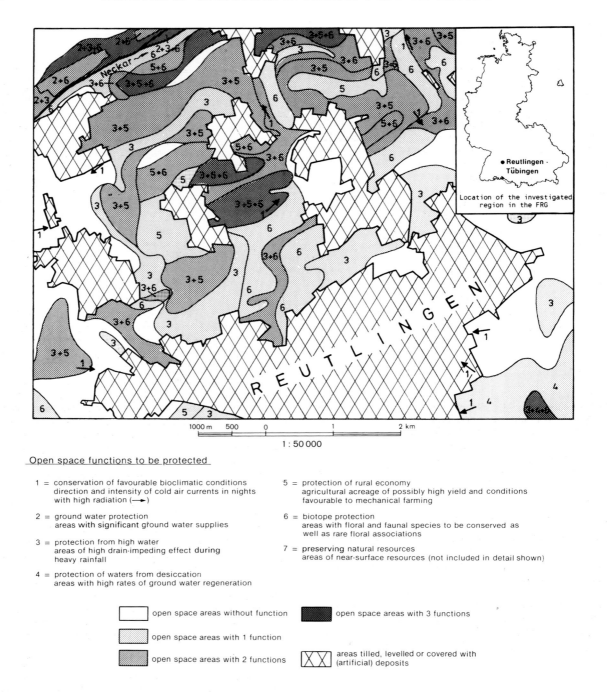

1 : 50 000

Open space functions to be protected

1 = conservation of favourable bioclimatic conditions
direction and intensity of cold air currents in nights
with high radiation (→)

2 = ground water protection
areas with significant ground water supplies

3 = protection from high water
areas of high drain-impeding effect during
heavy rainfall

4 = protection of waters from desiccation
areas with high rates of ground water regeneration

5 = protection of rural economy
agricultural acreage of possibly high yield and conditions
favourable to mechanical farming

6 = biotope protection
areas with floral and faunal species to be conserved as
well as rare floral associations

7 = preserving natural resources
areas of near-surface resources (not included in detail shown)

☐ open space areas without function

☐ open space areas with 1 function

▦ open space areas with 2 functions

■ open space areas with 3 functions

▨ areas tilled, levelled or covered with
(artificial) deposits

Fig. 6.6 Map of open-space functions to be protected in the associated districts of Reutlingen/Tübingen, Germany. Shading indicates number of functions to be considered while figures indicate the type of function.

Bavaria, in contrast to the soil map that covers only a small portion of the region. The basic soil inventory is complemented by a variety of standard soil or thematic maps, integrating measurements from stations that monitor pollution and erosion hazards. Another aspect of the Bavarian system is a network of some 1100 observation and control points and a specific long-term network of 230 stations monitored by three separate authorities, the Geological Survey of Bavaria and two institutions under the Ministry of Agriculture.

Also in Bavaria, a specific data bank for the Alpine region, called GEORISK, has been assembled for landslides. All

information on sites of potential instability is recorded. If possible, the information is checked in the field and supplemented by further data. Most recent mass movements concern areas where similar events have taken place before. The data are therefore of particular value for warning planners and engineers about the vulnerability of specific sites [86] [87].

In Italy, recently enforced laws and regulations have underlined the Geological Survey's function in the coordination of all public activities producing, collecting or using territorial data related to geosciences. On the basis of the existing group of geological data banks, the Integrated National Geological

Information System GEODOC is now being established. It is designed to give easy access to highly detailed information [129].

The Dutch National Inventory of Clay for the ceramic industry and embankment covers is an example of a data bank of a more specific nature. In order to establish the future potential of this resource a nationwide inventory has been completed. It was based on descriptions of boreholes held by the Geological Survey and of all relevant near-surface clay beds, and includes a calculation of resources. Some resources were excluded on the basis of environmental considerations. Maps were prepared by means of a GIS on the basis of the digital data. In The Netherlands several inventory studies on near-surface construction materials have been made by the Geological Survey, because the required expertise of regional geology, together with the management and maintenance of a vast file of field data, are not available outside the Survey [141].

To handle the overwhelming amount of data, most data banks have to be based on Electronic Data Processing (EDP) systems. The digital data stored in the EDP databases have the advantage over analogue data that they can be updated easily and combined according to actual needs. Both data banks and EDP systems are constantly updated.

Conclusion

The development of computer science, and especially its graphic component, has resulted in the rigid distinction between maps and data banks breaking down. Already, and to an increasingly greater extent in the future, the tendency will be to supply a variety of à la carte and up-to-date presentations. This leads to much easier access to data within the Geological Surveys in a more useful format.

6.3.2. Managing conflicting interests

In tackling environmental problems, conflicting interests commonly come to light. It is the task of the relevant authorities to find solutions which will go some way towards satisfying all the conflicts involved. The best way of handling conflicting interests is by creating an atmosphere of mutual confidence amongst the professionals involved. Conflicts related to land-use often create strong feelings and tend to escalate if agreement on a common basis of information is not established. So, a strategy, policy, and methodology in the handling of conflicting interests, based on best available information, will usually lead to the best results.

Management of resources and development

A most important limiting factor in development planning is the sterilization of resources by other human activities. In some cases the sterilization is final. It is out of the question, for example, to produce bulk minerals from historical sites such as the Forum Romanum in Rome and from most other urban areas. In other cases the degree or period of sterilization may be varied by certain conditions of development. If an urban area is demolished, raw materials may be extracted before the ground is reconstituted and used for other purposes. In planning, the ideal is to identify the final long-term use of a given land or sea area, and to make a comprehensive, multipurpose plan for the successive temporary exploitation of the area to achieve the maximum use of the available natural resources as a whole. There is a degree of idealism in such a concept, but,

without doubt, the trend of planning and development must be in that direction.

A case-history from Aldridge-Brownhills near Birmingham (UK) demonstrates some of the many limiting factors in the exploitation of natural resources. Though largely urban, this area still contains large reserves of sand and gravel, brick clay, and coal in areas of open country, but these are difficult to exploit because of conflicting existing developments. In some places pillar and stall workings in Wenlock Limestone are a hazard, as the potential risk of subsidence is high. An additional problem is that the abandoned shallow workings are not all shown on plans. Elsewhere, subsidence resulting from coal mining is an important factor in land-use planning. Other restrictive factors are a major, controlled, chemical waste-disposal facility established in abandoned coal mines and old marl quarries, and the numerous boreholes for public water supply. The latter factors tend to set limits for the future production of brick clay in part of the area because of the danger of breaking the clay seal between waste and drinking water. In planning the working of the valuable resources it is obvious that the documentation provided by the British Geological Survey from its archives is of critical value [223].

In urban areas land is a most important resource. Land is needed for buildings and construction works, and for facilities for transportation in the widest possible sense. It is well known that information on the exact location of sewers, cables, water, gas, and heat pipes is often unsatisfactory and that this lack of knowledge creates considerable problems for the maintenance of the facilities, and for the development of new construction works. The same is often the case with regard to certain types of geoscientific information which are essential to land-use planning, such as knowledge of the existence of raw materials, groundwater reserves, waste dumps, and special types of soil or rock.

Commonly, conflicts exist between the exploitation of raw materials and urbanization. The typical case is that of a small town which develops close to a new industry. As time goes by the town expands, while at the same time the mines, quarries, and pits of the industry extend towards the town-limits in the search for still more raw materials for exploitation. The problem then arises of balancing the rate and extent of the commercial development, the potential unemployment, and acceptable living conditions for the population. In such conflicts it is crucial to get the best possible knowledge of all relevant factors related to the quantity and quality of the resources, and their location in and relation to the environment, so that options may be assessed and the best possible plan evolved.

In East Tyrone and at Ballymoney in Northern Ireland, two large deposits of lignite were found by an exploration programme undertaken by the Geological Survey of Northern Ireland. Each of the sites raised distinct socio-economic or environmental problems in relation to any exploitation. Part of the deposits are close to an area where wildlife is especially protected and another area is fringed by significant housing developments along road-sides. In East Tyrone local opposition prevented further exploitation. At Ballymoney it was decided that the thick multiple lignite sequence about 70 metres beneath the town itself should not be exploited, and that the bulk of the deposit beyond the town be designated as an area where future extraction can occur [224] [225] [226] (Fig. 6.7).

The case of Madrid illustrates the conflicts between the needs of a large city for natural aggregates and environmental problems arising out of their exploitation. Mining in the Madrid Autonomous Region is limited to exploitation of non-metallic minerals and quarry products. Based on consumption and production statistics for cement, the production of sand and gravel in 1984 amounted to

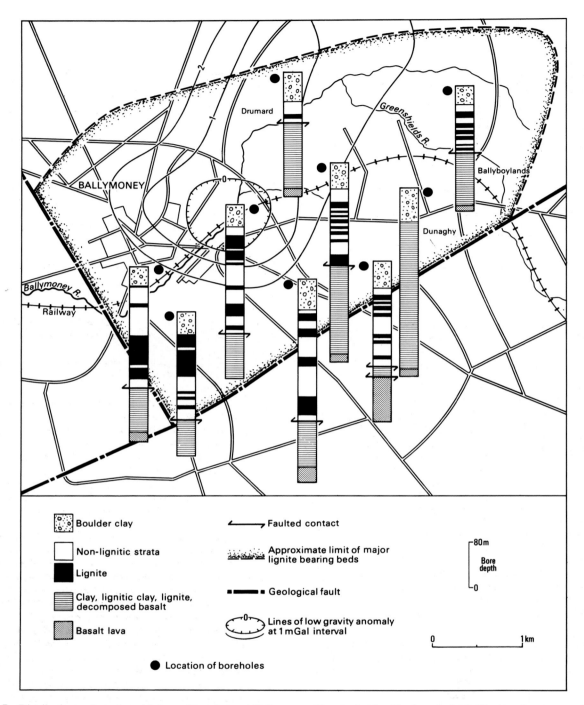

Fig. 6.7 Distribution and stratigraphic logs of boreholes at Ballymoney, County Antrim, Northern Ireland. The logs demonstrate that the thickest lignite deposits are situated in the urban area.

about 6–7 million tonnes. Assuming stagnant demand, over 100 million tonnes will be required by the end of the century and it can be assumed that by 2050 there will be problems concerning the supply of aggregate in the Region. The exploitation of aggregates is not dependent only on the actual and limited reserves and the agreement of the owners of the land. Environmental constraints are many. The largest amount of development and reserves of the resources is found in a complex of terraces along the Lower Jarama River which constitutes one of the most fertile areas of the Region. As is often the case with sand and gravel deposits, they form an important aquifer which supplies water to the area's small industries and farms. Exploitation of the materials below the water table creates large depressions and artificial lakes which cannot be filled up again, a well known situation for many European rivers (the Meuse, the Seine, and the Rhine for example). This means a steady loss of area of fertile, arable land. In the Madrid Region, moreover, the lakes suffer from progressive salinization and eutrophication that gradually degrades the quality of the water. Another impact is the progressive lowering of the water table over a radius of several kilometres, which leads to decline of water-level, salinization, and

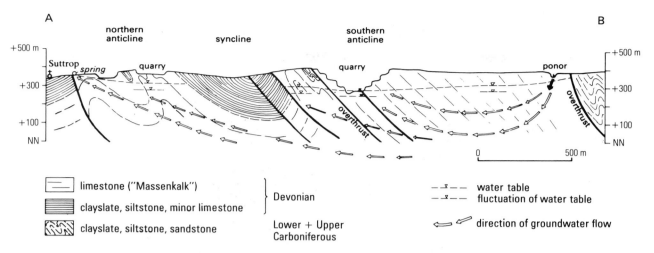

Fig. 6.8 Exploitation of limestone and extraction of groundwater in a karstic region. The quarrying is being carried out related to the groundwater fluctuations. 'Ponor' is a sinkhole (Warstein, Germany).

pollution of water wells in the affected area. Also, the river is affected by alterations to its course and fluvial dynamics. This, in turn, has a negative impact on the river's vegetation and fish population. Considering all of these effects it is essential that the reserves of the resource are evaluated so as to ascertain the availability of sand and gravel to the construction industry in the long term. At the same time, resource and environmental planning should be carried out urgently to set standards, give guidelines, and demonstrate where and how future exploitation should take place. In addition, it would be wise to begin to plan for the use of products which can replace natural aggregates eventually in the construction industry of the city [152] [162].

Exploitation of a raw material deposit can have a strong impact on a groundwater reservoir. This is the case not only for sand and gravel deposits but also for many other types of soil and rock. Limestones are often important aquifers and, in some cases, even crystalline rocks and clay serve as aquifers of lesser potential.

The conflict of interest between exploitation of limestone and the extraction of groundwater can be illustrated by a case-history from the surroundings of Warstein, east of the Ruhr area in Northrhine-Westphalia, Germany.

The karst limestone region of Warstein consists of an isolated deposit of massive reef limestones of Devonian age surrounded by slates, shales, greywackes, and quartzites. The deposit has a complicated internal structure due to being folded, faulted, and overthrust. The limestone is of equal potential as a highly productive aquifer and as a source of raw material for the limestone and cement industries, with a production of about 4 million tonnes per year. There is an increasing conflict between groundwater abstraction for human consumption and the quarrying activity. The minimum groundwater outflow from the karst complex is 22 million cubic metres a year from an area of about 11 square kilometres and tracer tests have revealed that the main flow of groundwater is through fractures. Water abstraction has been given priority over limestone quarrying and legally enforced restrictions mean that base level of the quarries must be at least two metres above the highest groundwater levels. Also, a regional drawdown of the karst water level as a result of pumping from wells is not allowed as this would result in the springs used for water supply becoming dry. One suggested solution to the conflict is to deepen the quarries during the annual period of low groundwater level and to refill with rock debris as soon as the water level rises again. This use of backfill would allow

the intermittent extraction of high-quality limestone and yet safeguard the abstraction of groundwater [105] (Fig. 6.8).

The selected examples presented above represent classical problems associated with the exploitation of mineral resources. It must be emphasized that the negative impacts reach their maximum in the exploitation of such as industrial minerals, when large areas of land are involved.

Other aspects of conflicting interests concern other types of resources. Two examples are given below, one concerning underground space, the other dealing with national parks and protected areas.

Human activities through centuries have transformed the strata underlying many urban areas into a sponge-like state. The existence of unknown or forgotten prehistoric or historic mines, cellars, buildings, and monuments has caused many problems and attracted much publicity. Such cases are well known from many locations and have been reported from the construction of subways in most European capitals. The unexpected discovery of remains from the past is just as interesting to the general public and the media as are the technical or economic consequences. Historic relics are of such importance to society that they should be protected, if possible, but at the same time, delays in construction and extra costs should be avoided. For this reason archaeologists and geologists in many countries are asked to work together to provide clearance before the excavators start working. In this way geological investigations can contribute to establishing the existence of the relics of former developments and to the implications for developers and planners in the future. One example is the case of the 'caves' of the city of Notthingham in England.

A large part of the City of Nottingham (UK) is built over the outcrop of the Triassic Nottingham Castle Sandstone Formation, or is underlain by it at shallow depth. It is a weakly cemented low- to medium-strength sandstone, which is easily excavated with hand tools. This ease of working has resulted in many cellars, store rooms, dwellings, and passages being dug beneath the city since the twelfth century at least (Fig. 6.9). These voids are known locally as 'caves'. Although the rock is relatively weak many of the 'caves' have remained unchanged for hundreds of years due, in part, to the very wide (over 1 m) spacing of discontinuities in the rock. However, changes in conditions and loads imposed by new development or redevelopment may affect this apparent stability. The large number

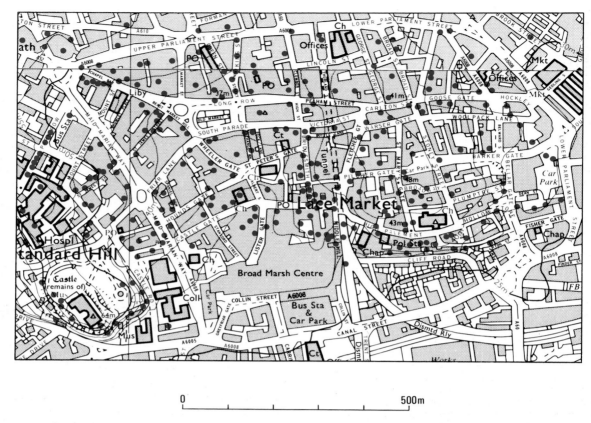

0   500m

Fig. 6.9 'Caves' cut in sandstone beneath Nottingham (UK) are a problem in town planning. ● Location of underground void (not to scale).

of voids beneath the city presents a valuable cultural and archaeological asset; but they also constitute a potential subsidence hazard to existing buildings and to future constructions. In 1988 the British Geological Survey, funded by the Department of the Environment, produced a register and location map of the known 'caves' describing their geology, history, extent, and possible hazardous implications. This documentation is an important source of information for future development planning [200] [212] [221].

The so-called 'green' movements have accelerated the interest and concern of the general public about what may be termed the 'soft' resources of the natural environment. The resulting local opposition to development has itself become an important sterilising factor in the exploitation of mineral resources, groundwater resources and land for development. Plans to change the environment by the construction of hydroelectric plants and reservoirs, for example, are now commonly opposed by the public. Now, the increased awareness among the public and politicians of the importance of protecting our natural environment has led to the establishment of an increasing number of national parks and other protected areas.

Many case histories demonstrate that these areas are, indeed, effectively protected now. In Norway, for instance, in Anarjokka National Park in the north, an application from a private company to carry out prospecting for nickel resources was turned down. In the area of Jostedalen National Park, central west Norway, the Norwegian Geological Survey had insufficient information to make realistic assessments of the mineral potential. An application to the Ministry of the Environment to carry out more detailed studies in the area,

even before the park was established, was turned down as well [149].

Integrated management

In the examples presented in the preceding section, the conflicts that arise between the various potential uses of an area have been considered. Such cases occur frequently. Commonly a spectrum of interconnected uses have to be dealt with simultaneously and studied together. This is also true of successive and integrated uses concerning the same area. This type of problem has been dealt with in various ways depending on the period of time, the country, or the region. The examples presented below are just a few among many which testify to a diversity of approaches. The first case-history relates to a methodological trial in the perspective of long-term planning.

A model was set up by the Geological Survey of Austria and tested in the Mattigtal area in Upper Austria. The evaluation model takes into account the quality, quantity, and discernible vulnerability of environmental potentials and the interaction with other potentials. In the test area, about 300 square kilometres, groundwater and near-surface mineral deposits were evaluated as were the impacts of their use on the ecology and landscape, conservation, settlement and recreation areas, climate and air quality, and agriculture and forestry. The model identified positive conservation zones for further exploitation and use of the evaluated resources, and negative conservation zones where development should be restricted or forbidden. The result of the work is available in map and documentation form. The evaluation is based on a rating (none, weak, medium, strong) of the effects of selected elements, such as the exploitation of aggregate resources or groundwater abstraction on other land-uses, such as agriculture/soil, forestry/soil, biotopes, recreation, urbanization, infrastructure, waste disposal, and sewage disposal. It is expected that the provincial Government of Upper Austria will adopt the evaluation model as standard procedure after it has been refined to some extent. The model is at present in analogue form but the next step in the project is expected to be its conversion to digital form [1] [2] (Fig. 6.10).

The example of the development of the Seine Valley between Nogent-sur-Seine and Montereau (France) shows in a very concrete way, the requirements of an integrated regional development. The area in question contains the largest alluvial deposit and one of the most important aquifers of the region. The Geological Survey of France (BRGM) investigated the area in 1965 and in later hydrogeological assessments, in order to find the means of meeting the growing needs for drinking water in the Paris region. Investigations performed by many different techniques revealed that the Seine Valley is functioning as a sand filter capable of supplying 300–600 000 cubic metres of drinking water per day. So, the area, which in 1965 was still unpolluted and not yet developed, contained a considerable water resource which ought to be protected immediately. Alongside this situation, the growing demand for aggregates for construction and development in Paris made it necessary to open numerous sand-pits in the same area. The phased development of the area included new and extended industrial zones, the construction of a canal, the TGV Line from Paris to Lyon, a highway construction and a nuclear power plant. Feasibility studies on the geology and hydrogeology were performed by the Geological Survey throughout all phases of the development. Today the valley has been transformed from its undisturbed prehistoric state as an area with important biotopes and a river with occasional very low levels of water, into an area from which both aggregates and large amounts of drinking water are extracted. The river has been transformed into a canal in which the water level is regulated. Also, the water supply is now abstracted wholly as groundwater, thus protecting it from possible pollution from the nuclear power plant and other industries (Fig. 6.11) [29].

The quantitative and qualitative influence on groundwater of dredging aggregates has been experienced in Bavaria, Germany, where the demand for aggregates over many decades has led to a rather uncoordinated and harmful programme of extraction.

Now attempts are made increasingly to direct and monitor the exploitation by means of regional planning and administrative approval procedures. The problem is complex and includes all the well-known conflicts between the exploitation of aggregate and groundwater resources and, in some places, an almost complete destruction of the landscape. The creation of dredged lakes tends to interfere with the groundwater flow. The strong eutrophication of the lakes and the evaporation from the lake surfaces also create problems to which no solutions have been found as yet. Such problems should be tackled in a joint research programme by the water authorities and relevant scientists such as geologists, hydrologists, and meteorologists. As a first step, even a hydrogeological evaluation carried out prior to the creation of dredged lakes would help in preventing hydraulic and hydrochemical changes from having adverse effects on the surroundings [89].

A further example of geoscientific contributions to general regional planning and to managing conflicting interests, is the geoscientific zoning of an area for land-use planning in the Madrid Region, Spain.

In 1987, at the request of the Madrid Autonomous Region, the Geological Survey (ITGE) prepared a geoscientific atlas of the 'Natural Medium' of the region. The zoning was used as a tool for the analysis of the physical surroundings. It was established by

defining homogeneous integral units including all geological factors that fundamentally affect or condition human activity, and vice versa. Four levels of division were introduced: environment, sub-environment, system, and unit. Two large environmental domains were considered: mountain range and depression. Four sub-environments were defined by climatic and topographic features: high sierra, mountainsides, high surfaces and valleys; they, again, were divided into fifteen systems that included the large lithological and geomorphological domains, plant associations, and surface formations. At unit level, 131 units were defined mainly on the basis of lithology and plant associations. The maps and thematic cartograms of the atlas were found to be an important contribution to land-use planning in the Region. In addition to updating knowledge, the synthetic nature of the atlas has served to reveal aspects of the physical medium that were little known to the population, thus serving as a didactic tool for environmental education [152] [163].

Site investigation

The problems raised by conflicts of use and associated planning are not restricted to large areas only. Many problems relate to local site studies, even when the conclusion reached must be integrated into a larger regional perspective. Examples of such problems are found in Germany and Denmark.

A most important social problem is the disposal of hazardous waste into the natural environment. At the end of the 1970s, the Geological Survey of Lower Saxony, Germany, at the initiative of the Hannover District Administration, began to prepare maps of the distribution of clay deposits suitable for the development of special waste-disposal sites. A comprehensive interdisciplinary study on the suitability of claystone was conducted during 1981–86. During the investigation, many holes were drilled, field and laboratory tests were made, an evaluation scheme was developed, and the results of the evaluation were presented on maps. The study demonstrated the importance of the geological suitability of the substratum for any waste-disposal site. Pollution of the surroundings will result from sites developed in unsuitable strata. Failure of the surface seal and the bottom seal of a multibarrier surface repository of solid hazardous waste may be especially dangerous if the last barrier, the substratum, does not have the retention capability to prevent or at least hinder the escape of toxic materials. Extensive knowledge is needed for an assessment of the effectiveness of a barrier: convective and diffusive migration of substances, the mechanisms of precipitation, sorption, desorption, and degradation, as well as the physical, chemical, and mineralogical stability of the rocks. All this information is necessary for the mathematical models used for predicting the migration of pollutants from a waste-disposal site [92].

In Denmark field tests are being made of a surface seal for waste dumps based on capillary forces in fine sand underlain by coarser sand and superimposed by a soil layer. This so-called Capillary Barrier is capable of draining off most of the precipitation, thus preventing the formation of poisonous leacheate which can pollute the groundwater [11] (see also Chapter 5.5).

It would be expected that all site investigations would involve geoscientific input, but this is not always the case. It has been reported from Belgium, for instance, that many major roads and bridges have been constructed without any geological advice whatever. The foundation of a motorway in the coastal plain was made in places by the removal of one or two of the top layers of peat and other unconsolidated sediments, and their replacement with sands. This was done systematically to a certain depth without taking account of the

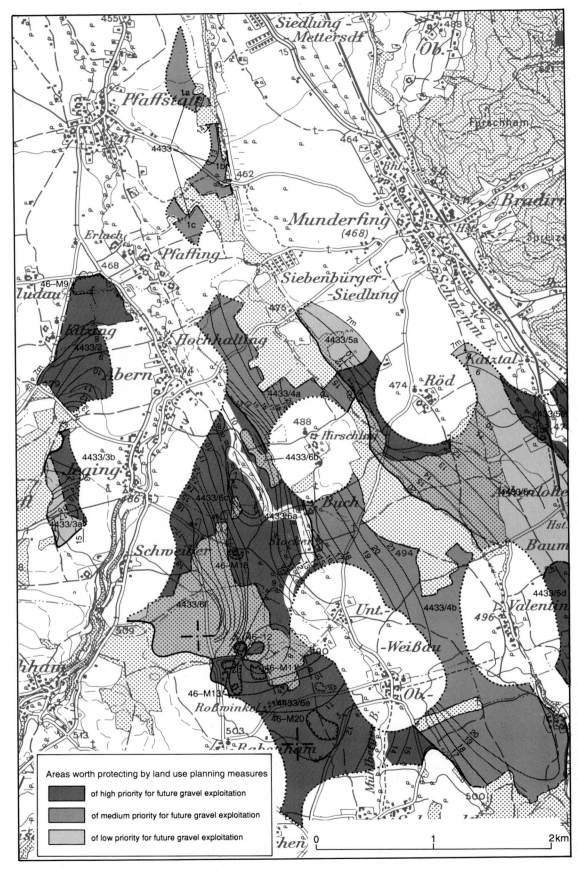

Fig. 6.10 Abstract from a map entitled 'Conservation worthiness of gravel/sand resources' at Mattigtal, Austria (originally at the scale of 1:20 000, Sheet 4433, Mattigtal).

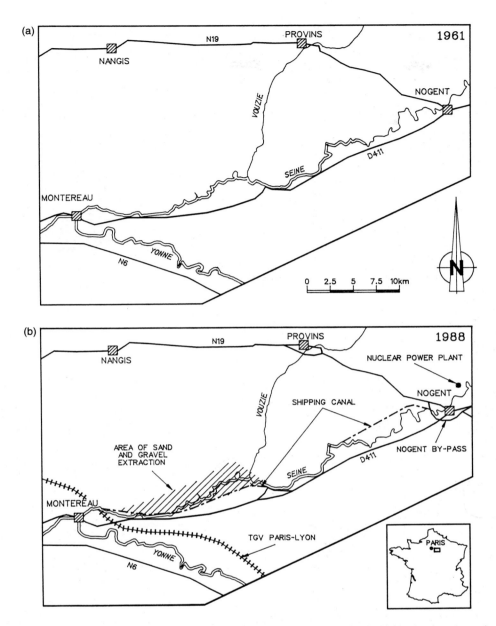

Fig. 6.11 The Seine Valley east of Paris: (a) the largely unspoilt state in 1961; (b) the highly-developed state in 1988.

nature of the underlaying sediments. In some places, sediments which were suitable for road construction, were replaced while in other places only part of a thick peat layer was removed. This has resulted in the uneven compaction of the deposits under the foundations of the road, which is now developing a typical washboard surface in some places [3] [6] (Fig. 6.12).

In such cases there have, without doubt, been conflicting interests, but it is not always clear what interests are involved. Society requires high-quality development at an affordable price, and this may be in conflict with social desires regarding the protection of natural resources and the environment, and the maintenance of a national construction industry. There is great advantage, therefore, in involving geological information and expertise at an early stage of planning when alternative solutions may be considered. Slight adjustments to the preliminary plans may save money and raw materials and may

also lead to better solutions from an environmental point of view.

6.3.3. Land conservation and after-use

A rather new aspect of land-use planning is the concern about the impact of our use of the environment and its resources on the socio-economic conditions of future generations. In some countries nuclear energy has been rejected. Denmark made such a decision because, at the time, no reliable and economic means of disposal of radioactive waste was available.

The recommendations made by the World Commission on Environment and Development, which urged the world community to have a greater concern for the environment, are generally accepted. The application of these principles in land-use planning would imply that the after-use must also be planned during the initial planning phase of any potential

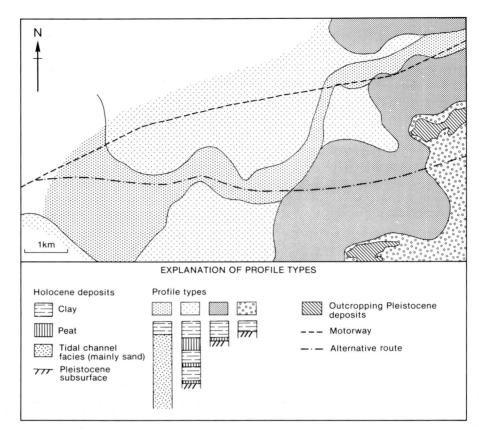

Fig. 6.12   Simplified profile type map indicating alternative route of a motorway in Belgium through areas of clay over sand or clay over Pleistocene sub-surface. The existing motorway is partly located in areas with a clay and peat sequence.

development. It is recognized, of course, that the use of any area may change naturally over time as mines are abandoned, as the need for certain resources disappear, as climate or sea level change, and so forth.

Past and future changes of function

All mines, quarries, and pits are eventually abandoned. All these cavities and voids in the surface layers of the Earth's crust are tempting places in which to dispose of the gigantic amounts of garbage and waste produced by man. Old mines are commonly used as repositories for hazardous industrial waste; clay pits are used for waste-disposal. Former sand and gravel pits have been used as waste dumps or as tank parks, neatly kept out of sight, but a potential threat to the quality of groundwater supply. Pits, quarries, and open mines situated under the level of the water-table often become lakes. Not all abandoned, or partly abandoned workings are used for other developments, however; some areas continue to deteriorate slowly or are simply abandoned, leaving geotechnical and environmental problems behind.

This was the situation for some years at Deeside in North Wales, UK, an industrial area for more than 100 years and once rich in coal, lead, and zinc. Up to 10 years ago it was a scene of industrial dereliction and unemployment, showing the ravages of mine shafts, coal tips, and disused quarries. It included much unnatural scenery and many man-made hazards. One large working quarry, surrounded by many abandoned uncapped shafts from lead mining, is now in conflict with a pleasant Country Park and a major site of

scientific interest. Now the area and the region are reviving and the county is doing its best to heal the scars of the past and to organize the after-use which includes continued production of bulk minerals and the conservation of a reasonable environment. The planners must act quickly to ensure that the county remains a scenically attractive gateway to Wales. The Geological Survey has actively supported the planning procedures by making available relatively simple, single-element Environmental Geology Maps and well-illustrated reports on the practical aspects of the local geology [189] [192] [203].

Former industrial areas may be used for urban development. Typically, they contain buried, hidden, and forgotten waste repositories. Urban areas themselves may also change; in former times they have sometimes been abandoned as a result of war or a major hazard, such as a fire, an earthquake, the raising or lowering of sea level, the lack of enough drinking water, or for many other reasons. Near Copenhagen, Denmark, for example, a forest and open-land resort was created some hundred years ago by removing a few villages and farms to satisfy the need for a hunting area for the Royal house. Later in this century, a hotel and a number of private houses were demolished, opening the beach to the public again and creating a green belt along the coast north of the capital.

Reclamation and restoration

It is obvious that Europe cannot tolerate the abandonment of areas temporarily used for industrial purposes without remedial treatment and preparation for the re-development.

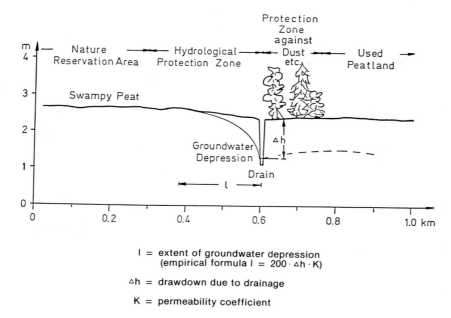

I = extent of groundwater depression
(empirical formula I = 200 · Δh · K)

Δh = drawdown due to drainage

K = permeability coefficient

Fig. 6.13 Concept of protection zones between farmland and a nature reserve area, Germany.

The traditional objective for reclamation and re-cultivation is to gain land for agriculture or other development. Today the creation of a natural environment including landforms, landscapes, vegetation, and animal wildlife is a new and important objective for reclamation all over the world. So, two different lines of action are now being followed in reclamation: either to find new and acceptable ways to exploit a given area when the first use has stopped, or, to bring the area back to its natural state.

Strip-mining techniques used in working lignite, limestone, and placer minerals are well known 'rolling' developments, which may leave the areas concerned more or less as they were, so long as precautions had been taken to avoid possible environmental problems such as acidification of water and pollution of water by iron compounds like ochre. Abandoned opencast mines and former waste dumps are sometimes reconstituted and re-cultivated successfully as agricultural areas as a result of relevant studies completed before the work started.

In Spain a study of the environmental issues relating to reclamation of areas degraded by strip-mining operations was performed by the Geological Survey. A cost estimate of reclaiming several areas of mining operations was made. The measures necessary to balance the environmental impact of abandoned mines were specified and a basic programme for reclamation was proposed [163] [162].

A typical conflict involving reclamation occurs in German peatlands, which are subject to special considerations concerning their future use because of their importance to the human environment in Germany. Peatlands cover about 4 per cent of the total land-area of the country and more than 9 per cent of the Länder Lower Saxony and Schleswig-Holstein. The question arises as to whether these areas should be used for peat production for fuel, horticultural, and medical purposes, or for pasture and farming, or should they be left as undisturbed natural biotopes and reserved areas for wildlife? For centuries most of the area of lowmoor has been used for agriculture while 20 per cent of the area of highmoor is in the natural state. The need for protection of the remaining peat-

land in a densely populated area is undisputed. In 1981 the laws regulating the exploitation of near-surface raw materials were integrated into the new Nature Protection Act. Since then, the primary objective has been the regeneration of peatlands after the peat has been cut. All upland bogs in Lower Saxony that were uncultivated or were in peat production in 1980 were mapped by the Geological Survey and classified according to value and ecological potential. Criteria for re-wetting and re-establishing a natural state were defined on the basis of long-term field tests. Since nutrient input in re-wetted peatlands is much higher than nutrient loss, a non-typical vegetation may develop. As a result, a concept of protection zones for groundwater and dust derived from adjacent farm lands has been developed [95] (Fig. 6.13).

The hydrological and other prerequisites for the re-wetting of peatlands after cutting have been researched so thoroughly (more than 100 studies) that predictions can be made on the basis of a survey of the area in question (e.g. depth, stratigraphy, field permeability, degree of peat decomposition).

Problems arise with respect to natural rehabilitation of peatlands after cutting in areas previously used for agriculture. The nutrient-rich surface layer resulting from fertilizing of crops is unsuitable for regenerating peatlands. Such areas are better used for cultivation after deep ploughing. This is suitable for intensive cultivation of many types of crops.

Another type of regeneration is the use of abandoned quarries and pits for leisure purposes, such as parks, bathing facilities, and lakes for water sports. In some countries reclamation and restoration is encouraged and awards for good restoration practice may be granted [14].

A special problem arises when landscape architects and geoscientists together try to plan the restoration of an area which has suffered industrial degradation. Should a restored landscape be made to look natural or to be clearly artificial? Examples are known of waste tips which have been shaped like natural geological landforms such as glacial eskers. In Norway a case study has been carried out into the feasibility of completely restoring a ridge which would be removed by the possible exploitation of an iron ore–apatite deposit [149].

Fig. 6.14   Cleaned exposure of Maastrichtian limestone. An example of geological conservation (Province of Limburg, The Netherlands).

Conservation

The conservation of sites of special geoscientific interest is an issue not only for geoscientists but also for naturalists, research workers, and teachers. Most sites are valuable elements of a beautiful or characteristic landscape, certainly of local and possibly of national importance, or they contain a biotope for plants and animal wild life and contribute to the conservation of a diverse fauna and flora. In broader terms, many organizations and authorities throughout Europe work for the protection of the environment generally. Sometimes conservation yields remarkable results of benefit to present and future generations. In some cases, however, the end-product is not helpful to the geologist. If a coastal cliff is protected too effectively, for example, erosion may stop completely and the preserved geoscientific profile may stabilize and the state of the surface may deteriorate. The cliff will be protected, certainly, but it will not be directly visible or available, and further studies of the geological section may be forbidden or impossible. This has happened already in many places. The same situation can arise in abandoned quarries and pits, when profiles are protected to the point where they cease to be useful for scientific study.

In planning the after-use it is sometimes better to adopt a deliberate policy of leaving a quarry or pit as it is and letting natural restoration pursue its course. But it must be stressed that leaving reclamation to nature itself must at all times be as a result of informed decision and not an excuse for doing nothing (Fig. 6.14).

To achieve the best possible basis for the planning of land-use, after-use, and conservation, it is vital to make use of the information on the landscape and its geoscientific contents which can be supplied by the Geological Surveys. Many examples from all parts of Europe demonstrate that the Geological Surveys contribute actively to this work. One example is from Schleswig-Holstein in Germany. This region has many geologically young (Pleistocene and Holocene) land-forms and is considered to be a classic area for the study of these geological–geomorphological phenomena. For the protection of the most valuable and special parts of this landscape, the Geological Survey has located, registered, and made an evaluation of relevant features. The Earth Science Conservation Map at the scale of 1:250 000 yields a quick and comprehensive view of the topographic position and dimensions of the features which are also listed, described and mapped in more detail in a special catalogue. This documentation is, of course, of much value for land-use planning and at the same time contributes to the adequate selection of areas which should be protected and preserved for succeeding generations [116].

An example from The Netherlands [142] demonstrates the successful cooperation and interaction between scientists and planners to the benefit of the environment. An earth science conservation policy was developed here on the basis of information on land-forms and associated deposits, their functions, and vulnerability. Land-forms have two important functions: they provide information about the geological processes which created the landscape, and they have a great influence on hydrological processes and the development of soils and vegetation. From 1965 to 1988 the inter-institutional working group GEA under the responsibility of the Research Institute of Nature Management, established an inventory of geological and pedological sites in each of the Dutch provinces. More than 800 earth science sites, ranging from geological sections and single land-forms to complete patterns of land-forms, were selected and reported upon. The documentation was widely based on the geomorphological map of The Netherlands at the scale of 1:50 000 which covers approximately 60 per cent of the land surface.

While this work was going on planners and decision makers did their part of the work. In 1982, an earth science conservation policy was incorporated into the Structural Scheme for Nature and Landscape Conservation. In the Soil Protection Law (1987) Dutch provinces were requested to protect valuable abiotic areas by ascribing high earth science values, and earth science conservation now plays a substantial part in the National Policy Scheme published in 1989.

Thus, it is advised that comprehensive plans made for land-use should include recommendations for the after-use. Relevant geoscientific input should be integrated into these plans. Cooperation between authorities and private enterprise should be encouraged and awards should be given to private enterprise for good environmental practices in excess of minimum statutory requirements. And last but not least, the general public should be encouraged to support the expenditure of national resources to conserve the natural environment, by being well informed and involved in planning processes.

6.4. FINAL REMARKS

The general understanding that there are limits to Nature's resources, and of the reaction of the environment to human interference is now growing rapidly. Natural resources are understood today in a much broader sense than ever before. They consist of much more than minerals and water, as they include all elements constituting the environment. All natural resources are finite. Natural landscapes once destroyed and extinct wildlife cannot be replaced. It makes sense, therefore, to conserve natural resources carefully and to use them in an informed way. This is basically a political task, but recourse to information and expertise held by Geological Surveys is essential.

The occurrence of all types of environmental resources should be identified and their relative or absolute value should be established, as far possible. The application of geoscientific data and expertise is essential to such a process.

The development of new tools for evaluation is of considerable importance to society. Such development will inevitably be complex and it is recommended that it should be made through close national and international cooperation between geoscientists and decision makers in order to reach an agreed common evaluation of trends and future needs.

Non-use of natural resources is not an option. Leaving resources undisturbed is merely to classify them as reserves. The conscious decision to sterilize resources should be taken only with the understanding of the possible environmental consequences and side-effects.

Natural hazards are and always have been present in the environment. The potential development of hazardous conditions can be predicted, to a large extent, by geologists. On the basis of knowledge of geological history and processes, the nature, magnitude, location, and possible impact of geological hazards can be described. Such evidence is essential to the development of global scenarios for the future. What the geologists can contribute especially is the evaluation of the geoscientific elements of the models, taking into account regional and local geological conditions. In this way, geologists assist in establishing the norm, in identifying significant variations from that norm, and in forecasting the possible effect on the natural environment of any future developments.

The science of prediction is certainly inadequate as yet, and further scientific research is essential. In the light of the serious economic and social impact of geological hazards, some governments have established forecasting systems with a view to understanding the incidence and impact of such events. These need to be extended and improved, in tandem with technological and scientific progress.

Land-use planning, including planning of the reconstitution and after-use of land, should benefit from the expertise of the Geological Surveys and from the application of the existing data collected and managed by them over many decades. Geoscientists, as well as planners and decision makers, should ensure that relevant information is better integrated into planning. To make this happen, geoscientists must ensure that the information is presented in a way that meets the needs of practical planning locally, as well as at a regional, national, and international level.

The more important conclusions of this chapter are that geoscientific information must be integrated more systematically into planning, and that Geological Surveys must be recognized as major contributors to decision-making. The sensible planning of resource development, the prevention of sterilization of resources, the protection of groundwater from pollution, the identification of criteria for the safe disposal of wastes on and in the crust of the Earth, and the recognition of hazards to development in the geological environment, are all issues for which full and comprehensive assessments of the available geological information are required prior to the making of planning decisions related to potential development.

It is recommended, therefore, that it becomes a statutory obligation of planning organizations to consult Geological Surveys and the data held by them, whenever appropriate. This is particularly necessary in the early stages of planning, as many case histories have demonstrated that early contact could have avoided later difficult problems and costly remedial actions. The interpretation and release of geoscientific information, as a basis for an open and democratic planning process, is important and must be the principal one of the many responsibilities of national Geological Surveys. This would provide the safeguard to society against inadequate, incorrect, or falsely interpreted information being an obstacle to the successful development of the environment.

RECOMMENDED FURTHER READING

Bender, F. (ed.) (1986). *Geo-resources and the environment—Proceedings of the fourth international symposium, Hannover, F.R. Germany, October 1985.* (E. Schweizerbart'sche, Stuttgart).

Culshaw, M. G., Bell, F. G., Cripps, J. C. and O'Hara, M. (eds) (1987). *Planning and engineering geology.* Geological Society Engineering Geology Special Publication No. 4, London.

McCall, G. J. H. and Marker, B. (eds) (1989). *Earth science mapping for planning, development and conservation.* (Graham and Trotman, London).

de Mulder, F. J. and Hageman, B. P. (eds) (1989). *Proceedings of the INQUA symposium on applied Quaternary studies. Ottawa, 6 August 1987.* (A. A. Balkema, Rotterdam).

Wolff, F. C. (ed.) (1987). *Geology for environmental planning. Proceedings of the International symposium 'Geological mapping in the service of environmental planning', Trondheim, Norway, 6–9 May 1986.* (NGU Special Publications Series, Trondheim).

Wolff, F. C. and Cendrero, A. (eds) (1990). *Geology and the environment.* Engineering Geology Special Issue vol. 29.4. (Elsevier, Amsterdam).

7

Proposals for the future

The Editorial Board

7.1. INTRODUCTION [210]

The preceding chapters have demonstrated the importance of applied geosciences for a rational use of our natural resources and the preservation of our environment. The future of the industrial, agricultural, and social infrastructure of nations depends on the adequate management of these resources and the protection of our environment.

Rock and soil provide the environment on and in which present and future generations will live; the contained minerals provide the basic substance for industry and commerce; fossil fuels and alternative energy resources will continue to provide vital sources of power; and groundwater is essential to the survival of existing or even greater levels of population.

In addition to natural hazards, such as landslides, seismic risk, volcanic risk, or flooding that must be prevented or minimized as much as possible, population increase leads to an ever greater pressure on resources and the environment. An acceptable future will be secured only by recognizing the problems inherent in such developments, and by careful management of the exploitation of natural resources. Although the whole population is responsible, only governments, politicians, and their agencies can ensure appropriate arrangements, and only scientists trained in the acquisition and analysis of relevant data can provide a solid basis of geology to service that process.

In that respect, applied geosciences, which are a major concern of the Geological Surveys, have an essential role to play. It is fundamental that this role should be developed with maximum efficiency at all levels (local, national, and international) in close association with the relevant decision makers.

It is essential for the protection and control of the environment, that the Geological Surveys are considered to be major components of a network of organizations, with access to and the ability to apply a great wealth of multidisciplinary expertise. In environmental geology the Geological Surveys must carry out two main functions:

(1) the provision of geological data and expertise for the study and management of the environment and its resources;

(2) the identification, management, and control of natural hazards, and of the interaction between the natural environment and human activities.

However, before considering any proposals for the building up of this European network of geological organizations, it is necessary to discuss briefly the problem of standards. The environmental problems in the countries of Western Europe may be similar, but the ways and means of tackling them show great diversity. Such regional response may arise out of obvious differences in geology and climate, but the historical fragmentation of Europe has played a role too. Today, it is clear that a unified approach, based on adequate scientific data, is essential.

7.2. THE GREAT SIMILARITY OF PROBLEMS IN THE COUNTRIES OF WESTERN EUROPE

7.2.1. Similarity of problem types

Many diverse environmental problems have been described already. Although the case studies concern specific areas in a given country, it is well known that similar examples and types of problem exist within the boundaries of other countries, and that their impact may be transnational. These similarities result from the uniformity of the physical environment, despite its apparent diversity, as a source for mineral resources, and as a body of varied solid components interacting with air and with water, both at the surface and underground.

In Western Europe the EEC has become an economic entity able to influence the environmental future of the whole globe, and Western Europe has become the example for the evolution of Eastern Europe with all its inherent environmental problems. Within this framework, geology is a basic and unifying factor as it relates to no national boundaries.

There are regions of hard rocks and softer rocks, mountains and plains, low-lying areas with poor natural drainage, strata containing large volumes of solid and liquid resources and strata with localized mineral resources, coastal areas with associated features of climate and erosion, and areas of extreme climate generating erosional factors which may be exaggerated by the relative instability of steep topography. All these factors, described in more detail in Chapter 2, transcend national boundaries. Environmental problems related to physical features or to man's interaction with them are, therefore, common to all countries.

For example, the coastal zones of the countries bordering the southern North Sea have similarities in the landscape, the climate, and the geology which lead to common problems in water supply, raw-materials extraction, shoreline stability, communication, navigation, and development in general. Problems of a local or more confined nature may not cross national boundaries, but their common geological basis is well recognized from country to country.

7.2.2. Natural hazards

Hazards arising out of natural geological phenomena are common to various countries. The most readily observed are problems related to instability, particularly landslides. These occur under various climatic conditions in areas with relatively unconsolidated strata, especially in mountainous areas where discontinuity of geological structures and high-angle slopes may coincide.

Low-lying areas and areas of gentle topography can be subject to sediment accumulation and flooding, with all the associated problems. In coastal zones the juxtaposition of deposits of mud and sand presents civil engineers with particular problems in construction projects.

As construction also takes place on the sea bed, properties of sea bed sediments, the incidence of slope instability, and problems related to the occurrence of shallow gas, are of particular importance. The presence of groundwater, although in itself a resource of great importance, presents problems when associated with soluble strata. The consequent creation of large underground cavities can lead to the occurrence of subsidence and surface instability. Instability of another kind, arising out of the presence of zones of weakness resulting from crustal dislocation and movement, presents problems in construction, particularly for underground facilities.

Earthquakes and volcanic eruptions provide the problem of the unexpected. The centres of greatest risk are well known in the southern fringes of the area and Iceland, but, although most of Western Europe is relatively stable, the current development of increasingly sophisticated structures demanding extremely stable foundations means that all countries have become involved in monitoring even minor tectonic activity in the course of controlling such risks.

In areas where the emission of radon gas is significant, many people are now at risk as a result of living in hermetically sealed houses, designed and built with insulation related to government-sponsored energy-saving policies and without reference to the underlying geology. This is government irresponsibility on a grand scale.

The physical and chemical properties of soils and rocks are highly variable. They have a bearing on problems in the field of construction, and also in the interactions which take place between any development work and the geological environment. One facet of the chemistry of soils and strata, the identification of the levels of occurrence of particular trace elements, has become of vital importance in understanding their relationships to disease in crops, animals and man. Even natural variations in the Earth's magnetic field and other magnetic disturbances affect man's activities, not the least in achieving accurate positioning and navigation. National boundaries are irrelevant when considering such problems related to natural geological phenomena, and all countries have had the experience in most of them.

7.2.3. Man-made hazards

Many environmental problems facing man are self-induced. Ever since man recognized the wealth of natural resources available in the Earth's crust, he has been exploiting them ruthlessly and with scant regard for the detrimental effects on the environment as a whole.

Exploitation of mineral resources

Coal mining has taken place over widespread areas of Western Europe. Traditionally, abstraction has been by deep mining, which has had negative effects on the environment in the form of subsidence and the disposal of large quantities of waste rock. The natural habitats of large areas have been destroyed in the course of opencast exploitation of near-surface deposits of coal, lignite, and peat, with the attendant problems of drainage and rehabilitation. Only recently has there been recognition of the secondary environmental effects of burning these fuels; the countries of northern Europe are now affected by 'acid rain', and, globally, man is now concerned with the so-called 'greenhouse effect'. Here, the past is the key to the present and the future. The geological evidence of past climatic changes, particularly of relatively rapid rates of change before, during, and after the last Ice Age, has a key role in determining the nature and extent of such problems.

The extraction of oil and gas also takes place over wide areas on land, but more so in the North Sea, with lesser problems of subsidence for the areas concerned, but major interference with naturally occuring groundwater regimes and major pollution problems arising from leakages and spillages. The impact on the environment of the storage of hydrocarbons has been considerable but it has been ameliorated to some extent, particularly in northern countries, by the development of techniques of storage underground, but not without cost to the environment with respect to the disposal of bulk waste and to the effects on groundwater.

Metallic minerals have been worked for many centuries. High-grade occurrences tend to be localized, and the mining methods used have caused the problems of waste dumps and tailings, and the migration of toxic effluents. Improved methods of abstraction from low-grade deposits have resulted in extensive and deep excavations, with the drainage and rehabilitation problems common to all opencast operations.

In addition, widespread large-scale excavations are necessary for the exploitation of bulk industrial minerals for the construction industry, such as sand, gravel, limestone, granite, and brickclay, and of materials for more specific uses such as ironstone and china-clay. They have been accompanied by all the attendant problems of large excavations for civil engineering development, such as foundations for buildings, road and bridge construction, tunnelling, and the creation of underground space. Some of these activities heighten the potential for induced land-slipping, and for the variation of groundwater regimes.

Waste disposal and groundwater protection

One of the greatest effects on groundwater, however, arises out of modern developments in waste disposal, particularly in the form of surface landfill. The indiscriminate dumping of domestic and industrial waste, some of which is highly toxic, is now a serious problem in many industrialized countries. The seepage of poisonous leachates pollutes groundwater locally, but such pollution can also reach rivers and reservoirs, and creates widespread problems even well beyond the shoreline in adjacent estuaries and open seas. The problems of the chemical interaction between the waste product and the environment are ubiquitous, as are those of the physico-chemical processes controlling the migration of toxic leachates, particularly with respect to the absorption and desorption of specific organic chemicals. Further problems arise from the generation and migration of toxic and explosive gases, with potentially dire consequences for nearby communities.

In all countries the development of intensive farming methods has led to the contamination of both ground and surface water. In regions of agriculture, where irrigation is widely practised, the effect on groundwater regimes can be significant. This must be evaluated from the economic viewpoint against the need for such irrigation.

In current specialist developments, such as the disposal of radioactive waste into the geological environment, the attendant problems are the subject of much research.

The foregoing account is not meant to be an exhaustive catalogue of all the environmental problems in the geological environment which man has to face and overcome, but rather to show the widespread nature and occurrence of such problems. The causes and the problems are similar in whichever country they occur, and all the countries of Western Europe have long and bitter experience of the actual and potential consequences of ignoring environmental geological factors.

7.3. DIVERSIFIED APPROACH TO RESOURCE AND RISK MANAGEMENT

7.3.1. Diversity of practices and tradition

In all the countries of Western Europe some statutory regulations apply to the use of the natural environment, to the abstraction of resources from the geological environment, to the disposal of potential pollutants in the form of industrial and household waste, and to agricultural beneficiation, onto and into that environment. But legislation, regulations, standards, and codes of practice vary from country to country.

In some countries the public, pressure groups, political organizations, authorities, and governments are more concerned with environmental problems than in other countries, where the extent of the problems and the potentially serious implications for mankind are only at the early stages of appreciation and understanding.

Traditions also vary and have a large part to play in the perception of problems and of the necessary steps to remove them. This point is well demonstrated by readily observed variations in agricultural, industrial, and mining practices.

In some cases the magnitude of the problem is now so great that the cost of removing it exceeds the resources available in the economy of the country concerned and, unless alternative solutions can be found, the population will have to bear with it in the long term. A classic example of this is the widespread existence of abandoned waste dumps and repositories containing hazardous and toxic materials. Commonly, the extent of these and their content, and the knowledge of chemical processes going on within and around them, is only partly known and documented. They should be removed or isolated from the environment, of course, but, in many countries, unless there is the incentive of financial gain from reclamation, this is 'a labour of Hercules', and much remains to be done to improve the situation.

Because regional and national environmental problems remain largely unsolved or continue to increase, they impinge detrimentally on the environment and the socio-economic situation generally, and they contribute to the growing awareness of neighbouring countries that, taken together, these problems constitute a threat of global proportions to the well-being of mankind. Consequently, governments are continuously at work on national and international agreements to control and mitigate the basic problems. Gradually, common agreements, regulations, and standards are developed to form the legal basis for the international protection of the environment, which is crucial for an acceptable existence for humans on planet Earth. Depending on the country, the Geological Survey will be involved to a greater or lesser extent in advising on such negotiations. Our case is that a national Geological Survey should always be involved: the geology beneath us, the biota, and the atmosphere cannot and should not be treated separately.

Similar international activity and cooperation is taking place at the technical level on geoscientific research and development. Here again national traditions and practices vary. Related particularly to the style of the dominant geological regimes and to traditional structures in educational systems, each country has developed strengths in experience and research in particular aspects of geoscience. The corollary is that significant weaknesses exist in some fields of the discipline. Likewise there are variations from country to country in the relationships between research establishments, in the grouping of associated activities, and in the lines of communication between research areas and implementing authorities.

The stimulus of concern, however, together with rapidly improving facilities for communication, are overcoming national and international parochialism. Significant improvements in international cooperation are now furthering new approaches, methods, materials, and instrumentation, and contributing to the broadening of political vision and the making and implementation of relevant agreements. The involvement of Geological Surveys in such matters, though it varies from country to country, is a key component.

7.3.2. A key concept: political and executive power at the appropriate scale of action

Complementing the diversity among European countries in the fields of regulations, traditions and R&D, one of the main differences between countries lies in a different administrative organization and, in particular, a different political and executive power at the various levels: national, regional (region, county, province, etc.), and local (commune). Without falling into the utopia of attempting to unify this administrative organization, it is obviously necessary that the problems of the environment be dealt with at the appropriate scale and at an adequate level of power, and that there should be an overriding concern for the collective interest.

It is also essential to remove another major dividing factor: the selfish attitude of institutes and organizations that consider themselves to be solely responsible for one specific sector under the supervision of their area of central government. On the other hand, it is also essential to set up the appropriate national arrangements for taking full advantage of the expertise and advice of such institutions.

Elsewhere in this book, the experience and case histories from the Geological Surveys and associated institutions of twenty-one countries of Western Europe were set out. In some countries, the Geological Survey is the national body funded by government to assemble the national geoscience database, to analyse it, and to advise on developments on and in the Earth's crust in support of mankind, and on problems arising from such activities. In every country the national Survey makes an important contribution to this process.

The general objectives are clear and common to all countries, and yet it is clear from the previous pages that, even though the scientists in these national Surveys are aware and cooperate globally, the end product, the ways in which environmental problems are handled by scientists and by the responsible authorities whom they advise, varies considerably from country to country.

The reasons for the variations may be found in history, in local tradition, in the financial support arranged by government, or in differences in natural conditions. Some organizations have developed quickly through strong involvement in the applied field, while others, perhaps as a result of inadequate funding or strong traditions in the research area, have shown some inertia in developing appropriate methodology, expertise, and involvement in the field of environmental geology. Part of the variation also relates to the demands for involvement made by governments and local authorities, and that, in turn, reflects on national policies, national economies, and national awareness.

Many differences between Geological Surveys and their national roles reflect the history of their foundation, or of the working conditions or organization relating to the principal functions they have carried out for most of their existence. Some have been in existence for more than 150 years, and were set up initially to survey the geology of the country, and to publish maps and accounts of the geology. Others, particularly those with a much shorter history, have been founded to search for natural resources and to advise on the relevance of geological factors to the national economy.

In countries where the development of heavy industry was related to readily available resources of coal and iron ore, the Geological Surveys have had an obvious, but somewhat restricted, role in socio-economic planning. Some surveys were established with this as their principal objective and were typically affiliated to government departments associated with industry and energy. For some, the function has been limited to surveying and geological interpretation in support of the exploration and development function carried out by another government or quasi-government agency. For others the role has been more extensive with greater responsibility in resource assessment.

In some countries that are heavily afflicted by landslides, volcanic eruptions, earthquakes, flooding, and other natural hazards, the emphasis of the Geological Surveys is much more towards specialist thematic surveys and more closely linked with planning. These Surveys tend to have strongly developed capabilities in the fields of environmental and geotechnical geology. In countries with high dependence on groundwater resources, the capability in hydrogeology is highly developed, and where the national economy is highly dependent on indigenous resources of metalliferous minerals, the Surveys have well-developed expertise in geochemistry.

In all countries the emphasis is changing rapidly towards the application of the geosciences to the environment, with an increasing tendency towards quantification of qualitative parameters. In recent years, even the most traditional Surveys have been forced to change and to become concerned with a much wider range of studies in the complex field of geoscience. Such developments, however, do not detract from the traditional Survey role of data acquisition and analysis, but rather underline the requirement for such basic functions to be developed to the highest possible standards.

In some cases the mistake was made of attempting to develop modern studies at the expense of the traditional, with the inevitable result that the advice in the newer areas of activity is founded on a less-than-adequate database.

The demand for prediction is increasing. Geologists traditionally have been able to identify potential hazards and to predict that they will occur sooner or later. The more sophisticated and developed society of today demands that such identification and prediction is more precise, so that potential catastrophes can be avoided or, at least, that problems can be dealt with at minimum expense. Such demands expose the fact that there is no simple means of predicting the timing of a hazardous natural phenomenon and underline the obvious need for a strengthening of basic research into geological processes and conditions, and for improved monitoring techniques and models.

All of these pressures have brought about reorientation and reorganization within most Surveys, and for some the change has been greater in that they have been transferred to government departments strictly concerned with environmental matters. For some, however, the direct links with government are remote, and involvement is less than the current situation and future trends demand. In such cases it can be difficult for the Survey to recognize where it fits in national environmental programmes.

The rate of change has also varied greatly from one Survey to another, commonly as a result of the different emphasis governments place on the environment, on the relevance of geosciences to environmental problems, and on identifying funding to support adequately the work of national geological services. For some, the potential of the new way of thinking, which incorporates traditional functions with new and comprehensive activities to provide a national and essential service on environmental geology of the highest quality, has not yet been realized.

These differences of approach, and variations in expertise and capacity, are regarded by some as being a weakness in the scientific resources of Western Europe which are available to maximize the potential of the geological environment and to address the problems arising naturally or as a result of human activity. On the other hand, great strength is available from the varied experience of tackling these problems, and from the wider range of available solutions based on alternative or new methods and techniques.

But society can benefit only if this breadth of expertise is recognized, shared, and applied appropriately regardless of national boundaries. To this end, good links have been formed amongst the Geological Surveys of Western Europe, and much work has been done already to develop common standards, guidelines, and codes of practice. So far, however, governments have shown little concern and produced little funding to ensure maximum cooperation and transfer of expertise to achieve the best possible availability of service for any development or problem anywhere in Western Europe to which environmental geology relates. Governments may find an even greater need to foster such coordination now that Western Europe is inevitably involved in the re-development of Eastern Europe, where there are serious environmental problems.

The position of Geological Surveys in the framework of government varies from country to country, but, in general, they are not in the mainstream of international programmes related to industry and trade. Without question they should play the leading role in the establishment of technical standards related to geology, but here again the situation varies, with Surveys having to act on their own varied initiatives without the impetus of government instruction or support.

The overall position is that all the countries of Western Europe have national Surveys, but these are much diversified in function, in expertise, in funding, and in their relation to government. Much work has to be done in capitalizing on what exists already, and in achieving for Western Europe a service in environmental geology of the highest quality which relates to the developments and problems of the future. The establishment of a strong interface with government and with international environmental protection agencies is essential. The Surveys can do much from existing capacities, but only the awareness of governments, the allocation of sufficient national resources and adequate intervention at the appropriate scale will achieve the desired objective.

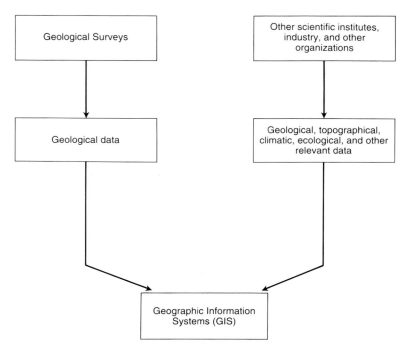

Fig. 7.1　Supply of geological data concerning the environment.

7.4. SHORT- AND LONG-TERM PROPOSALS

The Geological Surveys have two main functions in the field of environmental protection and management. Not only must the Surveys be the key suppliers of geological data concerning the environment, but they must also play a major role in the identification, management, and control of hazards to man and the environment. These actions must anticipate and respond to the needs of users at the appropriate level (local, national, and international) and be carried out in close connection with other institutions dealing with other aspects of the environment. As the distribution of tasks among relevant institutes and organizations varies from one country to the other, the necessity for uniformity in the technical approach must be emphasized.

7.4.1. Availability of geological data concerning the environment

The acquisition, processing, and making available of geological data are essential functions of the Geological Surveys. These data are stored in data banks, the organization and use of which were revolutionized by the development of computer support. However, remaining difficulties and problems should be urgently addressed, in particular those of data quality, comprehensiveness in acquisition, and availability to users.

The value of a data bank depends in the first place on the quality of the data. This requires precise geographic location, and the adequate choice and description of entry parameters. When those depend solely on the competence of Geological Surveys, such as the data contained in geological maps, the similarity of presentation and the control of their quality are adequate. When the data are provided by external bodies, such as drilling companies, sources of error are more numerous and require more specific control. Every company or organization should be fully aware of the necessity for high-quality data, such data being an essential part of the national archive. Apart from the quality, the data sets are commonly

incomplete. Again, using drilling as an example, it is known that many short holes drilled for geotechnical purposes are not recorded, and that data, often acquired at great expense, are lost forever.

Many data banks containing localized information (drill holes, sand and gravel deposits, seismic events, etc.) are now computerized, in a form sufficiently flexible to satisfy a wide range of users' requirements within the limits of confidentiality established in each country. Others types of information, map data for instance, are still mostly available in physical format. Geographic Information Systems (GIS), which provide geographically unified access to a variety of thematic fields (Fig. 7.1) are undergoing rapid development to meet the demand for easy access to data.

The Geological Surveys should be supported by their respective ministries to develop a leading position in the dialogue with users and to ensure the availability of adequate geological data. The main areas of concern are the type of data required, the level and timing of application, the structuring of access to data banks, and the form of data delivery.

In the case of geological mapping, most Western European countries publish their basic maps at the scale of 1:50 000, but three, Germany, Luxembourg, and Switzerland use 1:25 000. In each country the extent of coverage differs, as does the degree of computerization of map data. Although national requirements must be respected, international consideration must be given to identifying the best form of availability for users of geological mapping data. Important matters for consideration are:

(1) the balance between data on unconsolidated sediments and on underlying rocks;

(2) the scale of entry of data into data banks: a uniform scale for the whole country (1:50 000 scale for instance), with greater detail in specific areas (1:5000 to 1:10 000);

(3) the nature of the distribution network: centralized at the national level, with some regional outlets, or totally regional.

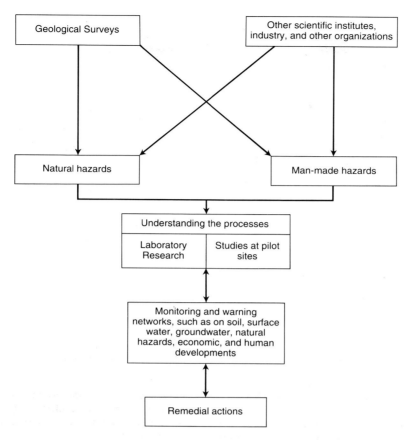

Fig. 7.2 Risks to the environment: research management and control.

More broadly, the question is raised of organizing integrated data management systems corresponding to the needs of users and decision makers in the field of land-use planning in the natural environment, and urban and rural development. To meet this goal, the following arrangements would be necessary:

(1) the setting up of appropriate networks (and associated systems) giving access to data provided by separate organizations, and so obviating the potential ineffectiveness of large national, centralized, multidisciplinary data banks;

(2) exhaustive data acquisition and the monitoring of data quality;

(3) decentralization of data collection, processing, and distribution, so that relevant data can be focused for use at the most appropriate technical and administrative levels;

(4) identification of the special needs of specific problems, such as ranking priorities and major local variations.

7.4.2. Risks and hazards to the environment: research, management, and control

Effective management and control of the environment is possible only through understanding the processes and mechanisms under which the different types of risk and hazard develop, whether they are natural hazards or risks deriving from economic development and population increase.

As shown in Fig. 7.2, the Geological Surveys are involved in studies within their own discipline and competence, in close collaboration with the other organizations which have complementary expertise. Full understanding of the complexities and problems of the environment is usually beyond the competence of a single organization, and is critically dependent on multidisciplinary R&D programmes.

Despite its diversity, this target requires a logical approach to the analysis of problems, starting from the understanding of processes through laboratory and pilot-site studies, to the installation of monitoring and warning networks with sound scientific bases, and reaching completion with the development of a predictive capacity and remedial actions.

The current situation in Western Europe can be summarized as follows:

1. Research into processes and studies at pilot sites is seldom coordinated, but are randomly dispersed through separate organizations, often on a parochial and regional basis. This commonly leads to a limited and partial approach to problems, with items such as soil, surface water, and groundwater being studied separately without reference to their wider significance.

2. Great diversity can exist in different countries, and even in different regions within the same country, in the design of monitoring networks, with respect to the density of stations, the periodicity of measurements and the validation and distribution of acquired data.

3. Different countries select different remedies for the same problem.

It is true to say that many separate issues relate to the physical environment and its socio-economic setting, but it is essential that the integrating factors are recognized and developed. Decisions are commonly based on insufficient technical and scientific knowledge. Geological Surveys, together with other organizations, must work to improve this situation, and public authorities must understand the necessity for supporting integrated action on these matters. Only then will there be the essential coordination of ideas to achieve the necessary standardization of monitoring techniques and impact studies for the benefit of man and his environment.

7.4.3. Geological Surveys and Environmental Impact Assessments

The assessment of dangers and the consideration of safety factors are standard procedure in development work. Risk analysis is also common with respect to the insurance and financing of development projects. In many countries these matters are covered by compulsory Environmental Impact Assessments (EIAs) for a wide range of projects.

For large-scale developments, such as major mining facilities, nuclear power plants, dams, tunnels, and bridges, it is obvious that each EIA should include reports and assessments of the geoscientific factors, made by professionally qualified geologists, who are capable not only of identifying all the relevant data, but also of carrying out the appropriate analyses of all the data. The results of such assessments must be communicated in a way that leaves those concerned with the decisions on the proposed development in no doubt about the salient geoscientific implications and any possible interactions with the advice of other disciplines.

It may be less obvious that such assessments are also necessary for minor development projects. Some of these may be less expensive and superficially less dangerous, but, because of their greater number, they have a cumulative impact on the local environment and require equally rigorous EIAs. At all levels of input, the effective integration of geoscientific data and analyses with those of other relevant disciplines is of paramount importance.

It is often frustrating for geologists to observe how the recognition of problems late in the history of, or even after the completion of a project, and the associated expenditure on unnecessary retrieval, have resulted from a failure to apply geological data and expertise at the appropriate time. It is particularly so when it is clear that the cost of a relevant, and even exhaustive, geological feasibility study or survey at the start of a project or during its planning phase, normally has no adverse impact on either the overall budget or the time-schedule. Even if the costs were significant, they would be offset by the early identification of potential problems with consequently even greater savings in expenditure and time. This seems to be so obvious and has been demonstrated so often by experience, that the reason for not involving geological expertise at the correct time, or at all in some cases, can have no other explanation than lack of knowledge of the relevance of geological data and expertise, and of the benefits which will accrue from their application.

So, it cannot be recommended too strongly that decision makers, planners, and industry must consult the national geological service whenever a development project is in prospect, and that governments ensure that EIAs become commonplace, with geology taking its rightful place alongside other contributing disciplines. Geological Surveys are in a position to supply information and expertise over a wide range of disciplines and, in some situations, they will be able to produce assessments and identify problems from existing resources, and without recourse to additional programmes of data acquisition and analysis. In those cases the cost–benefit ratio will be at its best and represents a real financial return on government investment in Geological Surveys.

Geologists understand fully that the possible increase in cost of a project, or even its delay or rejection, because of the identification of negative environmental impacts, can be unpleasant and difficult for the potential developer to accept. Geological Surveys, however, from their advisory position, recognize that national and international social responsibilities will have priority, and they feel confident that no responsible and adequately informed decision maker, planner, or developer will support a development if there is reasonable doubt about the potential disadvantageous or even dangerous environmental impact and consequences of a project. It is essential that all parties concerned use the assistance which Geological Surveys are prepared and able to give within their field of competence, to check for all possible geological risks and impacts on the environment.

7.4.4. Geological Surveys and the establishment of standards

The need for improving uniformity of geological data and methodology at universally acceptable levels is well recognized. National Geological Surveys and data banks have grown independently over many years, and reflect national programmes and resources. Now that a greater awareness of the environmental impact of development, a recognition that data and expertise may be available outside the country, and strong movements towards international cooperation on environmental problems are commonplace, it is clear that remedial actions must be based on a common understanding of the nature, magnitude, and consequences of the problems and the potential impact of such actions.

To facilitate agreement on such evaluations, it is crucial that all physical phenomena are described and entered into data banks in such a way that precise and unambiguous understanding is assured for any subsequent user. The Geological Surveys have the experience and expertise for this work. Traditionally they have cooperated in the establishment of common standards and methods for the description and evaluation of geological factors that relate to environmental problems. Amongst themselves they have cooperated on the development of minimum standards for the recording of data, for the construction of maps, and for procedures necessary for any involvement in identifying, quantifying, and solving these problems.

With the increase in the range of environmental problems, and of the awareness that remedial action is essential, the level of international cooperation on the development of necessary standards in land-use planning and environmental protection is rising. It is recommended that, in recognition of the relevance of data held in national archives, and of the expertise built up by long experience in associated fields of activity, governments ensure that Geological Surveys are given the mandate and resources to work on standards and methods which contribute to an official international standardization system.

7.4.5. Looking ahead

The Geological Surveys are in a key position to contribute to the protection and advantageous development of the environment. They make no pretence of being alone in this task, but, within the framework of data management systems and expertise, their contribution can be readily integrated with that of complementary disciplines, such as biology, ecology, and climatology. It is the task of public authorities to ensure the adequacy and quality of the various organizations involved, and to achieve the necessary integration of effort. This applies equally to the scientific, technical, and administrative aspects of the problems. If we do not want to be overwhelmed by environmental problems, the time has come for strong government action to utilize to best effect the fund of data and expertise accumulated over the years.

To achieve the objectives set out above, both for the Geological Surveys and for governments, will mean that in most countries the place of the Survey in the government machine, the mechanism for its management and application, and the levels of its staffing and funding will have to be considered at the highest levels. The contribution that an adequately resourced and properly used Geological Survey can make to the national economy of any nation is great. The integration of the Geological Surveys of Western Europe, along with their neighbours in Eastern Europe and similar organizations throughout the world, is an objective of fundamental importance. There is good reason now for dedicating more resources to strengthening such bonds and activities as one practical contribution to expanding international co-operation on ever more issues, including above all, the ever present environmental problems consequent upon man's activities.

Mankind has so much to lose if his relationship with the geological environment is not managed effectively. Decision makers in government cannot afford to ignore the considerable volume of geoscientific knowledge and expertise available for application towards achieving the maximum long-term benefit for all.

'There is a tide in the affairs of men,
Which, taken at the flood, leads on to fortune;
Omitted, all the voyage of their life
Is bound in shallows and in miseries.
On such a full sea are we now afloat;
And we must take the current when it serves,
Or lose our ventures.'
William Shakespeare
Julius Caesar, iv.3

Appendix 1: Addresses and Directors of the Western European Geological Surveys

(as at 7 March 1991)

AUSTRIA

Geologische Bundesanstalt
Rasumoffskygasse 23
Postfach 154
A-1030 Vienna
Director: Professor Dr T. E. Gattinger

BELGIUM

Service Géologique de Belgique
13 Jenner Street
B1040 Brussels
Head and Inspector General of Mines: Professor Dr J. Bouckaert

CYPRUS

Geological Survey Department
Nicosia
Director: Dr G. Constantinou

DENMARK

Danmarks Geologiske Undersøgelse
Thoravej 8
DK-2400 Copenhagen NV
Director: Dr O. W. Christensen

FINLAND

Geologian Tutkimuskeskus
Geological Survey of Finland
SF-2150 Espoo 15
Director: Professor L. K. Kauranne

FRANCE

Bureau de Recherches Géologiques et Minières
Service Géologique National
BP 6009
45060 Orléans Cedex 2
Director: Dr L. Le Bel

GERMANY

Bundesanstalt für Geowissenschaften und Rohstoffe
Postfach 51 01 53
D 3000 Hannover 51
President: Professor Dr M. Kürsten

GERMAN STATES represented by

Geologisches Landesamt Nordrhein-Westfalen
Postfach 1080
De-Greiff-Strasse 195
4150 Krefeld
President: Professor Dr Ing P. Neumann-Mahlkau

GREECE

Institute of Geology and Mineral Exploration
70 Messoghion Street
Athens 11526
Director General: Dr V. Andronopoulos

GREENLAND

Grønlands Geologiske Undersøgelse
Ostervoldgade 10
1350 Copenhagen K
Director: Dr M. Ghisler

ICELAND

Orkustofnun (National Energy Authority)
Grensasvegur 9
108 Reykjavik
Director: Dr G. Palmason

IRELAND

Geological Survey of Ireland
Beggars Bush
Haddington Road
Dublin 4
Director: Dr R. Horne

ITALY

Servicio Geologico d'Italia
Largo S Susanna 13
00187 Rome
General Director: Dr A. Todisco

LUXEMBOURG

Service Géologique du Luxembourg
43 Boulevard Grande Duchesse Charlotte
Luxembourg
Director: Ing J. Bintz

THE NETHERLANDS

Rijks Geologische Dienst
Richard Holkade 10
Postbus/PO Box 157
2000 AD Haarlem

Director: Dr C. Staudt

NORWAY

Norges Geologiske Undesøkelse
Liev Erikssons Vei 39
PO Box 3006 Lade
N-7002 Trondheim

Director General: Dr K. S. Heier

PORTUGAL

Serviços Geológicos de Portugal
Rua da Academia das Ciencias 19, 2
12 00 Lisbon

Director: Dr D. de Carvalho

SPAIN

Instituto Tecnológico Geominero de España
Rios Rosas 23
28003 Madrid

Director General: Mr E. Llorente

SWEDEN

Sveriges Geologiska Undersökning
Box 670
S-751 28 Uppsala

Director General: Mr J. O. Carlsson

SWITZERLAND

Service Hydrologique et Géologique National
3003 Berne

Chief and Deputy Director of the Federal Environment Protection Agency: Dr C. Emmenegger

TURKEY

Maden Tetkik ve Arama Genel Müdürlügü
Ankara

Director General: Professor Dr O. Baysal

UNITED KINGDOM

British Geological Survey
Keyworth
Nottingham NG12 5GG

Director: Dr P. Cook

Appendix 2: Contributors

All those who have made a scientific contribution of any kind are listed below by country and organization. Each has been given a number, and that number appears in square brackets [123] in the text at the point where acknowledgement is appropriate.

In addition, the important contribution of draughtsmen and photographers in many Geological Surveys is to be acknowledged. Particular credit is due to those in BGR, BGS, BRGM, DGU, and GLA/NW who were responsible for the final compilation and coordination of the illustrations.

The Directors also wish to acknowledge the extensive work of those in BRGM who prepared the final typescript of the whole book.

AUSTRIA

Geological Survey—Geologische Bundesanstalt (GBA):
1. G. Letouzé-Zezula; 2. H. Pirkl.

BELGIUM

Geological Survey—Service Géologique de Belgique (SGB):
3. C. Baeteman; 4. F. Bogemans; 5. H. Goethals; 6. R. Paepe.

CYPRUS

Geological Survey Department (GSD):
7. E. Christodoulou; 8. G. Constantinou; 9. A. Panayiotou; 10. G. Petrides.

DENMARK

Geological Survey—Danmarks Geologiske Undersøgelse (DGU):
11. L. J. Andersen; 12. A. Dinesen; 13. P. Gravesen; 14. E. Stenestad.

FINLAND

Geological Survey—Geologian Tutkimuskeskus (GSF):
15. G. Gáal; 16. K. Kauranne; 17. V. Knosmanen; 18. E. Lappalainen; 19. J. Niemalä; 20. Y. Pekkala; 21. B. Sattikoff; 22. V. Souminen; 23. H. Tanskanen; 24. M. Tontti.

Department of Material Science and Rock Engineering, Helsinki University of Technology
25. H. Niini.

FRANCE

Geological Survey—Bureau de Recherches Géologiques et Minières (BRGM):
26. J. P. Aste; 27. E. de Backer; 28. M. Barres; 29. G. Berger; 30. P. le Berre; 31. A. Boisdet; 32. R. Bouteloup; 33. M. Bouvet; 34. B. Cabrol; 35. J. Casanova; 36. J. J. Chateauneuf; 37. C. Chauby; 38. P. Ciron; 39. J. J. Collin; 40. B. Come; 41. S. Cottez; 42. S. Courbouleix; 43. M. Donsimoni; 44. G. Farjanel; 45. J. L. Garnier; 46. A. Gerard; 47. P. Godefroy; 48. M. Jaujou; 49. P. Lagny; 50. M. Lansiart; 51. P. Margron; 52. P. Massal; 53. P. Masure; 54. F. Maubert; 55. A. Menjoz; 56. E. Michalski; 57. P. Mouroux; 58. C. Oliveros; 59. J. E. Pasquet; 60. P. Peaudecerf; 61. J. Piraud; 62. G. Rampon; 63. J. C. Roux; 64. J. M. Sionneau; 65. G. M. Sustrac; 66. H. Traineau; 67. S. A. Chessy.

Institute of Global Physics, Paris:
68. P. Bernard.

Institute for Restoration of Mountain Land (RTM):
69. L. Besson.

National Agronomic Research Institute (INRA):
70. D. Baize; 71. G. Monnier.

National Research Centre for Farming Equipment, Farming Practices, Waters and Forests (CEMAGREF):
72. J. Guet.

Société Nationale Elf Aquitaine (Production):
73. T. Tanguy.

GERMANY

Geological Survey—Bundesanstalt für Geowissenschaften und Rohstoffe (BGR):
74. H. Aust; 75. M. Langer; 76. A. Pahl.

Cadastral Survey of City of Essen:
77. H. Dahm.

Institute of Geology and Mineralogy, Friedrich-Alexander-University, Erlangen:
78. G. Lüttig; 79. U. Mattig.

State Geological Surveys
Baden-Württemberg:
80. P. Hummel; 81. E. Rogowski; 82. W. Schloz; 83. E. Wallrauch; 84. F. Zwölfer.

Bavaria:
85. H. J. Baumann; 86. W. Grottenthaler; 87. A. v. Poschinger; 88. O. Wittman; 89. J.-P. Wrobel.

Hamburg:
90. A. Paluska.

Hesse:
91. A. Quadflieg.

Lower Saxony:
92. J. Drescher; 93. H. Geissler; 94. R. Hindel; 95. H. Kuntze; 96. W. Schäfer; 97. L. Schröder; 98. H. Sponagel.

Northrhine-Westphalia:
99. B. Alberts; 100. G. Drozdzewski; 101. H. Hager; 102. B. Jäger; 103. J. Kalterherberg; 104. K. Köwing; 105. G. Michel; 106. G. Milbert; 107. P. Neumann-Mahlkau;

108. W. Paas; 109. U. Pahlke; 110. F. K. Schneider;
111. G. Schollmayer; 112. H. P. Schrey; 113. H. Vogler;
114. P. Weber; 115. H. Wilder.

Schleswig-Holstein:
116. F. Grube.

GREECE

Geological Survey—Institute of Geology and Mineral Exploration (IGME):
117. T. Christodoulou; 118. G. Gioni.

GREENLAND

Geological Survey—Grønlands Geologiske Undersøgelse (GGU):
119. J. Bondam.

ICELAND

Orkustofnun (National Energy Authority) (NEA):
120. G. Palmason; 121. K. Saemundsson; 122. F. Sigurdsson.

IRELAND

Geological Survey of Ireland (GSI):
123. C. R. Aldwell; 124. D. Daly; 125. A. M. Flegg;
126. P. McArdle.

ITALY

Geological Survey—Servicio Geologico d'Italia (SGI):
127. F. Ferri; 128. G. Stefani; 129. A. Todisco.

Geological Survey of Emilia-Romagna:
130. L. di Bello; 131. R. Pignone.

Geological Survey of Lombardy:
132. D. Fossati; 133. G. Mannucchi; 134. L. Ottenziali.

Geological Survey of Piemonte:
135. V. Coccolo.

Vesuvius Observatory, Naples:
136. G. Luongo; 137. G. Tedesco.

LUXEMBOURG

Geological Survey—Service Géologique du Luxembourg (SGL):
138. J. Bintz; 139. R. Maquil.

THE NETHERLANDS

Geological Survey—Rijks Geologische Dienst (RGD):
140. S. Jelgersma; 141. E. J. F. de Mulder.

Research Institute for Nature Management, Leersum:
142. G. P. Gonggrijp.

NORWAY

Geological Survey—Norges Geologiske Undersøkelse (NGU):
143. H. Barkey; 144. B. Bergstrom; 145. L. H. Blikra;
146. G. Juve; 147. R. T. Otteson; 147a. A. Reite;
148. E. Sigmond; 149. F. C. Wolff.

PORTUGAL

Geological Survey—Serviços Geológicos du Portugal (SGP):
150. D. de Carvalho.

SPAIN

Geological Survey—Instituto Tecnólogico Geominero de España (ITGE):
151. R. Arteaga; 152. F. J. Ayala-Carcedo; 153. D. Barettino;
154. R. del Castillo Gonzalez; 155. D. B. Fraille; 156. E. Gallego;
157. L. I. Gonzales; 158. L. Lain; 159. J. A. Lopez-Geta;
160. F. X. Montserrat i Rebull; 161. P. Navarette;
162. L. Vadillo Fernandez; 163. E. G. Valcarce.

Prospeccion e Ingenieria, S A:
164. L. Gonzáles.

SWEDEN

Geological Survey—Sveriges Geologiska Undersökning (SGU):
165. S. Arvidsson; 166. C. L. Axelsson; 167. T. B. Fagerlind;
168. D. Fredrikson; 169. L. Karlqvist; 170. Th. Lundqvist;
171. J. Pousette; 172. L. Rudmark; 173. O. Selinus.

Mineconsult, Stockholm:
174. C. O. Morfeldt.

State Mining Property Commission:
175. G. Norblad.

Swedish Rock Engineering Research Foundation—BeFo, Stockholm:
176. A. Nordmark.

Uppsala Municipality Department of Water and Waste Water:
177. S. Ahlgren; 178. R. B. Bergström.

SWITZERLAND

Geological Survey—Service Hydrologique et Géologique National (SHGN):
179. L. Hauber; 180. A. Petrascheck; 181. B. Schädler;
182. J. P. Tripet.

TURKEY

Geological Survey—Maden Tetkik ve Avama Genel Müdürlüğü (MTA):
183. S. Ates; 184. E. Gürsoyfrak; 185. A. Oral;
186. H. Tahsin Aktimut.

UNITED KINGDOM

British Geological Survey (BGS):
187. N. Aitkenhead; 188. K. Ambrose; 189. R. A. B. Bazley;
190. C. W. A. Browitt; 191. M. A. E. Browne;
192. S. G. D. Campbell; 193. B. Cannell; 194. G. R. Chapman;
195. N. A. Chapman; 196. A. H. Cooper; 197. J. C. Cripps
(university collaborator); 198. M. G. Culshaw;
199. R. A. Downing; 200. A. Forster; 201. M. J. Gallagher;
202. T. P. Gostelow; 203. B. A Hains; 204. P. M. Harris;
205. D. J. Harrison; 206. R. T. Haworth; 207. J. H. Hull;
208. P. D. Jackson; 209. B. Kelk; 210. F. G. Larminie;
211. T. V. Loudon; 212. D. J. Lowe; 213. D. M. McCann;
214. A. A. McMillan; 215. B. S. P. Moorlock; 216. J. A. Plant;
217. D. W. Redmayne; 218. S. J. Stoker; 219. R. G. Thurrell;
220. A. J. Wadge; 221. J. C. Walsby; 222. G. M. Williams;
223. A. A. Wilson.

Geological Survey of Northern Ireland:
224. A. E. Griffith; 225. I. C. Legg; 226. W. I. Mitchell.

Ben Eagach Quarry, Scotland:
227. A. R. Burns.

Cabinet Office, Whitehall, London.
228. E. F. P. Nickless.

M I Great Britain Ltd, Scotland:
229. M. J. D. Butcher.

Petroleum Science and Technology Institute, Edinburgh:
230. S. Brown.

Southern Science, Worthing:
231. H. G. Headworth.

Appendix 3: Figures

Index

Bold entries denote case histories.
Entries in italics denote major entries.